INTRODUCTION TO MICROWAVE CIRCUITS

Radio Frequency and Design Applications

Robert J. Weber
Iowa State University

IEEE Microwave Theory and Techniques Society, *Sponsor*

IEEE Press Series on RF and Microwave Technology
Roger D. Pollard and Richard Booton, *Series Editors*

IEEE PRESS

The Institute of Electrical and Electronics Engineers, Inc., New York

This book and other books may be purchased at a discount
from the publisher when ordered in bulk quantities. Contact:

IEEE Press Marketing
Attn: Special Sales
445 Hoes Lane
P.O. Box 1331
Piscataway, NJ 08855-1331
Fax: +1 732 981 9334

For more information about IEEE Press products, visit the
IEEE Online Catalog & Store: http://www.ieee.org/store.

The author and publisher of this book have used their best efforts in preparing this book.
These efforts include the development, research, and testing of the theories and programs
to determine their effectiveness. The author and publisher shall not be liable in any event
for incidental or consequential damages with, or arising out of, the furnishing, performance,
or use of these theories or programs.

Printed in the United States of America.

10   9   8   7   6   5   4   3   2   1

ISBN 0-7803-4704-8
IEEE Order No. PC5758

Library of Congress Cataloging-in-Publication Data

Weber, Robert J., 1942-
    Introduction to microwave circuits : radio frequency and design applications / Robert J.
  Weber ; IEEE Microwave Theory and Techniques Society, sponsor.
      p. cm. -- (IEEE Press series on RF and microwave technology)
    ISBN 0-7803-4704-8
    1. Microwave circuits. I. IEEE Microwave Theory and Techniques. II. Title.
  III. Series.

TK7876.W42 2000
621.381'32--dc21

                                                                00-059672

Because He Lives

Contents

Preface

This book is meant to serve both as a textbook and a reference book for practicing engineers. Most undergraduates studying microwave engineering hope to shortly become practicing engineers. This book is meant to stay with the student when the student becomes a practicing engineer. During the approximately 25 years I spent at the former Collins Radio Company and Rockwell International, I had the privilege of working with some great microwave engineers, who were real practitioners. Much of what I accomplished in those years can be attributed to their patience and desire to train me in good experimental techniques in the microwave laboratory. They reinforced my conviction that good microwave design is built upon good characterization. Good characterization is proven in the laboratory. Although I have written well over 25,000 lines of Fortran code for microwave circuit analysis, I still believe in proving the circuit in the laboratory. Design cycle time and costs are very important to engineers in the microwave industry. The experimentalist knows the value of computer-aided design, but also knows when the intelligence contained in the program is not sufficient to mimic and model what occurs in the laboratory. First-time design success is the goal of many engineers and is required of many managers. One purpose of this book is to help engineers reach that goal.

Much of the material contained in Chapters 1 to 3, 5, 7 to 9, and 10 has been used in a lecture in a one-semester course at Iowa State University. The material in Chapter 6 is usually learned by those students in the concurrent laboratory assignments. At Iowa State University, the microwave student designs a 1-GHz oscillator, amplifier, and filter on a printed circuit board during the first two-thirds of the semester. During the last one-third of the semester, the boards are fabricated and the student assembles, tests, and analyzes the design and compiles a report on the performance of the small power module that was designed.

The material in Chapter 4 is meant for the engineer working in the mixed-mode analog-digital MMIC area. The mixed-mode three- and four-port scattering parameter measurement and analysis technique is covered in that chapter. The undergraduate student or practicing engineer who is interested in two-port circuits only can skip that chapter.

The material is arranged in an order that enables an engineer first to gain an appreciation of component parasitics, then to learn the scattering parameter technique and subsequently apply both of these knowledge bases to an amplifier design. The amplifier design procedure is then applied to an oscillator design. High-efficiency amplifier design is inserted between amplifiers and oscillators in order to provide a basis for optimum power generation in an oscillator. Filter design procedures follow oscillator design and give the engineer the tools to provide spectrum cleanup for oscillator outputs.

Noise, detection, and mixing (Chapters 10 and 11) follow these chapters. Portions of these chapters can be included in a one-semester course, but these chapters plus Chapters 12 and 13 provide a basis for a second-semester course in RF and microwave engineering. The material in Chapter 13 provides a beginning analysis for bridging the gap between analog and high-speed digital microwave circuits. Parts of the material in Chapter 4 could be used in a second-semester course as well.

The last two chapters of the book, Chapters 14 and 15, should be read along with Chapters 1 through 8. Chapter 15 contains a more detailed worked example of a 1-GHz oscillator, amplifier, and printed circuit board filter. The reader can easily use that chapter as a basis for a printed circuit board or monolithic microwave integrated circuit design. Chapter 14 contains some details about bias components and concerns that are often overlooked.

Material for the book was first compiled in order to teach an on- and off-campus course when I was an adjunct instructor at the University of Iowa while working at Rockwell International. That content has been used during the 13 years I have been on the faculty of Iowa State University. I thank all of my former students for their comments; some of the comments have been enlightening and others encouraging. My teaching assistants who helped in the laboratory deserve special thanks. I was privileged to take a six-month professional development leave from Iowa State University, which provided me with relief from pressing day-to-day duties and with an opportunity for further study in the microwave area, as well as the environment necessary for the final details of the book to be completed.

ACKNOWLEDGMENTS

My family deserves special thanks. Our daughter is a missionary doctor in Zambia. We communicate by email. She asks how many extensions of deadlines one can get while writing a book. Her pragmatism as well as the spiritual faith that we share has been a real encouragement to me. Our sons, now both engineers, one industrial, one civil, often said, "How's the book coming, Dad?" to encourage me along. Thanks. My wife, Joan, really deserves much of the credit. Without her love, patience, and encouragement this book wouldn't be done. Thanks, Joan.

Department of Electrical
and Computer Engineering
Iowa State University

1

Microwave Circuits

1.1 INTRODUCTION

The term *microwave circuits* means different things to different people. The prefix *micro* comes from the Greek μικρός (micros) and among its various meanings has the meaning of little or small. Little and small are relative words. *Microwave* means little or small waves. This often means that the wavelength is small with respect to the physical variations in a circuit or small with respect to a component size. To many people, microwave means those frequencies in the 1-to-10-gigahertz range. This is the range over which the wavelength in air varies from 30 centimeters to 3 centimeters. In dielectric materials, the wavelength is reduced from its value in air by a factor of the square root of the dielectric constant. In the 1-to-10-gigahertz frequency range, components with linear dimensions of 3 to 30 millimeters or approximately 1 to 20 millimeters for common dielectric materials become a tenth of a wavelength long. Other people like to think about microwave circuits as those for which distributed effects are important. For those circuits, wave propagation effects are important. Whether wave propagation effects are important depends again on the physical size or fabrication technology used to fabricate a circuit that is being described or characterized. Between the middle and the end of the twentieth century, microwave circuit techniques included printed circuit board (PCB), microwave integrated circuit (MIC), and monolithic microwave integrated circuit (MMIC) technologies in addition to coaxial and waveguide technologies. The circuit designer uses packaged devices roughly one centimeter in size for PCB circuits, beam-leaded devices roughly one millimeter in size for MIC circuits, and monolithic devices roughly ten to a hundred microns in size for MMIC circuits.

In this book, the term *microwave circuits* includes those circuits fabricated on printed circuit boards (PCB), microwave integrated circuits (MIC), or monolithic microwave integrated circuits (MMIC) for which the circuit parasitic elements form a significant portion of the circuit element values. A parasitic element might arise due to distributed circuit effects or arise from fabrication constraints. This definition includes but is not limited to those circuits for which distributed effects need to be considered. Printed circuit boards and microwave integrated circuits are in use in most avionics and wireless circuit applications. The relative amount of a parasitic element depends on the dielectric constant of the medium in which the circuit element is embedded, the distance a signal needs to travel from a source to a load, or the

1

distance between a component and a ground conductor or region. This concept can be used to describe circuits that are analog in nature, digital in nature, or mixed mode (analog and digital) in nature. Both parasitic elements and distributed effects need to be considered in microwave circuit design. For the digital designer, these parasitics affect the time delay and wave shape associated with voltage or current pulses traveling along a wire or conductor.

A model of a circuit element accounts for the electric, magnetic, and dissipated energies in the circuit element. A combination of lumped-constant elements consisting of resistors, capacitors, or inductors can often be used to model a circuit element. A resistor is used to characterize the power or energy lost in a circuit, while a capacitor and an inductor are used to characterize the electric and magnetic energy stored in a circuit respectively. In those cases where the circuit is large with respect to a wavelength, a transmission line model consisting of resistors, capacitors, and inductors will be used to characterize the distributed parameter effect for the circuit elements.

Giving the reader the necessary tools needed to analyze or synthesize microwave circuits is the objective of this book. Some of these tools will be developed in a sequence and other tools are developed in specific chapters. Some general analysis tools are developed and then the tools needed to design an amplifier, the tools needed to design an oscillator, and the tools needed to design filter circuits are developed. The scattering matrix formalism is developed and then measurement methods for obtaining the scattering parameters are described. For those readers who are interested in differential circuit characteristics, scattering parameters of active differential circuits are described before stability, gain, and match are described. For those readers who are interested only in single-ended circuits, Chapter 4 can be skipped and the reader can go directly to Chapter 5. Readers who are not interested in measuring scattering parameters and who wish to use scattering parameters directly may go from Chapter 2 to Chapter 5. Chapter 6 contains material that will give the reader several methods to match lumped- and distributed-constant circuits. Different modes of power amplification are discussed in Chapter 7 before oscillators are discussed in Chapter 8, in order to give the reader the design information needed to design oscillators for high power-conversion efficiency.

Chapters 2, 3, 5, 6, 7, and 8 provide the material necessary for designing an oscillator including any necessary phase shifters, an attenuator, an amplifier, and a filter. Material from this set of chapters has been used for over a decade in a one-semester beginning microwave circuits course for senior and first-year-graduate students at Iowa State University. The thrust of that course has been to look at the circuit's aspect of microwave circuit design. The distributed nature of the design is incorporated using de-embedding techniques and the scattering matrix for transmission line sections. The laboratory portion of the course consists of the design, fabrication, build, and test of a 1-gigahertz oscillator, followed by an attenuator, an amplifier, and a two-pole filter using microstrip printed circuit boards and leaded as well as chip parts for the discrete passive components. During the time the students were fabricating and testing those parts of a single-frequency power source, selected material from Chapters 10 and 11 was included in the lecture material.

Other topics in this book include noise in Chapter 10, and detection and frequency translation in Chapter 11. Special attention is given in Chapter 12 to the use of PIN diodes for switching applications for the reader who is incorporating sensitive receivers with high-power transmitters in a single antenna system. Chapter 12 contains a description of some microwave components that can be used by the microwave engineer to build a system. Fully developing that chapter in a college course could easily comprise a semester of learning and fill several books. However, sufficient material is given to allow the reader to use the components in a system. Components and connecting wires that are used in microwave circuits need to be characterized. Methods to measure, manipulate, and verify component values are discussed. A method for time domain analysis applicable to high-speed digital and mixed-mode digital-analog circuits is described in Chapter 13. Chapter 14 includes a discussion of bias circuit component effects and nonlinear phenomena associated with bias circuit configurations.

Chapter 15 is the final chapter of the book. It contains a worked example of a 1.25-GHz amplifier, a 1.25-GHz loop oscillator, a 1.25-GHz shunt oscillator, and a couple of 1.25-GHz filters. The reader should refer to that chapter while reading Chapters 2, 5, 6, 7, 8, and 9. Appendix D contains a laboratory procedure to measure the parasitic values of R, L, and C components. The procedures are described in terms of a test fixture with connectors; however, these procedures are applicable to MMIC versions of the components when using rf or microwave probes to connect to the test fixture.

1.2 CIRCUIT ELEMENTS

A circuit component consists of several simple circuit elements. As will be discussed in Chapter 2, reciprocal passive components we call resistors, or capacitors, or inductors each consist of various combinations of the three circuit elements—resistance, capacitance, and inductance. When these components are used, they are placed in space somewhere in relation to a ground plane or ground reference. Transmission lines are composed of these same three elements—resistance, capacitance, and inductance—but in a transmission line, these elements are distributed over a region of space rather than being identified with a point in space. In addition to these components, controlled current and voltage sources are used to describe the performance of active elements. Equivalent circuits for transistors and diodes are given assuming the reader has had some introduction to diode equations and transistor phenomena.

The author often tells his students that he has never seen any of the simple circuit elements, a resistor, a capacitor, or an inductor, much less a ground plane. This should become more evident in Chapter 2 when components are described. What people have seen is a component called a resistor, a capacitor, or an inductor but each of these components has varying amounts of other circuit element types associated with it.

1.2.1 Ground Planes

Typically the metal on the back of a printed circuit board or the back side metalization on a MMIC chip is called a *ground plane*. That is the term used in the industry and the literature. Calling something by the term, however, does not guarantee that the metal structure has the mathematical characteristics of a ground plane, i.e., that the potential is the same everywhere on the surface.

What is a ground plane? A ground plane that represents the mathematical description extends from minus infinity to plus infinity in both directions. It does not have any holes or cuts in it and it has zero resistance. The author has never seen one of these! The microwave circuit designer needs to appreciate what effect a ground conductor of finite size has on a microwave circuit and what effect drilling a hole or making a cut in the ground conductor has on the circuit. All conductors have a finite resistance. Even ground conductors formed out of superconductors have a finite resistance at high frequencies. Normal conductors have resistance at all frequencies. The performance of a microwave circuit depends on whether and where a surface of zero potential exists.

A ground plane is assumed to have a constant potential everywhere on it. Unbalanced transmission line analysis assumes that a ground plane exists. When a ground plane does not exist as assumed for transmission line analysis, then one needs to question how the results of transmission line analysis can be applied to real circuits. Often the ground plane is assumed to be at a potential of zero volts everywhere on it. A perfect ground plane can conduct microamperes of current or hundreds of amps of current and in either case has no voltage developed across it. What is commonly called the ground plane in the circuit board industry is an approximation for what a real ground conductor is. The ground plane approximation often gets the microwave engineer into difficulty when using the results of circuit or transmission line analysis based on an infinite ground plane assumption. Real ground conductors do have voltages across them even when they are formed from a continuous foil or sheet of material. This material is not infinite in extent and does have a finite resistance.

1.2.2 Linear *R, L, C* Circuit Elements

Some brief comments will be made about the simple linear circuit elements, the resistor, the capacitor, and the inductor. It is appreciated that the reader will likely know these circuit elements quite well. However, with regard to microwave circuit modeling, it is helpful to review how the circuit elements function in a circuit. Keep in mind that the mathematical description of a circuit element and not the function of a real-world component is being considered when these elements are discussed in this section.

Multiple terminal networks have separate currents and voltages into and across each set of terminal pairs. In microwave circuits, a port is a region through which energy flows. A port might be considered a terminal pair, a region between and around two wires, or it might be the opening into a tube called a *waveguide*, or further, it might just be an opening into some volume. A circuit element, either a resistor, a capacitor, or an inductor, in the absence of a ground plane, is a one-port network. It has only one terminal pair and only one port. When each terminal is separately connected somewhere but also in the presence of a ground plane or ground node then these elements become two-port networks.

A resistor of resistance value R is used to describe a linear circuit element for which the current through the resistor is in phase with the voltage across the resistor. All energy entering a resistor is dissipated in the resistor. The relationship between current through a resistor and voltage across a resistor is expressed as $V = RI$. This relationship can also be described as a conductor or a conductance. The expression is then rewritten as $I = GV$. Notice that although the relationships appear to be the same and $R = 1/G$, there is a significant difference in the equations. In the resistance expression, current, I, is the independent variable. In the conductance expression, voltage, V, is the independent variable. This does not appear to be very important for a simple resistor, but it is quite important when the linear relationships for circuit elements are extended to multiple terminal networks.

A capacitor of capacitance value C is used to describe the linear circuit element for which the current through the capacitor is said to lead the voltage across the capacitor. The capacitor does not dissipate energy and does not store magnetic energy, but does store or release electric energy. When current into the capacitor is the independent variable, 1 over C times the time integral of the current into the capacitor gives the voltage across a capacitor. When voltage across the capacitor is the independent variable, the current through the capacitor is equal to C times the time derivative of the voltage across the capacitor.

$$V(t) = \frac{1}{C} \int_{-\infty}^{t} I(t)\,dt$$

$$I(t) = C \frac{dV(t)}{dt}$$

An inductor of inductance value L is used to describe the linear circuit element for which the current through the inductor is said to lag the voltage across the inductor. The inductor does not dissipate energy and does not store electric energy, but does store or release magnetic energy. When current into the inductor is the independent variable, L times the time derivative of the current through the inductor gives the voltage across the inductor. When the voltage across the inductor is the independent variable, the current into an inductor is equal to 1 over L times the time integral of the voltage across the inductor.

$$V(t) = L\frac{dI(t)}{dt}$$

$$I(t) = \frac{1}{L}\int_{-\infty}^{t} V(t)\,dt$$

These well-known relationships should be kept in mind when MMIC and VLSI bias networks are designed. The inductance of the bias line will cause the voltage across a device to drop when that device's current changes rapidly. The MMIC circuit designer should attempt to incorporate enough capacitance close to an amplifier module to support at least one sinusoidal or pulse cycle of current that exists in the dc bias lead to that amplifier. These formulas are used to determine the effect of inductance or the value of capacitance needed.

Nonlinear circuit elements also exist. For the purposes of this book, unless specifically stated otherwise, all circuit values are assumed to be linear or the values used for circuit elements are those derived in a piecewise linear region of operation of the circuit. Those circuit values are often derived by differentiation of the immittance or transfer functions of the device around a bias point resulting in a small signal ac analysis. In the case of PIN diode analysis, the circuit values are derived by large signal linear ac analysis around a small dc signal that does not change appreciably over one large ac cycle.

1.2.3 Distributed Parameter Circuits

Transmission lines are used to model some circuits. Figure 1.1 shows a model of a lossless transmission line, Figure 1.2 shows a model of a transmission line with losses, and Figure 1.3 shows various types of transmission lines. In the transmission line model, R, L, C, and G values are given as per-unit-length values. In the metric system, transmission line resistance has units of ohms/m, inductance has units of henries/m, capacitance has units of farads/m, and conductance has units of siemens/m. Many microwave wireless systems, radar systems, and PCB circuits for computers use the microstrip type of construction.

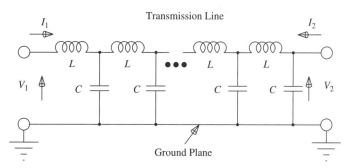

Figure 1.1 Model of a lossless transmission line.

Transmission lines are used to model distributed parameter circuits. Capacitance is distributed over a region of space between two conductors and inductance is distributed along the length of these conductors. One cannot identify a resistance, inductance, capacitance, or conductance with any single point on the conductors since the energies associated with these elements are distributed along the length of the conductors.

Two important parameters result from solving the differential equations for the uniform transmission line equivalent circuits shown in Figure 1.1 and Figure 1.2 [1]. If a transmission

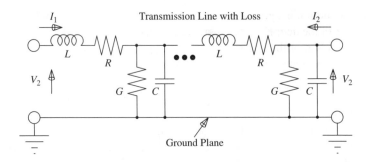

Figure 1.2 Model of a transmission line with losses.

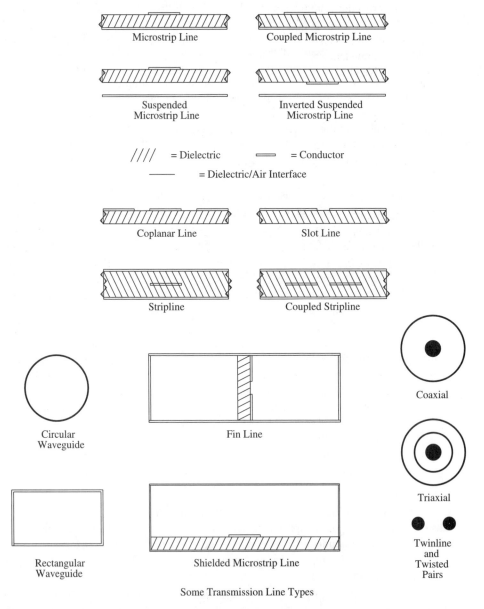

Some Transmission Line Types

Figure 1.3 Various types of transmission lines.

line has an infinite length, then the input impedance looking into the line is the characteristic impedance of the line. For a lossless line, the characteristic impedance is purely real. Consider one L-C section of the line shown in Figure 1.2. Terminate that section in a real impedance R. Set the impedance looking into this shunt C, series L, shunt R circuit equal to the same value of R.

$$\frac{\frac{(R+j\omega L \Delta x)}{j\omega C \Delta x}}{R + j\omega L \Delta x + \frac{1}{j\omega C \Delta x}} = R$$

$$R + j\omega L \Delta x = R + j\omega C \Delta x R^2 - \omega^2 LC(\Delta x)^2 R$$

$$\omega L \Delta x = \omega C \Delta x R^2 \Rightarrow R = \sqrt{\frac{L}{C}}$$

This procedure can be done for each LC section along the line. The input impedance at each spot along the line stays equal to the same value of R. The term containing the differential length squared has been ignored in the equation since it is much smaller than R. The value of the termination impedance R is called the characteristic impedance Z_0 of a lossless transmission line and is given by:

$$Z_0 = \sqrt{\frac{L}{C}}$$

The values in the equivalent circuit of a transmission line are differential values, Ldx, Cdx, etc. Therefore, the cutoff frequency of these transmission line sections becomes infinite as dx goes to zero.

$$f_{\text{cutoff}} \approx \frac{1}{2\pi \sqrt{LdxCdx}} = \frac{1}{2\pi \sqrt{LC}\,dx}$$

This might imply that the transmission line functions from dc all the way up to beyond light frequencies. If the model were correct this might be true. However, notice that there are no capacitors shown between nodes of the transmission line model. These capacitors do exist but are not shown for a TEM (transverse electromagnetic) transmission line. The energy contained in those capacitors for transmission lines operating in the TEM mode is negligible. This results from the assumption that important voltage differences exist only between the conductor and ground. The voltage difference between each of the nodes is zero since the electric field in the direction of propagation is zero. The small impedance of the series R-L circuit essentially shorts out any capacitance reactance between these nodes. When voltage differences between the distributed nodes become important (i.e., the wavelength along the line becomes short with respect to the distance between the conductor and ground), the TEM model shown is no longer valid. The capacitance that does exist between nodes on the transmission line then begins to have significant energy. In field analysis, the line is then said to support higher-order modes and the TEM assumption is violated.

Each section of the transmission line has a time delay associated with it. Recall that the dimension of the product LC is equal to time squared (the resonant frequency of an LC circuit is 1 over the square root of the product LC). This differential time delay for a small section of the line is:

$$\tau \Delta x = \sqrt{L\Delta x C \Delta x} = \sqrt{LC}\,\Delta x$$

When L and C are per-meter values, the time delay is also in seconds per meter. Therefore, the phase velocity v of wave propagation along a lossless transmission line is given by:

$$v = \frac{1}{\sqrt{LC}}$$

When lumped-constant equivalents for transmission lines are given later in this book for lines that have a finite length, the time delay for those circuits cannot be determined using this simple formula. Integration of the differential time delay of the transmission response needs to be done from dc to the frequency of interest to get the time delay at a given frequency for those networks as discussed in Chapter 13.

TEM transmission lines have both the electric and magnetic field lines contained in a transverse plane perpendicular to the direction of propagation. Other lines such as the microstrip and slot line have some components of the electric and magnetic field line in the direction of propagation. Those lines that have only a very small amount of the total energy in the line associated with these longitudinal components of the electromagnetic field are often termed *quasi-TEM lines*. At low frequencies, the microstrip line is a quasi-TEM line. In general, non-TEM lines have propagation constants and characteristic impedances that vary with frequency. This type of line is called *dispersive* since its propagation constant and therefore the phase velocity of propagation varies with frequency.

When a transmission line is propagating energy in a mode that is not TEM, the electric field lines are not all contained in the transverse plane. Therefore, all of the electrical field energy cannot be represented by a shunt capacitor. When the portion of the electric field transverse to the line varies with frequency, the shunt capacitance also varies with frequency. Some of the shunt capacitive energy will be in air and some in the dielectric for an air/dielectric microstrip line. The phase velocity of propagation for an air-filled line is:

$$v = \frac{1}{\sqrt{L_0 C_0}} = c$$

where C_0 and L_0 are the capacitance and inductance per unit length for an air-filled line and c is the speed of light. The phase velocity of propagation for a uniformly filled dielectric but nonmagnetic medium TEM transmission line is:

$$v = \frac{1}{\sqrt{\varepsilon_r}\sqrt{L_0 C_0}} = \frac{c}{\sqrt{\varepsilon_r}} \text{ for a TEM line}$$

where ε_r is the relative dielectric constant of the dielectric material in the line. When the cross-sectional area of the transmission line contains more than one value of dielectric constant, the capacitive energy is distributed among several dielectric constants. That distribution varies with frequency and the resultant electric and magnetic field distributions are not TEM. In this case the phase velocity of propagation for the nonmagnetic, non-TEM line is not given by equation above but by:

$$v = \frac{1}{\sqrt{\varepsilon_{\text{reff}}}\sqrt{L_0 C_0}} = \frac{c}{\sqrt{\varepsilon_{\text{reff}}}} \text{ for a non-TEM line}$$

where $\varepsilon_{\text{reff}}$ is the effective dielectric constant of the line. The effective dielectric constant accounts for the portion of electric energy that is distributed among the different materials with different dielectric constants. Approximate equations for the effective dielectric constant and characteristic impedance of a microstrip line are given in Appendix A. Equations for other transmission line configurations are given in the literature [2]. The approximate equations given in Appendix A give an effective dielectric constant and characteristic impedance but they do not account for any variation with frequency. However, the results are quite useful when used at frequencies for which that variation is small. Many computer-aided design programs perform computations that determine the variation of these parameters with frequency. One should remember that a microstrip line is not a TEM line and at higher frequencies when the line deviates from a TEM form, the variation of $\varepsilon_{\text{reff}}$ with frequency should be taken into account.

One period of a sinusoidal signal of frequency f is $1/f$. A wave with phase velocity v travels a distance of v/f in one period of the waveform. The wave also has a phase shift of two pi over this distance at a given point in time. A phase shift constant beta with dimensions of radians per unit length is defined such that beta times one wavelength is two pi.

$$\beta \frac{v}{f} = 2\pi$$

$$\beta = \frac{\omega}{v}$$

A transmission line of length D is often referred to as theta radians long with a phase shift constant of beta.

$$\beta D = \theta$$

1.2.4 Transmission Line Types

Figure 1.3 shows five types of microstrip lines. Whether one uses a suspended dielectric line or a dielectric line that is placed on a ground plane often depends on manufacturing requirements. Fin line is used for placing printed circuit components and circuits inside waveguides. Stripline is often used to seal the circuits for environmental and electromagnetic protection and for high-power operation. Coplanar circuits are often found on the top of integrated circuit chips. Coplanar circuits often also have a ground plane under the circuit. The top "ground" lines are often connected to the supporting ground plane. The effect of a bottom conductor on the characteristics of the top coplanar circuit needs to be considered. Even if the back side of the chip or board does not have a conductor, the proximity effect of a conductor somewhere in the vicinity of the back side needs to be considered. Both coaxial and waveguide circuitry are still used. They were used more often in the several decades before miniature manufacturing processes and devices were available.

Throughout this book, microstrip lines are referred to. However, the results of the various analyses in the text should not be limited to microstrip circuits and can be applied to different transmission line types. When a device, such as a transistor, is placed on a transmission line, a voltage and a current are assumed to exist on the conductor leading to the device. The type of transmission line type is not as important in applying the results of the microwave analysis given here as much as the fact that there is a point at which a voltage and a current exist. For devices that interact with a field over a surface or a volume and for which a current or a voltage for the device cannot be defined, the results of the analyses in this book have a more limited application. However, when the structure contains multiple modes, the reader can extend the analyses given in this book to some of those instances by using superposition of an analyses for each of the multiple modes that exist. The scattering parameter techniques given in this book do apply to distributed structures, waveguide structures, and, as indicated in Section 3.1.3, even to optical devices under certain conditions. The analyses are focused not on the characteristics of individual transmission line types but on developing techniques that apply to circuit analysis and synthesis that have applicability independent of the transmission line type or of whether the circuit is of a distributed nature.

EXERCISE

1-1 Calculate the characteristic impedance, phase velocity, effective dielectric constant, and phase shift constant at 1 GHz for a line that has 35 pF/m capacitance and 750 nH/m inductance. Assume the material medium is nonmagnetic.

2

Models, Modeling, and Characterization

2.1 MODELING AND CHARACTERIZATION

A significant amount of the effort expended designing MMICs (monolithic microwave integrated circuits), MICs (microwave integrated circuits), ASICs (application-specific integrated circuits), high-speed digital circuits, and PCBs (printed circuit boards) at high, microwave, and millimeter wave frequencies has to do with the engineering of parasitic elements of various components. Every circuit component is an R, L, and C circuit. The R part of the circuit has to do with the power dissipated in the circuit. The L part has to do with how the magnetic field energy of the circuit is accounted for, and the C part has to do with how the electric field energy of the circuit is accounted for. The equivalent circuit may contain various arrangements of the R's, L's, and C's in parallel or series depending on the circuit element. The formation of such a circuit is often called *modeling the circuit*. The equivalent circuit of a component also depends on how it is positioned with respect to a ground plane. The length of current flow through the part might be affected by the orientation of the part with respect to the ground plane. The length of current flow through the element has a direct bearing on the inductance of the element, and the position of the element with respect to a ground plane has a direct bearing on any parasitic capacitance of the element.

One additional parameter in the characterization of a device or circuit is the time delay through the circuit. Time delay is often modeled using a transmission line or sometimes with simply a time delay element. A transmission line contains stored magnetic and electric field energy. Some of this energy is dissipated in various loss elements. When constructing a model of a device using a transmission line, one needs to determine how much of the electric and magnetic field energy in a device is incorporated in the transmission line model and what field energy is not in the transmission line model. The same holds true for dissipated power. A model should account for all of the energy but should not account for any energy twice.

In addition to accounting for dissipated power and stored magnetic and electric energies of the circuit, the microwave engineer is often interested in how much of the available power is delivered to a load. Available power is the amount of power that can be obtained from a source. The microwave engineer is also interested in how the ratio of signal to noise varies as the signal progresses through a structure (the circuit). Several tools are developed in this book to help the microwave engineer model circuits and perform these calculations.

Mathematical modeling of circuits will be discussed first. A review is made of mathematical models of circuits using voltage and current as the variables since most devices used in microwave circuits have voltage-current characteristics. Then these models are extended to models that use one particular linear combination of voltage and current, the scattering parameter method. Scattering parameters are then related to ratios of traveling voltage waves on transmission lines. The review of one-port and two-port linear parameters is included here since their use is still important to the understanding and calculation of several aspects of microwave circuits. That discussion is followed by a brief treatment of loss in transmission lines and several sections that give models of three components: a *capacitor*, a *resistor*, and an *inductor*.

2.1.1 Two Terminal Components

It is helpful at this time to define a word to represent either impedance or admittance. The word *immittance* (*imp*edance + ad*mittance* = immittance) is used in circuit theory to mean either impedance or admittance. Immittance will be used to indicate impedance or admittance as necessary. Two terminal components can be either one-port or two-port networks. For one-port networks, when voltage is the independent variable and current is the dependent variable, $I = YV$ where Y is the admittance of the component. When current is the independent variable, then $V = ZI$, where Z is the impedance of the component. If Y is zero, then Z is undefined but the immittance is called an open circuit. Likewise, if Z is zero, then Y is undefined but the immittance is called a short circuit. Short-circuit and open-circuit immittances are used as important reference points when scattering parameters are discussed. When stubs are discussed later in this chapter, short or open immittances will also be assumed at a given point on a transmission line.

2.2 TWO-PORT CIRCUITS

A structure that has both an input and an output is often characterized by a linear two-port network. These two-port networks are described in terms of various sets of independent variables and dependent variables. Often the input of a circuit is assumed to be *fed* or *sourced* with a two-wire pair or with a wire and an infinite ground plane. Likewise the output of a circuit is *loaded* or *terminated* with a two-wire pair or with a wire and an infinite ground plane. Sometimes the two-wire input pair and the two-wire output pair have a *ground* conductor in common. In order to have an adequate characterization of the structure, the independent variables and the dependent variables need to be identified. For instance, if an input two-wire pair has a different ground than the ground for an output two-wire pair, it needs to be determined whether the voltage (and thus the field energy) between the input reference potential (input ground) and the output reference (output ground) is significant. If the energy that must exist between the input ground and the output ground affects the operation of the circuit, then the structure is no longer a two-port but is a three-port circuit or four-port circuit. Three- and four-port circuits are discussed in Chapter 3.

Linear two-port analysis assumes a ground point exists. A wire is usually at a zero potential at only a point. A ground plane would have zero potential at all points on it. However, if the ground plane is cut, or has a hole drilled in it, and so on, it is no longer a plane, and thus no longer a ground plane, and thus zero potential is no longer guaranteed everywhere. If the "ground" plane does not have zero resistivity, it is not a ground plane. The validity of two-port analysis depends on knowing where ground exists.

2.2.1 Two-Port Parameters

In circuit theory, including microwave circuit theory, impedance, admittance, and chain matrices are defined and used to describe two-port networks. This book makes use of four types of two-port parameters. These are the impedance, admittance, $ABCD$ chain, and scattering parameter sets of two-port parameters. The hybrid h parameters are discussed with regard to transistor parameters. Actual measurements in the range of frequencies associated with microwave frequencies are often made using scattering matrices but many of the circuit theories and derivations rely on the impedance, admittance, or chain matrix formalization. Hybrid parameters are often used for device characterization, including characterization at microwave frequencies. These parameters are defined in this chapter.

Consider the two-port shown in Figure 2.1. Various combinations of independent variables, the two-port parameter names, and the matrix formulations of these sets of parameters are given in the sections to follow. If there are four variables, there are six ways to select two independent variables and two dependent variables. When the two dependent variables are written in equation form, a choice could be made to put either one of the dependent variables in the first equation. This could generate twelve different sets of equations. However, common convention gives only six combinations. The dependent variable on port one is put in equation one and the dependent variable on port two is put in equation two. When both V and I from a single port are the dependent variables, then the voltage dependent variable equation is written first. It is possible to choose different linear combinations of two independent variables as new independent variables and different linear combinations of the dependent variables as new dependent variables. There are an infinite number of ways to do that. One particular combination is described in what are called the *scattering parameters*. Different combinations of scattering parameters, and thus different combinations of voltage and current variables, are used for three- and four-port scattering parameters in differential circuits. These are discussed in Chapter 3. In the following sections, both the individual equations are given as well as their matrix formulation.

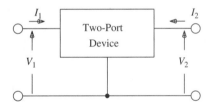

Figure 2.1 Voltage and current for a three-terminal two-port

2.2.2 Immittance Two-Port Parameters

The Z (impedance) parameters of a two-port relate I_1 and I_2 as independent variables to V_1 and V_2 as dependent variables. This relationship is defined as:

$$V_1 = Z_{11}I_1 + Z_{12}I_2$$
$$V_2 = Z_{21}I_1 + Z_{22}I_2$$
$$(V) = (Z)(I)$$

The Y (admittance) parameters of a two-port relate V_1 and V_2 as independent variables to I_1 and I_2 as dependent variables. This relationship is defined as:

$$I_1 = Y_{11}V_1 + Y_{12}V_2$$
$$I_2 = Y_{21}V_1 + Y_{22}V_2$$
$$(I) = (Y)(V)$$

It is readily seen from the matrix formulation, that (Z) is the inverse of (Y) and vice versa. However, if (Z) exists, (Y) may not exist and vice versa. This is similar to the open-circuit and short-circuit situation for a one-port. Linear combinations of voltage and current can give a more general description of a network. A network that does not have an immittance matrix might be described by a different choice of independent variables. The scattering matrix formulation that will be described in Section 2.3 uses a linear combination of the V and I variables. Impedance matrices are used when two-port networks are put in series and admittance matrices are used when two-port networks are put in parallel [1].

2.2.3 Chain Two-Port Parameters

The *ABCD* parameters of a two-port relate V_2 and I_2 as independent variables to V_1 and I_1 as dependent variables. In older literature, these parameters are sometimes referred to as "a" parameters. Those "a" parameters are not to be confused with the variable a used in the scattering matrix formulation. The *ABCD* relationship is defined as:

$$V_1 = AV_2 - BI_2$$
$$I_1 = CV_2 - DI_2$$
$$\begin{pmatrix} V_1 \\ I_1 \end{pmatrix} = \begin{pmatrix} A & B \\ C & D \end{pmatrix} \begin{pmatrix} V_2 \\ -I_2 \end{pmatrix}$$

where the minus sign in the I_2 term results from the convention that the current under consideration is into a two-port. When *ABCD* chain matrices were first defined, it was assumed that the current was into port one and out of port two. Thus a minus sign is used when the current is assumed to be into port two. The *ABCD* matrix is often called the chain matrix because the overall result of a cascade of circuits can easily be determined by *chaining* the matrices from the individual circuits together. The T parameters to be defined later with respect to S parameters have properties similar to the *ABCD* parameters. From Figure 2.2:

$$\begin{pmatrix} V_1 \\ I_1 \end{pmatrix} = \begin{pmatrix} A & B \\ C & D \end{pmatrix} \begin{pmatrix} V_2 \\ -I_2 \end{pmatrix}$$

$$= \begin{pmatrix} A & B \\ C & D \end{pmatrix} \begin{pmatrix} \hat{V}_1 \\ \hat{I}_1 \end{pmatrix}$$

$$= \begin{pmatrix} A & B \\ C & D \end{pmatrix} \begin{pmatrix} \hat{A} & \hat{B} \\ \hat{C} & \hat{D} \end{pmatrix} \begin{pmatrix} \hat{V}_2 \\ -\hat{I}_2 \end{pmatrix}$$

and:

$$\begin{pmatrix} \tilde{A} & \tilde{B} \\ \tilde{C} & \tilde{D} \end{pmatrix} = \begin{pmatrix} A & B \\ C & D \end{pmatrix} \begin{pmatrix} \hat{A} & \hat{B} \\ \hat{C} & \hat{D} \end{pmatrix}$$

$$= \begin{pmatrix} A\hat{A} + B\hat{C} & A\hat{B} + B\hat{D} \\ C\hat{A} + D\hat{C} & C\hat{B} + D\hat{D} \end{pmatrix}$$

The overall circuit matrix is the product of the "chained" matrices for the individual circuits. The product can be continued for additional circuits in cascade.

The *EFGH* parameters of a two-port relate V_1 and I_1 as independent variables to V_2 and I_2 as dependent variables. These parameters are sometimes referred to as "b" parameters in older literature. Those "b" parameters are not to be confused with the variable b used in the

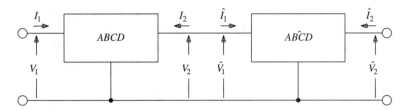

Figure 2.2 Two two-ports in cascade.

scattering matrix formulation. This relationship is defined as:

$$V_2 = EV_1 - FI_1$$
$$I_2 = GV_1 - HI_1$$

$$\begin{pmatrix} V_2 \\ I_2 \end{pmatrix} = \begin{pmatrix} E & F \\ G & H \end{pmatrix} \begin{pmatrix} V_1 \\ -I_1 \end{pmatrix}$$

The *EFGH* parameters are not used very often but can be used to perform chain matrix analysis of circuits also.

2.2.4 Hybrid Two-Port Parameters

The *h* parameters (sometimes called hybrid parameters although there are two types of hybrid parameters) of a two-port relate I_1 and V_2 as independent variables to V_1 and I_2 as dependent variables. This relationship is defined as:

$$V_1 = h_{11}I_1 + h_{12}V_2$$
$$I_2 = h_{21}I_1 + h_{22}V_2$$

$$\begin{pmatrix} V_1 \\ I_2 \end{pmatrix} = \begin{pmatrix} h_{11} & h_{12} \\ h_{21} & h_{22} \end{pmatrix} \begin{pmatrix} I_1 \\ V_2 \end{pmatrix}$$

The *h* parameters are often used to describe the operation of transistors. For bipolar transistors, the *h* parameter equations are rewritten as:

$$V_1 = h_{ix}I_1 + h_{rx}V_2$$
$$I_2 = h_{fx}I_1 + h_{ox}V_2$$

$$\begin{pmatrix} V_1 \\ I_2 \end{pmatrix} = \begin{pmatrix} h_{ix} & h_{rx} \\ h_{fx} & h_{ox} \end{pmatrix} \begin{pmatrix} I_1 \\ V_2 \end{pmatrix}$$

where the first subscript *i* stands for the input parameter, *r* stands for the reverse parameter, *f* stands for the forward parameter, and *o* stands for the output parameter. The second subscript *x* is equal to *e* for a common emitter part, is equal to *b* for a common base part, and is equal to *c* for a common collector part. A similar set of parameters can be described for a field effect transistor using the gate, source, or drain as a common terminal.

The *g* parameters (also sometimes called hybrid parameters) of a two-port relate V_1 and I_2 as independent variables to I_1 and V_2 as dependent variables. This relationship is defined as:

$$I_1 = g_{11}V_1 + g_{12}I_2$$
$$V_2 = g_{21}V_1 + g_{22}I_2$$

$$\begin{pmatrix} I_1 \\ V_2 \end{pmatrix} = \begin{pmatrix} g_{11} & g_{12} \\ g_{21} & g_{22} \end{pmatrix} \begin{pmatrix} V_1 \\ I_2 \end{pmatrix}$$

The hybrid h parameter matrices of two networks can be added to describe networks that are in series at their input and in parallel at their output. The hybrid g parameter matrices of two networks can be added to describe networks that are in parallel at their input and in series at their output.

2.3 S PARAMETERS (SCATTERING PARAMETERS— VOLTAGE REFERENCED)

Scattering parameters also represent a linear description of a two-port. However, the independent variables are neither a voltage nor a current. They are a linear combination of voltages and currents. Under some conditions, this linear combination of voltage and current can be related to the traveling waves on a transmission line with a specified characteristic impedance. Scattering parameters are a very useful tool for a microwave engineer. Optimal power transfer from a source to a load is used in this chapter to derive a set of S parameters or scattering parameters. In this text, rms amplitudes are used for terminal and port quantities. Scattering parameters are normalized values. The normalization impedance choice depends on the user or depends on the measurement system used for making these measurements. When measurement techniques are considered in Chapter 3, the independent variables used in the scattering parameter relationships will be related to traveling voltage waves on a transmission line for those cases where the scattering parameters are normalized to a transmission line's characteristic impedance. Scattering parameters are often measured and referenced to the characteristic impedance of a transmission line by measuring the different forward and reverse traveling voltage waves on a transmission line that has a specified characteristic impedance. However, scattering parameters and matching and gain relationships that result from their use are very often applied when the connections between different components are not a transmission line of that value. Then the traveling waves that actually exist on the interconnecting transmission lines that have different characteristic impedances are not the waves that were used in the original measurement. In order to avoid confusion between the actual traveling waves that exist on the interconnections and the waves that exist on a transmission line having a characteristic impedance equal to the normalization impedance, a more general approach is taken for the scattering parameter development. The application of the scattering parameter method to voltage traveling waves on a transmission line will be given in Chapter 3.

2.3.1 S Parameters—One-Port

Consider the one-port described in the Figure 2.3. The phase reference is assumed to be equal to the phase of E. The maximum power that can be delivered by the source occurs when the load is a conjugate match of the source.

$$Z_S = R_S + jX_S$$
$$Z_L = R_L + jX_L$$
$$Z_L = Z_S^*$$

The maximum power is defined to be the product aa^* and is given as:

$$P_{\max} = \frac{EE^*}{4R_s} = aa^*$$
$$= \frac{(V + IZ_s)(V + IZ_s)^*}{4R_s}$$
$$\sqrt{P_{\max}} = \frac{|E|}{2\sqrt{R_s}} = |a|$$

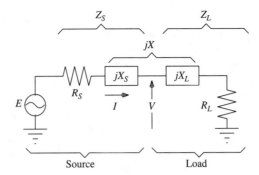

Figure 2.3 One-port used to derive scattering parameters.

The dimension of the magnitude of a is the square root of power. There are an infinite number of ways to choose the phase reference for a in the equation above. The choice of phase reference for a as defined below is the usual choice for voltage-referenced scattering parameters:

$$a = \frac{V + IZ_S}{2\sqrt{R_S}} = \frac{E}{2\sqrt{R_S}}$$

The quantity a is a linear variable since V and I are linear variables. The quantity a is often casually called the *incident power* but it is actually a square root of power quantity, has dimensions of square root of power, and is a linear variable. The quantity a could also have been defined by using the conjugate portion of the above equation. It has sometimes been defined that way and care needs to be exercised when reading older literature to be sure one knows what definition is being used. The quantity aa^* is incident power using either definition. Let P_{max} be called the incident power and then let $b = \Gamma a$ where bb^* stands for a reflected power quantity. Then:

$$bb^* = P_{ref}$$

$$P_{delivered} = P_{del} = P_{max} - P_{ref} = P_{inc} - P_{ref}$$

Substituting:

$$bb^* = P_{ref} = P_{max} - P_{del}$$

$$= \frac{II^*(Z_L + Z_S)(Z_L^* + Z_S^*)}{2(Z_S + Z_S^*)} - \frac{II^*(Z_L + Z_L^*)}{2}$$

$$= \frac{II^*(Z_L - Z_S^*)(Z_L^* - Z_S)}{2(Z_S + Z_S^*)}$$

$$= \frac{(V - IZ_S^*)(V^* - I^*Z_S)}{4R_S}$$

The quantity b has a magnitude equal to the square root of the above equation. The phase of b can be referenced to any quantity one desires just as the phase of a could be referenced to any quantity as a reference. The quantity b has not been uniquely defined in the literature; however, the choice of b that follows is the usual choice.

$$b = \frac{V - IZ_S^*}{2\sqrt{R_S}}$$

Using these definitions, a reflection coefficient Γ is defined by:

$$\Gamma = \frac{b}{a} = |\Gamma|e^{j\phi} = \frac{V - IZ_S^*}{V + IZ_S} = \frac{IZ_L - IZ_S^*}{IZ_L + IZ_S}$$

$$\frac{b}{a} = \frac{Z_L - Z_S^*}{Z_L + Z_S}$$

Voltage traveling waves on a transmission line are related to a and b but are not the same as a and b. Voltage traveling waves are voltage quantities while a and b are square root of power quantities. When the source and load impedances are conjugates of each other, the quantity b divided by a is zero as expected. Usually the source impedance is a real number and then the reflection coefficient given by the above ratio is the same as a voltage reflection coefficient ratio.

$$\Gamma = \frac{b}{a} = |\Gamma|e^{j\phi} = \left.\frac{V - IR_S}{V + IR_S}\right|_{X_S=0} = \frac{IZ_L - IR_S}{IZ_L + IR_S}$$

$$\frac{b}{a} = \frac{Z_L - R_S}{Z_L + R_S}$$

If the source resistance value is the same as the characteristic impedance of a transmission line, then the reflection coefficient ratio is equal to the voltage reflection coefficient on the transmission line. The a and b quantities are then related to the forward and reverse voltage waves on the transmission line by a multiplicative constant.

As an example, determine the ratio of b to a for an impedance of 10 k ohms in series with a 100 fF capacitance at 100 MHz. Do the calculation using a normalization impedance (source impedance) of 50 ohms and using a normalization impedance of 10 k ohms.

$$Z_L = 10^4 - j\frac{1}{2\pi 10^8 10^{-13}} = 10^4(1 - j1.59)$$

$$\Gamma|_{R_S=50} = \frac{9950 - j15900}{10050 - j15900} = 0.9972\angle -0.258°$$

$$\Gamma|_{R_S=10k} = \frac{0 - j15900}{20000 - j15900} = 0.622\angle -51.5°$$

It would be extremely difficult to accurately measure this impedance value using a 50-ohm measurement system. The magnitude of the reflection coefficient is -0.024 dB in a 50-ohm system. The magnitude error in a measurement system would have to be better than 0.001 dB to make an accurate measurement. It would take a phase error of only slightly more than a quarter degree to mistakenly give an inductive impedance rather than a capacitive impedance. When using a 10 k-ohm normalization impedance it is much easier to measure the load, and measurements require much less precision. However, it is not always practical to use normalization impedances of 10 k ohms. This is discussed further in Chapter 3.

2.3.2 *S* Parameters—Multiport

For a network with more than one port, a and b become column vectors and the reflection coefficients Γ_{ij} form a square matrix (S). Then:

$$(b) = (S)(a)$$

Consider the two-port network shown in Figure 2.4. Note that the values of S_{ij} are not unique but depend on the reference impedances Z_i chosen for each port.

$$b_1 = S_{11}a_1 + S_{12}a_2$$

$$b_2 = S_{21}a_1 + S_{22}a_2$$

$$a_i = \frac{V_i + I_iZ_i}{2\sqrt{|\mathrm{Re}(Z_i)|}}$$

$$b_i = \frac{V_i - I_iZ_i^*}{2\sqrt{|\mathrm{Re}(Z_i)|}}$$

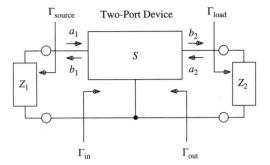

Figure 2.4 Two-port described in terms of scattering parameters.

The value of the reference impedance often depends on the impedance level used by the instrument that is used to make the scattering parameter measurements. By convention, when Z_1 and Z_2 are not specified, it is assumed that they are each 50 ohms. These impedances, Z_1 and Z_2, are generally real but in some instances can be complex. The subscripts for the (S) matrix are in normal convention. For S_{ij}, the i subscript denotes the port at which a response is measured and the j subscript denotes the port where a source is connected for a measurement. By substituting load and source impedances in the equations, one can determine what the input reflection coefficient is in terms of the load reflection coefficient or one can determine the output reflection coefficient as a function of the source reflection coefficient. These equations are:

$$\Gamma_{in} = \frac{S_{11} - \Delta_S \Gamma_{load}}{1 - S_{22} \Gamma_{load}}$$

$$\Gamma_{out} = \frac{S_{22} - \Delta_S \Gamma_{source}}{1 - S_{11} \Gamma_{source}}$$

$$\Delta_S = S_{11} S_{22} - S_{21} S_{12}$$

These are important relationships and will be used when two-port stability is considered. Notice that the first equation assumes that one is looking to the right for both the input and the load in Figure 2.4 while the second equation assumes that one is looking to the left for both the output and the source. It is important to point out that Γ_{in} is not equal to S_{11} and that Γ_{out} is not equal to S_{22}. They have equality only when the opposite port is terminated in a zero reflection coefficient immittance or when $S_{12}S_{21} = 0$. The quantities a and b shown in Figure 2.4 are not to be identified with the top input conductor but with the ports of the network identified by (S). These quantities are often shown schematically in the literature as drawn in Figure 2.5. The lines with arrows in Figure 2.5 stand for vectors and not wires or conductors. In terms of terminal voltages and currents, S_{ij} are given by:

$$S_{11} = \frac{V_1 - I_1 Z_1^*}{V_1 + I_1 Z_1} \bigg|_{V_2 = -I_2 Z_2}$$

$$S_{21} = \frac{V_2 - I_2 Z_2^*}{V_1 + I_1 Z_1} \sqrt{\frac{|Re(Z_1)|}{|Re(Z_2)|}} \bigg|_{V_2 = -I_2 Z_2}$$

$$S_{12} = \frac{V_1 - I_1 Z_1^*}{V_2 + I_2 Z_2} \sqrt{\frac{|Re(Z_2)|}{|Re(Z_1)|}} \bigg|_{V_1 = -I_1 Z_1}$$

$$S_{22} = \frac{V_2 - I_2 Z_2^*}{V_2 + I_2 Z_2} \bigg|_{V_1 = -I_1 Z_1}$$

Figure 2.5 Two-port diagram in terms of a and b quantities.

In terms of a and b quantities, S_{ij} are given by:

$$S_{11} = \left.\frac{b_1}{a_1}\right|_{a_2=0}$$

$$S_{21} = \left.\frac{b_2}{a_1}\right|_{a_2=0}$$

$$S_{12} = \left.\frac{b_1}{a_2}\right|_{a_1=0}$$

$$S_{22} = \left.\frac{b_2}{a_2}\right|_{a_1=0}$$

Considering the definitions of a_1 and a_2, $a_2 = 0$ implies that $V_2 = -I_2 Z_2$ and $a_1 = 0$ implies that $V_1 = -I_1 Z_1$. When $a_2 = 0$, then $\Gamma_2 = 0$ and when $a_1 = 0$, then $\Gamma_1 = 0$. Using circuit variables V and I, the circuit given in the top part of Figure 2.6 is used to determine S_{11} and S_{21}, while the circuit given in the bottom part of Figure 2.6 is used to determine S_{12} and S_{22}. Note that the circuit termination conditions are different between the two circuits. One has a source on the left side and one has a source on the right side. Those using the (S) formulation for the first time often make the mistake of trying to determine all four S_{ij} from just one of the circuits. Two circuits and terminations are needed to satisfy the circuit assumptions.

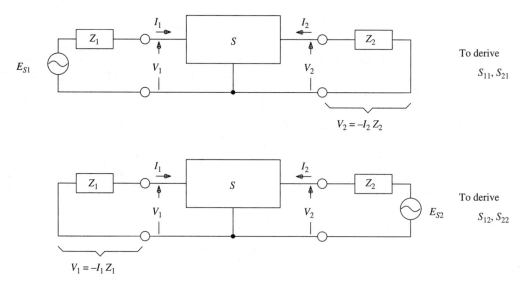

Figure 2.6 Two-port diagram used to measure scattering parameters using V and I.

The schematic given in the top part of Figure 2.7 is used to determine S_{11} and S_{21} using a and b quantities. The schematic given in the bottom part of Figure 2.7 is used to determine S_{12} and S_{22} using a and b quantities. The source is shown as a voltage source with an impedance. It could also be shown as a vector source.

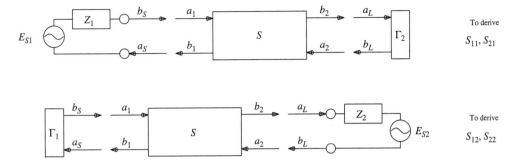

Figure 2.7 Two-port diagram used to measure scattering parameters using *a* and *b*.

Note that the S matrix is defined in terms of measurable terminal quantities, V_i and I_i, plus two reference impedances Z_1 and Z_2. Generally, published S parameters with no impedance reference data mentioned are by convention S parameters referenced to $Z_1 = Z_2 = 50 + j0$. Note that 1 k ohm or 1 M ohm or even $(1 + j2)$ ohm reference impedance S parameters can be defined. The reference impedance for port one can be different than the reference impedance for port two.

Normalized impedances are often used in reflection coefficient equations. Impedance is normalized if it is divided by a reference impedance. Normalized impedance is dimensionless. Admittances can be normalized as well, but admittance is divided by the normalization admittance. The convention for printing normalized impedances varies with different authors. In this book, lowercase y and z will usually be used to designate normalized admittance and impedance respectively. Uppercase Y and Z will be usually be used to designate nonnormalized immittances. Therefore:

$$Z = R + jX$$
$$Y = G + jB$$
$$z = r + jx = \frac{R + jX}{R_n} = \frac{Z}{R_n}$$
$$y = g + jb = \frac{G + jB}{G_n} = \frac{Y}{G_n} = \frac{R_n}{Z} = \frac{1}{z}$$

where R_n is the normalization impedance and G_n is the normalization conductance. These relationships are often used in reflection coefficient equations as follows.

$$Z = R + jX$$
$$Y = G + jB$$
$$\Gamma = \frac{Z - Z_0}{Z + Z_0} = \frac{\frac{Z}{Z_0} - 1}{\frac{Z}{Z_0} + 1} = \frac{z - 1}{z + 1}$$

2.3.3 Relationships between *S* Parameters and Other Parameters

The two-port S-parameter matrix may be related to the system matrix in circuit theory and then to the Z and Y matrices as shown in the following steps. The other two-port parameters can then be determined by transforming between the two-port parameters.

$$b_1 = S_{11}a_1 + S_{12}a_2$$

$$b_2 = S_{21}a_1 + S_{22}a_2$$

$$\frac{V_1 - I_1 Z_1^*}{2\sqrt{|\text{Re}(Z_1)|}} = S_{11}\frac{V_1 + I_1 Z_1}{2\sqrt{|\text{Re}(Z_1)|}} + S_{12}\frac{V_2 + I_2 Z_2}{2\sqrt{|\text{Re}(Z_2)|}}$$

$$\frac{V_2 - I_2 Z_2^*}{2\sqrt{|\text{Re}(Z_2)|}} = S_{21}\frac{V_1 + I_1 Z_1}{2\sqrt{|\text{Re}(Z_1)|}} + S_{22}\frac{V_2 + I_2 Z_2}{2\sqrt{|\text{Re}(Z_2)|}}$$

Multiplying through the equations by 2 and grouping together the voltage and current vectors gives:

$$(0) = \begin{pmatrix} S_{11} - 1 & S_{12} \\ S_{21} & S_{22} - 1 \end{pmatrix} \begin{pmatrix} \frac{1}{\sqrt{R_1}} & 0 \\ 0 & \frac{1}{\sqrt{R_2}} \end{pmatrix} \begin{pmatrix} V_1 \\ V_2 \end{pmatrix}$$

$$+ \begin{pmatrix} S_{11}Z_1 + Z_1^* & S_{12}Z_1 \\ S_{21}Z_1 & S_{22}Z_1 + Z_2^* \end{pmatrix} \begin{pmatrix} \frac{1}{\sqrt{R_1}} & 0 \\ 0 & \frac{1}{\sqrt{R_2}} \end{pmatrix} \begin{pmatrix} I_1 \\ I_2 \end{pmatrix}$$

$$R_i = \text{Re}(Z_i)$$

or in general for an N-port network:

$$(0) = ((S) - (U)) \left(\frac{1}{2\sqrt{R_i}}\right)(V) + ((S)(Z_i) + (Z_i)^*) \left(\frac{1}{2\sqrt{R_i}}\right)(I)$$

$$(0) = ((S) - (U)) \left(\frac{1}{\sqrt{R_i}}\right)(V) + ((S)(Z_i) + (Z_i)^*) \left(\frac{1}{\sqrt{R_i}}\right)(I)$$

where (U) is the identity matrix and (Z_i) is a diagonal matrix. In order to determine the (Z) matrix, the general equation is rearranged to give:

$$(0) = [(U) - (S)] \left(\frac{1}{\sqrt{R_i}}\right)(-U)(V) + ((S)(Z_i) + (Z_i)^*) \left(\frac{1}{\sqrt{R_i}}\right)(I)$$

$$(0) = (-U)(V) + (\sqrt{R_i})[(U) - (S)]^{-1}((S)(Z_i) + (Z_i)^*) \left(\frac{1}{\sqrt{R_i}}\right)(I)$$

and since:

$$(0) = (-U)(V) + (Z)(I)$$

$$(Z) = (\sqrt{R_i})((U) - (S))^{-1}((S)(Z_i) + (Z_i)^*) \left(\frac{1}{\sqrt{R_i}}\right)$$

if (Z) exists (i.e., if $[(U) - (S)]^{-1}$ exists). If (Y) exists, inverting the equation for (Z) or rearranging the starting equation gives (Y) as:

$$(Y) = (\sqrt{R_i})((S)(Z_i) + (Z_i)^*)^{-1}[(U) - (S)] \left(\frac{1}{\sqrt{R_i}}\right)$$

When the normalization impedance is real, then the immittance matrices are related to the scattering matrices as follows.

$$(Z) = (\sqrt{R_i})((U) - (S))^{-1}((S) + (U))(\sqrt{R_i})$$
$$(Y) = (\sqrt{G_i})((S) + (U))^{-1}((U) - (S))(\sqrt{G_i})$$
$$(S) = ((\sqrt{G_i})(Z)(\sqrt{G_i}) - (U))((\sqrt{G_i})(Z)(\sqrt{G_i}) + (U))^{-1}$$
$$(S) = ((U) - (\sqrt{R_i})(Y)(\sqrt{R_i}))((U) + (\sqrt{R_i})(Y)(\sqrt{R_i}))^{-1}$$

If either one of the immittance matrices is known, then the scattering matrix normalized to a different set of resistances can be derived. For instance, if a two-port scattering matrix is measured using a high-normalization impedance for port one and a low-normalization impedance for port two, the immittance matrix can be found. A new scattering matrix can then be calculated from that immittance matrix, normalizing the scattering matrix to 50 ohms. It might not be feasible to measure high-impedance devices with a 50-ohm impedance but once the device is measured with a high-impedance, 50-ohm scattering parameters can be calculated. The number of digits that must be retained in the calculation depends on the transformation ratio needed.

For $Z_i = Z_0$, and Z_0 real:

$$(Z) = Z_0[(U) - (S)]^{-1}[(U) + (S)]$$
$$(Y) = Y_0[(U) + (S)]^{-1}[(U) - (S)]$$
$$(Y) = (Z)^{-1}$$

Starting from the two-port definitions, the following relationships between the two-port parameters definitions can be shown in terms of the immittance parameters. These equations assume the normalization impedance is the same for port one and for port two.

$$S_{11} = \frac{(Y_0 - Y_{11})(Y_0 + Y_{22}) + Y_{12}Y_{21}}{(Y_0 + Y_{11})(Y_0 + Y_{22}) - Y_{12}Y_{21}} = \frac{(Z_{11} - Z_0)(Z_{22} + Z_0) - Z_{12}Z_{21}}{(Z_{11} + Z_0)(Z_{22} + Z_0) - Z_{12}Z_{21}}$$

$$S_{12} = \frac{-2Y_{12}Y_0}{(Y_0 + Y_{11})(Y_0 + Y_{22}) - Y_{12}Y_{21}} = \frac{+2Z_{12}Z_0}{(Z_{11} + Z_0)(Z_{22} + Z_0) - Z_{12}Z_{21}}$$

$$S_{21} = \frac{-2Y_{21}Y_0}{(Y_0 + Y_{11})(Y_0 + Y_{22}) - Y_{12}Y_{21}} = \frac{+2Z_{21}Z_0}{(Z_{11} + Z_0)(Z_{22} + Z_0) - Z_{12}Z_{21}}$$

$$S_{22} = \frac{(Y_0 + Y_{11})(Y_0 - Y_{22}) + Y_{12}Y_{21}}{(Y_0 + Y_{11})(Y_0 + Y_{22}) - Y_{12}Y_{21}} = \frac{(Z_{11} + Z_0)(Z_{22} - Z_0) - Z_{12}Z_{21}}{(Z_{11} + Z_0)(Z_{22} + Z_0) - Z_{12}Z_{21}}$$

For the hybrid and chain parameters the relationships are as follows.

$$S_{11} = \frac{+A + BY_0 - CZ_0 - D}{+A + BY_0 + CZ_0 + D} = \frac{(h_{11}Y_0 - 1)(h_{22}Z_0 + 1) - h_{12}h_{21}}{(h_{11}Y_0 + 1)(h_{22}Z_0 + 1) - h_{12}h_{21}}$$

$$S_{12} = \frac{2(AD - BC)}{+A + BY_0 + CZ_0 + D} = \frac{+2h_{12}}{(h_{11}Y_0 + 1)(h_{22}Z_0 + 1) - h_{12}h_{21}}$$

$$S_{21} = \frac{2}{+A + BY_0 + CZ_0 + D} = \frac{-2h_{21}}{(h_{11}Y_0 + 1)(h_{22}Z_0 + 1) - h_{12}h_{21}}$$

$$S_{22} = \frac{-A + BY_0 - CZ_0 + D}{+A + BY_0 + CZ_0 + D} = \frac{(h_{11}Y_0 + 1)(1 - h_{22}Z_0) + h_{12}h_{21}}{(h_{11}Y_0 + 1)(h_{22}Z_0 + 1) - h_{12}h_{21}}$$

The reverse transformations are as follows.

$$Y_{11} = \frac{(1 - S_{11})(1 + S_{22}) + S_{12}S_{21}}{(1 + S_{11})(1 + S_{22}) - S_{12}S_{21}} Y_0; \qquad Z_{11} = \frac{(1 + S_{11})(1 - S_{22}) + S_{12}S_{21}}{(1 - S_{11})(1 - S_{22}) - S_{12}S_{21}} Z_0$$

$$Y_{12} = \frac{-2S_{12}}{(1 + S_{11})(1 + S_{22}) - S_{12}S_{21}} Y_0; \qquad Z_{12} = \frac{+2S_{12}}{(1 - S_{11})(1 - S_{22}) - S_{12}S_{21}} Z_0$$

$$Y_{21} = \frac{-2S_{21}}{(1 + S_{11})(1 + S_{22}) - S_{12}S_{21}} Y_0; \qquad Z_{21} = \frac{+2S_{21}}{(1 - S_{11})(1 - S_{22}) - S_{12}S_{21}} Z_0$$

$$Y_{22} = \frac{(1 + S_{11})(1 - S_{22}) + S_{12}S_{21}}{(1 + S_{11})(1 + S_{22}) - S_{12}S_{21}} Y_0; \qquad Z_{22} = \frac{(1 - S_{11})(1 + S_{22}) + S_{12}S_{21}}{(1 - S_{11})(1 - S_{22}) - S_{12}S_{21}} Z_0$$

For the hybrid and chain parameters, the reverse transformations are as follows.

$$A = \frac{(1 + S_{11})(1 - S_{22}) + S_{12}S_{21}}{2S_{21}}; \qquad h_{11} = \frac{(1 + S_{11})(1 + S_{22}) - S_{12}S_{21}}{(1 - S_{11})(1 + S_{22}) + S_{12}S_{21}} Z_0$$

$$B = \frac{(1 + S_{11})(1 + S_{22}) - S_{12}S_{21}}{2S_{21}} Z_0; \qquad h_{12} = \frac{+2S_{12}}{(1 - S_{11})(1 + S_{22}) + S_{12}S_{21}}$$

$$C = \frac{(1 - S_{11})(1 - S_{22}) - S_{12}S_{21}}{2S_{21}} Y_0; \qquad h_{21} = \frac{-2S_{21}}{(1 - S_{11})(1 + S_{22}) + S_{12}S_{21}}$$

$$D = \frac{(1 - S_{11})(1 + S_{22}) + S_{12}S_{21}}{2S_{21}}; \qquad h_{22} = \frac{(1 - S_{11})(1 - S_{22}) - S_{12}S_{21}}{(1 - S_{11})(1 + S_{22}) + S_{12}S_{21}} Y_0$$

When using these transformations, it is important to note where the characteristic impedance or admittance is used.

2.3.4 The Series and the Shunt Impedance Circuits

Two two-ports are shown in Figure 2.8. One consists of a series immittance connected between port one and port two with no connections to ground. That two-port does not have a Z matrix since I_1 and I_2 are not independent ($I_1 = -I_2$) but does have a Y matrix. Likewise, the other two-port that consists of port one connected to port two with a wire in addition to having an immittance to ground does not have a Y matrix since V_1 and V_2 are not independent ($V_1 = V_2$), but does have a Z matrix. Each of these circuits has an S matrix. It is left to the reader to show that the scattering parameters for these circuits are as given here.

$$(S)_{Z_{SE}} = \begin{pmatrix} \dfrac{Z_{SE}}{2Z_0 + Z_{SE}} & \dfrac{2Z_0}{2Z_0 + Z_{SE}} \\[3mm] \dfrac{2Z_0}{2Z_0 + Z_{SE}} & \dfrac{Z_{SE}}{2Z_0 + Z_{SE}} \end{pmatrix}$$

$$(S)_{Y_{SH}} = \begin{pmatrix} \dfrac{-Y_{SH}}{2Y_0 + Y_{SH}} & \dfrac{2Y_0}{2Y_0 + Y_{SH}} \\[3mm] \dfrac{2Y_0}{2Y_0 + Y_{SH}} & \dfrac{-Y_{SH}}{2Y_0 + Y_{SH}} \end{pmatrix} = \begin{pmatrix} \dfrac{-Z_0}{Z_0 + 2Z_{SH}} & \dfrac{2Z_{SH}}{Z_0 + 2Z_{SH}} \\[3mm] \dfrac{2Z_{SH}}{Z_0 + 2Z_{SH}} & \dfrac{-Z_0}{Z_0 + 2Z_{SH}} \end{pmatrix}$$

2.3.5 Measuring the Series *R* of a Series *R-L-C* Circuit

If a device such as a diode has a model consisting of a series *R-L-C* circuit in parallel with an impedance, the value of the resistance can be readily found by measuring the

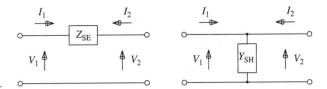

Figure 2.8 Two simple two-ports.

shunt impedance across a transmission line at series resonance [2]. This technique is often referred to as the *DeLoach method*. A schematic of the series *R-L-C* circuit in shunt with a transmission line is shown in Figure 2.9. The measurement equipment is calibrated with a phase reference plane right at the junction of the device and the transmission line. The package or device parallel capacitance has been ignored in the figure since this capacitance is predominately shunted out by the series resonant circuit at resonance. If that assumption is not valid then the shunt capacitance needs to be considered. One can set the instrument to read out a value for S_{21} equal to one-half of the value of the characteristic impedance of the transmission lines when there is no device in the fixture. When R is equal to infinity in the equation for transmission across a shunt *R-L-C* circuit, S_{21} would be equal to one. Setting the instrument to read a value of one-half the value of the characteristic impedance calibrates the equipment with some scale factor K. The instrument reference value indicated is now $K S_{21}$. A device is then inserted and the frequency is changed either manually or automatically until resonance is achieved. If R is small with respect to Z_0 then the indicated value of $K S_{21}$ is approximately equal to R if the instrument has a reference value of $K = Z_0/2$. The value of R can be used to calculate Q of the device if the value of C or L is known.

Let $K = Z_0/2$

$$K S_{21\text{ref}} = K \left. \frac{2Z_{SH}}{Z_0 + 2Z_{SH}} \right|_{Z_{SH}=\infty} = \frac{Z_0}{2}$$

Let $R \ll Z_0$ and $\omega L = 1./\omega C$

$$K S_{21} = \frac{Z_0}{2} \left. \frac{2R_{SH}}{Z_0 + 2R_{SH}} \right|_{Z_{SH}=R\ll Z_0} \approx \frac{Z_0}{2}\frac{2R}{Z_0} \approx R$$

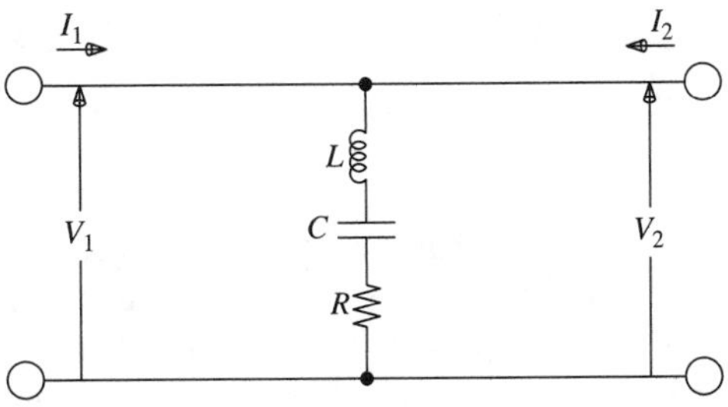

Figure 2.9 Two-port of a series *R-L-C* in shunt.

2.4 *T* PARAMETERS

Transmission matrices using a and b quantities for cascaded networks will now be described. These relationships are called the *T* parameters. *T* is an abbreviation for chain transmission parameters or chain transfer parameters. Refer to Figure 2.10 and set the variables flowing

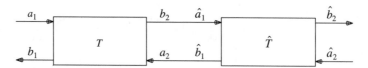

Figure 2.10 Two two-ports in cascade using T parameters.

between the two networks equal to each other as shown in the following equation.

$$b_2 = \hat{a}_1$$

$$a_2 = \hat{b}_1$$

$$\frac{V_2 - I_2 Z_2^*}{2\sqrt{R_2}} = \frac{\hat{V}_1 + \hat{I}_1 \hat{Z}_1}{2\sqrt{\hat{R}_1}}$$

$$\frac{V_2 + I_2 Z_2}{2\sqrt{R_2}} = \frac{\hat{V}_1 - \hat{I}_1 \hat{Z}_1^*}{2\sqrt{\hat{R}_1}}$$

The port voltage and currents are related to each other as given in the following equation.

$$V_2 = \hat{V}_1$$

$$I_2 = -\hat{I}_1$$

This results in the following two simultaneous equations:

$$\frac{V_2 - I_2 Z_2^*}{2\sqrt{R_2}} = \frac{\hat{V}_1 + \hat{I}_1 Z_2^*}{2\sqrt{R_2}} = \frac{\hat{V}_1 + \hat{I}_1 \hat{Z}_1}{2\sqrt{\hat{R}_1}}$$

$$\frac{V_2 + I_2 Z_2}{2\sqrt{R_2}} = \frac{\hat{V}_1 - \hat{I}_1 Z_2}{2\sqrt{R_2}} = \frac{\hat{V}_1 - \hat{I}_1 \hat{Z}_1^*}{2\sqrt{\hat{R}_1}}$$

Subtracting the first equation from the second equation shows that the normalization resistance at port two of the first network must be equal to the normalization resistance at port one of the second network.

$$\hat{R}_1 = R_2$$

Substituting this result back into either of the two equations gives the following result.

$$\hat{Z}_1^* = Z_2$$

If purely real characteristic impedances are used, the output normalization impedances of two ports that are tied together must be equal to each other in order to use cascaded transfer matrices. If one wants to insert the networks together in any order, then the normalization impedances must be real and be the same on all ports. If the normalization impedances are not real, they must be conjugates of each other on the ports that are tied together and then care must be taken in the order the networks are cascaded.

Two definitions of T parameters exist in the literature. The first type uses the a_1 as the dependent variable in the first equation while the other type uses b_1 as the dependent variable in the first equation. Relationships for both sets will be given [3, 4, 5]. Real and equal port normalization impedances are assumed. These relationships can be found by a simple manipulation of the two-port scattering parameter equations. For the type described by Gonzalez, a subscript G will be attached to the matrix. For the type described in the Hewlett-Packard application notes, a subscript H will be attached to the matrix. The defining equations are as given here.

$$\begin{pmatrix} a_1 \\ b_1 \end{pmatrix} = \begin{pmatrix} T_{11} & T_{12} \\ T_{21} & T_{22} \end{pmatrix}\bigg|_G \begin{pmatrix} b_2 \\ a_2 \end{pmatrix} \qquad \begin{pmatrix} b_1 \\ a_1 \end{pmatrix} = \begin{pmatrix} T_{11} & T_{12} \\ T_{21} & T_{22} \end{pmatrix}\bigg|_H \begin{pmatrix} a_2 \\ b_2 \end{pmatrix}$$

$$(T)|_G = \begin{pmatrix} \dfrac{1}{S_{21}} & -\dfrac{S_{22}}{S_{21}} \\ \dfrac{S_{11}}{S_{21}} & -\dfrac{\Delta_S}{S_{21}} \end{pmatrix} \qquad (T)|_H = \begin{pmatrix} -\dfrac{\Delta_S}{S_{21}} & \dfrac{S_{11}}{S_{21}} \\ -\dfrac{S_{22}}{S_{21}} & \dfrac{1}{S_{21}} \end{pmatrix}$$

$$(S) = \begin{pmatrix} \dfrac{T_{21}}{T_{11}} & +\dfrac{\Delta_T}{T_{11}} \\ \dfrac{1}{T_{11}} & -\dfrac{T_{12}}{T_{11}} \end{pmatrix}_G \qquad (S) = \begin{pmatrix} \dfrac{T_{12}}{T_{22}} & +\dfrac{\Delta_T}{T_{22}} \\ \dfrac{1}{T_{22}} & -\dfrac{T_{21}}{T_{22}} \end{pmatrix}_H$$

The T matrices defined by each set are related but are different! Sometimes in older literature the T matrices are defined with the second port variable a_2 away from the port and b_2 toward the port. This is like the differences in definition in the literature for I_2 when using $ABCD$ matrices. Be sure you know which set is being using when trying to decipher an article. Just like the chain matrix using voltage and current variables, the net T matrix for two networks that are cascaded is the product of the individual matrices as given here.

$$(\tilde{T}) = (T)(\hat{T})$$

2.5 THE SMITH® CHART[1]

Appendix B gives some brief complex variable facts that are helpful in reading this section if the reader is not familiar with the description of circular loci in the complex plane.

2.5.1 Bilinear Transformations and the Reflection Coefficient Plane

The relationship that was derived between reflection coefficients and immittance in Section 2.3 is a bilinear transformation. The reflection coefficient chart as defined by Smith [6] is a layout or mapping of the impedance plane onto the reflection coefficient plane. The reflection coefficient plane is labeled with impedance plane values. The formula used for this mapping is:

$$\Gamma = \frac{Z - Z_0}{Z + Z_0} = \frac{z - 1}{z + 1}; \qquad z = \frac{Z}{Z_0}$$

The inverse transformation is:

$$Z = Z_0 \frac{1 + \Gamma}{1 - \Gamma}$$

The voltage standing wave ratio (simply S or VSWR) on a transmission line is defined as the maximum voltage on the line divided by the minimum voltage on the line. If Γ is real, there are two cases to be considered. One is for the magnitude of Γ greater than one and the

---

[1] SMITH® is a Registered Trademark of Analog Instrument Co., Box 950, New Providence, NJ 07974

other for the magnitude of Γ less than one.

$$\text{VSWR} = \frac{1+\Gamma}{1-\Gamma} = \frac{Z}{Z_0}\bigg|_{Im(\Gamma)=0; 0<\Gamma<1} \qquad Z = Z_0(\text{VSWR})|_{Im(\Gamma)=0; 0<\Gamma<1}$$

$$\text{VSWR} = \frac{1-\Gamma}{1+\Gamma} = \frac{Z_0}{Z}\bigg|_{Im(\Gamma)=0; -1<\Gamma<0} \qquad Z = \frac{Z_0}{\text{VSWR}}\bigg|_{Im(\Gamma)=0; -1<\Gamma<0}$$

In the impedance plane, a point position is determined by going a distance x from the origin in the x direction and then a distance y from x axis point in the y direction. The same thing is done in the transformed coordinate system that is labeled with impedance plane values. To plot a point in the mapped coordinate system, first go a distance x in curvilinear coordinates from the mapped impedance plane origin ($z = 0 + j0$), ($\Gamma = -1 + j0$), and then go a distance y in curvilinear coordinates in y direction. The Smith® Chart is drawn on the reflection coefficient plane but labeled with impedance coordinates. The center of the Smith® Chart is $\Gamma = 0$, and using the equation above, this is:

$$Z = Z_0\frac{1+0}{1-0} = Z_0$$

If the chart is normalized, then the center of chart is labeled as $z = Z_0/Z_0 = 1$. If the chart is not normalized, then the center of the chart is labeled as Z_0. The reflection coefficient plane is often called the *gamma plane*.

2.5.2 Straight Lines in the Impedance Plane on the Reflection Coefficient Chart

The mapping of straight lines from the impedance plane to the gamma plane to form the reflection coefficient chart is given next. Pictorially the mapping results as shown in Figure 2.11. For a vertical line in the impedance plane, the resistance is constant. Let that constant be R_1. Then for an impedance $Z = R_1 + jX$:

$$Z + Z^* = 2R_1$$

$$z + z^* = 2r_1$$

$$\frac{1+\Gamma}{1-\Gamma} + \frac{1+\Gamma^*}{1-\Gamma^*} = 2r_1$$

Rearranging and grouping terms:

$$\Gamma\Gamma^*(2r_1 + 2) - 2r_1\Gamma - 2r_1\Gamma^* = 2 - 4r_1$$

$$\Gamma\Gamma^* - \frac{r_1}{1+r_1}\Gamma - \frac{r_1}{1+r_1}\Gamma^* + \frac{r_1^2}{(1+r_1)^2} = \frac{1}{(1+r_1)^2}$$

Therefore:

$$\Gamma_c = \frac{r_1}{1+r_1} \qquad \rho = \frac{1}{|1+r_1|}$$

For a horizontal line, the reactance is constant. Then for an impedance $Z = R + jX_1$:

$$Z - Z^* = 2jX_1$$

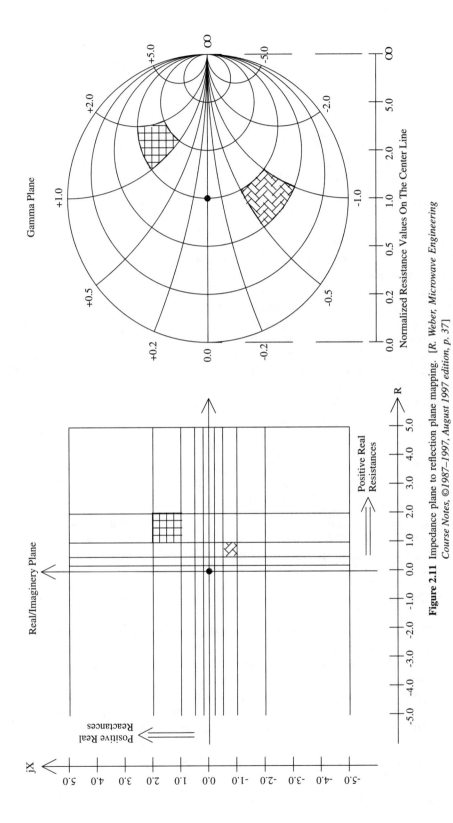

Figure 2.11 Impedance plane to reflection plane mapping. [*R. Weber, Microwave Engineering Course Notes, ©1987–1997, August 1997 edition, p. 37*]

Gamma Plane

Normalized Resistance Values On The Center Line

Real/Imaginery Plane

Positive Real Resistances

Positive Real Reactances

Substituting for Z:

$$Z - Z^* = 2jX_1$$

$$z - z^* = 2jx_1$$

$$\frac{1+\Gamma}{1-\Gamma} - \frac{1+\Gamma^*}{1-\Gamma^*} = 2jx_1$$

$$\left(\frac{1+\Gamma}{1-\Gamma}\right)\left(\frac{1-\Gamma^*}{1-\Gamma^*}\right) - \left(\frac{1+\Gamma^*}{1-\Gamma^*}\right)\left(\frac{1-\Gamma}{1-\Gamma}\right) = 2jx_1$$

$$\Gamma\Gamma^* - \left(\frac{x_1-j}{x_1}\right)\Gamma - \left(\frac{x_1+j}{x_1}\right)\Gamma^* + \left(\frac{1+x_1^2}{x_1^2}\right) = \frac{1}{x_1^2}$$

Therefore:

$$\Gamma_c = \frac{x_1+j}{x_1} \qquad \rho = \frac{1}{|x_1|}$$

The horizontal line $Im(Z) = jZ_0$ in the impedance plane maps to a circle in the reflection coefficient plane with center $1 + j$ and radius equal to 1. The vertical line $Re(Z) = Z_0$ in the impedance plane maps to the circle in the reflection coefficient plane with center $0.5 + j0$ and radius of one-half. Performing this mapping for several vertical and horizontal lines allows one to draw a chart that is the mapping of the impedance plane onto the reflection coefficient plane. Specific maps of this type are called Smith® Charts.

Since admittance is a bilinear transformation of impedance, an admittance reflection co-efficient chart (the mapping of the positive real admittance plane onto the reflection coefficient plane) can also be drawn. Admittances can be read off of an impedance reflection coefficient chart by noting that:

$$\Gamma = \frac{Z - Z_0}{Z + Z_0} = \frac{\frac{1}{Y} - \frac{1}{Y_0}}{\frac{1}{Y} + \frac{1}{Y_0}} = -\frac{Y - Y_0}{Y + Y_0}$$

$$\frac{Y}{Y_0} = \frac{1-\Gamma}{1+\Gamma} = \frac{1+(-\Gamma)}{1-(-\Gamma)}$$

This implies that the normalized admittance can be read from a reflection coefficient chart by plotting $-\Gamma$ on the chart and reading the result as normalized admittance. The quantity $= -\Gamma$ is a rotation of Γ by 180 degrees on the reflection plane. In reading older literature, it needs to be determined whether admittance is plotted on an impedance reflection coefficient chart or an admittance reflection coefficient chart.

2.5.3 The *Z-Y* Chart

The Z-Y chart is derived by first rotating the admittance reflection coefficient chart by 180 degrees and then superimposing the rotated admittance reflection coefficient chart onto the impedance reflection coefficient chart. The step described above of reflection by 180 degrees across the origin of the reflection coefficient chart interchanges the capacitance and inductive sides of the chart. However, the subsequent operation (rotation by 180 degrees) then changes the inductive and capacitive sides of the chart back so that the upper part of the resultant chart is still inductive and the bottom of the resultant chart is still capacitive. The rotated admittance chart is then superimposed on the impedance chart as shown in Figure 2.12 to yield the Z-Y impedance-admittance chart. Notice that the top part of the Z-Y chart is a positive reactance and a negative susceptance. The bottom of the Z-Y chart is a negative reactance and a positive susceptance. An inductive admittance has a negative susceptance. A capacitive admittance has

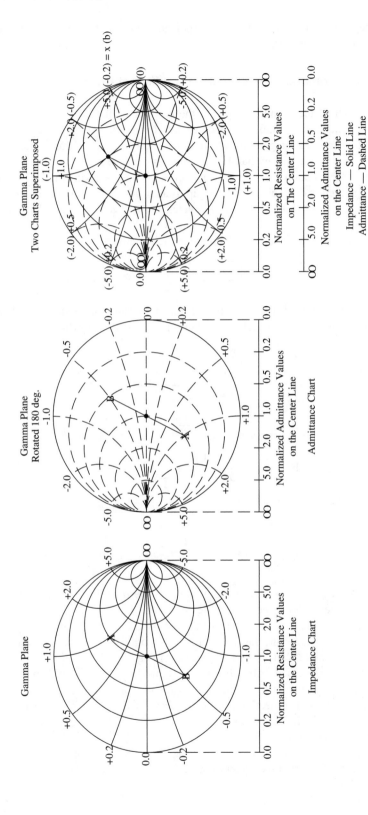

Figure 2.12 Impedance and admittance planes. [*R. Weber, Microwave Engineering Course Notes, ©1987–1997, August 1997 edition, p. 36*]

a positive susceptance. One can use this chart to perform sequential impedance and admittance manipulations on the same chart.

2.5.4 Extended (Compressed) Reflection Coefficient Chart

The right half of the immittance plane, the half containing all the positive resistance or conductance values, maps into the inside of the unit circle in the reflection coefficient plane. Using the same mapping as used to map the right half of the immittance plane into the inside of the unit circle, the left half of the immittance plane maps into the region outside the unit circle. When stability of two-port networks is considered in Chapter 5, the region outside the unit reflection coefficient circle is important in addition to the region inside the unit reflection coefficient circle because unstable devices might have negative real input or output immittances.

Substituting $Z = -Z_0$ into the reflection coefficient equation:

$$|\Gamma| = \left|\frac{Z - Z_0}{Z + Z_0}\right|\Bigg|_{Z \to -Z_0} \to \infty$$

Consider a mapping of a straight line in the impedance plane that lies close to the $R = -Z_0$ line onto the reflection coefficient plane. If $z = -(1 + \delta)$, where δ is real:

$$z = -(1 + \delta)$$

$$\Gamma_c = \frac{-(1 + \delta)}{1 - 1 - \delta} = \frac{1 + \delta}{\delta} = 1 + \frac{1}{\delta}$$

$$\rho = \left|\frac{1}{1 - 1 - \delta}\right| = \frac{1}{|\delta|}$$

$$\Gamma = \frac{-(1 + \delta) - 1}{-(1 + \delta) + 1} = \frac{-2 - \delta}{-\delta} = 1 + \frac{2}{\delta}$$

If $z = -1 + jx$, where x is the value of the imaginary part of z:

$$z = -1 + jx$$

$$\Gamma = \frac{-1 + jx - 1}{-1 + jx + 1} = \frac{x^2 + 2jx}{x^2} = 1 + \frac{2j}{x}$$

Notice that for $z = -1 + jx$, the point is on the vertical line at a position $2/x$ above or below the real line in the reflection coefficient plane. For $z = -(1 + \delta)$, when $\delta > 0$, the negative resistance magnitude is greater than the magnitude of Z_0 and the magnitude of Γ is larger than one. This means that the points to the right of a vertical line through $(1, 0)$ in the reflection coefficient plane have a magnitude of negative R greater than the magnitude of the characteristic impedance. All points to the right of the Z equal to the negative of the characteristic impedance line in the impedance plane then map to the left of a vertical line running through $(1, 0)$ in the reflection coefficient plane. This is important when it necessary to determine whether the reflection coefficient of a network is stable when using a source impedance equal to the normalization impedance.

The mapping for gamma larger than 1 in magnitude is shown in Figure 2.13. The unit circle is shown just to the left of the center of the figure. All the positive real immittance values lie within the unit circle.

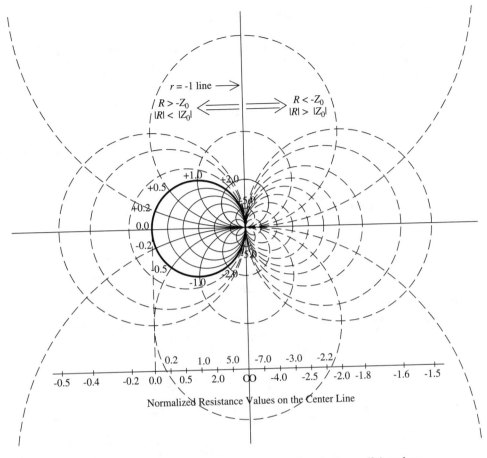

Figure 2.13 Negative real impedance mapped onto the reflection coefficient plane.

Another technique can be used to plot negative real values on the reflection coefficient chart. Consider an impedance with a negative real part:

$$Z = -|R| + jX$$

$$Z^* = -|R| - jX = -(|R| + jX)$$

$$\Gamma = \frac{Z - Z_0}{Z + Z_0} \qquad \Gamma^* = \frac{Z^* - Z_0}{Z^* + Z_0}$$

$$\frac{1}{\Gamma^*} = \frac{Z^* + Z_0}{Z^* - Z_0} = \frac{(-|R| + jX) - Z_0}{(-|R| + jX) + Z_0}$$

This implies that if $1/\Gamma^*$ is plotted, resistance values shown as positive in the standard positive real immittance reflection coefficient chart can be read as negative values, but the imaginary values are read from that same chart with the correct sign and magnitude value. In other words, inductance is still on the top side of the chart and capacitance is still on the bottom part of the chart. The whole left half immittance plane has been mapped inside the unit circle. This mapping is very helpful when plotting the output impedance of oscillator circuits.

2.5.5 Normalization Impedance Change

When changing the position of points or loci plotted on a normalized reflection coefficient chart when changing from one normalization impedance to another normalization impedance consider:

$$z_{old} = \frac{Z}{Z_{0old}}; \quad Z = z_{old} Z_{0old}; \quad z_{new} = \frac{Z}{Z_{0new}}; \quad Z = z_{new} Z_{0new}; \quad z_{new} = z_{old} \frac{Z_{0old}}{Z_{0new}}$$

The lowercase letters are the normalized values plotted on a normalized reflection coefficient chart.

2.5.6 The Carter Chart

An impedance with constant magnitude in the impedance (or admittance) plane lies on a circle centered about the origin in the impedance (or admittance) plane. This circle can be transformed by the reflection coefficient formula to yield a circle in the reflection coefficient plane. The rays of constant angle for the impedance are straight lines and they will also transform into circles in the reflection coefficient plane. A chart made up of these constant impedance magnitude circles and constant angle circles is called a *Carter Chart* [7].

The transformation for constant impedance magnitude $K Z_0$ is:

$$\Gamma_c = \frac{K^2 + 1}{K^2 - 1} \quad \rho = \frac{2K}{|K^2 - 1|}$$

The transformation for a ray through the origin of slope m in the impedance plane is:

$$\Gamma_c = \frac{-j}{m} \quad \rho = \frac{\sqrt{1 + m^2}}{m}$$

Figure 2.14 shows several of these circles plotted with dashed lines on the Z-Y chart in the reflection coefficient plane. Notice that to the left of a vertical line running through the center of the chart, the impedance magnitude is less than Z_0. Recall from the negative impedance chart discussed previously that to the left of a vertical line placed at $(1, 0)$ in the reflection coefficient plane, the magnitude of the negative real part of impedance is less than Z_0. Lines of constant phase angle in the impedance plane are related to Q. Lines of constant Q will be discussed further when resonators are discussed.

2.6 TRANSMISSION LINE MODEL

The equations given in this section are used to model a transmission line in terms of scattering parameters. They are based on traveling voltage waves on a transmission line.

2.6.1 *S* Matrix of a Transmission Line

The scattering matrix of a transmission line of length D and normalized to an impedance that is the same as the characteristic impedance of the line can be derived from considering the traveling waves existing on the transmission line as shown in Figure 2.15. When the line is terminated on the right side with its characteristic impedance, the voltage waves traveling

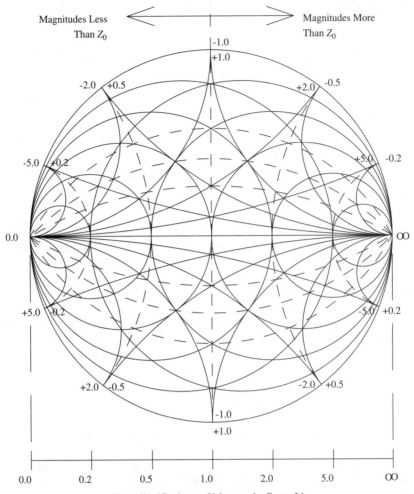

Figure 2.14 Diagram used on a Carter chart (dashed lines).

to the left are zero ($a_2 = 0$ and $b_1 = 0$).

$$b_2 = a_1 e^{-j\theta} = a_1 e^{-j\beta D}$$

$$a_1 = \frac{V_1 + Z_0 I_1}{2\sqrt{Z_0}} = \frac{E}{2\sqrt{Z_0}}$$

$$b_2 = \frac{V_2 - Z_0 I_2}{2\sqrt{Z_0}} = \frac{E}{2\sqrt{Z_0}} e^{-j\theta}$$

$$S_{11} = \left.\frac{b_1}{a_1}\right|_{a_2=0} = 0 = S_{22}$$

$$S_{21} = \left.\frac{b_2}{a_1}\right|_{a_2=0} = e^{-j\theta} = S_{12}$$

$$(S) = \begin{pmatrix} 0 & e^{-j\theta} \\ e^{-j\theta} & 0 \end{pmatrix}$$

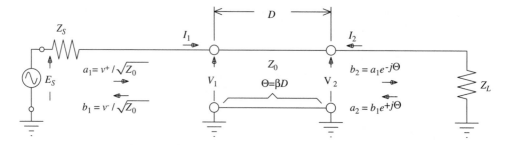

Figure 2.15 Relationship between voltage waves and voltage and current. [*R. Weber, Microwave Engineering Course Notes, ©1987–1997, August 1997 edition, p. 21*]

Notice that the phase shift is negative for a positive length line. This is important when reflection coefficient chart manipulations are considered for a transmission line. It is also necessary to emphasize that this S matrix is valid only when the characteristic impedance of the transmission line is the same as the normalization impedance.

2.6.2 The *S* Matrix of a Lossy Transmission Line

When the transmission line has loss, the quantity θ in the above equations has both a real and an imaginary part. The scattering matrix still has zeroes on the diagonal since the scattering matrix is normalized to the characteristic impedance of the line. The scattering matrix for a lossy line is given by:

$$(S) = \begin{pmatrix} 0 & e^{-\alpha D} e^{-j\beta D} \\ e^{-\alpha D} e^{-j\beta D} & 0 \end{pmatrix}$$

where α is the attenuation constant for the transmission line and β is the phase constant for the transmission line.

The S matrix given for the lossy transmission line above was normalized to the characteristic impedance of the transmission line. The S matrix of a transmission line that has a characteristic impedance of Z_0 but normalized to a different characteristic impedance Z_n is:

$$S_{11} = S_{22} = \frac{\left[\left(\frac{Z_o}{Z_n}\right)^2 - 1\right]\sinh(\gamma D)}{2\left(\frac{Z_o}{Z_n}\right)\cosh(\gamma D) + \left[\left(\frac{Z_o}{Z_n}\right)^2 + 1\right]\sinh(\gamma D)}$$

$$S_{12} = S_{21} = \frac{2\left(\frac{Z_o}{Z_n}\right)}{2\left(\frac{Z_o}{Z_n}\right)\cosh(\gamma D) + \left[\left(\frac{Z_o}{Z_n}\right)^2 + 1\right]\sinh(\gamma D)}$$

where:

$$\gamma D = \alpha D + j\beta D$$

and:

$$\sinh(\gamma D) = \sinh((\alpha + j\beta)D)$$
$$= \sinh(\alpha D)\cos(\beta D) + j\cosh(\alpha D)\sin(\beta D)$$
$$\cosh(\gamma D) = \cosh((\alpha + j\beta)D)$$
$$= \cosh(\alpha D)\cos(\beta D) + j\sinh(\alpha D)\sin(\beta D)$$

where α is the attenuation constant in nepers per meter and β is the propagation constant in radians per meter. This S matrix can be used to describe a lossless line for $\alpha = 0$ and will reduce to the familiar form of the S matrix for $Z_0 = Z_n$. In terms of the transmission line elements R, L, C, and G, the quantities α and β are given as:

$$\gamma = \alpha + j\beta$$
$$= \sqrt{(R + j\omega L)(G + j\omega C)}$$

$$Z_0 = \sqrt{\frac{R + j\omega L}{G + j\omega C}}$$

The $ABCD$, Y, and Z matrices of a lossless transmission line of length D with a characteristic impedance Z_0 are:

$$\begin{pmatrix} A & B \\ C & D \end{pmatrix} = \begin{pmatrix} \cos(\beta D) & jZ_0 \sin(\beta D) \\ jY_0 \sin(\beta D) & \cos(\beta D) \end{pmatrix}$$

$$(Y) = \begin{pmatrix} -jY_0 \cot(\beta D) & jY_0 \csc(\beta D) \\ jY_0 \csc(\beta D) & -jY_0 \cot(\beta D) \end{pmatrix}$$

$$(Z) = \begin{pmatrix} -jZ_0 \cot(\beta D) & -jZ_0 \csc(\beta D) \\ -jZ_0 \csc(\beta D) & -jZ_0 \cot(\beta D) \end{pmatrix}$$

The $ABCD$, Y, and Z matrices of a lossy transmission line of length D with a characteristic impedance Z_0 are:

$$\begin{pmatrix} A & B \\ C & D \end{pmatrix} = \begin{pmatrix} \cosh(\gamma D) & Z_0 \sinh(\gamma D) \\ Y_0 \sinh(\gamma D) & \cosh(\gamma D) \end{pmatrix}$$

$$(Y) = \begin{pmatrix} Y_0 \coth(\gamma D) & -Y_0 \operatorname{csch}(\gamma D) \\ -Y_0 \operatorname{csch}(\gamma D) & Y_0 \coth(\gamma D) \end{pmatrix}$$

$$(Z) = \begin{pmatrix} Z_0 \coth(\gamma D) & Z_0 \operatorname{csch}(\gamma D) \\ Z_0 \operatorname{csch}(\gamma D) & Z_0 \coth(\gamma D) \end{pmatrix}$$

2.6.3 Input Reflection Coefficient Seen Looking into a Terminated Two-Port

Consider a network as shown in Figure 2.16.

$$\frac{b_L}{a_L} = \frac{a_2}{b_2} = \Gamma_L$$

$$b_1 = S_{11}a_1 + S_{12}\Gamma_L b_2$$

$$b_2 = S_{21}a_1 + S_{22}\Gamma_L b_2$$

$$\Gamma_{\text{in}} = \frac{b_1}{a_1} = \frac{S_{11} - \Delta_S \Gamma_L}{1 - S_{22}\Gamma_L}$$

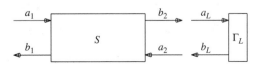

Figure 2.16 Two-port with a load only.

The scattering matrix of a lossless line of length D with characteristic impedance Z_0 equal to the normalization impedance as given in Section 2.6 is:

$$(S) = \begin{pmatrix} 0 & e^{-j\beta D} \\ e^{-j\beta D} & 0 \end{pmatrix}$$

The input reflection coefficient into this length of transmission line as a function of the load reflection coefficient is:

$$\Gamma_{in} = \frac{b_1}{a_1} = \frac{0 - (-e^{-j\theta}e^{-j\theta})\Gamma_L}{1 - 0} = e^{-j2\theta}\Gamma_L$$

where θ is the phase shift constant equal to β times the length of the line. Note that the phase is negative for a positive θ. This means that if the input reflection coefficient Γ_{in} is plotted on a Smith® Chart, the radius of Γ_{in} is the same as the radius of Γ_L. The argument (angle) of Γ_{in} is equal to the angle of Γ_L added to twice the negative of the angular length (-2θ) of the transmission line. Note that the rotation is in the negative angular direction. This is clockwise in the reflection coefficient plane since mathematically positive angles are in the counterclockwise direction. The rotation angle is twice the angular length of the line. This allows one to use the reflection coefficient chart to easily determine the input reflection coefficient and input immittances looking into a transmission line of length θ equal to βD that is terminated in an arbitrary impedance Z_L. A microwave engineer often refers to this as rotating the impedance down a transmission line.

2.6.4 Shift of Reference Plane on a Transmission Line

Often measurements are made at one reference plane on a transmission line but the reflection coefficients are desired at another reference plane or point on that transmission line. How does the S matrix change when the reference point is shifted? Suppose, as shown in Figure 2.17, that the original S matrix is measured at points "B" but that the S matrix referenced to a different spot, either points "A" or "C", is desired. Further assume that the transmission lines of length D_1, D_2, D_3, and D_4 all have a characteristic impedances equal to the normalization impedances of their respective ports. Using either T parameters or S parameters, the incident and reflected waves on each side at the three points "A", "B", and "C" are:

$$a_{1\text{"C"}} = a_{1\text{"B"}}e^{-j\beta D_2} \qquad a_{2\text{"C"}} = a_{2\text{"B"}}e^{-j\beta D_4}$$
$$a_{1\text{"B"}} = a_{1\text{"A"}}e^{-j\beta D_1} \qquad a_{2\text{"B"}} = a_{2\text{"A"}}e^{-j\beta D_3}$$
$$b_{1\text{"C"}} = b_{1\text{"B"}}e^{+j\beta D_2} \qquad b_{2\text{"C"}} = b_{2\text{"B"}}e^{+j\beta D_4}$$
$$b_{1\text{"B"}} = b_{1\text{"A"}}e^{+j\beta D_1} \qquad b_{2\text{"B"}} = b_{2\text{"A"}}e^{+j\beta D_3}$$

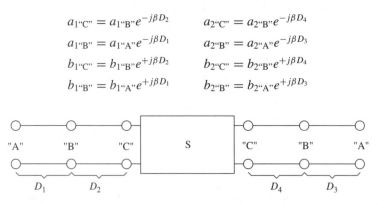

Figure 2.17 Reference plane shifts for scattering parameters.

Moving the reference planes from "B" to "C" on both sides gives the following relationships.

$$\begin{pmatrix} b_1 \\ b_2 \end{pmatrix}_{\text{"B"}} = \begin{pmatrix} e^{-j\beta D_2} & 0 \\ 0 & e^{-j\beta D_4} \end{pmatrix} \begin{pmatrix} b_1 \\ b_2 \end{pmatrix}_{\text{"C"}}$$

$$\begin{pmatrix} a_1 \\ a_2 \end{pmatrix}_{\text{"B"}} = \begin{pmatrix} e^{+j\beta D_2} & 0 \\ 0 & e^{+j\beta D_4} \end{pmatrix} \begin{pmatrix} a_1 \\ a_2 \end{pmatrix}_{\text{"C"}}$$

$$\begin{pmatrix} b_1 \\ b_2 \end{pmatrix}_{\text{"B"}} = \begin{pmatrix} e^{-j\beta D_2} & 0 \\ 0 & e^{-j\beta D_4} \end{pmatrix} \begin{pmatrix} b_1 \\ b_2 \end{pmatrix}_{\text{"C"}}$$

$$= \begin{pmatrix} S_{11} & S_{12} \\ S_{21} & S_{22} \end{pmatrix}_{\text{"B"}} \begin{pmatrix} a_1 \\ a_2 \end{pmatrix}_{\text{"B"}} = \begin{pmatrix} S_{11} & S_{12} \\ S_{21} & S_{22} \end{pmatrix}_{\text{"B"}} \begin{pmatrix} e^{+j\beta D_2} & 0 \\ 0 & e^{+j\beta D_4} \end{pmatrix} \begin{pmatrix} a_1 \\ a_2 \end{pmatrix}_{\text{"C"}}$$

$$\begin{pmatrix} b_1 \\ b_2 \end{pmatrix}_{\text{"C"}} = \begin{pmatrix} e^{+j\beta D_2} & 0 \\ 0 & e^{+j\beta D_4} \end{pmatrix} \begin{pmatrix} S_{11} & S_{12} \\ S_{21} & S_{22} \end{pmatrix}_{\text{"B"}} \begin{pmatrix} e^{+j\beta D_2} & 0 \\ 0 & e^{+j\beta D_4} \end{pmatrix} \begin{pmatrix} a_1 \\ a_2 \end{pmatrix}_{\text{"C"}}$$

$$\begin{pmatrix} S_{11} & S_{12} \\ S_{21} & S_{22} \end{pmatrix}_{\text{"C"}} = \begin{pmatrix} e^{+j\beta D_2} & 0 \\ 0 & e^{+j\beta D_4} \end{pmatrix} \begin{pmatrix} S_{11} & S_{12} \\ S_{21} & S_{22} \end{pmatrix}_{\text{"B"}} \begin{pmatrix} e^{+j\beta D_2} & 0 \\ 0 & e^{+j\beta D_4} \end{pmatrix}$$

$$= \begin{pmatrix} e^{+j2\beta D_2} S_{11B} & e^{+j\beta(D_2+D_4)} S_{12B} \\ e^{+j\beta(D_2+D_4)} S_{21B} & e^{+j2\beta D_4} S_{22B} \end{pmatrix}$$

Using the same procedure, changing the reference plane from "Bfs" to "A" gives the following relationship between the scattering matrices.

$$\begin{pmatrix} S_{11} & S_{12} \\ S_{21} & S_{22} \end{pmatrix}_{\text{"A"}} = \begin{pmatrix} e^{-j\beta D_1} & 0 \\ 0 & e^{-j\beta D_3} \end{pmatrix} \begin{pmatrix} S_{11} & S_{12} \\ S_{21} & S_{22} \end{pmatrix}_{\text{"B"}} \begin{pmatrix} e^{-j\beta D_1} & 0 \\ 0 & e^{-j\beta D_3} \end{pmatrix}$$

$$= \begin{pmatrix} e^{-j2\beta D_1} S_{11B} & e^{-j\beta(D_1+D_3)} S_{12B} \\ e^{-j\beta(D_1+D_3)} S_{21B} & e^{-j2\beta D_3} S_{22B} \end{pmatrix}$$

Change-of-reference-plane equations are useful when it is not possible to measure a circuit right at the plane where the immittance parameters are desired. Specific note should be made that as a move is made farther away from the device for the desired reference plane, the exponent is negative. As a move is made closer to the device, the exponent is positive. It is readily seen that a move could be made toward the device on one side and away from the device on the other side using a negative distance in the equations, and the equations would be easily modified to accommodate that.

2.7 COMPONENTS—RESISTOR, CAPACITOR, INDUCTOR, AND STUBS

In this section some models of a resistor, a capacitor, and an inductor are considered. Transmission line sections called *stubs* are defined. These stubs are often used in place of a capacitor or an inductor.

Each of the three components to be described will have voltage across them. Therefore, electric energy will exist across the component and will be represented by a capacitance.

Current will flow through each component and therefore magnetic energy will exist around the component and will be represented by an inductance. Each component will have energy loss associated with it. Resistances will be used to account for this loss. The relative value of each element used in the model of a component will depend on whether it is intended to be a resistor, capacitor, or inductor. It should be pointed out that a capacitor appears to be an inductive element at very high frequencies just as an inductor appears to be a capacitive element at very high frequencies. The microwave engineer will become accustomed to which reactive component values are to be used in which frequency ranges. Appendix D contains laboratory exercises that can be used to help characterize resistors, capacitors, and inductors.

2.7.1 Resistor Components

A resistor mounted on a PCB is shown in Figure 2.18. An equivalent circuit for it is shown in Figure 2.19. Components are often fabricated symmetrically. Therefore, the circuit should show mirror symmetry. The shunt capacitors to ground in the circuit represent the capacitance of the body of the resistor to a ground plane. Therefore, these values will change as a function of conductor-to-ground-plane distance. The capacitor from input to output is used to represent the electric field energy that exists across the resistor. The voltage across a resistor is often significant and therefore that capacitive energy cannot be ignored. If the resistor is fabricated on the top of a semiconductor in an MMIC, the primary capacitive parasitic might be a capacitance from input to output rather than from one of the leads to ground. The inductors represent the magnetic energy occurring from current flowing in the leads as well as from the current flowing through the length of the resistor. Leadless resistors such as are fabricated on MMICs will still exhibit inductance. The current still must travel from input to output and thus create a magnetic field and an equivalent inductance. The resistor element is shown in the center. Many resistors are fabricated with the resistive element in the center. Some resistors are fabricated from resistive spirals of wire or etched film around a mandrel. These resistors might require an equivalent circuit similar to an inductor that has a large amount of loss. An inductor circuit is considered later. The amount of lead inductance depends on where the resistor is attached to a transmission line or contact in a circuit. Chip resistors have a small amount of lead inductance, whereas a cylindrical resistor with axial leads usually has a larger amount of inductance.

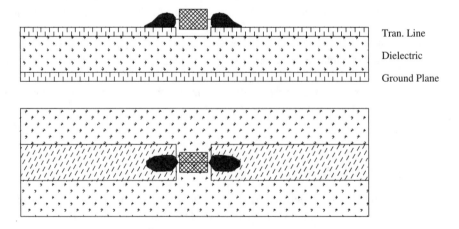

Figure 2.18 Cross section of a chip resistor on a PCB. [P.S.]

In order to understand why some resistors have an inductance reactance and other resistors have a capacitance reactance when they are measured, consider the circuit model of a

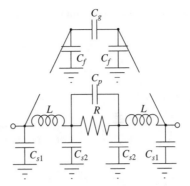

Figure 2.19 Equivalent circuit of a resistor on a PCB. [P.S.]

chip resistor shown in Figure 2.20. This configuration assumes that one end of the resistor is grounded. The analysis is slightly different when the other end that is currently grounded is instead terminated in a transmission line. The ground shorts out some of the shunt capacitance associated with the resistor. The input admittance looking into the resistor is:

$$Y = Y_t + Y_b = \frac{1}{R + j\omega L} + j\omega C_b + j\omega C_t$$

$$= \frac{R + j\omega[(\omega^2 L^2 + R^2)(C_t + C_b) - L]}{R^2 + \omega^2 L^2}$$

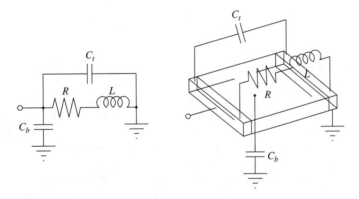

Figure 2.20 Equivalent circuit of a chip resistor grounded at one end.

The imaginary term of the admittance must be negative for the resistor to display inductance. This condition is:

$$\omega^2 < \frac{1}{L(C_t + C_b)} - \left(\frac{R}{L}\right)^2$$

Since the left-hand side of the equation must be positive for real frequencies, the smallest value of frequency for which the condition can be met is when the frequency is equal to zero. Then means that the resistance must be less than:

$$R < \sqrt{\frac{L}{C_t + C_b}}$$

The value of the lower capacitance C_b depends on what the termination impedance on the far end of the resistor is. When the far end is shorted to ground, the values for R, L, and

C_b can be derived from the input admittance for a shorted, shunt, lossy stub. This is:

$$Y_{in} = Y_0 \coth(\gamma \text{ length})$$

$$\approx Y_0 \left(\frac{1}{\gamma \, (\text{length})} + \frac{\gamma \, (\text{length})}{3} \right)$$

$$\approx \sqrt{\frac{\hat{G} + j\omega\hat{C}}{\hat{R} + j\omega\hat{L}}} \left(\frac{1}{\sqrt{(\hat{R} + j\omega\hat{L})(\hat{G} + j\omega\hat{C})}(\text{length})} + \frac{\sqrt{(\hat{R} + j\omega\hat{L})(\hat{G} + j\omega\hat{C})}(\text{length})}{3} \right)$$

$$\approx \frac{1}{(\hat{R} + j\omega\hat{L})(\text{length})} + \frac{(\hat{G} + j\omega\hat{C})(\text{length})}{3}$$

where the tilde quantities are the per-length parameters of the transmission line formed by the resistor, substrate, and ground plane. The value of the capacitance C_b, therefore, is one-third of the lower-side capacitance or:

$$C_b = \frac{Y_0\sqrt{\varepsilon_{\text{eff}}}\,(\text{length})}{3c}$$

The upper-side capacitance is the coupling capacitance across the resistor. The resistance and the inductance are equal to:

$$L = \frac{Z_0\sqrt{\varepsilon_{\text{eff}}}\,(\text{length})}{c}$$

$$R = R_\square \#_{\text{square}}$$

The number of squares times the ohms-per-square of the film determines the resistance. The characteristic impedance and admittance used in the calculations above are approximately the characteristic impedance one would get from a transmission line that is as wide as the resistor and made up of a two-level dielectric—the dielectric of the printed circuit substrate and the substrate of the resistor. If the resistor is relatively thick, the characteristic impedance is somewhat smaller than this impedance due to a thick line effect. If the resistor is mounted with the film side downward, the characteristic impedance used is approximately the same as the characteristic impedance of a transmission line of that width on the substrate. When C_t is zero, then the resistance would have to be less than the square root of three times the characteristic impedance of this effective transmission line to have a net inductive component.

$$R < \sqrt{\frac{L}{C_b}}$$

$$< \sqrt{\frac{\frac{Z_0\sqrt{\varepsilon_{\text{eff}}}\,(\text{length})}{c}}{\frac{Y_0\sqrt{\varepsilon_{\text{eff}}}\,(\text{length})}{3c}}}$$

$$< \sqrt{3}Z_0$$

For large-value resistors that have no resonant frequency, the immittance has a capacitive component at all frequencies while smaller-value resistors are inductive up to the resonance frequency.

When the far end of the resistor is terminated in an open circuit, the resistor appears as an open-circuited stub. The input impedance into the open-circuited stub can similarly be shown to be:

$$Z_{in} = Z_0 \coth(\gamma \text{ length})$$

$$\approx Z_0 \left(\frac{1}{\gamma(\text{length})} + \frac{\gamma(\text{length})}{3} \right)$$

$$\approx \sqrt{\frac{\hat{R} + j\omega\hat{L}}{\hat{G} + j\omega\hat{C}}} \left(\frac{1}{\sqrt{(\hat{R} + j\omega\hat{L})(\hat{G} + j\omega\hat{C})}(\text{length})} + \frac{\sqrt{(\hat{R} + j\omega\hat{L})(\hat{G} + j\omega\hat{C})}(\text{length})}{3} \right)$$

$$\approx \frac{1}{(\hat{G} + j\omega\hat{C})(\text{length})} + \frac{(\hat{R} + j\omega\hat{L})(\text{length})}{3}$$

Notice that this equation can be used to model the loss and therefore the Q of an open-circuited stub that has some resistive loss. The total resistance and inductance of the top electrode is divided by three. As shown earlier, when the resistor is treated as a short-circuited stub, the capacitance is divided by three. Since the lumped model of the resistor depends on its termination impedance, a more accurate model of the resistor is to model the resistor as a lossy transmission line. The transmission line model of the resistor is most appropriate for chip resistors. Leaded resistors can also be treated this way if the leaded section and the resistor body are treated as separate components.

A resistor in series on a printed circuit board was shown in Figure 2.18. One possible lumped-constant equivalent circuit for the resistor mounted on a printed circuit board or substrate was shown in Figure 2.19. Other equivalent circuits using more components can also be used to model the resistor. Whether C_{s1} or C_{s2} best models the resistor will depend on the physical configuration of the resistor and the mounting arrangement. For each different mounting configuration, the best match to measured data should be used. The equivalent circuit in Figure 2.19 was chosen because each component is readily identifiable with a physical portion of the resistor. Each component in the model approximates and represents some physical aspect of the actual resistor in the microstrip circuit of Figure 2.18. The C_g and C_f capacitances represent the gap-coupling capacitance and the gap-fringing capacitance respectively.

Figure 2.21 shows the calculated and measured responses of a 1206-size chip resistor. A drawing of a 1206-size chip is given in Appendix E. The resistive element was assumed to be 2.2 mm (0.085 inch) long and a 70-ohm transmission line was assumed for the lossy line. The leakage capacitance across the resistor was based on measuring several values and was set equal to 55 fF. A lead inductance of 0.2 nH was assumed at each end to account for the current flowing from the transmission line to the resistive metal end contact. The resistor was assumed to be mounted on a printed circuit board that was 0.75 mm (0.030 inch) thick with a dielectric constant equal to 3.38. Actual measurements on resistors were quite close to the calculated curves except that at the high-frequency portions, the immittances had somewhat more capacitance. This chart shows the trend that a low-value chip resistor is inductive and higher-value units are capacitive. Keep in mind that these resistors were characterized as mounted on a 50-ohm transmission line. Their response would be different if one end were shorted or opened.

2.7.2 Capacitor Components

A chip capacitor mounted on a PCB is shown in Figure 2.22. One equivalent circuit is shown in Figure 2.23. Notice that there are series resistances to represent the energy lost due to conduction of current through the leads as well as a parallel resistance to account for energy loss in the dielectric. The value of the parallel resistance may be a function of frequency for some dielectrics. This is the model for a single pair of plates. If the capacitor is constructed

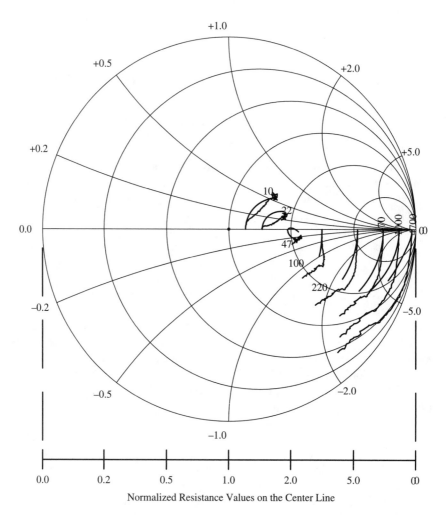

Gamma Plane

Normalized Resistance Values on the Center Line

Figure 2.21 Calculated vs. measured values for a 1206 set of chip resistors dc-6GHz.

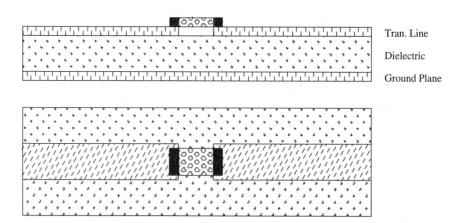

Tran. Line

Dielectric

Ground Plane

Figure 2.22 Cross section of a chip capacitor on a PCB. [P.S.]

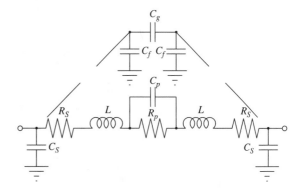

Figure 2.23 Equivalent circuit of a capacitor on a PCB. [P.S.]

from a multiple pair of plates, then the model for the capacitor depends on how the part is attached to a board. If the capacitor is mounted as shown in Figure 2.24, then the inductance for the current that travels to the upper pairs of electrodes is greater than the inductance for the current that travels to the lowest pair of electrodes. Note that the current cannot travel straight upward from underneath the capacitor to the upper plates. The current will travel on the surface of the metal and will travel up the side of the contact and then in between the layers. This results in different values of inductance and therefore results in an equivalent circuit as shown in Figure 2.25. Even though the electromagnetic fields in the region are coupled, the coupling factor may be small enough to allow each individual current path to exhibit its own resonance. The circuit consists of several series R-L-C circuits in parallel. Figure 2.25 shows only three sections in parallel. There may be more or less than three sections in any particular capacitor. In a small region near series resonance, the circuits with the largest inductance will be inductive while the circuits with the smallest inductance will be capacitive. The result will be a parallel resonance at some frequency between the two serial resonant frequencies. This will cause current to *idle* between a pair of plates. This will reduce the effective capacitance of the component. If the capacitor has only two plate pairs, the result will be nearly an open circuit. The current will idle between the plate pairs and this current circulating through the small plate resistances creates significant power dissipation. Halford [8] suggested that multiple-plate chip capacitors should be mounted on their edge as shown in Figure 2.26 to give each plate pair the same path length. Then the R-L-C elements for each plate pair are essentially the same and one can use the equivalent circuit shown in Figure 2.23. If this capacitor is placed onto

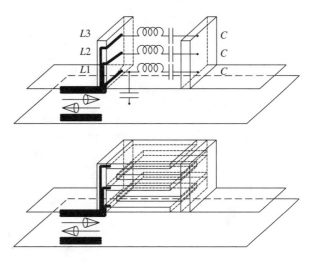

Figure 2.24 Circuit consideration for a chip capacitor.

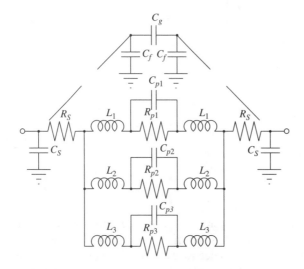

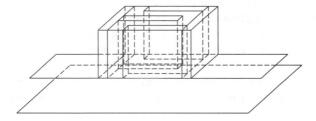

Figure 2.25 Equivalent circuit for a three-layer chip capacitor.

Figure 2.26 Chip capacitor mounted on its edge to minimize parasites.

a transmission line with its plates vertical, the current paths are not only similar but they are shorter. If the capacitor has an even number of plates and it is installed with its plates vertical, then the capacitor is almost symmetrical. When the inductance and capacitance of each layer of the chip are the same, then each layer has the same resonant frequency and the parallel combination of these series L-C circuits results in a single series L-C for the equivalent circuit for the internal portion of the capacitor.

Capacitors used for low-frequency bypass are sometimes formed from two electrodes wound up in a spiral. The equivalent circuits for those capacitors depend on how the current enters the conductors forming their plates.

There are several other capacitor models that can be used. A distributed circuit model is shown in Figure 2.27 for a single-layer capacitor. This model works well for some MMIC capacitors formed from two metal layers with an intervening dielectric layer. When several layers are put one on top of the other, the model is somewhat different. The bold lines indicate an edge view of the current path. The magnitude of the current flowing on the electrodes between the top and bottom plates decreases as the distance from the input end increases. If one assumes that the current goes to zero at the far end, then the single stub model can be used and is a good approximation for many different capacitors. It is important to note from Figure 2.24 that it does make a difference which end of the capacitor goes to which part of the circuit. The parasitic inductor and the length of the transmission line can affect the operation of the capacitor depending on how it is placed in a circuit. If the capacitor is placed on its edge with its plates vertical, then the capacitor becomes more symmetrical. There are also several lumped-constant equivalent circuit models for a capacitor. If leads are added to the capacitor, then the inductance of the leads needs to be added to the equivalent circuit. Whether to consider a lead as a lumped inductor or a distributed transmission line depends on the length of the lead. The equivalent circuit shown in Figure 2.23 is a symmetric equivalent circuit.

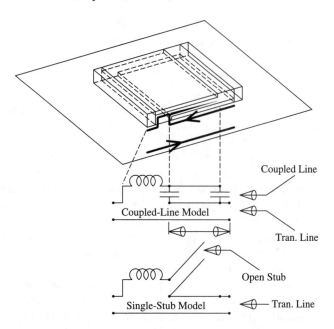

Figure 2.27 Distributed model of a chip capacitor.

The C_g and C_f capacitances represent the gap-coupling capacitance and the gap-fringing capacitance respectively. Other possible equivalent circuit models that might be more valid in specific applications are shown in Figure 2.28. The primary difference between each of these models is where the shunt capacitor is placed. In some applications, the bottom electrode of the capacitor has a significant amount of shunt capacitance to ground. Then the model shown

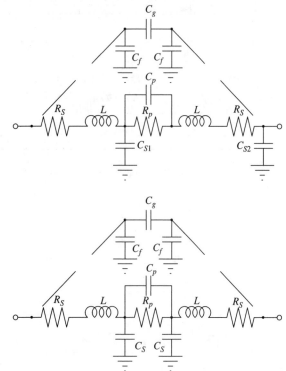

Figure 2.28 Alternate models of capacitors.

in the left-hand side of Figure 2.28 would be used. At other times, a symmetrical circuit model as shown in the right-hand side of Figure 2.28 might better represent the capacitor.

2.7.3 Inductor Components

A simple one-layer wound inductor is shown mounted on a PCB in Figure 2.29. An equivalent circuit of this component is shown in Figure 2.30. If an inductor is formed from multiple layers of windings, then a different equivalent circuit that includes the mutual coupling between layers needs to be used. The single-layer circuit is also applicable to chip inductors that are formed from etched windings on small circular or rectangular mandrels. The primary difference between chip inductors and single-layer wire wound inductors is in the relative magnitude of the parallel capacitance. The capacitor across the inductance is due to the capacitance that exists between turns of the inductor, and this causes the inductor to have a parallel resonant frequency. The parallel resonant frequency is an important parameter of the inductor. The inductor does not give an inductive reactance at frequencies above its parallel resonant frequency. In fact, it is more of a capacitor above its resonant frequency. Instead of presenting higher impedances with frequency it presents lower impedances with frequency. The parallel capacitance accounts for the electric field energy that exists between each turn of the inductor. If there is minimal next-nearest-neighbor coupling between the turns of an inductor, the capacitance between nonadjacent turns can be ignored just as was done in the transmission line model. Under these conditions, the model consists of numerous sections of

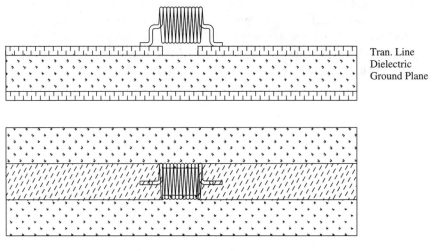

Figure 2.29 Cross section of an inductor mounted on a PCB. [P.S.]

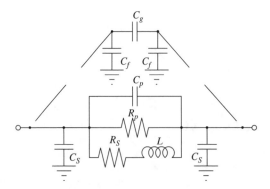

Figure 2.30 Equivalent circuit of an inductor mounted on a PCB. [P.S.]

parallel L-C sections in series. If all of these sections have the same element values, they can be merged into a single L-C as shown in the model. The simple one-layer inductor can be fabricated on a magnetic core. This will increase the inductance but also increase the core loss. The loss in the core must be accounted for. A parallel resistor as shown across the inductor model accounts for this loss. For an air wound inductor, the parallel resistance does not exist except for radiation from the winding. If the core has its own skin depth, then two or more parallel L-C-R sections are put in series to model the inductor component across a wide frequency range as shown in Figure 2.31. Each parallel L-C-R section then accounts for different magnetic core skin effect regions. The parallel resistor is put across the part of the winding that is represented by that part of the core. The parallel capacitor accounts for that amount of electric energy that is in the same region as the magnetic field energy. Some of the magnetic field energy does not penetrate the core at all and is represented by leakage inductance in series with the elements of the component. The skin effect resistance of the winding conductor is accounted for by a series resistance in series with the elements of the components. The series resistor is not a constant due to the variation in skin depth with frequency. In Chapter 12, a transformer will be described and the model of the inductor will be extended to account for mutual coupling between two windings.

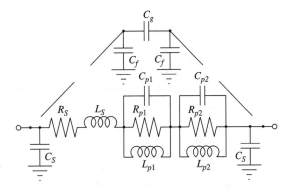

Figure 2.31 Alternate equivalent circuit of an inductor.

The inductance of an air-wound inductor is given approximately by [9]:

$$L = N^2 \frac{R^2}{229R + 254D}$$

where L is in microhenries, R is in mm, and D is in mm. N is the number of turns and R and D are as shown in Figure 2.32. The value of the inductance is said to be accurate to about 1%. As is well known, the total length of the wire must be shorter than a small portion of a wavelength

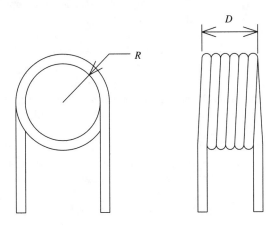

Figure 2.32 Drawing of a single-layer wound inductor.

for this formula to hold, and the parasitic capacitance of the windings will dictate how high a frequency for which the coil acts solely like an inductor. When the turns are closely packed a conformal or enamel coating keeps the windings from shorting and the inductance will be at its maximum. The inductance decreases when the inductor is stretched, pulling the turns apart. This procedure is often used to tune an inductor. The inductance of the inductor will decrease if the space inside the winding (the core) is partially filled with a highly conductive metal and will increase if the space is filled with a high-permeability material. Tunable inductors are often tuned by using a screw made out of a magnetic material such as ferrite to partially fill the volume inside the windings. The screw is used to fill a variable portion of the core. This in effect puts two inductors in series, one primarily air filled and one primarily ferrite filled.

2.7.3.1 MMIC Inductors.

When an inductor is fabricated on top of a semiconductor chip, the structure is almost always planar. The cross-sectional area of the inductor is normal to the surface of the chip. This causes the magnetic field to come out of the surface of the wafer and to go down into the surface of the chip. This magnetic field induces current in the semiconductor and the back mounting surface of chip. Various types of square, polygonal, and circular spirals have been tried for the conductors. Other variations have to do with the width of the conductor versus distance from the center of the spiral and how far into the center of the spiral the conductors go. Still other variations have to do with the thickness of the metal and whether the metal is fabricated on top of the dielectric on top of the chip or supported by air-bridge structures of the surface of the dielectrics. Some examples are given in [10,11]. In order to characterize these inductors, the same circuit diagram used for chip and PCB inductors is used. However, when silicon or other low-resistance semiconductors are used, the shunt capacitors that exist from the inductor to ground must have a resistance in series with them to account for loss due to the current induced in the surface of the low-resistance chip. For semi-insulating GaAs, that resistor is not as much a portion of the circuit as it is for silicon. The metal in the spiral has a resistivity typically in the tens of milli-ohms per square. This results in a fairly large resistance in series with the inductance of the model. Since the inductor is fabricated so close to the surface of a dielectric medium, the substrate, the parallel capacitance between turns is high. This causes the inductor to have a low parallel resonant frequency.

2.7.4 Stubs

Capacitive and inductive circuit elements can often be made from a length of transmission line. These lengths of line are used as one-port elements and are called stubs. Often the length of line is short with respect to a wavelength. The length of line may be terminated at the far end in an impedance. If the impedance at the far end of the stub is a short circuit, the length of line is called a short-circuited stub, shorted stub, or short stub. If the impedance at the far end of the stub is an open circuit, the length of line is called an open-circuited stub or open stub. Examples of some of these lines are shown in Figure 2.33.

Notice that the stubs can either be put in shunt or in series with another transmission line, or they can be put in parallel or series with other circuit components. At the point the stub is connected to a circuit, the total admittance changes if the stub is inserted in parallel while the total impedance changes if the stub is inserted in series. This effect is the same as the change in admittance or impedance resulting when a resistor, a capacitor, or an inductor is inserted at a point. The immittance changes by the amount of admittance or impedance one calculates using the formula for the component—a resistor, a capacitor, or an inductor. The reader will recognize that a stub has an immittance just as a capacitor or an inductor has an immittance.

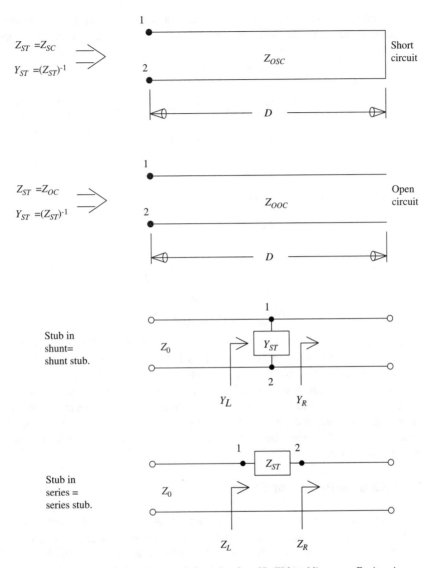

Figure 2.33 Series shunt, open and shorted stubs. [*R. Weber, Microwave Engineering Course Notes, ©1987–1997, August 1997 edition, p. 43*]

Thus in Figure 2.33, for a shunt stub:

$$Y_L = Y_{st} + Y_R$$

and for a series stub:

$$Z_L = Z_{st} + Z_R$$

Short (in length) open stubs can be used for capacitors and short (in length) shorted stubs can be used as inductors. They can be used either in series or in shunt. Just as one uses an equation to calculate the immittance of an inductor or a capacitor, one uses an equation to calculate the immittance of a stub. The input immittance into stubs can be found from the input reflection coefficient equation in Section 2.3.2 and the scattering matrix of a transmission line normalized to its own characteristic impedance given in Section 2.6.1. The input reflection coefficient looking into an open-circuited stub of length L and characteristic impedance Z_0 is:

$$\Gamma_{oc} = \frac{0 + e^{-2j\beta L}}{1 - 0} = e^{-2j\beta L}$$

$$Z_{oc} = Z_0 \frac{1 + \Gamma_{oc}}{1 - \Gamma_{oc}} = Z_0 \frac{1 + e^{-2j\beta L}}{1 - e^{-2j\beta L}}$$

$$Z_{oc} = -jZ_0 \cot(\beta L)$$

$$Y_{oc} = +jY_0 \tan(\beta L)$$

Likewise for a short-circuited stub:

$$\Gamma_{sc} = \frac{0 - e^{-2j\beta L}}{1 - 0} = -e^{-2j\beta L}$$

$$Z_{sc} = Z_0 \frac{1 + \Gamma_{sc}}{1 - \Gamma_{sc}} = Z_0 \frac{1 - e^{-2j\beta L}}{1 + e^{-2j\beta L}}$$

$$Z_{sc} = +jZ_0 \tan(\beta L)$$
$$Y_{sc} = -jY_0 \cot(\beta L)$$

where $\beta = \omega/v = 2\pi f/v$, Z_0 is the characteristic impedance of the transmission line, and v is the velocity of propagation on the transmission line. A single-stub circuit uses one stub, a double-stub circuit uses two stubs, a triple-stub circuit uses three stubs, etc. The immittance presented by a stub can also be found from a reflection coefficient chart. If the stub is a shorted stub, one starts at the short part of the chart and moves clockwise twice the physical length of the line expressed in degrees. After rotating the short, the resulting point on the reflection coefficient chart represents the stub immittance. Similarly for an open stub, one starts at the open point and rotates clockwise.

2.7.5 Stubs in Parallel or Series

Figure 2.34 shows two open stubs placed in parallel at the same point on a transmission line. Generally in circuits that contain more than one stub, there is a length of line between each of the stub insertion points. However, sometimes a shunt or parallel open stub is used to provide a low-rf impedance at a certain point in a circuit while maintaining a dc open. Since the input impedance of an open-circuited stub is low whenever βL is close to $\pi/2$, users are tempted to put two of the stubs in parallel to reduce that impedance even further.

Top View of Microstrip Lines

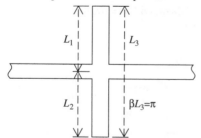

Figure 2.34 Two open stubs in parallel at one point. [R. Weber, *Microwave Engineering Course Notes*, ©1987–1997, August 1997 edition, p. 44]

Suppose that the two stubs differ in length by a small amount ε. Let the lengths L_1 and

L_2 be designated by $L_2 = L_1 + \varepsilon$ and let $L_3 = L_1 + L_2$ where $\beta L_3 = \pi$. Then the two stub admittances Y_1 and Y_2 add, giving $Y_{\text{total}} = Y_1 + Y_2$. Therefore:

$$Y_1 = jY_0 \tan(\beta L_1) = jY_0 \tan(\beta[L_3 - L_2])$$

$$Y_2 = jY_0 \tan(\beta L_2)$$

$$Y_{\text{total}} = jY_0 \tan(\beta[L_3 - L_2]) + jY_0 \tan(\beta L_2)$$

or:

$$\frac{Y_{\text{total}}}{Y_0} = jY_0 \tan(\beta[L_3 - L_2]) + jY_0 \tan(\beta L_2)$$

$$= \left[j \frac{\tan(\beta L_3) - \tan(\beta L_2)}{1 + \tan(\beta L_2)\tan(\beta L_3)} + j\tan(\beta L_2) \right]_{\beta L_2 \neq \frac{\pi}{2}}$$

$$= \left[j \tan(\beta L_3) \frac{1 + \tan^2(\beta L_2)}{1 + \tan(\beta L_2)\tan(\beta L_3)} \right]_{\beta L_2 \neq \frac{\pi}{2}}$$

At $\beta L_3 = \pi$ and $\beta L_2 \neq \pi/2$:

$$\frac{Y_{\text{total}}}{Y_0} = \left[j \tan(\beta L_3) \frac{1 + \tan^2(\beta L_2)}{1 + \tan(\beta L_2)\tan(\beta L_3)} \right]_{\beta L_2 \neq \frac{\pi}{2}; \beta L_3 = \pi} = 0$$

Therefore, if $\beta L_2 \neq \pi/2$, the use of parallel stubs that have a total electrical length of π yields an open circuit, not a low impedance! One should be very careful when using two open-circuited stubs in parallel. The total length may be well below an electrical length of π at the frequency of use but may be at an electrical length of π at a harmonic or other interfering frequency. Two parallel stubs will then not present a low-impedance bypass at the harmonic or interfering frequency. The dual result will be found for the series combination of two shorted series stubs. When the total length of the two series stubs is π, then the total impedance is zero rather than an expected higher impedance. The parallel combination of two open stubs or the series combination of two shorted stubs constitutes a resonator as will be shown in Chapter 9. A resonator has a defined passband.

As given above, the input immittance into an open-circuited stub of length D is:

$$Z_{\text{in}} = -jZ_0 \cot(\beta D)$$

$$Y_{\text{in}} = +jY_0 \tan(\beta D)$$

and the input immittance into a short circuited stub is:

$$Z_{\text{in}} = +jZ_0 \tan(\beta D)$$

$$Y_{\text{in}} = -jY_0 \cot(\beta D)$$

These stubs can be used in place of capacitors and inductors. Let the length of the stub be length D. If βD is small, then for a shorted stub:

$$Z_{\text{in}} = +jZ_0 \tan(\beta D) = j\omega L_{\text{eq}}$$

$$\approx jZ_0 \beta D = jZ_0 \frac{\omega}{v}D = j\omega L_{\text{eq}}$$

and for an open stub:

$$Y_{\text{in}} = +jY_0 \tan(\beta D) = j\omega C_{\text{eq}}$$

$$\approx jY_0 \beta D = jY_0 \frac{\omega}{v}D = j\omega C_{\text{eq}}$$

Therefore a short-shorted stub can represent an equivalent inductance of:

$$\omega L_{\text{eq}} \approx Z_0 \frac{\omega D}{v} \Rightarrow L_{\text{eq}} \approx Z_0 \frac{D}{v}$$

and a short-open stub can represent an equivalent capacitance of:

$$\omega C\text{eq} \approx Y_0 \frac{\omega D}{v} \Rightarrow C\text{eq} \approx Y_0 \frac{D}{v}$$

where v is the velocity of propagation in the stub.

Since $v = 3 \times 10^{10}$ cm per second in air and since the characteristic impedance of a wire in air is on the order of 120 ohms, the inductance of a cm of wire in air is approximately:

$$L_{\text{eq}} \approx \frac{Z_0 D}{v} = \frac{120 \text{ ohms} \times 1 \text{ cm}}{3 \times 10^{10} \text{ cm per second}} \approx 4 \text{ nH per cm} \approx 10 \text{ nH per inch}$$

Likewise:

$$C_{\text{eq}} \approx \frac{Y_0 D}{v} = \frac{0.01 \text{ mhos} \times 1 \text{ cm}}{3 \times 10^{10} \text{ cm per sec}} \approx 0.3 \text{ pF per cm} \approx 0.75 \text{ pF per inch}$$

These equivalent values work well as an estimate in most applications. Keep in mind that these estimates are based on a characteristic impedance for the wire of 120 ohms. If the wire is quite close to a ground plane, the characteristic impedance might be lower. Remember that each transmission line has some inductance and some capacitance. In filter designs, the amount of capacitive energy in a shorted stub or the amount of inductive energy in an open stub must be accounted for at frequencies near circuit-resonant frequencies. These approximations may not work well when the bandwidth of the resonant circuits is a consideration.

2.7.6 Input Immittance into Lossy Stubs

When stubs are considered in Chapter 9 on filter design, it will be important to know the input impedance of a stub that has loss. The input impedance of a short-circuited stub at either an open-circuit ($\beta D = n\pi + \pi/2$) or a short-circuit ($\beta D = n\pi$) resonance is given by:

$$Z_{\text{in}}|_{\beta D=\frac{\pi}{2}+n\pi} = Z_0 \tanh(\gamma D)|_{\beta D=\frac{\pi}{2}+n\pi} = Z_0 \coth(\alpha D)$$

$$Z_{\text{in}}|_{\beta D=n\pi} = Z_0 \tanh(\gamma D)|_{\beta D=n\pi} = Z_0 \tanh(\alpha D)$$

Likewise, the input impedance of an open-circuited stub at either an open-circuit ($\beta D = n\pi$) or short-circuit ($\beta D = n\pi + \pi/2$) resonance is given by:

$$Z_{\text{in}}|_{\beta D=n\pi} = Z_0 \coth(\gamma D)|_{\beta D=n\pi} = Z_0 \coth(\alpha D)$$

$$Z_{\text{in}}|_{\beta D=\frac{\pi}{2}+n\pi} = Z_0 \coth(\gamma D)|_{\beta D=\frac{\pi}{2}+n\pi} = Z_0 \tanh(\alpha D)$$

Figure 2.35 shows a drawing of the voltage and current along a quarter-wavelength-long stub. The distributed capacitors and inductors are shown dashed to emphasize that the electric energy and magnetic energy varies on reactively terminated lines. The voltage is small near the short circuit and thus the capacitive energy is small, but the inductive energy is large there since current is flowing in all of the inductors. Likewise, near the open circuit, the distributed capacitors and inductors are shown dashed again to emphasize that the electric and magnetic (capacitive and inductive) energy varies as a function of position from the open circuit. The inductive energy is small near the open end since the current must go to zero at the open circuit. However, the voltage is high and thus the capacitive energy is high near the open circuit. When filter resonators are discussed, the effect of this energy variation will be important in filter design and filter loss.

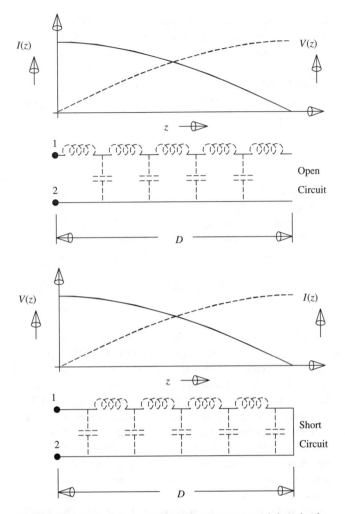

Figure 2.35 Voltage and current distribution along open and shorted stubs.

2.7.7 Foreshortening Stubs

As discussed in the previous section, the capacitive energy near a shorted end of a stub is nearly zero. When a shorted stub is put on a printed circuit board, the effect of getting the shorted end of the stub to ground must be considered. The current has to get to ground and current travel is inductance. When a piece of metal is used to get the shorted end of the stub to ground, that metal adds length to the stub. Since the current in the part of the metal going to ground from the top of the board is at right angles to the current in the stub on the top of the board, the fields from the two portions do not interact. Since there is very little capacitive energy to consider at that point, the stub can be considered as starting at the back side of the board rather than at the starting point on the top side of the board. To the first approximation, the length of the stub is the drawn length on the top side of the board plus the length through the thickness of the board to the ground plane. In filter designs and other narrowband circuits, if this length is ignored, the circuit will be mistuned.

A similar argument holds for the open portion of a stub. The stub appears longer than the drawn length of the stub. This can be determined to a first-order approximation by plotting the characteristic admittance of a transmission line versus its width. Consider the open-circuited piece of transmission line shown in Figure 2.36. A dashed line is drawn along the edge of the transmission line to indicate that the effective width of the line is wider than it really is. The characteristic admittance of a transmission line is related to the capacitance per unit length of the line as follows:

$$Y_0 = \frac{c}{\sqrt{\varepsilon_{\text{eff}}}} C = vC$$

where c is the speed of light, v is the velocity of light on the transmission line, ε_{eff} is the effective dielectric constant, and C is the capacitance per unit length of the line. In Figure 2.36, the transmission line fringing capacitance makes the line appear wider than it really is. The characteristic admittance can be expressed in terms of the capacitance calculated from the parallel plate capacitance plus the fringing capacitance as follows.

$$Y_0 = vC \approx v\left(\frac{\varepsilon_r \varepsilon_0 W \cdot 1 \text{ meter}}{H} + C_{\text{fringe}}\right) = \frac{\varepsilon_{\text{eff}} \varepsilon_0 (W + 2x) \cdot 1 \text{ meter}}{H}$$

Notice that on the end of the line, the dashed line is drawn the same distance away from the end of the line. The charges that accumulate on the edge of the line to make the line look wider than it really is also accumulate on the end of the line and make it look longer than it really is. How much is that length? Plotting the characteristic admittance versus width can be used to determine how much wider a line is than it appears. That width is $2x$. The line appears x longer than it really is. That amount of length must be taken off the length of the line when it is put on a printed circuit board. Most of the computer-aided design tools compensate for that automatically if the user calls out an open stub. However, if the user just uses a transmission line of given length for the stub and leaves the line open, the actual length of the line must be made shorter than the calculated length by the amount x. For a printed circuit board with a relative dielectric constant equal to 3.38, a plot of Y_0 versus width is shown in Figure 2.37. The line appears to be wider by approximately $1.7 \, W/H$. This implies that x/H is equal to approximately 0.85. The line must be shortened by that amount to account for fringing off the end of the line. This approximate calculation ignores the fringing off the corner of the line but if the line is wide, as it likely would be for a filter section or open stub, that amount of fringing capacitance would be small compared to the fringing capacitance at the end of the stub.

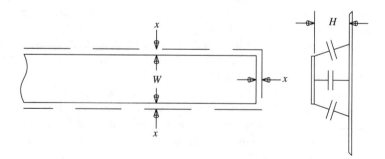

Figure 2.36 Top and end view of an open circuit stub end.

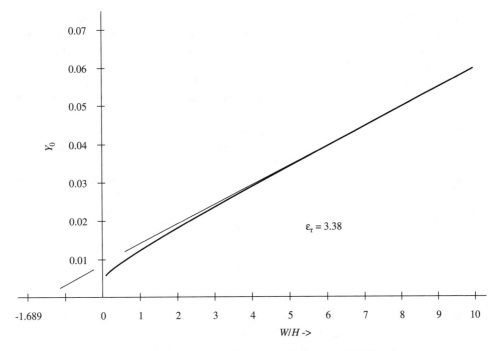

Figure 2.37 Characteristic admittance for a microstrip line and 3.38 dielectric constant material.

2.8 TEST FIXTURE CONSIDERATION

When components are characterized, the test fixture that they are measured in should reflect the environment that they will be used in. The test fixture also needs to have adequate shorts, opens, and through transmission lines.

Figure 2.38 shows a drawing of a test fixture that has been used to characterize leaded and chip resistors, capacitors, and inductors. This particular test fixture was made to characterize 1206 components and therefore has a 2.54-mm (0.1-inch) space between the transmission lines for the component. Similar fixtures have been used with smaller gaps to characterize smaller components. The test fixture shown in Figure 2.38 was made on board material 0.75 mm (0.03 inch) thick and having a dielectric constant of 3.38. Notice that there is a through line on the

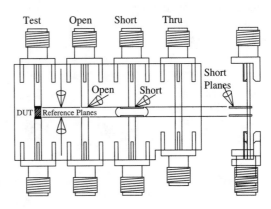

Figure 2.38 Drawing of a PCB test fixture for measuring components.

fixture. The fixture is notched down so that the distance between the connectors on the through line is equal to the sum of the distances of the transmission lines from the connectors to the component under test. The short is shown as protruding above the board. A simple conductor from the transmission line to ground will not act as a short. The electric fields project out from the transmission line in the area above the line and to the sides of the line. The short is wider than the line and extends above the board to provide a short circuit "plane" for these fields. When that short circuit plane does not exist, the currents must go to the ground plane through a length of metal to reach a plane of zero potential. This metal conductor is an inductor and not a ground. The position of the short would be markedly altered if one simply puts a piece of metal strip or wire the same width of the transmission line from the transmission line to ground. The open is also used for calibration of the test fixture. The fixture is calibrated on the vector network analyzer using the on-board open, on-board short, and board through. The component under test is soldered across the open space between the two transmission lines and the full two-port scattering parameters are measured on it.

2.9 TRANSMISSION LINE DISCONTINUITIES

When considering the effect of transmission line discontinuities, an examination of the structure will often guide one to the form of the equivalent circuit for the additional reactive components that must be put in to account for the discontinuities.

Consider the current flowing down a transmission line such as shown in Figure 1.1. Notice that the current must move along the transmission line. When this current moves a little way, it creates magnetic field energy, and thus the section of the conductor through which it moves represents an inductance. This moving charge is separated from an equal and opposite charge moving along the ground conductor. The electric energy associated with this charge separation is capacitive energy. The equivalent circuit of the lossless transmission line is thus as shown in Figure 1.1, where the dots in the middle stand for many more L-C sections. If two transmission lines of different sizes are connected together, parasitic capacitance and parasitic inductance resulting from this connection must be added to the overall circuit to adequately incorporate all of the inductive and capacitive energy storage. Thus the model for the connection between two transmission lines would look similar to that shown in Figure 2.39. Current must move from the edge of one transmission line to the edge of the other transmission line and this movement represents inductance. The capacitance of the charge that exists along the lateral direction across the line to ground is represented by the capacitors in the gap model. The two inductors and the one capacitor in a tee formation for the parasitic section might also be modeled in a pi formation with two capacitors and one inductor or one might just have a two-element el section depending on the physical configuration of the connection. The important thing to remember is to incorporate all of the parasitic energy, but do not duplicate

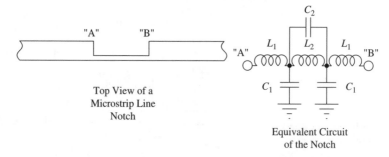

Top View of a
Microstrip Line
Notch

Equivalent Circuit
of the Notch

Figure 2.39 A notch example of transmission line discontinuities.

the energy included in another part of the model. Keep in mind that the energy associated with each of the transmission lines is included in the transmission line circuit model. However, the transmission line model is for an infinite length of line. Whenever the line is truncated, one has to determine what parasitic capacitance or inductance results from the truncation. The end nodes, points "A" and "B" in Figure 2.39, previously did not have any capacitance acting on them on them except directly down to the ground plane. Now charges there can look elsewhere, resulting in parasitic capacitance from those nodes to different parts of a circuit or to the ground plane.

EXERCISES

2-1 A chip resistor has a lumped-constant model with a total inductance of 1 nH, a capacitance across the part of 0.2 pF, and a resistance 100 ohms. The total shunt capacity from the part to ground is 100 fF. Determine the input reflection coefficient S_{11} of the resistor when it is measured in series with a 50-ohm line and when it is placed at the end of a 5-ohm line. Determine S_{21} of the resistor when it is placed in series with a 50-ohm line.

2-2 Given $h_{11} = 130$; $h_{12} = -j0.1$; $h_{21} = -j4$, and $h_{22} = 0$, find the S matrix.

2-3 Plot the reflection coefficient on a reflection coefficient chart normalized to 50 ohms when the following components are placed on a 50-ohm transmission line terminated with 50 ohms.

> A 100-ohm shunt shorted stub that is one-eighth of a wavelength long.
> A 50-ohm series open stub that is three-eighths of a wavelength long.
> A 10-nH series inductor at 1 GHz.
> A 15-pF series capacitor at 1 GHz.
> A 100-nH shunt capacitor at 1 GHz.
> A 10-nH shunt inductor at 1 GHz.
> A 100-ohm series resistor.
> A 10-ohm shunt resistor.

2-4 A 10-pF capacitor is placed in shunt with a 50-ohm resistor. Determine the locus of points traced by the impedance on a reflection coefficient chart normalized to 50 ohms from 1 MHz to 5 GHz.

2-5 A 5-nH inductor is placed in series with a 50-ohm resistor. Determine the locus of points traced by the impedance on a reflection coefficient chart normalized to 50 ohms from 1 MHz to 5 GHz.

2-6 Start with a 50-ohm resistor. Use a reflection coefficient chart that is normalized to the appropriate impedance. The normalization impedance may change as different components are added. Add the following components in sequence, starting with the 50 ohm resistor. Work at 1.59155 GHz.

> A 2-pF capacitor in series.
> A 10-nH inductor in shunt.
> A quarter-wavelength 31.62-ohm transmission line.
> A one-eighth wavelength 10-ohm shorted stub in series.
> A 25-ohm resistor in shunt.
> A 50-ohm resistor in series.
> A 100-ohm, one-quarter wavelength shorted stub in shunt.

Determine the input impedance looking into the 100-ohm shorted stub.

2-7 A perfect two-port reverse isolator circuit has $S_{11} = 0$, $S_{12} = 1$, $S_{21} = 0$, $S_{22} = 0$. Determine the Z, Y, and $ABCD$ matrices if they exist. If one of the sets of two-port matrices does not exist, explain why. Redo the problem by interchanging S_{12} and S_{21} for a perfect forward isolator circuit.

3

S-Parameter Measurement
Methods

3.1 *S*-PARAMETER MEASUREMENT METHODS

If one can measure terminal voltages and currents while imposing required termination conditions on the ports, scattering parameters can be measured. For networks that do not have terminals for which voltages and current can be measured, then traveling waves related to the equivalent voltages and currents at the various ports can be measured. There are commercially available network analyzers and vector voltmeters that measure the magnitude and phase of one signal in reference to another signal. Various types of couplers and samplers are used to form a sample of the parameters necessary for the measurements. The phase and magnitude of a low-frequency signal can be displayed on an oscilloscope. These measurements can then be used to calculate the scattering parameters of networks at low frequencies. Scattering parameters at very low frequencies, for example, 10 Hz or below (subaudible), all the way up to very high frequencies, for example, 500 THz and beyond (light and beyond), can be measured. All that is necessary is that the equipment respond to the voltage and current (electric field and magnetic field) magnitude and phase relationships. The parameters can be measured as continuous waves (cw) or during a pulsed signal.

There are two fundamental types of network analyzers. The vector network analyzer measures the amplitude and phase of a signal with respect to a reference signal. The scalar network analyzer measures only the amplitude of the waves with respect to the amplitude of a reference. There are several kinds of vector network analyzers. There are two-detector, three-detector, and four-detector versions. If a signal and a reference are measured, then two detectors are required. The signal and reference detectors can be switched between the various spots in the test circuit to complete the measurements required. However, when the detectors are switched from one part of the circuit to another, some error occurs in the switches and leakage exists in the switches, allowing some of one signal to leak into the measured signal of another part, and so on.

3.1.1 *S*-Parameter Measurement Using an Oscilloscope

Scattering parameters are the ratio of two linear variables under specified terminal termination conditions. Using a setup similar to that shown in Figure 3.1, the voltage of the source E_s can be displayed on one channel of a dual-channel oscilloscope. Other channels of

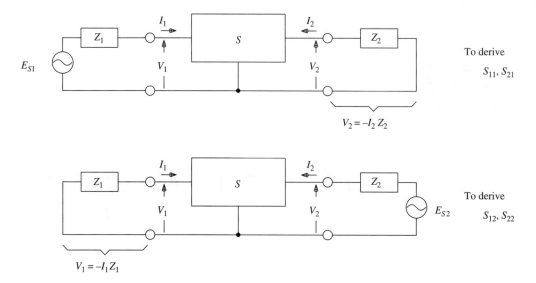

Figure 3.1 Setup to measure scattering parameters using voltage and current.

the oscilloscope can be used to measure V_1 and V_2. As long as the time base of the oscilloscope is referenced to the voltage source E_s, the phase and magnitude relationship between the terminal voltages and the reference can be established. The currents can be determined from these voltages if Z_1 and Z_2 are known. It is important to recognize that both the top and bottom circuits are needed to determine all four scattering parameters of a two-port network. If the circuit has a higher number of ports, then a multichannel oscilloscope can be used to make all of the measurements or a probe can be used to sequentially measure the port voltages. The source needs to be put sequentially on all the ports as well. This enables one to make measurements on differential circuits using the mixed-mode analysis of Chapter 4 as well as regular scattering parameter measurements described in this chapter. If a phase meter is available, the phase difference between the two signals can be measured with better accuracy than reading the difference between zero crossings on an oscilloscope. In other techniques that use a network analyzer, a very-high-frequency signal is down converted (translated to a lower frequency) for magnitude and phase determination at a lower frequency. This process preserves the amplitude and phase of the signal. Down conversion will be described when mixing phenomena is discussed in Chapter 11.

3.1.2 *S*-Parameter Measurement Using Traveling Waves

Many measurement methods for measuring scattering parameters rely on the measurement of traveling waves on a transmission line. The incident and reflected a and b quantities are related to traveling waves on a transmission line. A lossless ideal transmission line has a real characteristic impedance. The scattering parameters can be referenced to any value of impedance. However, if the quantities a and b are referenced to an impedance Z_0 that is also the characteristic impedance of a transmission line, a simple relationship exists between the traveling voltage waves v^+ and v^- on the transmission line and the a and b quantities. This relationship is shown in Figure 3.2. The characteristic impedance Z_0 must be real for the relationship to hold. The voltage V and the current I at any point on a lossless transmission

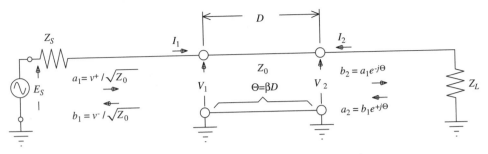

Figure 3.2 Relationship between voltage waves and voltage and current.

line are:

$$V = v^+ + v^-$$

$$I = \frac{v^+}{Z_0} - \frac{v^-}{Z_0}$$

where V is the total voltage at a point on the line and I is the total current in the positive direction (often but not always to the right) at this point on the line, and the lowercase v's are the positive and negative voltage waves on the line. When using voltage referenced S parameters, if:

$$Z_i = Z_i^* = Z_0$$

$$R_S = Z_0$$

then:

$$a + b = \frac{V}{\sqrt{Z_0}} = \frac{v^+ + v^-}{\sqrt{Z_0}}$$

$$a - b = I\sqrt{Z_0} = \frac{v^+ - v^-}{Z_0}\sqrt{Z_0} = \frac{v^+ - v^-}{\sqrt{Z_0}}$$

Notice that these a and b quantities are related to the traveling waves by a multiplicative constant. If the voltage traveling waves are measured and if the reference impedances are chosen equal to the characteristic impedance of the transmission line, then the S parameters are easily determined from the traveling waves.

The above relationship relating traveling waves on a uniform transmission line to scattering matrix variables holds only when the reference impedance Z_0 of a port of the scattering matrix is equal to the transmission line characteristic impedance. Using these assumptions:

$$b_1 = S_{11}a_1 + S_{12}a_2$$

$$b_2 = S_{21}a_1 + S_{22}a_2$$

$$\sqrt{Z_0}b_1 = S_{11}\sqrt{Z_0}a_1 + S_{12}\sqrt{Z_0}a_2 \Rightarrow v_1^- = S_{11}v_1^+ + S_{12}v_2^+$$

$$\sqrt{Z_0}b_2 = S_{21}\sqrt{Z_0}a_1 + S_{22}\sqrt{Z_0}a_2 \Rightarrow v_2^- = S_{21}v_1^+ + S_{22}v_2^+$$

Figure 3.3 shows a setup for measuring S parameters using transmission line couplers to sample the traveling waves and thus the a and b quantities. A transmission line coupler samples part of one of the traveling waves on the line and puts that sample on the sampling port of the coupler. Couplers are described in Chapter 12. The scattering parameter S_{ij} is determined by:

$$S_{ij} = \frac{b_i}{a_j}\bigg|_{a_k=0}$$

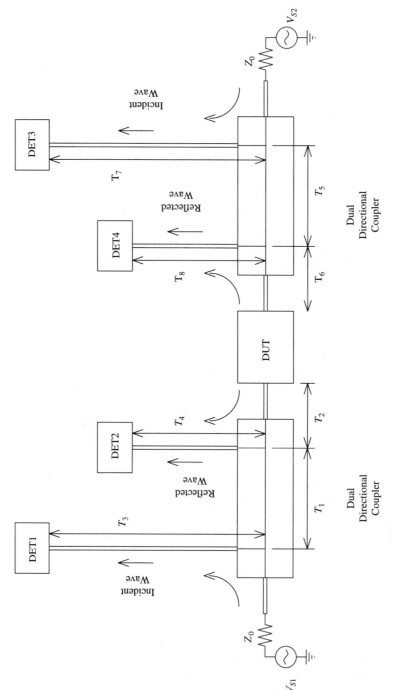

Figure 3.3 Basic network analyzer system.

where k takes on all possible values except j. In Figure 3.3, this condition is met by having the device terminated in the characteristic impedance of the respective transmission lines. Recall that when a transmission line is terminated in a resistance equal to its characteristic impedance, the reflected voltage wave on that transmission line is zero.

In order to get one-half of the parameters, the source is put on the output of the device. This gives accurate parameters for the device only when the bias of the device does not depend on the input drive and when the part is linear. A Class C circuit, and many Class B circuits, do not met these criteria. Class A, B, and C circuits are discussed in Chapter 7.

Most rf and microwave measurement network analysis equipment measures the vector relationship between either v^+ or v^- and a reference signal. The following relationships hold if the reference impedances are real and if the traveling waves measured are measured right at or referenced to the device ports.

$$S_{11} = \left. \frac{\frac{v_1^-}{\sqrt{Z_{01}}}}{\frac{v_1^+}{\sqrt{Z_{01}}}} \right|_{v_2^+=0} \qquad S_{21} = \left. \frac{\frac{v_2^-}{\sqrt{Z_{02}}}}{\frac{v_1^+}{\sqrt{Z_{01}}}} \right|_{v_2^+=0} \qquad S_{12} = \left. \frac{\frac{v_1^-}{\sqrt{Z_{01}}}}{\frac{v_2^+}{\sqrt{Z_{02}}}} \right|_{v_1^+=0} \qquad S_{22} = \left. \frac{\frac{v_2^-}{\sqrt{Z_{02}}}}{\frac{v_2^+}{\sqrt{Z_{02}}}} \right|_{v_1^+=0}$$

3.1.3 Measuring the *S* Parameters with a Slotted Line

This section is not necessary for understanding the rest of the book. It is included only for its historical significance. It shows how to measure scattering parameters with an inexpensive piece of equipment, the slotted line. A slotted line is a piece of equipment that allows the total voltage at a point on the line to be sampled. Figure 3.4 shows a sketch of the voltage along a line terminated with a $\Gamma = 0.5$ at 135 degree. The probe is moved along the line and the total voltage is plotted as a function of position on the line. Since there are two voltage waves traveling on the transmission line, using linear superposition, the total voltage on the line is the sum of these two voltage waves. A voltage probe will sample the magnitude of the total voltage on the line. A wave traveling in the positive direction has a complex voltage that varies with distance z as:

$$v^+ = v_f e^{-j\beta(z-z_o)}$$

where β is the propagation constant and z_0 is a reference point on the line where the wave phase is equal to zero. A wave traveling in the negative direction has a complex voltage that varies with distance z as:

$$v^- = v_r e^{j\beta(z-z_o)} e^{j\phi_0}$$

where ϕ_0 is the phase shift in the reverse wave at the forward phase reference point. The constants v_f and v_r are real quantities. At some point, the minimum magnitude of the voltage will be:

$$v_{\min} = |v_f - v_r|$$

At some other point, the maximum magnitude of the voltage will be:

$$v_{\max} = v_f + v_r$$

The phase shift can be determined from the position of the minimum voltage. At the position of the minimum voltage:

$$\phi_0 = -2\beta(z - z_o) + \pi$$

If the forward wave v_f is set equal to the reference wave v_1^+ and the reverse wave v_r is set equal to v_1^-, v_2^+, and v_2^- one at a time using a coupler or other sampling network, then the

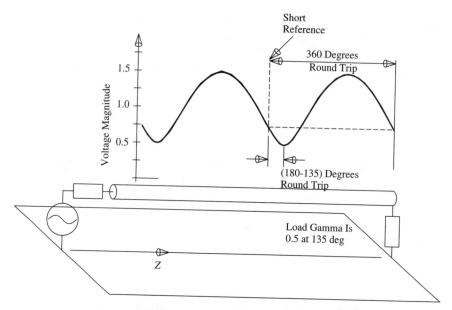

Figure 3.4 Sketch of a standing wave on a transmission line.

full set of two-port scattering parameters (both magnitude and phase) can be measured with the slotted line setup. Measuring v_{max} allows the magnitude of v_r to be determined when v_f is known. The magnitude of v_f is determined by terminating the line in a matched load. Using a device like a directional coupler allows the four traveling waves associated with the device to be sampled. The following set of conditions can then be used to determine the scattering parameters. The quantities $a_1, b_1, a_2,$ and b_2 represent samples of the $a_1, b_1, a_2,$ and b_2 actually on the device. Reference planes and throughs are established using short circuits and through lines in the same way they are used with other techniques.

Connect the input and the output of the slotted line

between ports a_1 and b_1 to measure S_{11},

between ports b_2 and a_1 to measure S_{21},

between ports b_1 and a_2 to measure S_{12}, and

between ports b_2 and a_2 to measure S_{22}.

The author used this technique in the late sixties to measure S parameters of high-power devices under pulse operation. Since an optical equivalent of a slotted line is an interferometer, the two-port S parameters (both magnitude and phase) of optical devices can also be measured using the same connections whenever the optical device operates on a wavefront (single mode) of optical energy. This includes to the first order the gain and phase characteristics of a laser or an LED. The results also apply to waveguide analysis. For optical and waveguide analysis, superposition of modes can be included but each mode set will have its own set of scattering parameters and the instrumentation will have to be adapted to separate and measure each mode set.

3.1.4 One-Plus-Γ Method of Measuring Scattering Parameters

A method exists for measuring scattering parameters using power dividers and vector voltage measuring equipment [1]. A matched symmetrical power divider is connected to a source as shown in Figure 3.5. The reference output port of the power divider is terminated

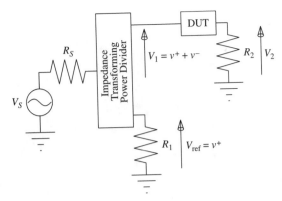

Figure 3.5 Block diagram of a one-plus-gamma system

with a load equal to the reference impedance value. The other output port of the power divider is terminated by the input of a two-port under test. The output port of the two-port under test is terminated in a load equal to the second port reference impedance. Under this set of conditions, the voltage across the input of the device is $v_1^+ + v_1^-$. The voltage across port two is v_2^+. There is no v_2^- since the output port is terminated in a matched load. The quantity v_1^+ is determined by measuring the voltage at the reference output load of the power divider. At that port, there is no reflected wave since the port is terminated in the reference impedance and therefore the voltage is just the incident voltage. The conditions seen by the two-port under test are the conditions for determining S_{11} and S_{21}. The ratio of the input voltage on the two-port to the reference voltage (from the first output port of the power divider) is:

$$\frac{v_1^+ + v_1^-}{v_1^+} = 1 + \frac{v_1^-}{v_1^+} = (1 + S_{11})|_{v_2^-=0}$$

The ratio of the output voltage of the two-port to the reference voltage is:

$$\frac{v_2^+}{v_1^+} = S_{21}|_{v_2^-=0}$$

In order to determine S_{12} and S_{22}, the device is turned end for end in the same manner as making normal scattering parameter measurements. S_{22} is determined from:

$$\frac{v_2^+ + v_2^-}{v_2^+} = 1 + \frac{v_2^-}{v_2^+} = (1 + S_{22})|_{v_1^-=0}$$

and S_{12} is determined from:

$$\frac{v_1^+}{v_2^+} = S_{12}|_{v_1^-=0}$$

This measurement technique can be used to measure high- or low-impedance scattering parameters. The input divider network can be designed to provide a high-impedance source for the device. For instance, at frequencies in the low tens of megahertz, the input power divider can be designed to provide a one- or ten-kilo-ohm impedance reference. This allows one to make high-impedance scattering parameter measurements of high-impedance devices such as CMOS or junction FETs at low frequencies. Figure 3.6 shows a photo of a fixture used for measuring FETs in the tens-of-megahertz range. Figure 3.7 shows a schematic of Figure 3.6. The parasitic capacitance and inductance of the resistor power divider network (resistors and mounting structure) determines the limit on the useful frequency range of the fixture.

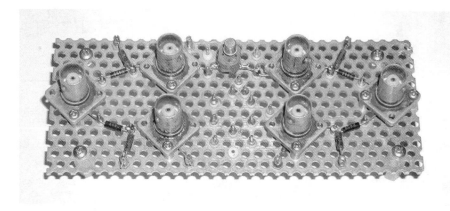

Figure 3.6 Photo of a low-frequency, high-impedance one-plus-gamma system.

The voltages at ports one and two are measured by probing. The probe impedance needs to be considered part of the device or subtracted from the measurements. If the probe impedance is insignificant with respect to the device impedance level, it can be ignored.

High-impedance measurements above a couple of hundred ohms are impractical using transmission line couplers and so on. A simple calculation of the practical impedance levels for a coaxial line will give one a feel for these impedance levels. Determine the outer diameter necessary for a 500-ohm coaxial air line that has a 1/4-mm inner diameter, using:

$$500 = 60 \ln (D2/D1)$$
$$D2 = .25 e^{(500/60)} \approx 1 \text{ m}$$

One can almost walk through such a line. For low-frequency, high-impedance measurements, the outer diameter of the coax could be as large as about 10 mm. Most commercial coaxial lines have a plastic dielectric. Depending on the material used, the dielectric constant value might be somewhere between 2.1 and 2.3. When the center conductor is 1.27 mm in diameter and the dielectric constant is 2.07, the outer diameter of the dielectric is about 4.2 mm for a 50-ohm line. If the impedance required is 100 ohms, then the outer diameter would go up to 14 mm. In high-impedance lines, the space between the inner and outer conductor is often filled with a spiral of dielectric. The line would have an effective dielectric constant of approximately 1.5. A cable with a 10-mm outer diameter and a 0.25-mm inner diameter would have a characteristic impedance of approximately 180 ohms.

On the other hand, for a 5-ohm characteristic impedance, the outer diameter for an inner diameter of 0.25 mm in air is:

$$5 = 60 \ln (D2/D1)$$
$$D2 = .25 e^{(5/60)} \approx 0.272 \text{ mm}$$
$$(0.272 - 0.25)/2 = 0.011 \text{ mm}$$

The space between the inner conductor and the outer conductor is about 11 microns—not quite the distance of two red blood cells. A 10-ohm line with a 0.64-mm inner diameter and a dielectric constant of 3 would have a dielectric thickness of approximately 0.1 mm. This type of line can be fabricated. From this brief calculation, it is seen that most practical transmission lines have impedance somewhere between 5 or 10 ohms and 200 ohms. Higher-impedance lines are available but they have helixes for center conductors. In planar

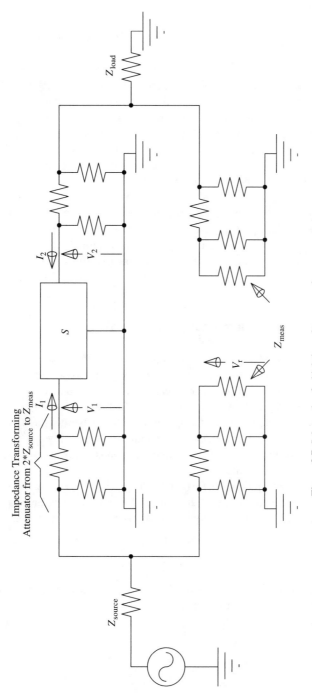

Figure 3.7 Schematic of a high-impedance one-plus-gamma system.

structures, lower impedances are possible but the width of the lines soon gets to be an appreciable portion of a wavelength and then the TEM approximation breaks down due to the large line width.

3.1.5 Low-Impedance Scattering Parameter Measurements

High-power rf and microwave devices operate at a low-impedance level. The impedance level is often under 1 ohm. In the high-impedance setup described in Section 3.1.4, the important parasitic is the shunt capacity. In a low-impedance setup, the important parasitic is series inductance. Voltage and current measurements can be made in the hundreds-of-megahertz range using resistor voltage dividers and transformers. For impedance levels of 1 ohm, a 100-ohm resistor is large enough to use for a voltage tap. The current can be measured with a transformer. Voltage is measured right at the device port using a resistive voltage divider. Current is measured using a transformer. The current measured is the sum of the device current plus the current in the voltage divider. This technique has been used at frequencies approaching 500 megahertz [2]. Figure 3.8 shows a diagram of a low-impedance measurement setup. The scattering parameters are calculated from the defining formulas. The termination impedances on the ports need to be accurately determined when the device is removed and the load is placed directly on the measurement port. If the parasitics of the transformer used for sampling the current are too large, then a load alone can be used on the load ports.

3.1.6 Scattering Parameters Using Load Pull Methods

If a device develops its bias from the input rf signal, then the reverse parameters cannot be measured by using the source on the output. The part bias would change and the reverse parameters would be invalid. Class C parts can be characterized in a linear region of operation that often exists around a given input drive level. Stability calculations in this linear region of operation can be made if the scattering parameters can be measured. Figure 3.9 shows a load pull setup for determining the scattering parameters when only the load is varied [3]. The source remains on the input side. Two separate algorithms can be used to extract the scattering parameters from the measurements.

In the following formulas, let B stand for b values that are divided by a reference value. Usually the reference value is a_1. Let A stand for a values that are divided by this same reference value. The scattering parameter equations can then be written as follows.

$$B_1 = S_{11} + S_{12}A_2$$

$$B_2 = S_{21} + S_{22}A_2$$

Put two different loads Γ_{L1} and Γ_{L2} on the part. Add a second subscript to the A and B quantities to differentiate these two loads.

$$\Gamma_L = \frac{a_2}{b_2} = \frac{A_2}{B_2}$$

$$\Gamma_{L1} = \frac{A_{21}}{B_{21}} \qquad \Gamma_{L2} = \frac{A_{22}}{B_{22}}$$

$$B_{11} = S_{11} + S_{12}A_{21} \qquad B_{21} = S_{21} + S_{22}A_{21}$$

$$B_{12} = S_{11} + S_{12}A_{22} \qquad B_{22} = S_{21} + S_{22}A_{22}$$

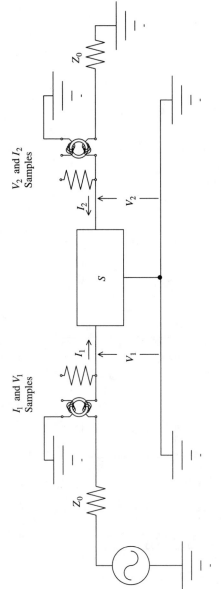

Figure 3.8 Schematic of a low-impedance voltage and current measurement system. [*Robert J. Weber, "Oscillator Design Techniques Using Calculated and Measured S Parameters," Copyright ©1991, Short Course, 45th Annual Symposium on Frequency Control, May 1991, Los Angeles, CA.*]

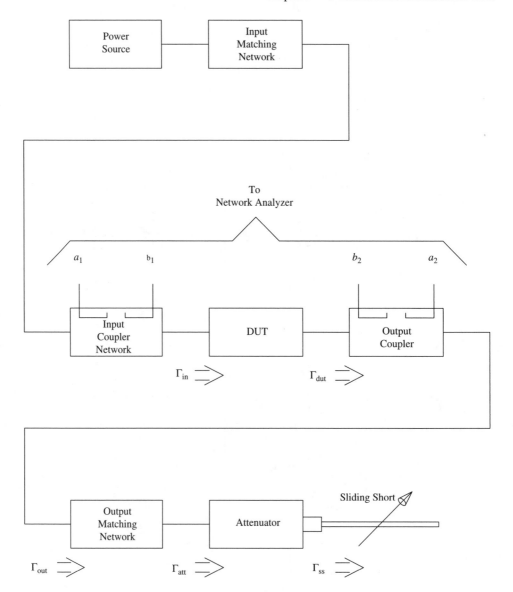

Figure 3.9 Schematic of a load pull method of measuring scattering parameters. [*Robert J. Weber, "Oscillator Design Techniques Using Calculated and Measured S Parameters," Copyright ©1991, Short Course, 45th Annual Symposium on Frequency Control, May 1991, Los Angeles, CA.*]

Subtracting the two sets of equations and then substituting the result back into the first of the sets of equations gives the scattering parameters.

$$B_{11} = S_{11} + S_{12}A_{21} \qquad B_{21} = S_{21} + S_{22}A_{21}$$
$$B_{12} = S_{11} + S_{12}A_{22} \qquad B_{22} = S_{21} + S_{22}A_{22}$$

$$S_{12} = \frac{B_{11} - B_{12}}{A_{21} - A_{22}} \qquad S_{22} = \frac{B_{21} - B_{22}}{A_{21} - A_{22}}$$

$$S_{11} = \frac{A_{21}B_{12} - A_{22}B_{11}}{A_{21} - A_{22}} \qquad S_{21} = \frac{A_{21}B_{22} - A_{22}B_{21}}{A_{21} - A_{22}}$$

All four of the scattering parameters can be measured by measuring the A and B ratios under two load conditions. These load conditions can be chosen close to the optimum load for the chosen operating conditions of the part. When saturation of voltage or current is taking place, the loads are chosen to minimize the change in saturation (i.e., the locus of change should be at right angles to the path of maximum change in saturation). The load has been pulled over two different load values.

This procedure can be extended using a circular locus for a load pull. Consider the setup shown in Figure 3.9. A sliding short will give a reflection coefficient plane locus of magnitude one but the phase changes over two pi as the short is pulled along a transmission line. Automated tuners are available commercially to allow specified loci to be placed on the device as well. Those tuners that place a capacitive slug near a transmission line and then move that slug along the line present a circular load locus to the system. This circular locus facilitates the determination of the scattering parameters by load pull.

The input reflection coefficient locus into the load network is a bilinear transformation of the output locus of the load as it is pulled along a circular locus. When an attenuator (a circuit that reduces the magnitude of the traveling waves by a fixed amount) is placed on a sliding short, the circular locus is transformed to a smaller circular locus at the input to the attenuator. When a reflection coefficient that has a circular locus is placed on the output of an output matching network, the input reflection coefficient into that matching network is also circular. Continuing back through the circuit in Figure 3.9, at each point the reflection coefficient has a circular locus as the sliding short is pulled. Figure 3.10 depicts what this might look like. Figure 3.11 shows a blowup of B_1, B_2, and A_2 all plotted on the same reflection coefficient chart.

Since the various parameters are related by bilinear transformations, if the loci are not circles, then the device is not operating regionally linear and therefore the S parameters are not valid. Harmonic generation will vary as the load is pulled if the load locus variation is too large, and this can cause the loci to depart from a circular shape. A valid set of S parameters must be obtained from the device operating at the same condition when determining each of the parameters. The author has used different control networks on the input power of a Class C device to either given constant collector current or constant output power in order to obtain circles. If some control is not done, the circles appear as one-sided flattened ellipses due to saturation effects. These saturation effects are discussed in Chapter 7 for power devices. When using small circular loci, the input drive does not vary very much and the part remains under essentially the same operating conditions.

The *linear* S parameters are derived from regionally linear operation of a non-linear device using the following derivation.

$$B_1 = S_{11} + S_{12}A_2 \qquad\qquad B_2 = S_{21} + S_{22}A_2$$

$$= S_{11} + S_{12}A_{2c} + S_{12}A_{2\rho}e^{j(\phi_{A_{2\rho 0}}+\phi_k)} \qquad = S_{21} + S_{22}A_{2c} + S_{22}A_{2\rho}e^{j(\phi_{A_{2\rho 0}}+\phi_k)}$$

$$= B_{1c} + B_{1\rho}e^{j(\phi_{B_{1\rho 0}}+\phi_k)} \qquad\qquad = B_{2c} + B_{2\rho}e^{j(\phi_{B_{2\rho 0}}+\phi_k)}$$

$$B_{1c} = S_{11} + S_{12}A_{2c} \qquad\qquad B_{2c} = S_{21} + S_{22}A_{2c}$$

$$B_{1\rho} = |S_{12}|A_{2\rho} \qquad\qquad B_{2\rho} = |S_{22}|A_{2\rho}$$

$$S_{12} = \frac{B_{1\rho}}{A_{2\rho}}e^{j(\phi_{B_{1\rho 0}}-\phi_{A_{2\rho 0}})} \qquad\qquad S_{22} = \frac{B_{2\rho}}{A_{2\rho}}e^{j(\phi_{B_{2\rho 0}}-\phi_{A_{2\rho 0}})}$$

$$S_{11} = B_{1c} - S_{12}A_{2c} \qquad\qquad S_{21} = B_{2c} - S_{22}A_{2c}$$

3.1.7 A General Relationship for Determining Scattering Parameters

As was shown in the previous sections, the scattering matrix for a linear *n*-port can be determined without the need for matched terminations. For a general relationship, consider the

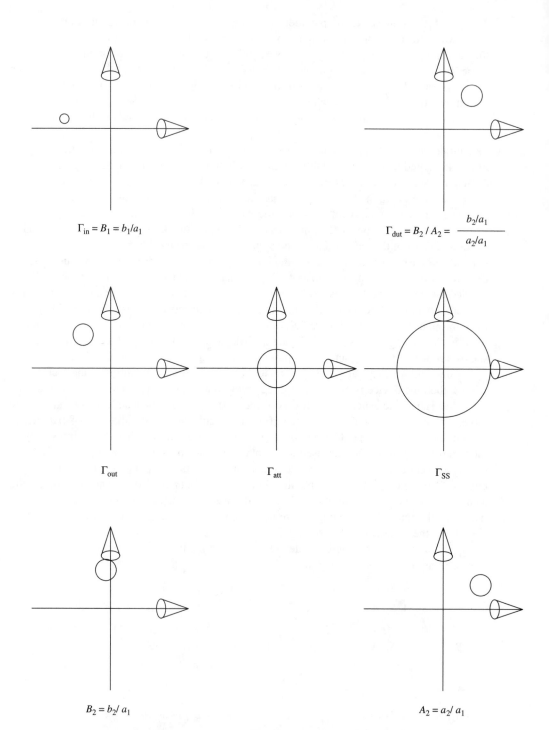

Figure 3.10 Reflection plane responses from a load pull method. [*Robert J. Weber, "Oscillator Design Techniques Using Calculated and Measured S Parameters," Copyright ©1991, Short Course, 45th Annual Symposium on Frequency Control, May 1991, Los Angeles, CA.*]

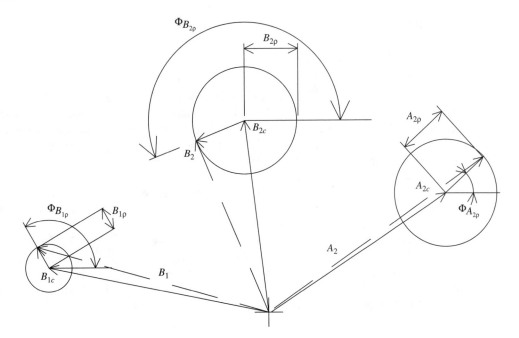

Figure 3.11 Expanded view of individual circles from a load pull method.

following relationship where the vector (b) is related to the vector (a) by the transformation determined by (S).

$$(b) = (S)(a)$$

Let new matrices be defined for several sets of vectors (b) and vectors (a). Consider each of these sets an individual measurement. Arrange these measurement vectors into an $n \times n$ matrix. Then there are n sets of (b) and (a) vectors filling a square b matrix and a square a matrix.

$$(b)_n = (S)_n (a)_n$$

A subscript n is used to designate the square state measurement matrix. For a 2-port network the matrices would be as given below.

$$\begin{pmatrix} b_{11} & b_{12} \\ b_{21} & b_{22} \end{pmatrix}_n = \begin{pmatrix} S_{11} & S_{12} \\ S_{21} & S_{22} \end{pmatrix} \begin{pmatrix} a_{11} & a_{12} \\ a_{21} & a_{22} \end{pmatrix}_n$$

If the determinant of the square a measurement matrix is not zero, then the inverse of the square a measurement matrix exists. Then:

$$\begin{pmatrix} S_{11} & S_{12} \\ S_{21} & S_{22} \end{pmatrix} = \begin{pmatrix} b_{11} & b_{12} \\ b_{21} & b_{22} \end{pmatrix}_n \begin{pmatrix} a_{11} & a_{12} \\ a_{21} & a_{22} \end{pmatrix}_n^{-1}$$

The existence of the determinant means that each individual measurement set is linearly independent from each of the others. In the above equation, if a_{21} and a_{12} are set equal to zero, then the same equation for each of the scattering parameter results as given earlier for determining scattering parameters for matched loads.

In subscript notation, the first subscript in the measurement matrix will be used for the port designation and the second subscript will be used for the measurement number. The columns of both the a measurement matrix and the b measurement matrix could be interchanged together and the result would give the same scattering matrix. For a 2-port network, the scattering matrix of the network is given by the following relationship.

$$\begin{pmatrix} S_{11} & S_{12} \\ S_{21} & S_{22} \end{pmatrix} = \frac{1}{a_{11}a_{22} - a_{21}a_{12}} \begin{pmatrix} b_{11}a_{22} - b_{12}a_{21} & b_{12}a_{11} - b_{11}a_{12} \\ b_{21}a_{22} - b_{22}a_{21} & b_{22}a_{11} - b_{21}a_{12} \end{pmatrix}$$

The above equation will be called the difference equation for determining the scattering matrix. The scattering parameters for a 2-port can be determined by making measurements of the independent and dependent variable for any two different port conditions (load match or source match). Notice that when a_{11} and a_{12} are set equal to one, this equation gives the same formula as was given the previous section for a two point measurement.

A similar technique can be used to determine the immittance, chain, or hybrid two port parameters from two independent sets of measurements using the dependent and independent variables and can be used for an n-port network to determine the immittance or scattering matrices from n independent measurements.

This technique can be applied to the circular locus load pull method of the previous section. Using a c subscript to designate the center of a circular locus, a rho subscript to designate the radius of the circle, and an angular quantity phi to designate the position on the circular locus, the a and b variables can then be expressed as follows. The running parameter phi in each circle equation varies the same in each of the variables.

$$b_1 = b_{1c} + b_{1\rho}e^{j(\phi + \phi_{b1})}$$
$$b_2 = b_{2c} + b_{2\rho}e^{j(\phi + \phi_{b2})}$$
$$a_1 = a_{1c} + a_{1\rho}e^{j(\phi + \phi_{a1})}$$
$$a_2 = a_{2c} + a_{2\rho}e^{j(\phi + \phi_{a2})}$$

Let the running variable (phi without a subscript) be given two values. This would be for measurements at two positions around the loci. The $n \times n$ matrix relationship would be given as follows.

$$\begin{pmatrix} b_{1c} + b_{1\rho}e^{j(\phi_1 + \phi_{b1})} & b_{1c} + b_{1\rho}e^{j(\phi_2 + \phi_{b1})} \\ b_{2c} + b_{2\rho}e^{j(\phi_1 + \phi_{b2})} & b_{2c} + b_{2\rho}e^{j(\phi_2 + \phi_{b2})} \end{pmatrix}$$
$$= \begin{pmatrix} S_{11} & S_{12} \\ S_{21} & S_{22} \end{pmatrix} \begin{pmatrix} a_{1c} + a_{1\rho}e^{j(\phi_1 + \phi_{a1})} & a_{1c} + a_{1\rho}e^{j(\phi_2 + \phi_{a1})} \\ a_{2c} + a_{2\rho}e^{j(\phi_1 + \phi_{a2})} & a_{2c} + a_{2\rho}e^{j(\phi_2 + \phi_{a2})} \end{pmatrix}$$

The determinant of the a matrix is:

$$\Delta_a = (e^{j\phi_2} - e^{j\phi_1})(a_{1c}a_{2\rho}e^{j\phi_{a2}} - a_{2c}a_{1\rho}e^{j\phi_{a1}})$$

If the measurements are made at two points around the loci, the determinant is not equal to zero. Then the scattering matrix can be determined by inverting the a matrix giving the following result.

$$\begin{pmatrix} S_{11} & S_{12} \\ S_{21} & S_{22} \end{pmatrix}$$
$$= \frac{1}{\left(a_{1c}a_{2\rho}e^{j\phi_{a2}} - a_{2c}a_{1\rho}e^{j\phi_{a1}}\right)} \begin{pmatrix} b_{1c}a_{2\rho}e^{j\phi_{a2}} - a_{2c}b_{1\rho}e^{j\phi_{b1}} & a_{1c}b_{1\rho}e^{j\phi_{b1}} - b_{1c}a_{1\rho}e^{j\phi_{a1}} \\ b_{2c}a_{2\rho}e^{j\phi_{a2}} - a_{2c}b_{2\rho}e^{j\phi_{b2}} & a_{1c}b_{2\rho}e^{j\phi_{b2}} - b_{2c}a_{1\rho}e^{j\phi_{a1}} \end{pmatrix}$$

The scattering matrix can be determined by using several methods. If the circular loci are known, then the centers and radii are known and these can be substituted into the above equation. The value for the subscripted phi's are taken from one particular measurement. If two points on the loci are known, those two points can be used in the difference equations. Actually any two points can be used for linear circuits.

3.2 SCATTERING PARAMETER CALCULATIONS USING SPICE

In a program like Spice, one way to determine the S parameters is to use the circuit in Figure 3.12. The voltages and currents on the left side of the device under test (DUT) are:

$$V_1 = 2E_s - I_1 Z_0$$

$$S_{11} = \left. \frac{V_1 - I_1 Z_0}{V_1 + I_1 Z_0} \right|_{a_2=0} = \frac{2E_s - I_1 Z_0 - I_1 Z_0}{2E_s - I_1 Z_0 + I_1 Z_0} = \frac{E_s - I_1 Z_0}{E_s} = \frac{V_{m1}}{E_s}$$

If $E_s = 1$, then only $E_s - I_1 Z_0$ needs to be measured. This is equal to the voltage V_{m1}. Likewise:

$$S_{21} = \left. \frac{V_2 - I_2 Z_0}{V_1 + I_1 Z_0} \right|_{a_2=0} = \frac{2V_2}{2E_s - I_1 Z_0 + I_1 Z_0} = \frac{V_2}{E_s}$$

If $E_s = 1$, then one need only measure V_2 to determine S_{21}. In short, with the above circuit, measuring V_m gives S_{11} and measuring V_2 gives S_{21}. For the other parameters, the circuit in the bottom part of Figure 3.12 is used. The reverse parameters are:

$$S_{22} = \left. \frac{V_2 - I_2 Z_0}{V_2 + I_2 Z_0} \right|_{a_1=0} = \frac{V_{m2}}{E_s}$$

$$S_{12} = \left. \frac{V_1 - I_1 Z_0}{V_2 + I_2 Z_0} \right|_{a_1=0} = \frac{V_1}{E_s}$$

In ac analysis, E_s is set equal to one volt. Then the magnitude and phase of the four voltages indicated by the equation above give the magnitude and phase of the scattering parameters. In transient analysis, the value of each input voltage source may need to be more or less than one volt depending on the dynamic range of the part. The maximum magnitude is determined by the largest signal allowed before some part of the circuit goes nonlinear. With two equal-valued sine wave voltages in the input of the circuit, each of the measured values is divided by the value of one of the voltage sources. For instance, if one needs 20 volts (2 times 10 V), two 10-volt sources are used and $\frac{1}{10}$ of the value of the measured values give the magnitudes of the scattering parameters. Often devices saturate at levels much less than one

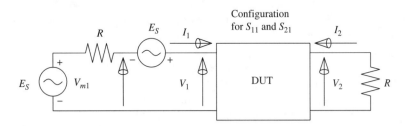

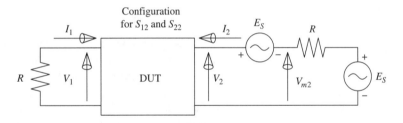

Figure 3.12 Scattering parameter measurements using spice programs.

volt. Then the voltage sources need to be set below one volt. If the device is not linear, just as in making measurements in the laboratory, one cannot turn the device around since the drive at the input needs to stay constant. When this condition exists, a load pull circuit is used in the spice analysis. Using a variable-length transmission line to form the circles allows S_{22} and S_{12} to be determined as discussed in Section 3.1.6.

3.3 NETWORK ANALYZER CALIBRATION

The amplitude and phase measurement equipment needs to be calibrated to make accurate scattering parameter measurements. The couplers, attenuators, and sampling circuits need to be accurately characterized if the resultant scattering parameters are to be accurate. Phase measurements using vector voltmeter equipment can present a difficulty if the time delays of the measurement setup are not carefully considered. Computer-based equipment can do virtual time delay compensation. Before a time delay analysis is given, a brief analysis of calibration of a network analyzer will be given.

In Figure 3.3, the input and output couplers are four-port networks. The input and output sources shown on the outside of the input and output couplers in most systems are not both active at once. In some load pull systems, both sources are active at the same time. For the purposes of this discussion, the sources are considered to be active only one at a time. If a single source is switched between the input and output, then there is some isolation that needs to be taken into account, because essentially then both sources are on at the same time with one source much larger than the other. The diagram in Figure 3.3 also ignores any leakage that could exist between the various detectors. The discussion of calibration given here ignores the leakage terms. Many network analyzers also have an option to ignore isolation or leakage terms in the calibration procedure.

The detectors are assumed to be perfectly matched to the couplers. Any mismatch that actually exists is attributed to the coupler and not to the detector. Therefore, there are no reflected terms that are taken into account from reflections off of the detectors. V_{di} will be used to designate the voltage detected at the ith detector. If the source is not perfectly matched to the input of the directional coupler, then reflections that come back from the device under test and the input coupler will re-reflect back into the system. The input coupler treats that as a part of the input wave it measures. However, when there is source mismatch, the detector level in detector one varies more than it would when there is no source mismatch and the instrument then has a larger dynamic range.

By assumption, the four-port circuit on the input has only one signal into the coupler from the source and one signal from the device under test. What is desired in the first coupler is the ratio of the reflected signal to the incident signal. The ratio of the reflected signal to the incident signal is related by a bilinear transformation to the incident and reflected wave as discussed in the load pull section. That relationship has three independent variables. The equation given has four constants but one can divide through the equation by any one of the constants (if it is not zero), leaving three independent constants.

$$\Gamma_{\text{MEAS}} = \frac{V_{d2}}{V_{d1}} = \frac{A\Gamma_{\text{DUT}} + B}{C\Gamma_{\text{DUT}} + D}$$

If the constants, A, B, C, and D can be determined, then the device reflection coefficient can be determined from the measured reflection coefficient by the inverse of the transformation above. If a matched load is placed on the circuit, the ratio of B to D is known. Likewise, if two other known reflection coefficients are put on the network the two other measured ratios are known. Three measured ratios using three known values are enough to solve the three linear equations in three unknown ratios. These values are often stored in an error matrix equation that is then used to correct the measured values. In the equation shown above, the quantity A is often one of the larger in value and can be used to divide through the equation to produce three calibration constants.

$$\Gamma_{MEAS} = \frac{V_{d2}}{V_{d1}} = \frac{\Gamma_{DUT} + (B/A)}{(C/A)\Gamma_{DUT} + (D/A)} = \frac{\Gamma_{DUT} + \hat{B}}{\hat{C}\Gamma_{DUT} + \hat{D}}$$

The three constants can be determined from the three known loads and measurements from the following matrix.

$$\begin{pmatrix} \hat{B} \\ \hat{C} \\ \hat{D} \end{pmatrix} = \begin{pmatrix} -1 & \Gamma_{MEAS1}\Gamma_{DUT1} & \Gamma_{MEAS1} \\ -1 & \Gamma_{MEAS2}\Gamma_{DUT2} & \Gamma_{MEAS2} \\ -1 & \Gamma_{MEAS3}\Gamma_{DUT3} & \Gamma_{MEAS3} \end{pmatrix}^{-1} \begin{pmatrix} \Gamma_{DUT1} \\ \Gamma_{DUT2} \\ \Gamma_{DUT3} \end{pmatrix}$$

Using the same procedure for the output coupler generates another three measured ratios that are used to calibrate the output coupler. There is another constant that is needed. This is the ratio between the input and output coupler values or the difference between detector one and detector four. The relationship between these detectors and the other two detectors can be determined using the six constants just obtained. The seventh constant is obtained by putting a through transmission line in at the device under test plane and measuring the ratio between detector one and four. A through should give a unity value but phase offsets and amplitude variations will cause that number to deviate than unity. A redundant set of measurements giving the differences between detector two and detector three can be taken to help reduce calibration error, minimizing the computation necessary to determine the ratio, for example, of detector three to detector four. This procedure assumes, as indicated above, that there are no leakage signals between the detectors and between the two sources shown. If there is significant leakage, that leakage term will change when the source is switched between position one and position two and when the device under test is changed. Extra terms are needed for that type of calibration. The calibration just described is a seven-term (eight-term if the reverse transmission term is used) calibration for a network analyzer. Currently vector network analyzers perform this calibration inside the equipment with high-speed computation for each frequency under consideration but within the vector network analyzer. Older systems performed this computation in a computer connected to the vector network analyzer via a data bus and the raw measurements were sent to the computer for correction and presentation in the peripheral data processor.

If the couplers in Figure 3.3 have known scattering parameters, then the scattering parameters for the unknown are easily determined. Let a source be put on port 1, the DUT on port 2, a reference detector on port 3, and a second detector on port 4 of the first coupler as shown in Figure 3.3. Let the input coupler have the following known scattering matrix.

$$(S) = \begin{pmatrix} S_{11} & S_{12} & S_{13} & S_{14} \\ S_{21} & S_{22} & S_{23} & S_{24} \\ S_{31} & S_{32} & S_{33} & S_{34} \\ S_{41} & S_{42} & S_{43} & S_{44} \end{pmatrix}$$

Then the following relationships hold under the assumption the detectors on ports 2 and 4 are matched to the port impedances. This assumption is especially useful when the above scattering matrix has been measured using the same detector that will be used to perform the unknown measurement.

$$\begin{pmatrix} b_1 \\ b_2 \end{pmatrix} = \begin{pmatrix} S_{11} & S_{12} \\ S_{21} & S_{22} \end{pmatrix} \begin{pmatrix} a_1 \\ a_2 \end{pmatrix}$$

$$\begin{pmatrix} b_3 \\ b_4 \end{pmatrix} = \begin{pmatrix} S_{31} & S_{32} \\ S_{41} & S_{42} \end{pmatrix} \begin{pmatrix} a_1 \\ a_2 \end{pmatrix}$$

$$\begin{pmatrix} a_1 \\ a_2 \end{pmatrix} = \begin{pmatrix} S_{31} & S_{32} \\ S_{41} & S_{42} \end{pmatrix}^{-1} \begin{pmatrix} b_3 \\ b_4 \end{pmatrix} = \begin{pmatrix} E_1 & F_1 \\ C_1 & D_1 \end{pmatrix} \begin{pmatrix} b_3 \\ b_4 \end{pmatrix}$$

$$\begin{pmatrix} b_1 \\ b_2 \end{pmatrix} = \begin{pmatrix} S_{11} & S_{12} \\ S_{21} & S_{22} \end{pmatrix} \begin{pmatrix} S_{31} & S_{32} \\ S_{41} & S_{42} \end{pmatrix}^{-1} \begin{pmatrix} b_3 \\ b_4 \end{pmatrix} = \begin{pmatrix} G_1 & H_1 \\ A_1 & B_1 \end{pmatrix} \begin{pmatrix} b_3 \\ b_4 \end{pmatrix}$$

$$\begin{pmatrix} b_2 \\ a_2 \end{pmatrix} = \begin{pmatrix} A_1 & B_1 \\ C_1 & D_1 \end{pmatrix} \begin{pmatrix} b_3 \\ b_4 \end{pmatrix}$$

$$\frac{b_2}{a_2} = \frac{A_1 b_3 + B_1 b_4}{C_1 b_4 + D_1 b_4} = \frac{A_1 + B_1 \frac{b_4}{b_3}}{C_1 + D_1 \frac{b_4}{b_3}}$$

A similar analysis holds for the second coupler giving the following relationship.

$$\begin{pmatrix} b_6 \\ a_6 \end{pmatrix} = \begin{pmatrix} A_2 & B_2 \\ C_2 & D_2 \end{pmatrix} \begin{pmatrix} b_7 \\ b_8 \end{pmatrix}$$

The resulting data from ports 2 and 6 are then used to determine the scattering parameters of the DUT.

3.3.1 Basic Network Analyzer and a Phase-Adjustment Procedure

A basic network analyzer diagram, based on using traveling waves, was shown in Figure 3.3. The figure shows two sources and four detectors. Various arrangements of this basic diagram are available in commercial network analyzers. Most analyzers use switches so that only one source is needed. In addition, some analyzers use only two detectors, some three detectors, and some four detectors to do two-port analysis. Two detectors are necessary if one needs amplitude and phase information. One detector sets the amplitude and phase reference and the other detector determines the amplitude and phase at that detector with respect to the reference detector. Switches can be used to switch the detectors between the various detector ports to minimize the number of physical detectors needed. Each of the different systems requires a different calibration routine. Calibration as discussed in the previous section is dependent on the match of a detector to a measurement port. If a detector is switched between measurement ports, the match varies depending on the switch position.

In order to make the physical portion of a network analyzer system phase track (i.e., produce the same phase difference between any two detectors), the time delay from a common point in the system where two quantities split to the separate detectors must be the same. If the phase difference is set equal to zero at any given frequency, there is no guarantee that the phase will be zero at any other frequency since phase is only measured to within two pi. In order to have a setup that phase tracks over a frequency range, there are time delay requirements to be met. Second-order effects such as coupler directivity and re-reflections will create smaller differences in phase versus frequency. However, in the basic network analyzer diagram shown in Figure 3.3, if $T_1 + 2T_2 + T_4$ is made equal to T_3, the analyzer will track on the first-order basis over frequency for an input reflection measurement. Likewise, $T_6 + T_8$ should be made equal to $T_2 + T_4$ for a forward scattering parameter measurement. Setting $T_6 + T_8$ equal to

$T_2 + T_4$ is often done by making the setup symmetrical. Similarly, if $T_5 + 2T_6 + T_8$ is the same as T_7 and $T_2 + T_4$ is the same as $T_6 + T_8$ then the measurements made with the source placed at the output also phase track. Early network analyzer setups used a line stretcher to vary the time delay of T_3 to satisfy these relationships. More modern network analyzers are able to accomplish this using a virtual delay using different computational algorithms within their detector instruments.

The basic network analyzer drawing in Figure 3.3 conceptually depicts four-port coupling devices as directional couplers. Actually, these devices can be any four-port networks that have directivity. Lossless, matched four-port networks have inherent directivity [4] and can be used for scattering parameter measurements.

Physically calibrating the phase of the system (instruments usually read phase difference rather than differential time delay) can present a difficulty. It needs to be determined if the displayed error is an error in time delay or an error in the zero of phase. Many older instruments had a vernier or adjustment for setting the zero of phase difference. A simple procedure allows one to tell whether the error is in time delay or meter reading. Let the phase at detector DET1 be $\phi 1$. Let the phase at detector DET2 be $\phi 2$. The difference in phase measured is $\phi m1 = \phi 2 - \phi 1 + \phi e$ where ϕe is the error in the meter reading. If one interchanges the two detectors (by swapping connectors, etc.), then the measured phase is $\phi m2 = \phi 1 - \phi 2 + \phi e$. The sum of $\phi m2$ and $\phi m2$ is $\phi m2 + \phi m1 = 2\phi e$. The average value between the two readings should be zero phase. Subtracting an amount equal to this average value from the meter reading or adjustment corrects the phase error of the meter. If an open circuit is used at the DUT plane, then the rest of the error is in the line stretcher and the line stretcher is varied until the meter reads zero phase. If a transmission through is being measured, and the network is symmetrical, then the phase difference between detector DET1 and DET4 should also be zero. When the meter is adjusted for its zero phase, the rest of the phase offset in either case is in the line stretcher. The line stretcher is adjusted to bring the meter reading to zero. If the meter reading varies when the frequency is changed, then the line stretcher may be off by a multiple of a wavelength. The line stretcher is then adjusted for a different multiple of a wavelength. When the time delays are equal, the analyzer has first-order calibration for phase and the phase correction does not substantially vary with frequency. Higher-order deviations from phase linearity arise from mismatches in connectors, transitions, and other terminations.

EXERCISES

3-1 A component consists of 1-uF capacitor in series with a resistance of 20 ohms. It is to be measured at 1 kHz. Using a 600-ohm voltage source of 10 volts at 1 kHz, sketch a measurement setup and determine what the voltage waveforms would look like to measure the scattering parameters of this series R-C when the component is measured on a dual-channel oscilloscope. Determine this for both S_{11} and S_{21}.

3-2 An active device has an input impedance of approximately 100 k ohms in parallel with 10 pF. The output impedance of the part is 100 ohms in parallel with 100 pF. The part is to be measured at 10 MHz. Magnitude ratios can be measured to no better than 0.1 dB. Determine a measurement fixture that can be used to measure the scattering parameters of the device. *Hint:* Consider the system shown in Figure 3.7 with a high impedance on the input side and a low impedance on the output side. The scattering parameters are derived with a high-impedance normalization on port one and a lower-impedance normalization on port two. Show why a 50-ohm system is not adequate using an assumption of 0.1-dB measurement accuracy for magnitude ratios.

3-3 Draw a circuit that can be used with Spice to calculate the scattering parameters of a 2N2222 (*npn* general-purpose transistor) at 100 MHz. The circuit should contain biasing sources with the necessary dc and ac components to bias the circuit at 10 V for Vce and 20 mA Ic. Determine the scattering parameters of a 2N2222 at 100 MHz using a Spice modeling program.

3-4 Two identical four-port directional couplers are used for a network analyzer system as shown in Figure 3.3. The couplers have the following scattering parameters: $S_{11} = S_{22} = 0.05$ at -30 degrees, $S_{12} = 0.95$ at 80 degrees, $S_{13} = S_{24} = 0.095$ at 20 degrees, $S_{14} = S_{23} = 0.009$ at -100 degrees, $S_{33} = S_{44} = 0.03$ at 135 degrees, and $S_{34} = 0.005$ at -135 degrees. The coupler is reciprocal, i.e., $S_{ij} = S_{ji}$. Determine two calibration matrices for the network analyzer. Ignore isolation.

3-5 Consider one direction coupler as specified in Exercise 3.4. It is used to measure a unity reflection coefficient magnitude. The angle of the reflection coefficient varies from zero degrees to 360 degrees. Determine the amplitude variation in dB of S_{11} that result in the measurement of the load with a system that has no calibration. *Hint:* Determine the bilinear transformation shown in Section 3.3 and map the unit circle into the measured reflection coefficient. Determine the minimum and maximum magnitude of the measured reflection coefficient.

4

Multiport and Differential-Mode Scattering Parameters

4.1 THREE-PORT AND FOUR-PORT SCATTERING PARAMETERS

In this chapter, different three- and four-port circuits are considered. The mixed-mode scattering parameter characterization method is discussed as well as looking at indefinite three-port scattering parameter matrices. Stability of multiport matrices will be covered in Chapter 5 along with stability of two-port networks.

4.2 TWO-PORT TO THREE-PORT AND THREE-PORT TO TWO-PORT CONVERSIONS

Often a three-terminal device is measured as a two-port with one port grounded. Feedback is then inserted in the common terminal for stabilization reasons or to use the device in an oscillator circuit. Oscillator circuits using this technique are discussed in Chapter 8.

4.2.1 Two-Port to Three-Port Conversion

If the device is measured as a two-port, the device parameters cannot always be converted to three-port parameters accurately. If one adds a common terminal impedance in series with the ground of a two-port, the process is easily described using the Z matrix. Let the grounded terminal be called port three when it has an impedance inserted at that spot. The current out of the third terminal when there are no hidden parasitics is the negative of the sum of the current into terminals one and two. Using Z parameters, the new Z matrix without parasitics is:

$$V_3 = -(I_1 + I_2)Z_3$$

$$V_1 - V_3 = Z_{11}I_1 + Z_{12}I_2 \quad V_1 = (Z_{11} + Z_3)I_1 + (Z_{12} + Z_3)I_2$$

$$V_2 - V_3 = Z_{21}I_1 + Z_{22}I_2 \quad V_2 = (Z_{21} + Z_3)I_1 + (Z_{22} + Z_3)I_2$$

$$(Z) = \begin{pmatrix} Z_{11} + Z_3 & Z_{12} + Z_3 \\ Z_{21} + Z_3 & Z_{22} + Z_3 \end{pmatrix}$$

This immittance matrix needs to be checked to determine whether it is valid. Figure 4.1 is shown in terms of the Y matrix. A two-port has four parameters. A three-port has nine

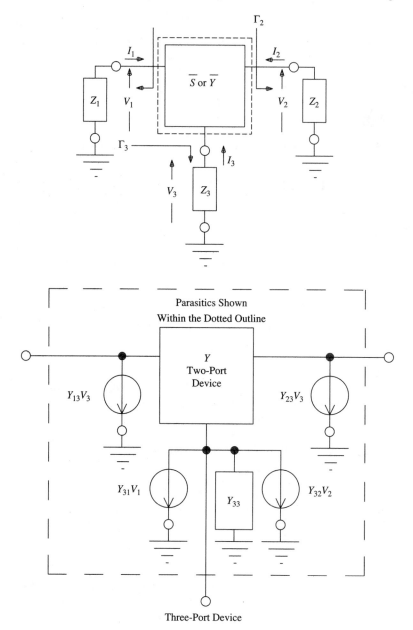

Figure 4.1 Three-port Y matrix considerations. [*Robert J. Weber, Oscillator Design Techniques Using Calculated and Measured S Parameters, Copyright ©1991, Short Course*, 45th Annual Symposium on Frequency Control, *May 1991, Los Angeles, CA.*]

parameters. There are five parameters of the device that are not measured when the device is measured as a two-port. For instance, if the device is a FET device, the back gating terms are not measured when the device is measured with the back terminal grounded to the body of the FET. The five terms not originally in the two-port matrix are shown in Figure 4.1. Notice that when V_3 is zero, the two current sources on the third port, $Y_{31}V_1$ and $Y_{32}V_2$, are shorted out and do not enter the device. In addition, the current from the two current sources, $Y_{13}V_3$ and $Y_{23}V_3$, on ports one and two has zero value. The admittance Y_{33} is also shorted out when V_3

is equal to zero. However, when the third port is lifted from ground, these current sources are in the circuit and affect the circuit operation. A careful analysis of the circuit should be made to determine whether the two-port to three-port conversion is valid.

4.2.2 Two-Port to Three-Port and Three-Port to Two-Port Conversion for Three-Terminal Networks

The two-port to three-port conversion for three-terminal devices will be derived next. Keep in mind that the conversion does not always incorporate the five hidden terms discussed above.

In the derivation to follow, the quantities with a caret on them are the two-port quantities. From Figure 4.1, then:

$$I_1 = \hat{I}_1$$
$$I_2 = \hat{I}_2$$
$$V_1 = \hat{V}_1 + V_3$$
$$V_2 = \hat{V}_2 + V_3$$
$$I_3 = -I_1 - I_2$$

The relationship between these terminal quantities and the scattering matrix quantities is determined as follows.

$$a_1 + b_1 = \frac{V_1}{\sqrt{Z_0}}; \ a_2 + b_2 = \frac{V_2}{\sqrt{Z_0}}; \ a_3 + b_3 = \frac{V_3}{\sqrt{Z_0}}$$

$$a_1 - b_1 = I_1\sqrt{Z_0}; \ a_2 - b_2 = I_2\sqrt{Z_0}; \ a_3 - b_3 = I_3\sqrt{Z_0}$$

$$\hat{a}_1 = \frac{V_1 - V_3 + I_1 Z_0}{2\sqrt{Z_0}} = a_1 - \frac{a_3}{2} - \frac{b_3}{2}$$

$$\hat{b}_1 = \frac{V_1 - V_3 - I_1 Z_0}{2\sqrt{Z_0}} = b_1 - \frac{a_3}{2} - \frac{b_3}{2}$$

$$\hat{a}_2 = \frac{V_2 - V_3 + I_2 Z_0}{2\sqrt{Z_0}} = a_2 - \frac{a_3}{2} - \frac{b_3}{2}$$

$$\hat{b}_2 = \frac{V_2 - V_3 - I_2 Z_0}{2\sqrt{Z_0}} = b_2 - \frac{a_3}{2} - \frac{b_3}{2}$$

Therefore:

$$V_1 = +\hat{V}_1 + V_3 \Rightarrow a_1 + b_1 = +\hat{a}_1 + \hat{b}_1 + a_3 + b_3$$
$$V_2 = +\hat{V}_2 + V_3 \Rightarrow a_2 + b_2 = +\hat{a}_2 + \hat{b}_2 + a_3 + b_3$$
$$I_3 = -I_1 - I_2 \Rightarrow a_3 - b_3 = -a_1 + b_1 - a_2 + b_2$$

Substituting for the scattering parameters for the two-port the following set of equations results.

$$a_1 + b_1 = \hat{a}_1 + \hat{b}_1 + a_3 + b_3 = \hat{a}_1 + \hat{S}_{11}\hat{a}_1 + \hat{S}_{12}\hat{a}_2 + a_3 + b_3$$
$$a_2 + b_2 = \hat{a}_2 + \hat{b}_2 + a_3 + b_3 = \hat{a}_2 + \hat{S}_{21}\hat{a}_1 + \hat{S}_{22}\hat{a}_2 + a_3 + b_3$$
$$a_3 - b_3 = -a_1 + b_1 - a_2 + b_2$$

or:

$$a_1 + b_1 = (1 + \hat{S}_{11})\hat{a}_1 + \hat{S}_{12}\hat{a}_2 + a_3 + b_3$$

$$a_2 + b_2 = \hat{S}_{21}\hat{a}_1 + (1 + \hat{S}_{22})\hat{a}_2 + a_3 + b_3$$

$$a_3 - b_3 = -a_1 + b_1 - a_2 + b_2$$

Substituting in for the caret variables and rearranging the result as a matrix equation gives:

$$\begin{pmatrix} 2 & 0 & (-1+\hat{S}_{11}+\hat{S}_{12}) \\ 0 & 2 & (-1+\hat{S}_{21}+\hat{S}_{22}) \\ 1 & 1 & 1 \end{pmatrix} \begin{pmatrix} b_1 \\ b_2 \\ b_3 \end{pmatrix} = \begin{pmatrix} 2\hat{S}_{11} & 2\hat{S}_{12} & (1-\hat{S}_{11}-\hat{S}_{12}) \\ 2\hat{S}_{21} & 2\hat{S}_{22} & (1-\hat{S}_{21}-\hat{S}_{22}) \\ 1 & 1 & 1 \end{pmatrix} \begin{pmatrix} a_1 \\ a_2 \\ a_3 \end{pmatrix}$$

$$\begin{pmatrix} b_1 \\ b_2 \\ b_3 \end{pmatrix} = \begin{pmatrix} 2 & 0 & (-1+\hat{S}_{11}+\hat{S}_{12}) \\ 0 & 2 & (-1+\hat{S}_{21}+\hat{S}_{22}) \\ 1 & 1 & 1 \end{pmatrix}^{-1} \begin{pmatrix} 2\hat{S}_{11} & 2\hat{S}_{12} & (1-\hat{S}_{11}-\hat{S}_{12}) \\ 2\hat{S}_{21} & 2\hat{S}_{22} & (1-\hat{S}_{21}-\hat{S}_{22}) \\ 1 & 1 & 1 \end{pmatrix} \begin{pmatrix} a_1 \\ a_2 \\ a_3 \end{pmatrix}$$

If the starting two-port S matrix was originally derived from a full three-port S matrix (one with all of the parasitics included), the above equation does not give the original three-port S matrix back since information is lost by the shorting operation. The five terms from the equivalent Y matrix have been deleted in the conversion to a two-port matrix when the third port is grounded. The three-port immittance matrix that results when the common terminal of a three-terminal network is lifted from ground to form a three-port network is called an *indefinite matrix*. There is no path to ground for the input and output currents. There would be if the parasitics had been included and then the matrix would not be indefinite. There are an infinite number of three-terminal devices that have the same two-port matrix. However, the above equation will be helpful if one already knows which of those components have been deleted. When that is the case, the Y matrix is found from the S matrix and the parasitic components are added back in. Then a new, correct three-port S matrix could be derived from the modified three-port Y matrix.

The inverse procedure, going from a three-port to a two-port, is given now. When one has a three-port S matrix and one of the ports is terminated in a known impedance the resulting two-port can be determined. Let the termination be placed on the third port. The matrix can always be rearranged to make that port be port three. In addition, let the three-port scattering matrix be arranged so that port one is the input and port two is the output. There is no loss in generality in this procedure. Let the termination on the third port be Γ_3. Then:

$$b_1 = S_{11}a_1 + S_{12}a_2 + S_{13}a_3$$

$$b_2 = S_{21}a_1 + S_{22}a_2 + S_{23}a_3$$

$$b_3 = S_{31}a_1 + S_{32}a_2 + S_{33}a_3 = \Gamma_3^{-1}a_3$$

$$a_3 = \frac{S_{31}a_1 + S_{32}a_2}{\Gamma_3^{-1} - S_{33}}$$

Substituting the last equation into the first two gives:

$$b_1 = \left(S_{11} + \frac{S_{13}S_{31}}{\Gamma_3^{-1} - S_{33}} \right) a_1 + \left(S_{12} + \frac{S_{13}S_{32}}{\Gamma_3^{-1} - S_{33}} \right) a_2$$

$$b_2 = \left(S_{21} + \frac{S_{23}S_{31}}{\Gamma_3^{-1} - S_{33}} \right) a_1 + \left(S_{22} + \frac{S_{23}S_{32}}{\Gamma_3^{-1} - S_{33}} \right) a_2$$

If port three is terminated in a short circuit (i.e., grounded), then $\Gamma_3 = -1$ and:

$$b_1 = \left(S_{11} - \frac{S_{13}S_{31}}{1 + S_{33}} \right) a_1 + \left(S_{12} - \frac{S_{13}S_{32}}{1 + S_{33}} \right) a_2$$

$$b_2 = \left(S_{21} - \frac{S_{23}S_{31}}{1 + S_{33}} \right) a_1 + \left(S_{22} - \frac{S_{23}S_{32}}{1 + S_{33}} \right) a_2$$

4.2.3 Mutual Inductance Considerations

When the parasitic mutual inductance between leads is important, similar considerations concerning parasitics and their effect on two-port to three-port and three-port to two-port conversions need to be taken into account. This is especially important if one of the ports is terminated in an open circuit or a high impedance during a two-port measurement. Figure 4.2 shows a three-port with three inductors all coupled together. If a port is left open for a two-port measurement and then later connected to lower impedances in a circuit, the effect of mutual inductance between the terminal leads would need to be considered. An application where this needs to be considered is in mixer circuits. The third port might be the intermediate frequency port. The termination on that port is important in mixer analysis but is also important for matching considerations due to mutual inductance considerations.

4.3 THREE-PORT AND FOUR-PORT SCATTERING PARAMETERS

There are several types of four-port devices. Directional couplers are four-port devices. Differential amplifiers with two inputs and two outputs are four-port devices. Differential amplifiers with two inputs and a single output are three-port devices. Just as in the case of two-port networks, the choice of the independent and dependent variables determines the form of the matrix that relates the variables. There have been several different choices made in the literature in this regard. This section introduces coupled-line analysis. Several types of directional couplers and their application are discussed in more detail in Chapter 12 dealing with microwave components.

4.3.1 Multiport Analysis

When there are four ports there are four voltage and four current variables. In some applications (e.g., differential amplifier circuits or directional coupler networks), two of the ports can be identified as input ports and two of the ports can be identified as output ports. Further, each pair of ports can be excited with equal voltages, differential voltages, equal currents, or differential currents. Other sets of excitations might be important in any different application. If any two sets of excitation are linearly independent, those two can be used to form a basis for a complete set of excitations by linear addition. There have been several choices for the two sets made in the literature. Two such choices will be considered here. Since scattering parameters result from a linear superposition of voltage and current variables, a particular set of excitations can be used to form a given scattering parameter matrix for use in describing the behavior of a circuit.

A four-port network is shown in Figure 4.3. Which two ports are the input and which two ports are the output make a difference only in the mathematics but it is important to know the choice. Different authors have made different choices. In this section, ports one and two will be considered input ports and ports three and four will be considered output ports.

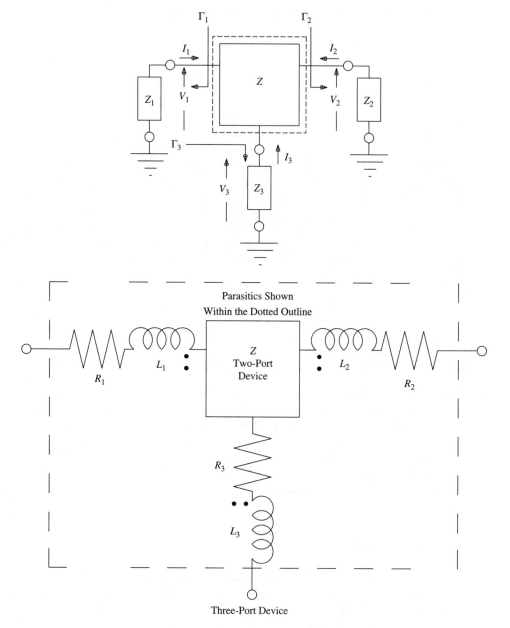

Figure 4.2 Three-port Z matrix considerations.

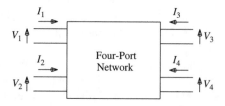

Figure 4.3 Four-port network.

4.3.2 Even- and Odd-Mode Analysis

Coupled-strip transmission line behavior has been described using even- and odd-mode analysis [1,2]. A four-port network can be analyzed by considering two of the four ports as inputs and the other two of the four ports as outputs. The inputs can be excited in phase and also can be excited out of phase. Likewise, the outputs can be measured in phase and also out of phase. If the network is symmetrical about some partition or line of symmetry between the inputs and the outputs, then when the two inputs to the network are excited in phase, the output signals will be in phase. Likewise if the inputs are excited out of phase, the outputs will be out of phase. Networks that do not exhibit symmetry will have in-phase and out-of-phase outputs for an even-phase input, and so on. Even- and odd-mode analysis consists of solving the four-port first when two input ports are excited in phase (even-mode analysis) and then solving the four-port when the input ports are excited out of phase (odd-mode analysis). The linear sum of the two solutions is a general solution for a general excitation. When the network is symmetrical about some line between the two input ports, then the analysis of the even and odd modes reduces to the analysis of a much simpler circuit. In the referenced papers [1,2] there was one transmission line between ports one and four and the other transmission line between ports two and three. This is an arbitrary choice of port designations and the choice is not consistent in the literature. The following set of variables was used in their analysis (their nomenclature is slightly different):

$$I_1 = I_{Ie} + I_{Io} \quad V_1 = V_{Ie} + V_{Io} \quad I_{Ie} = \frac{I_1 + I_2}{2} \quad V_{Ie} = \frac{V_1 + V_2}{2}$$

$$I_2 = I_{Ie} - I_{Io} \quad V_2 = V_{Ie} - V_{Io} \quad I_{Io} = \frac{I_1 - I_2}{2} \quad V_{Io} = \frac{V_1 - V_2}{2}$$

$$I_3 = I_{Oe} - I_{Oo} \quad V_3 = V_{Oe} - V_{Oo} \quad I_{Ie} = \frac{I_4 + I_3}{2} \quad V_{Oe} = \frac{V_4 + V_3}{2}$$

$$I_4 = I_{Oe} + I_{Oo} \quad V_4 = V_{Oe} + V_{Oo} \quad I_{Io} = \frac{I_4 - I_3}{2} \quad V_{Oo} = \frac{V_4 - V_3}{2}$$

where the first subscript is I for the input, O for the output, and the second subscript is e for even and o for odd.

It is also possible to describe a four-port $ABCD$ matrix where the input voltage and current variables are the dependent variables and the output voltage and current variables are the independent variables. Zysman and Johnson [2] describe the four-port Y and Z matrices for coupled lines in inhomogeneous dielectric mediums using these designations and then use those matrices and the four-port $ABCD$ matrix to describe the behavior of various coupled-line configurations.

A four-port $ABCD$ matrix can be derived from the Z or Y matrix or vice versa. This is done using partitioned matrix manipulations. The Z matrix is derived here but the Y matrix can also be derived similarly. The inverse transformation can also be easily derived.

$$\begin{pmatrix} \begin{pmatrix} V_1 \\ V_2 \end{pmatrix} \\ \begin{pmatrix} V_3 \\ V_4 \end{pmatrix} \end{pmatrix} = \begin{pmatrix} \begin{pmatrix} Z_{11} & Z_{12} \\ Z_{21} & Z_{22} \end{pmatrix} & \begin{pmatrix} Z_{13} & Z_{14} \\ Z_{23} & Z_{24} \end{pmatrix} \\ \begin{pmatrix} Z_{31} & Z_{32} \\ Z_{41} & Z_{42} \end{pmatrix} & \begin{pmatrix} Z_{33} & Z_{34} \\ Z_{43} & Z_{44} \end{pmatrix} \end{pmatrix} \begin{pmatrix} \begin{pmatrix} I_1 \\ I_2 \end{pmatrix} \\ \begin{pmatrix} I_3 \\ I_4 \end{pmatrix} \end{pmatrix}$$

$$\begin{pmatrix} \begin{pmatrix} V_1 \\ V_2 \end{pmatrix} \\ \begin{pmatrix} I_1 \\ I_2 \end{pmatrix} \end{pmatrix} = \begin{pmatrix} A & B \\ C & D \end{pmatrix} \begin{pmatrix} \begin{pmatrix} V_3 \\ V_4 \end{pmatrix} \\ \begin{pmatrix} -I_3 \\ -I_4 \end{pmatrix} \end{pmatrix}$$

$$(A) = \begin{pmatrix} Z_{11} & Z_{12} \\ Z_{21} & Z_{22} \end{pmatrix} \begin{pmatrix} Z_{31} & Z_{32} \\ Z_{41} & Z_{42} \end{pmatrix}^{-1}$$

$$(B) = \left(\begin{pmatrix} Z_{11} & Z_{12} \\ Z_{21} & Z_{22} \end{pmatrix} \begin{pmatrix} Z_{31} & Z_{32} \\ Z_{41} & Z_{42} \end{pmatrix}^{-1} \begin{pmatrix} Z_{33} & Z_{34} \\ Z_{43} & Z_{44} \end{pmatrix} - \begin{pmatrix} Z_{13} & Z_{14} \\ Z_{23} & Z_{24} \end{pmatrix} \right)$$

$$(C) = \begin{pmatrix} Z_{31} & Z_{32} \\ Z_{41} & Z_{42} \end{pmatrix}^{-1}$$

$$(D) = \begin{pmatrix} Z_{31} & Z_{32} \\ Z_{41} & Z_{42} \end{pmatrix}^{-1} \begin{pmatrix} Z_{33} & Z_{34} \\ Z_{43} & Z_{44} \end{pmatrix}$$

Coupled-line analysis will be discussed further when directional couplers are discussed. Four-port circuits that are not symmetric can also be analyzed in terms of even- and odd-mode excitations. The superposition of voltage and current to form new variables is not limited to symmetrical networks. However, the derivation of the immittance matrices is much easier when the networks are symmetrical.

4.3.3 Scattering Parameters for Differential- and Common-Mode Circuits

Differential circuits are often used in VLSI and MMIC design. If the circuits are totally differential (i.e., there is no common ground return immittance), then the Y matrix for the circuit is an indefinite matrix. Indefinite Y matrices will come up again when oscillator devices are considered. Differential analysis will also be considered when stability is considered for circuits defined by scattering parameters. In order to characterize differential circuits, a new set of scattering parameters was defined [3]. This set of scattering parameters is one of an infinite number of sets that could be defined. Consider the four-port networks shown in Figure 4.4. In order to describe the left-hand four-port, there are many linear combinations of V and I that can be used. Two of the most common are the impedance Z matrix and the admittance Y matrix. However, other combinations of V and I such as were used to develop the scattering parameters or the four-port $ABCD$ matrix described earlier can be used. As has been shown, even the conventional choice of linear combinations of V and I for the two-port scattering matrix is not unique.

The left-hand side of the networks in Figure 4.4 is considered side one and the right-hand side is considered side two. In order to describe the differential behavior of differential amplifiers, the following choice has been made by Bockelman and Eisenstadt [3].

$$I_1 = \left(\frac{I_{C1}}{2} + I_{D1} \right) \quad V_1 = \left(V_{C1} + \frac{V_{D1}}{2} \right) \quad I_{C1} = (I_1 + I_2) \quad V_{C1} = \frac{(V_1 + V_2)}{2}$$

$$I_2 = \left(\frac{I_{C1}}{2} - I_{D1} \right) \quad V_2 = \left(V_{C1} - \frac{V_{D1}}{2} \right) \quad I_{D1} = \frac{(I_1 - I_2)}{2} \quad V_{D1} = (V_1 - V_2)$$

$$I_3 = \left(\frac{I_{C2}}{2} + I_{D2} \right) \quad V_3 = \left(V_{C2} + \frac{V_{D2}}{2} \right) \quad I_{C1} = (I_3 + I_4) \quad V_{C2} = \frac{(V_3 + V_4)}{2}$$

$$I_3 = \left(\frac{I_{C2}}{2} - I_{D2} \right) \quad V_3 = \left(V_{C2} - \frac{V_{D2}}{2} \right) \quad I_{D2} = \frac{(I_3 - I_4)}{2} \quad V_{D2} = (V_3 - V_4)$$

V_{D1} and V_{D2} are the differential-mode voltages on the left-hand side and right-hand side, respectively. Likewise, V_{C1} and V_{C2} are the common-mode voltages on the left-hand side and right-hand side respectively. If V_{D1} is zero, then $V_1 = V_2$ and V_{C1} is equal to V_1 or V_2. This choice of linearly independent variables is different from those described earlier for couplers. Ports three and four are also switched as compared with the earlier definition. The

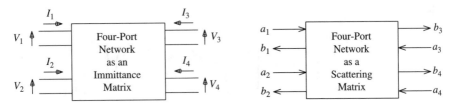

Figure 4.4 Four-port networks V and I linear parameters and a and b linear parameters.

common-mode voltage is defined the same as the even-mode voltage for the coupler and the differential-mode current is the same as the odd-mode current for the coupler. However, the differential-mode voltage is the total difference between ports one and two, not one-half as in the coupler case. The common-mode current is the total current into both ports on the input or output, not the average as in the coupler case.

These are another set of linearly independent variables. Other definitions are possible. Reference impedances can be defined for each port. These are only necessary when considering scattering matrices. In order to develop immittance matrices, we only need to consider the above relationships. Just as when scattering matrices were developed, the above variables can be considered either independent or dependent variables in circuit analysis.

If the new voltages are considered independent variables, then the new currents would be considered dependent variables. One could also choose some new voltages and some new currents as independent variables with the rest of the new voltages and new currents as dependent variables. The admittance parameters are given below. The impedance parameters can be developed in a similar manner.

$$I_1 = y_{11} V_1 + y_{12} V_2 + y_{13} V_3 + y_{14} V_4$$
$$I_2 = y_{21} V_1 + y_{22} V_2 + y_{23} V_3 + y_{24} V_4$$
$$I_3 = y_{31} V_1 + y_{32} V_2 + y_{33} V_3 + y_{34} V_4$$
$$I_4 = y_{41} V_1 + y_{42} V_2 + y_{43} V_3 + y_{44} V_4$$

Substituting the differential- and common-mode relationships in these equations gives the following set of equations.

$$+I_{D1} + \frac{I_{C1}}{2} = y_{11}\left(V_{C1} + \frac{V_{D1}}{2}\right) + y_{12}\left(V_{C1} - \frac{V_{D1}}{2}\right) + y_{13}\left(V_{C2} + \frac{V_{D2}}{2}\right) + y_{14}\left(V_{C2} - \frac{V_{D2}}{2}\right)$$

$$-I_{D1} + \frac{I_{C1}}{2} = y_{21}\left(V_{C1} + \frac{V_{D1}}{2}\right) + y_{22}\left(V_{C1} - \frac{V_{D1}}{2}\right) + y_{23}\left(V_{C2} + \frac{V_{D2}}{2}\right) + y_{24}\left(V_{C2} - \frac{V_{D2}}{2}\right)$$

$$+I_{D2} + \frac{I_{C2}}{2} = y_{31}\left(V_{C1} + \frac{V_{D1}}{2}\right) + y_{32}\left(V_{C1} - \frac{V_{D1}}{2}\right) + y_{33}\left(V_{C2} + \frac{V_{D2}}{2}\right) + y_{34}\left(V_{C2} - \frac{V_{D2}}{2}\right)$$

$$-I_{D2} + \frac{I_{C2}}{2} = y_{41}\left(V_{C1} + \frac{V_{D1}}{2}\right) + y_{42}\left(V_{C1} - \frac{V_{D1}}{2}\right) + y_{43}\left(V_{C2} + \frac{V_{D2}}{2}\right) + y_{44}\left(V_{C2} - \frac{V_{D2}}{2}\right)$$

With the new voltages as independent variables and the new currents as dependent variables, a new differential/common-mode admittance matrix can be defined. The above set of equations can be rearranged to give a new matrix. The matrix will be called $(Y)_{DC}$. [*Note:* notice the difference between the notation $(Y)_{DC}$ for the total matrix and (Y_{DC}) for one submatrix of $(Y)_{DC}$.] The relationship between the old (Y) matrix and the new matrix $(Y)_{DC}$

is:

$$\begin{pmatrix} I_{D1} \\ I_{D2} \\ I_{C1} \\ I_{C2} \end{pmatrix} = (Y)_{DC} \begin{pmatrix} V_{D1} \\ V_{D2} \\ V_{C1} \\ V_{C2} \end{pmatrix} = \begin{pmatrix} (Y_{DD}) & (Y_{DC}) \\ (Y_{CD}) & (Y_{CC}) \end{pmatrix} \begin{pmatrix} V_{D1} \\ V_{D2} \\ V_{C1} \\ V_{C2} \end{pmatrix}$$

$$= \begin{pmatrix} Y_{DD11} & Y_{DD12} & Y_{DC11} & Y_{DC12} \\ Y_{DD21} & Y_{DD22} & Y_{DC21} & Y_{DC22} \\ Y_{CD11} & Y_{CD12} & Y_{CC11} & Y_{CC12} \\ Y_{CD21} & Y_{CD22} & Y_{CC21} & Y_{CC22} \end{pmatrix} \begin{pmatrix} V_{D1} \\ V_{D2} \\ V_{C1} \\ V_{C2} \end{pmatrix}$$

$$\begin{pmatrix} \begin{pmatrix} I_{D1} \\ I_{D2} \end{pmatrix} \\ \begin{pmatrix} I_{C1} \\ I_{C2} \end{pmatrix} \end{pmatrix} = \begin{pmatrix} (Y_{DD}) & (Y_{DC}) \\ (Y_{CD}) & (Y_{CC}) \end{pmatrix} \begin{pmatrix} \begin{pmatrix} V_{D1} \\ V_{D2} \end{pmatrix} \\ \begin{pmatrix} V_{C1} \\ V_{C2} \end{pmatrix} \end{pmatrix}$$

$$= \begin{pmatrix} \begin{pmatrix} Y_{DD11} & Y_{DD12} \\ Y_{DD21} & Y_{DD22} \end{pmatrix} & \begin{pmatrix} Y_{DC11} & Y_{DC12} \\ Y_{DC21} & Y_{DC22} \end{pmatrix} \\ \begin{pmatrix} Y_{CD11} & Y_{CD12} \\ Y_{CD21} & Y_{CD22} \end{pmatrix} & \begin{pmatrix} Y_{CC11} & Y_{CC12} \\ Y_{CC21} & Y_{CC22} \end{pmatrix} \end{pmatrix} \begin{pmatrix} \begin{pmatrix} V_{D1} \\ V_{D2} \end{pmatrix} \\ \begin{pmatrix} V_{C1} \\ V_{C2} \end{pmatrix} \end{pmatrix}$$

where the submatrices are given below.

$$(Y_{DD}) = \begin{pmatrix} \dfrac{y_{11} - y_{12} - y_{21} + y_{22}}{4} & \dfrac{y_{13} - y_{14} - y_{23} + y_{24}}{4} \\ \dfrac{y_{31} - y_{32} - y_{41} + y_{42}}{4} & \dfrac{y_{33} - y_{34} - y_{43} + y_{44}}{4} \end{pmatrix}$$

$$(Y_{DC}) = \begin{pmatrix} \dfrac{y_{11} + y_{12} - y_{21} - y_{22}}{2} & \dfrac{y_{13} + y_{14} - y_{23} - y_{24}}{2} \\ \dfrac{y_{31} + y_{32} - y_{41} - y_{42}}{2} & \dfrac{y_{33} + y_{34} - y_{43} - y_{44}}{2} \end{pmatrix}$$

$$(Y_{CD}) = \begin{pmatrix} \dfrac{y_{11} - y_{12} + y_{21} - y_{22}}{2} & \dfrac{y_{13} - y_{14} + y_{23} - y_{24}}{2} \\ \dfrac{y_{31} - y_{32} + y_{41} - y_{42}}{2} & \dfrac{y_{33} - y_{34} + y_{43} - y_{44}}{2} \end{pmatrix}$$

$$(Y_{CC}) = \begin{pmatrix} \dfrac{y_{11} + y_{12} + y_{21} + y_{22}}{1} & \dfrac{y_{13} + y_{14} + y_{23} + y_{24}}{1} \\ \dfrac{y_{31} + y_{32} + y_{41} + y_{42}}{1} & \dfrac{y_{33} + y_{34} + y_{43} + y_{44}}{1} \end{pmatrix}$$

This new admittance matrix can be converted to a new scattering matrix using the standard admittance-to-scattering matrix formula. The transformation is:

$$(S) = ((U) - (\sqrt{R_i})(Y)(\sqrt{R_i}))((U) + (\sqrt{R_i})(Y)(\sqrt{R_i}))^{-1}$$

where R_i is the normalization impedance for the ith port. The matrix (U) is the identity matrix. The term containing the square root of R_i is a diagonal matrix. As in all scattering parameter matrices, the choice of the normalization impedance is somewhat arbitrary. The industry convention at this time is to use $2Z_0$ for the differential ports and $0.5Z_0$ for the common-mode ports. This corresponds to the measurement circuits shown in Figure 4.5 and the scattering quantities to be derived shortly. The normalization impedance matrix using this convention is:

$$(\sqrt{R_i}) = \begin{pmatrix} \sqrt{2Z_0} & 0 & 0 & 0 \\ 0 & \sqrt{2Z_0} & 0 & 0 \\ 0 & 0 & \sqrt{Z_0/2.} & 0 \\ 0 & 0 & 0 & \sqrt{Z_0/2.} \end{pmatrix}$$

The new scattering matrix looks similar to the admittance matrix. It is:

$$\begin{pmatrix} \begin{pmatrix} b_{D1} \\ b_{D2} \end{pmatrix} \\ \begin{pmatrix} b_{C1} \\ b_{C2} \end{pmatrix} \end{pmatrix} = \begin{pmatrix} (S_{DD}) & (S_{DC}) \\ (S_{CD}) & (S_{CC}) \end{pmatrix} \begin{pmatrix} \begin{pmatrix} a_{D1} \\ a_{D2} \end{pmatrix} \\ \begin{pmatrix} a_{C1} \\ a_{C2} \end{pmatrix} \end{pmatrix}$$

$$= \begin{pmatrix} \begin{pmatrix} S_{DD11} & S_{DD12} \\ S_{DD21} & S_{DD22} \end{pmatrix} & \begin{pmatrix} S_{DC11} & S_{DC12} \\ S_{DC21} & S_{DC22} \end{pmatrix} \\ \begin{pmatrix} S_{CD11} & S_{CD12} \\ S_{CD21} & S_{CD22} \end{pmatrix} & \begin{pmatrix} S_{CC11} & S_{CC12} \\ S_{CC21} & S_{CC22} \end{pmatrix} \end{pmatrix} \begin{pmatrix} \begin{pmatrix} a_{D1} \\ a_{D2} \end{pmatrix} \\ \begin{pmatrix} a_{C1} \\ a_{C2} \end{pmatrix} \end{pmatrix}$$

The measurement setup for determining immittance or scattering parameters can be configured in several ways. In the top of Figure 4.5, the voltage sources appear in a manner to show the differential and common-mode source voltages discretely. The tee formation will give equal common or differential-mode voltages on the ports only if the four-port is symmetrical or if the series source resistor is zero. In the bottom of Figure 4.5, the voltage sources are shown as two separate sources with variable phase shifts. When $\Phi_1 = \Phi_2$, then the setup is the same as $V_D = 0$. When $\Phi_1 = \Phi_2 + \pi$, then the setup is the same as $V_C = 0$. When $V_D = 0$, then the two source resistors on the left are connected in parallel for an effective common-mode source

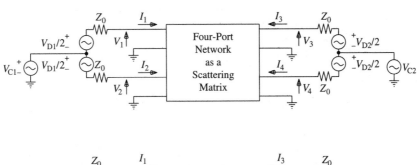

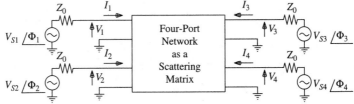

Figure 4.5 Two different measurement setups for a four-port network.

impedance of $Z_0/2$. When $V_C = 0$, then the two source resistors are connected in series (for symmetrical networks) with a source of value V_D. The load voltage in the common mode is $(V_3 + V_4)/2$. The load voltage in the differential mode is $V_3 - V_4$. It is important to determine which of the two voltages on the output serve as the differential voltage reference. There will be a difference of 180 degrees in the differential voltage between the two choices. Using the choices shown above, normalization impedances for the S_{DC} matrix of $2Z_0$ for the differential mode and $0.5Z_0$ for the common mode are appropriate for the measurement circuits shown.

The immittance parameters can be measured from the circuit described above. The scattering parameters can also be determined directly from measurements.

Starting from the differential and common-mode voltage and current definitions, a and b quantities for the differential and common-mode parameters can be determined. For the common mode:

$$a_{C1} = \frac{\dfrac{V_1 + V_2}{2} + (I_1 + I_2)Z_{0C}}{2\sqrt{Z_{0C}}}$$

$$= \frac{V_1 + 2I_1 Z_{0C}}{4\sqrt{Z_{0C}}} + \frac{V_2 + 2I_2 Z_{0C}}{4\sqrt{Z_{0C}}}$$

$$b_{C1} = \frac{\dfrac{V_1 + V_2}{2} - (I_1 + I_2)Z_{0C}}{2\sqrt{Z_{0C}}}$$

$$= \frac{V_1 - 2I_1 Z_{0C}}{4\sqrt{Z_{0C}}} + \frac{V_2 - 2I_2 Z_{0C}}{4\sqrt{Z_{0C}}}$$

If one substitutes $2Z_{0C} = Z_0$ in the above equations, the result is:

$$a_{C1} = \frac{V_1 + I_1 Z_0}{2\sqrt{2}\sqrt{Z_0}} + \frac{V_2 + I_2 Z_0}{2\sqrt{2}\sqrt{Z_0}} = \frac{a_1}{\sqrt{2}} + \frac{a_2}{\sqrt{2}}$$

$$b_{C1} = \frac{V_1 - I_1 Z_0}{2\sqrt{2}\sqrt{Z_0}} + \frac{V_2 - I_2 Z_0}{2\sqrt{2}\sqrt{Z_0}} = \frac{b_1}{\sqrt{2}} + \frac{b_2}{\sqrt{2}}$$

Likewise for the differential mode:

$$a_{D1} = \frac{V_1 - V_2 + \dfrac{(I_1 - I_2)}{2}Z_{0D}}{2\sqrt{Z_{0D}}}$$

$$= \frac{V_1 + I_1 \dfrac{Z_{0D}}{2}}{2\sqrt{Z_{0D}}} - \frac{V_2 + I_2 \dfrac{Z_{0D}}{2}}{2\sqrt{Z_{0D}}}$$

$$b_{D1} = \frac{V_1 - V_2 - \dfrac{(I_1 - I_2)}{2}Z_{0D}}{2\sqrt{Z_{0D}}}$$

$$= \frac{V_1 - I_1 \dfrac{Z_{0D}}{2}}{2\sqrt{Z_{0D}}} - \frac{V_2 - I_2 \dfrac{Z_{0D}}{2}}{2\sqrt{Z_{0D}}}$$

If one substitutes $Z_{0D} = 2Z_0$ in the above equations, the result is:

$$a_{D1} = \frac{V_1 + I_1 Z_0}{2\sqrt{2}\sqrt{Z_0}} - \frac{V_2 + I_2 Z_0}{2\sqrt{2}\sqrt{Z_0}} = \frac{a_1}{\sqrt{2}} - \frac{a_2}{\sqrt{2}}$$

$$b_{D1} = \frac{V_1 - I_1 Z_0}{2\sqrt{2}\sqrt{Z_0}} - \frac{V_2 - I_2 Z_0}{2\sqrt{2}\sqrt{Z_0}} = \frac{b_1}{\sqrt{2}} - \frac{b_2}{\sqrt{2}}$$

The new scattering parameters can be determined from the circuit shown in Figure 4.5 by measuring the standard scattering parameters but with two sources. Note that in the above equations, the phase reference is on V_1. Therefore, when V_1 and V_2 have the same phase, the differential mode tends to cancel (cancellation also depends on the currents) and when V_1 and V_2 are out of phase, the common mode tends to cancel. They only actually cancel when the network is symmetrical.

These equations can be further rearranged for addition by superposition. Using superposition, only one source (e.g., V_{si}) is on and all the others are off; the voltage at each port will be V_{ki}. Here the subscript k indicates which port the voltage is being summed at and i indicates which port has the source activated. The superposition quantities will be shown for excitation on the input by $i = 1$ and $j = 2$ and excitation on the output by $i = 3$ and $j = 4$. The total voltage at a port k is then $V_k = V_{ki} + V_{kj}$. This gives the following relationships:

$$a_{Ci} = \frac{V_{jm} + I_{jm}Z_0}{2\sqrt{2}\sqrt{Z_0}} + \frac{V_{jn} + I_{jn}Z_0}{2\sqrt{2}\sqrt{Z_0}} + \frac{V_{km} + I_{km}Z_0}{2\sqrt{2}\sqrt{Z_0}} + \frac{V_{kn} + I_{kn}Z_0}{2\sqrt{2}\sqrt{Z_0}}$$

$$= \frac{a_{jm}}{\sqrt{2}} + \frac{a_{jn}}{\sqrt{2}} + \frac{a_{km}}{\sqrt{2}} + \frac{a_{kn}}{\sqrt{2}}$$

$$b_{Ci} = \frac{V_{jm} - I_{jm}Z_0}{2\sqrt{2}\sqrt{Z_0}} + \frac{V_{jn} - I_{jn}Z_0}{2\sqrt{2}\sqrt{Z_0}} + \frac{V_{km} - I_{km}Z_0}{2\sqrt{2}\sqrt{Z_0}} + \frac{V_{kn} - I_{kn}Z_0}{2\sqrt{2}\sqrt{Z_0}}$$

$$= \frac{b_{jm}}{\sqrt{2}} + \frac{b_{jn}}{\sqrt{2}} + \frac{b_{km}}{\sqrt{2}} + \frac{b_{kn}}{\sqrt{2}}$$

$$a_{Di} = \frac{V_{jm} + I_{jm}Z_0}{2\sqrt{2}\sqrt{Z_0}} + \frac{V_{jn} + I_{jn}Z_0}{2\sqrt{2}\sqrt{Z_0}} - \frac{V_{km} + I_{km}Z_0}{2\sqrt{2}\sqrt{Z_0}} - \frac{V_{kn} + I_{kn}Z_0}{2\sqrt{2}\sqrt{Z_0}}$$

$$= \frac{a_{jm}}{\sqrt{2}} + \frac{a_{jn}}{\sqrt{2}} - \frac{a_{km}}{\sqrt{2}} - \frac{a_{kn}}{\sqrt{2}}$$

$$b_{Di} = \frac{V_{jm} - I_{jm}Z_0}{2\sqrt{2}\sqrt{Z_0}} + \frac{V_{jn} - I_{jn}Z_0}{2\sqrt{2}\sqrt{Z_0}} - \frac{V_{km} - I_{km}Z_0}{2\sqrt{2}\sqrt{Z_0}} - \frac{V_{kn} - I_{kn}Z_0}{2\sqrt{2}\sqrt{Z_0}}$$

$$= \frac{b_{jm}}{\sqrt{2}} + \frac{b_{jn}}{\sqrt{2}} - \frac{b_{km}}{\sqrt{2}} - \frac{b_{kn}}{\sqrt{2}}$$

where i stands for either 1 for the input parameters or 2 for the output parameters. The subscripts j and k would be 1 and 2 for $i = 1$ and 3 and 4 for $i = 2$ respectively. The subscripts m and n would be 1 and 2 if the excitation is on the input and 3 and 4 if the excitation is on the output. Whenever there is not a source on a particular node, the a term for that node goes to zero since the standard a and b quantities are normalized to Z_0. Likewise, when there is no source on a particular node, the b term numerator goes to 2 times the voltage associated with that term. These terms allow the differential and common-mode scattering parameters to be determined by superposition. Note that the phase reference for all of the voltages is still the phase of the input voltage on port one for input voltages. For output voltages, the phase can be referenced to either node 3 or node 4. However, to make the differential-mode parameters similar to single-ended scattering parameters, the phase reference should be on node 3 for excitations on the outputs.

Using the above equations and noting that $b_{D2} = b_{33} + b_{34} - b_{43} - b_{44}$, and so on, the following relationships between the differential/common-mode scattering parameters and the standard scattering parameters exist.

$$2S_{DD11} = S_{11} - S_{12} - S_{21} + S_{22}$$
$$2S_{DD12} = S_{13} - S_{23} - S_{14} + S_{24}$$
$$2S_{DD21} = S_{31} - S_{41} - S_{32} + S_{42}$$
$$2S_{DD22} = S_{33} - S_{43} - S_{34} + S_{44}$$
$$2S_{DC11} = S_{11} + S_{12} - S_{21} - S_{22}$$
$$2S_{DC12} = S_{13} + S_{14} - S_{23} - S_{24}$$
$$2S_{DC21} = S_{31} + S_{32} - S_{41} - S_{42}$$
$$2S_{DC22} = S_{33} + S_{34} - S_{43} - S_{44}$$
$$2S_{CD11} = S_{11} - S_{12} + S_{21} - S_{22}$$
$$2S_{CD12} = S_{13} - S_{14} + S_{23} - S_{24}$$
$$2S_{CD21} = S_{31} - S_{32} + S_{41} - S_{42}$$
$$2S_{CD22} = S_{33} - S_{34} + S_{43} - S_{44}$$
$$2S_{CC11} = S_{11} + S_{12} + S_{21} + S_{22}$$
$$2S_{CC12} = S_{13} + S_{14} + S_{23} + S_{24}$$
$$2S_{CC21} = S_{31} + S_{32} + S_{41} + S_{42}$$
$$2S_{CC22} = S_{33} + S_{34} + S_{43} + S_{44}$$

The reverse transformation is:

$$2. * S_{11} = +S_{DD11} + S_{CD11} + S_{DC11} + S_{CC11}$$
$$2. * S_{12} = -S_{DD11} - S_{CD11} + S_{DC11} + S_{CC11}$$
$$2. * S_{13} = +S_{DD12} + S_{CD12} + S_{DC12} + S_{CC12}$$
$$2. * S_{14} = -S_{DD12} - S_{CD12} + S_{DC12} + S_{CC12}$$
$$2. * S_{21} = -S_{DD11} + S_{CD11} - S_{DC11} + S_{CC11}$$
$$2. * S_{22} = +S_{DD11} - S_{CD11} - S_{DC11} + S_{CC11}$$
$$2. * S_{23} = -S_{DD12} + S_{CD12} - S_{DC12} + S_{CC12}$$
$$2. * S_{24} = +S_{DD12} - S_{CD12} - S_{DC12} + S_{CC12}$$
$$2. * S_{31} = +S_{DD21} + S_{CD21} + S_{DC21} + S_{CC21}$$
$$2. * S_{32} = -S_{DD21} - S_{CD21} + S_{DC21} + S_{CC21}$$
$$2. * S_{33} = +S_{DD22} + S_{CD22} + S_{DC22} + S_{CC22}$$
$$2. * S_{34} = -S_{DD22} - S_{CD22} + S_{DC22} + S_{CC22}$$
$$2. * S_{41} = -S_{DD21} + S_{CD21} - S_{DC21} + S_{CC21}$$
$$2. * S_{42} = +S_{DD21} - S_{CD21} - S_{DC21} + S_{CC21}$$
$$2. * S_{43} = -S_{DD22} + S_{CD22} - S_{DC22} + S_{CC22}$$
$$2. * S_{44} = +S_{DD22} - S_{CD22} - S_{DC22} + S_{CC22}$$

Once the circuit is measured in one of the manners described above, the standard admittance matrix can be determined from the measurements. The S_{DC} matrix can be first converted to a Y_{DC} matrix with the usual S matrix–to–Y matrix transformation or the differential/common-mode scattering parameters can be first converted to regular scattering parameters and then transformed back to admittance parameters. The transformation is:

$$(Y) = (\sqrt{R_i})^{-1}((U) + (S))^{-1}((U) - (S))(\sqrt{R_i})^{-1}$$

The terms in the equation are the same as given in the Y-to-S transformation. Once Y_{DC} is found, then the standard admittance matrix is found from the differential/common-mode admittance matrix by the following equation.

$$\begin{pmatrix} y_{11} & y_{12} \\ y_{21} & y_{22} \end{pmatrix}$$

$$= \begin{pmatrix} \dfrac{+4y_{DD11} + 2y_{DC11} + 2y_{CD11} + y_{CC11}}{4} & \dfrac{-4y_{DD11} + 2y_{DC11} - 2y_{CD11} + y_{CC11}}{4} \\[3mm] \dfrac{-4y_{DD11} - 2y_{DC11} + 2y_{CD11} + y_{CC11}}{4} & \dfrac{+4y_{DD11} - 2y_{DC11} - 2y_{CD11} + y_{CC11}}{4} \end{pmatrix}$$

$$\begin{pmatrix} y_{13} & y_{14} \\ y_{23} & y_{24} \end{pmatrix}$$

$$= \begin{pmatrix} \dfrac{+4y_{DD12} + 2y_{DC12} + 2y_{CD12} + y_{CC12}}{4} & \dfrac{-4y_{DC12} + 2y_{DC12} - 2y_{CD12} + y_{CC12}}{4} \\[3mm] \dfrac{-4y_{DD12} - 2y_{DC12} + 2y_{CD12} + y_{CC12}}{4} & \dfrac{+4y_{DD12} - 2y_{DC12} - 2y_{CD12} + y_{CC12}}{4} \end{pmatrix}$$

$$\begin{pmatrix} y_{31} & y_{32} \\ y_{41} & y_{42} \end{pmatrix}$$

$$= \begin{pmatrix} \dfrac{+4y_{DD21} + 2y_{DC21} + 2y_{CD21} + y_{CC21}}{4} & \dfrac{-4y_{DD21} + 2y_{DC21} - 2y_{CD21} + y_{CC21}}{4} \\[3mm] \dfrac{-4y_{DD21} - 2y_{DC21} + 2y_{CD21} + y_{CC21}}{4} & \dfrac{+4y_{DD21} - 2y_{DC21} - 2y_{CD21} + y_{CC21}}{4} \end{pmatrix}$$

$$\begin{pmatrix} y_{33} & y_{34} \\ y_{43} & y_{44} \end{pmatrix}$$

$$= \begin{pmatrix} \dfrac{+4y_{DD22} + 2y_{DC22} + 2y_{CD22} + y_{CC22}}{4} & \dfrac{-4y_{DD22} + 2y_{DC22} - 2y_{CD22} + y_{CC22}}{4} \\[3mm] \dfrac{-4y_{DD22} - 2y_{DC22} + 2y_{CD22} + y_{CC22}}{4} & \dfrac{+4y_{DD22} - 2y_{DC22} - 2y_{CD22} + y_{CC22}}{4} \end{pmatrix}$$

Notice the symmetry in the subscripts. All the same ij terms are grouped in a separate submatrix. All the differential terms have a 4 multiplier, all the cross-differential to common-mode terms have a 2 multiplier, and the common-mode terms have a unity multiplier. All the terms are then divided by 4.

These terms can be written in standard matrix notation as follows.

$$\begin{pmatrix} y_{11} & y_{12} \\ y_{21} & y_{22} \end{pmatrix} = \begin{pmatrix} \dfrac{+4\hat{y}_{11} + 2\hat{y}_{13} + 2\hat{y}_{31} + \hat{y}_{33}}{4} & \dfrac{-4\hat{y}_{11} + 2\hat{y}_{13} - 2\hat{y}_{31} + \hat{y}_{33}}{4} \\ \dfrac{-4\hat{y}_{11} - 2\hat{y}_{13} + 2\hat{y}_{31} + \hat{y}_{33}}{4} & \dfrac{+4\hat{y}_{11} - 2\hat{y}_{13} - 2\hat{y}_{31} + \hat{y}_{33}}{4} \end{pmatrix}$$

$$\begin{pmatrix} y_{13} & y_{14} \\ y_{23} & y_{24} \end{pmatrix} = \begin{pmatrix} \dfrac{+4\hat{y}_{12} + 2\hat{y}_{14} + 2\hat{y}_{32} + \hat{y}_{34}}{4} & \dfrac{-4\hat{y}_{12} + 2\hat{y}_{14} - 2\hat{y}_{32} + \hat{y}_{34}}{4} \\ \dfrac{-4\hat{y}_{12} - 2\hat{y}_{14} + 2\hat{y}_{32} + \hat{y}_{34}}{4} & \dfrac{+4\hat{y}_{12} - 2\hat{y}_{14} - 2\hat{y}_{32} + \hat{y}_{34}}{4} \end{pmatrix}$$

$$\begin{pmatrix} y_{31} & y_{32} \\ y_{41} & y_{42} \end{pmatrix} = \begin{pmatrix} \dfrac{+4\hat{y}_{21} + 2\hat{y}_{23} + 2\hat{y}_{41} + \hat{y}_{43}}{4} & \dfrac{-4\hat{y}_{21} + 2\hat{y}_{23} - 2\hat{y}_{41} + \hat{y}_{43}}{4} \\ \dfrac{-4\hat{y}_{21} - 2\hat{y}_{23} + 2\hat{y}_{41} + \hat{y}_{43}}{4} & \dfrac{+4\hat{y}_{21} - 2\hat{y}_{23} - 2\hat{y}_{41} + \hat{y}_{43}}{4} \end{pmatrix}$$

$$\begin{pmatrix} y_{33} & y_{34} \\ y_{43} & y_{44} \end{pmatrix} = \begin{pmatrix} \dfrac{+4\hat{y}_{22} + 2\hat{y}_{24} + 2\hat{y}_{42} + \hat{y}_{44}}{4} & \dfrac{-4\hat{y}_{22} + 2\hat{y}_{24} - 2\hat{y}_{42} + \hat{y}_{44}}{4} \\ \dfrac{-4\hat{y}_{22} - 2\hat{y}_{24} + 2\hat{y}_{42} + \hat{y}_{44}}{4} & \dfrac{+4\hat{y}_{22} - 2\hat{y}_{24} - 2\hat{y}_{42} + \hat{y}_{44}}{4} \end{pmatrix}$$

where the y parameters with a caret on top are the differential y parameters. The standard y parameters are shown related to submatrices of the standard y matrix in order to get the equations on the page.

Differential mode/common-mode scattering parameters are used to determine differential mode–to–common mode and common mode–to–differential mode conversion. Differential-mode stability and common-mode stability are determined from the two-port submatrices in the same manner stability will be considered for a two-port scattering matrix. Even- and odd-mode stability will be discussed later as well.

Some circuits are differential on the input and single-ended on the output, while other circuits are single-ended on the input and differential on the output. Circuits with only one set of differential ports and one single-ended port can be characterized using a three-port y matrix.

For a differential input and a single-ended output as shown in Figure 4.6, the transformation from a standard Y matrix to the Y_{DC} matrix is given below. Note that V_{C1} and V_3 are common-mode variables. V_3 is common mode only while V_1 has a common-mode and a differential-mode component. A subscript 3 is assigned to the output port to avoid confusion since there is not a second differential pair of ports.

$$\begin{pmatrix} I_{D1} \\ I_{C1} \\ I_3 \end{pmatrix} = (Y)_{DC} \begin{pmatrix} V_{D1} \\ V_{C1} \\ V_3 \end{pmatrix} = \begin{pmatrix} (Y_{DD}) & (Y_{DC}) \\ (Y_{CD}) & (Y_{CC}) \end{pmatrix} \begin{pmatrix} V_{D1} \\ V_{C1} \\ V_3 \end{pmatrix} = \begin{pmatrix} Y_{DD11} & Y_{DC11} & Y_{DC13} \\ Y_{CD11} & Y_{CC11} & Y_{CC13} \\ Y_{CD31} & Y_{CC31} & Y_{CC33} \end{pmatrix} \begin{pmatrix} V_{D1} \\ V_{C1} \\ V_3 \end{pmatrix}$$

$$\begin{pmatrix} (I_{D1}) \\ \begin{pmatrix} I_{C1} \\ I_3 \end{pmatrix} \end{pmatrix} = \begin{pmatrix} (Y_{DD}) & (Y_{DC}) \\ (Y_{CD}) & (Y_{CC}) \end{pmatrix} \begin{pmatrix} (V_{D1}) \\ \begin{pmatrix} V_{C1} \\ V_3 \end{pmatrix} \end{pmatrix} = \begin{pmatrix} (Y_{DD11}) & (Y_{DC11} \quad Y_{DC13}) \\ \begin{pmatrix} Y_{CD11} \\ Y_{CD31} \end{pmatrix} & \begin{pmatrix} Y_{CC11} & Y_{CC13} \\ Y_{CC31} & Y_{CC33} \end{pmatrix} \end{pmatrix} \begin{pmatrix} (V_{D1}) \\ \begin{pmatrix} V_{C1} \\ V_3 \end{pmatrix} \end{pmatrix}$$

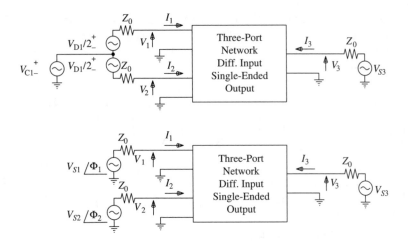

Figure 4.6 Two different measurement setups for a differential input three-port circuit.

$$(Y_{DD}) = \left(\frac{y_{11} - y_{12} - y_{21} + y_{22}}{4} \right)$$

$$(Y_{DC}) = \left(\frac{y_{11} + y_{12} - y_{21} - y_{22}}{2} \quad \frac{y_{13} - y_{23}}{2} \right)$$

$$(Y_{CD}) = \left(\begin{array}{c} \dfrac{y_{11} - y_{12} + y_{21} - y_{22}}{2} \\[2mm] \dfrac{y_{31} - y_{32}}{2} \end{array} \right)$$

$$(Y_{CC}) = \left(\begin{array}{cc} \dfrac{y_{11} + y_{12} + y_{21} + y_{22}}{1} & \dfrac{y_{13} + y_{23}}{1} \\[2mm] \dfrac{y_{31} + y_{32}}{1} & \dfrac{y_{33}}{1} \end{array} \right)$$

The normalization matrix for converting the differential input/single-ended output Y_{DC} matrix to the S_{DC} matrix is:

$$(\sqrt{R_i}) = \begin{pmatrix} \sqrt{2Z_0} & 0 & 0 \\ 0 & \sqrt{Z_0/2.} & 0 \\ 0 & 0 & \sqrt{Z_0} \end{pmatrix}$$

when measuring the circuit with all the ports terminated in the same characteristic impedance.

The reverse transformation from differential/common-mode y parameters to standard y parameters for a differential input to single-ended output is:

$$\begin{pmatrix} y_{11} & y_{12} & y_{13} \\ y_{21} & y_{22} & y_{23} \\ y_{31} & y_{32} & y_{33} \end{pmatrix} =$$

$$\begin{pmatrix} \dfrac{+4y_{DD11} + 2y_{DC11} + 2y_{CD11} + y_{CC11}}{4} & \dfrac{-4y_{DD11} + 2y_{DC11} - 2y_{CD11} + y_{CC11}}{4} & \dfrac{+4y_{DC13} + 2y_{CC13}}{4} \\[3mm] \dfrac{-4y_{DD11} - 2y_{DC11} + 2y_{CD11} + y_{CC11}}{4} & \dfrac{-4y_{DD11} - 2y_{DC11} + 2y_{CD11} + y_{CC11}}{4} & \dfrac{-4y_{DC13} + 2y_{CC13}}{4} \\[3mm] \dfrac{+4y_{CD31} + 2y_{CC31}}{4} & \dfrac{-4y_{CD31} + 2y_{CC31}}{4} & \dfrac{4y_{CC33}}{4} \end{pmatrix}$$

For a single-ended input and a differential output as shown in Figure 4.7, the transformation from a standard Y matrix to the Y_{DC} matrix is given below. Note that V_1 and V_{c2} are common-mode variables. V_1 is common mode only while V_{c2} has a common-mode and a differential-mode component.

$$\begin{pmatrix} I_{D2} \\ I_{C2} \\ I_1 \end{pmatrix} = (Y)_{DC} \begin{pmatrix} V_{D2} \\ V_{C2} \\ V_1 \end{pmatrix} = \begin{pmatrix} (Y_{DD}) & (Y_{DC}) \\ (Y_{CD}) & (Y_{CC}) \end{pmatrix} \begin{pmatrix} V_{D2} \\ V_{C2} \\ V_1 \end{pmatrix} = \begin{pmatrix} Y_{DD22} & Y_{DC22} & Y_{DC21} \\ Y_{CD22} & Y_{CC22} & Y_{CC21} \\ Y_{CD12} & Y_{CC12} & Y_{CC11} \end{pmatrix} \begin{pmatrix} V_{D2} \\ V_{C2} \\ V_1 \end{pmatrix}$$

$$\begin{pmatrix} (I_{D2}) \\ \begin{pmatrix} I_{C2} \\ I_1 \end{pmatrix} \end{pmatrix} = \begin{pmatrix} (Y_{DD}) & (Y_{DC}) \\ (Y_{CD}) & (Y_{CC}) \end{pmatrix} \begin{pmatrix} (V_{D2}) \\ \begin{pmatrix} V_{C2} \\ V_1 \end{pmatrix} \end{pmatrix} = \begin{pmatrix} (Y_{DD22}) & \begin{pmatrix} Y_{DC22} & Y_{DC21} \end{pmatrix} \\ \begin{pmatrix} Y_{CD22} \\ Y_{CD12} \end{pmatrix} & \begin{pmatrix} Y_{CC22} & Y_{CC21} \\ Y_{CC12} & Y_{CC11} \end{pmatrix} \end{pmatrix} \begin{pmatrix} (V_{D2}) \\ \begin{pmatrix} V_{C2} \\ V_1 \end{pmatrix} \end{pmatrix}$$

$$(Y_{DD}) = \left(\frac{y_{22} - y_{23} - y_{32} + y_{33}}{4} \right)$$

$$(Y_{DC}) = \left(\frac{y_{22} + y_{23} - y_{32} - y_{33}}{2} \quad \frac{y_{21} - y_{31}}{2} \right)$$

$$(Y_{CD}) = \begin{pmatrix} \dfrac{y_{22} - y_{23} + y_{32} - y_{33}}{2} \\ \dfrac{y_{12} - y_{13}}{2} \end{pmatrix}$$

$$(Y_{CC}) = \begin{pmatrix} \dfrac{y_{22} + y_{23} + y_{32} + y_{33}}{1} & \dfrac{y_{21} + y_{31}}{1} \\ \dfrac{y_{12} + y_{13}}{1} & \dfrac{y_{11}}{1} \end{pmatrix}$$

Notice the rows and columns appear to be interchanged. This is a result of keeping the differential-mode portion of the matrix in the upper left and the common-mode portion in the lower right. This interchanges the positions of the input and output variables.

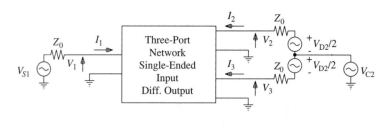

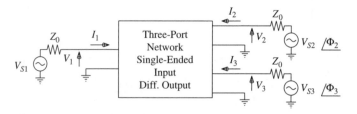

Figure 4.7 Two different measurement setups for a differential output three-port circuit.

The normalization matrix for converting the single-ended input/differential output Y_{DC} matrix to the S_{DC} matrix is:

$$(\sqrt{R_i}) = \begin{pmatrix} \sqrt{2Z_0} & 0 & 0 \\ 0 & \sqrt{Z_0/2.} & 0 \\ 0 & 0 & \sqrt{Z_0} \end{pmatrix}$$

when measuring the circuit with all the ports terminated in the same characteristic impedance.

The reverse transformation from differential/common-mode y parameters to standard y parameters for a single-ended input to differential output is:

$$\begin{pmatrix} y_{11} & y_{12} & y_{13} \\ y_{21} & y_{22} & y_{23} \\ y_{31} & y_{32} & y_{33} \end{pmatrix} =$$

$$\begin{pmatrix} \dfrac{4y_{CC11}}{4} & \dfrac{+4y_{CD12} + 2y_{CC12}}{4} & \dfrac{-4y_{CD12} + 2y_{CC12}}{4} \\[2ex] \dfrac{+4y_{DC21} + 2y_{CC21}}{4} & \dfrac{+4y_{DD11} + 2y_{DC21} + 2y_{DC22} + y_{CC22}}{4} & \dfrac{-4y_{DD11} + 2y_{DC21} - 2y_{DC22} + y_{CC22}}{4} \\[2ex] \dfrac{+4y_{DC21} - 2y_{CC21}}{4} & \dfrac{-4y_{DD11} - 2y_{DC21} + 2y_{DC22} + y_{CC22}}{4} & \dfrac{+4y_{DD11} - 2y_{DC21} - 2y_{DC22} + y_{CC22}}{4} \end{pmatrix}$$

Although the differential mode–common mode scattering parameter analysis has been done based on admittance matrices, the same procedure can be followed starting from impedance matrices. Impedance matrix analysis would be more appropriate when circuits are put in series rather than in parallel.

4.4 FOUR-PORT MIXED-MODE SCATTERING PARAMETER EXAMPLE

An example will be used to illustrate the use of the formulas given in Section 4.3. The circuit shown in Figure 4.8 contains a differential circuit. The two devices are slightly different and there is a cross-coupling resistor between the two input ports. The scattering parameters for the top and the bottom device were calculated in Spice. These parameters were then converted to admittance parameters with the S-to-Y conversion equation using 50 ohms for each port normalization resistance. The two-port S matrices are given below.

$$(S)_{top} = \begin{pmatrix} 0.06836 & @83.48 \text{ deg} & 0.1092 & @87.25 \text{ deg} \\ 3.818 & @69.60 \text{ deg} & 0.5505 & @25.15 \text{ deg} \end{pmatrix}$$

$$(S)_{bot} = \begin{pmatrix} 0.5520 & @158.5 \text{ deg} & 0.03635 & @70.22 \text{ deg} \\ 5.239 & @71.02 \text{ deg} & 0.3827 & @38.63 \text{ deg} \end{pmatrix}$$

These scattering parameters are converted to admittance parameters. The resistance is added to the admittance matrix by having $y_{12} = -0.001$ and adding 0.001 to y_{11} and y_{22}, giving a net admittance matrix of:

$$\begin{pmatrix} +0.01239 - j3.096x10^{-6} & -0.00100 + j0.0 & +0.0002410 - j0.002248 & +0.0 + j0.0 \\ -0.00100 + j0.0 & +0.04325 - j0.01359 & +0.0 + j0.0 & -0.0006476 - j0.001630 \\ -0.01580 - j0.07744 & +0.0 + j0.0 & +0.0003953 + j0.004606 & +0.0 + j0.0 \\ +0.0 + j0.0 & -0.09005 - j0.2362 & +0.0 + j0.0 & +0.003776 + j0.008981 \end{pmatrix}$$

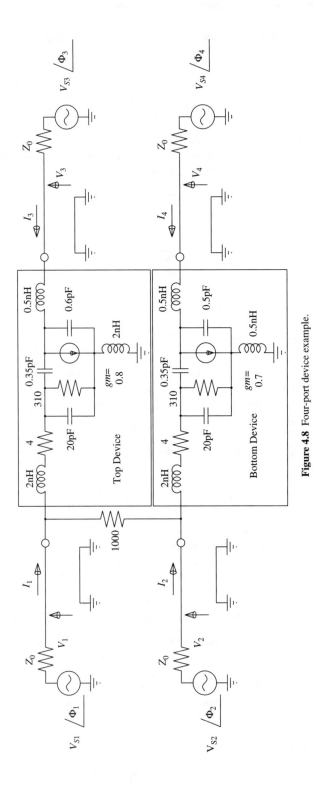

Figure 4.8 Four-port device example.

The admittance matrix was converted to a mixed-mode admittance matrix. The result is:

$$\begin{pmatrix} +0.01441 - j0.003398 & -0.0001017 - j0.0009693 & -0.01543 + j0.006792 & +0.0004443 - j0.000309 \\ -0.02646 - j0.07840 & +0.001043 + j0.003397 & +0.03712 + j0.07936 & -0.001690 - j0.002188 \\ -0.01543 + j0.006792 & +0.0004443 - j0.000309 & +0.05365 - j0.01359 & -0.0004066 - j0.003877 \\ +0.03712 + j0.07936 & -0.001690 - j0.002188 & -0.1058 - j0.3136 & +0.004171 + j0.01359 \end{pmatrix}$$

This matrix was then converted to a mixed-mode scattering matrix with reference impedances of 100 ohms for the differential "ports" and 25 ohms for the common-mode "ports." The result is:

$$\begin{pmatrix} 0.3059@155.8\text{ deg} & 0.06957@+82.5\text{ deg} & 0.2595@-14.83\text{ deg} & 0.03625@95.02\text{deg} \\ 4.365@70.05\text{ deg} & 0.4792@-30.55\text{ deg} & 0.7684@-106.6\text{ deg} & 0.1076@0.2214\text{ deg} \\ 0.2595@-14.83\text{ deg} & 0.03677@96.06\text{ deg} & 0.2902@+151.9\text{ deg} & 0.07169@83.04\text{ deg} \\ 0.6870@-105.5\text{ deg} & 0.09744@2.02\text{ deg} & 4.537@70.40\text{ deg} & 0.4623@-30.73\text{ deg} \end{pmatrix}$$

The circuit was then simulated on Spice as a four-port. The four-port scattering matrix is:

$$\begin{pmatrix} 0.06688@104.4\text{ deg} & 0.01283@26.08\text{ deg} & 0.1065@87.16\text{ deg} & 0.0008848@73.70\text{ deg} \\ 0.01283@26.08\text{ deg} & 0.5547@159.1\text{ deg} & 0.001387@109.5\text{ deg} & 0.03592@69.94\text{ deg} \\ 3.725@69.51\text{ deg} & 0.04848@91.82\text{ deg} & 0.5606@-25.12\text{ deg} & 0.003344@139.4\text{ deg} \\ 0.1275@74.50\text{ deg} & 5.177@70.74\text{ deg} & 0.01379@157.9\text{ deg} & 0.3872@-38.63\text{ deg} \end{pmatrix}$$

The four-port scattering parameters were used to calculate a mixed-mode scattering matrix using the conversion equations given in Section 4.3.3. The results were virtually the same as those obtained from the mixed-mode admittance matrix converted to scattering parameter.

Notice from the mixed-mode scattering matrix that there are significant amounts of differential- to common-mode conversion as well as common-mode to differential-mode conversion. Those two conversion submatrices are similar but not the same.

EXERCISES

4-1 A two-port S matrix has $S_{11} = -0.5$, $S_{12} = 0.01$, $S_{21} = 2.$, $S_{22} = 0.8$. Determine the three-port indefinite S matrix when the common terminal is lifted from ground. Assume a 50-ohm normalization.

4-2 Derive the four-port mixed-mode S matrix for a differential amplifier formed from the following two devices.

$$h_{11} = 100, h_{12} = 0.0001, h_{21} = 0.010, h_{22} = 0.01$$
$$h_{11} = 100, h_{12} = 0.0001, h_{21} = 0.008, h_{22} = 0.01$$

Hint: Method 1. Convert the h parameters to admittance parameters. Then form the four-port admittance matrix from the two-port admittance matrices. Then form the mixed-mode y matrix. Then form the mixed-mode scattering matrix.

Hint: Method 2. Write the voltage and current relationships for the network using the h parameters. Then calculate the mixed-mode scattering parameters from the defining relationships using the appropriate terminations on the devices.

4-3 Determine the gain of the two-port consisting of the two amplifiers in 4-2, first in parallel and then in the differential mode.

4-4 Determine the differential-mode to common-mode conversion of the amplifier in 4-2.

5

Stability, Stabilization, and Gain

5.1 STABILITY AND GAIN CONSIDERATIONS

In Chapter 2, two-port modeling parameters were considered. These allow one to determine the input and output characteristics of the networks given an output load or input source. In Chapter 4, some multiport characterization was done. This chapter addresses how much amplification gain is produced from these networks with specific loads, and equally important, what loads allow the networks to be stable from self-excited oscillations. This chapter addresses only linear phenomena. Chapter 14 of this book discusses some of the nonlinear phenomena that result in oscillations as a result of nonlinear effects.

Chapter 8 on oscillators contains criteria for a sustained self-excited steady-state oscillation from linear devices. This chapter is concerned not so much with sustained steady-state oscillations but whether there is instability. Such a circuit or network supports a signal that increases with time. The following list gives some of the types of instabilities that can be expected from circuits.

Types of Instability and Oscillations
- *Linear*
  1. Linear even mode
  2. Linear odd mode
  3. "Linear" nonlinear (e.g., Sampling Circuits)
  4. Signal generation, then translation in frequency, then amplification, and then translation back in frequency
- *Nonlinear*
  5. Squegging and motorboating
      Nonlinear components
  6. Parametric
      Subharmonic
      Idlers

Only the first two linear types of stability are discussed in this chapter. These are the linear even-mode and odd-mode stability conditions. Even-mode stability can be determined and controlled from the two-port terminals. Odd-mode stability cannot be determined from the two-port terminals and requires a change of the internal circuitry of the two-port to control it if it is a problem.

5.2 TWO-PORT REFLECTION COEFFICIENTS

In Chapter 2, input and output reflection coefficients were briefly discussed. The concept of (1) a source reflection coefficient, (2) an input reflection coefficient, (3) an output reflection coefficient, and (4) a load reflection coefficient is key to understanding even-mode two-port stability. These four reflection coefficients are separate and distinct.

5.2.1 Input, Load, Source, and Output Reflection Coefficients

There are four arrows shown in Figure 5.1. In addition to these four distinct reflection coefficients, the concept that, for instance, Γ_{in} is not equal to S_{11} is important to understanding stability. With reference to Figure 5.1, consider the arrows that point to the right. When the load Z_2 is changed, $\Gamma_2 = \Gamma_{load}$ changes. Changing load Z_2 does not affect the output reflection coefficient Γ_{out} or the source reflection coefficient Γ_{source}. It does, however, affect the input reflection coefficient Γ_{in}. Notice that the input reflection coefficient Γ_{in} and the load reflection coefficient Γ_{load} have arrows pointing to the right. In like manner, the source reflection coefficient $\Gamma_1 = \Gamma_{source}$ does not affect the input reflection coefficient Γ_{in} or the load reflection coefficient. The source reflection coefficient does affect the output reflection coefficient Γ_{out}. Notice that the source and the output reflection coefficients have arrows pointing toward the left. They are related or linked. The relationship can be easily determined as follows.

$$a_2 = \Gamma_{load}b_2$$

$$b_1 = S_{11}a_1 + S_{12}a_2 = S_{11}a_1 + S_{12}\Gamma_{load}b_2$$

$$b_2 = S_{21}a_1 + S_{22}a_2 = S_{21}a_1 + S_{22}\Gamma_{load}b_2$$

$$b_2 = \frac{S_{21}a_1}{1 - S_{22}\Gamma_{load}}$$

$$b_1 = S_{11}a_1 + S_{12}\Gamma_{load}\frac{S_{21}a_1}{1 - S_{22}\Gamma_{load}}$$

$$\Gamma_{in} = \frac{b_1}{a_1} = \frac{S_{11} - \Delta_S\Gamma_{load}}{1 - S_{22}\Gamma_{load}}$$

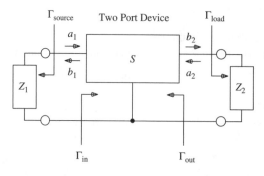

Figure 5.1 Four separate reflection coefficients.

The input reflection coefficient is a bilinear transformation of the load reflection coefficient. The transformation can be reversed and the load reflection coefficient that produces a given input reflection coefficient is given by the following relationship.

$$\Gamma_{\text{load}} = \frac{\Gamma_{\text{in}} - S_{11}}{\Gamma_{\text{in}} S_{22} - \Delta_S}$$

Care needs to be taken with the use of this last equation. It is not valid when S_{12} is zero. When S_{12} is zero, there is no relationship between the input reflection coefficient and the load reflection coefficient since when S_{12} is zero, the input reflection coefficient is always equal to S_{11}. When S_{12} is not zero, then the input reflection coefficient is not equal to S_{11}. As expected, this last equation is also a bilinear transformation. By a simple change of subscripts, the output reflection coefficient is related to the source reflection coefficient. This relationship is as follows.

$$\Gamma_{\text{out}} = \frac{b_2}{a_2} = \frac{S_{22} - \Delta_S \Gamma_{\text{source}}}{1 - S_{11} \Gamma_{\text{source}}}$$

$$\Gamma_{\text{source}} = \frac{\Gamma_{\text{out}} - S_{22}}{\Gamma_{\text{out}} S_{11} - \Delta_S}$$

The same care needs to be taken when using the above equation. Since there are four separate reflection coefficients, there are four separate reflection coefficient planes to plot them on. Each of the mappings is a bilinear transformation. The four reflection coefficient planes need to be kept separate and distinct to understand two-port stability. Figure 5.2 shows these four reflection coefficient planes. The top row of charts emphasizes the input and output reflection coefficients of the device. The left chart represents the source reflection coefficient plane. This plane is mapped into the third chart on the top row where it appears in the output reflection coefficient plane. These two charts represent the two arrows shown looking to the left in Figure 5.1. The second and fourth charts on the top row represent the two arrows looking to the right in Figure 5.1. The second chart shows the load reflection coefficient plane mapped back into the input reflection coefficient plane. The top row of charts is used to visualize the reflection coefficient chart when maximum gain is determined. In the bottom row of charts, the left chart shows the locus of source reflection coefficients that map into the unit circle in the output reflection coefficient plane. The rightmost chart in the bottom row shows the locus of the load reflection coefficients that map back into the unit circle in the input reflection coefficient plane. The bottom row of charts is used to visualize stability relationships for a two-port network. Notice that on the top row, the mapping is from the terminations to the input and output reflection coefficients. In the bottom row, the mapping is from the input and output reflection coefficients to the load and source termination reflection coefficients, respectively. Said in another way, the top row of charts sees the reflection coefficient plane on the opposite port looking through the network. The bottom row of charts shows the locus required on the opposite port that gives a unity reflection coefficient on the port one is looking into.

5.3 STABILITY

Stability criteria need to be evaluated at every frequency. It is sometimes possible for a two-port device to have a negative real immittance looking into a port at some frequency when the opposite port is terminated in some positive real immittance. This means that with positive real immittance port terminations, the circuit in which the device is embedded could become unstable, resulting in ever-increasing amplitudes for voltage and current with time. The device is then termed *potentially unstable*. If at a given frequency there is no negative real immittance at a port when the opposite port is terminated in any positive real immittance, then the device is called *unconditionally stable* at that frequency.

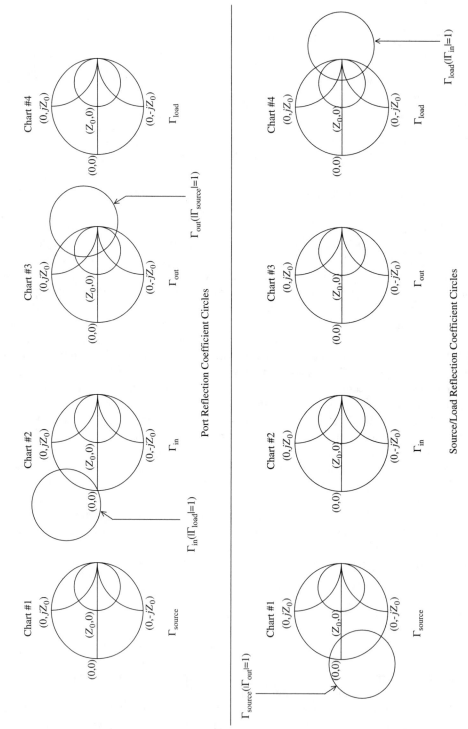

Figure 5.2 Four reflection coefficient planes for stability analysis.

5.3.1 Even-Mode and Odd-Mode Considerations

There are two modes of operation that have to be considered. First, the mode that is called the *even mode* of operation of a two-port will be considered. This is the mode that has current flowing in or out of the ports under consideration. Because of the current flowing in the port termination impedance, there is also a voltage at the two-ports under consideration. The other mode of operation, called the *odd mode,* has no current flowing in or out of the ports under consideration but has internal current and voltages. Since there is no current flowing in or out of the ports and since the ports can be terminated in general impedances, there is no voltage at the ports either. In the even mode, placing a short at a port would give zero voltage at that port but not zero current at that port. The mode of operation where zero voltage and zero current exist for a nonzero impedance termination at that port is called the odd mode of operation. The even and odd modes of operation have separate stability requirements. Figure 5.3 shows the currents and voltages for even and odd modes under consideration for stability analysis. Notice that the voltages for each node are the same on each half of the circuit but that the currents into each half of the circuit can be markedly different. The circuit shown shows the separate parts connected in parallel. The dual situation, the same current in the ports, but equal and opposite voltages might exist when two networks are put in series. This analysis will be carried out for the parallel case because it is the more common method by which circuits are combined. However, in some instances, circuits are put in series and a similar analysis using this procedure but using Z matrices can be carried out.

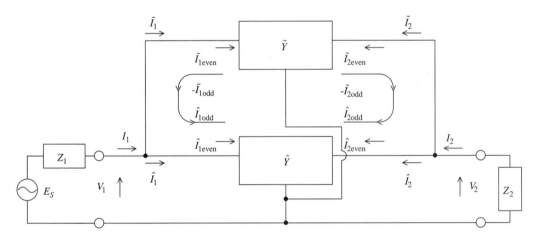

Figure 5.3 Configuration for considering odd-mode stability. [*Figure 1, Robert J. Weber,* "*Even Mode versus Odd Mode Stability,*" 40th Midwest Symposium on Circuits and Systems, 1997, pp. 607–610, August 1997.]

For linear circuits, the port currents can be summed as follows.

$$I_1 = \hat{I}_1 + \tilde{I}_1$$
$$= \hat{I}_{1\text{even}} + \hat{I}_{1\text{odd}} + \tilde{I}_{1\text{even}} + \tilde{I}_{1\text{odd}}$$
$$= \hat{I}_{1\text{even}} + \tilde{I}_{1\text{even}} + \hat{I}_{1\text{odd}} - \hat{I}_{1\text{odd}}$$
$$= I_{1\text{even}} + I_{1\text{odd}}$$

Identifying the even and odd parts of the input current gives the following equations.

$$I_{1\text{even}} = \hat{I}_{1\text{even}} + \tilde{I}_{1\text{even}}$$
$$I_{1\text{odd}} = 0$$

A similar set of equations holds for the second port. If the two networks are identical then the even-mode current would be the same into both networks and the even-mode input current would be double the input current into each network. The following analysis does not depend on the network being symmetrical. It depends only on the odd-mode current being equal and opposite into the top and bottom networks. Notice that when using superposition, the odd mode is not coupled to the "outside" of the circuit. The external odd-mode current is zero and therefore the odd-mode voltage is zero, yet there can be current flowing inside the circuit and therefore voltages exist across internal nodes. This condition is related to the eigenfunctions and eigenvalues of the Y matrix. The even mode of operation will be considered first.

5.3.2 Two-Port Even-Mode Stability Considerations

In order to investigate stability criteria for the two-port, consider the bottom line of reflection coefficient charts in Figure 5.2. If the unit circle of the Γ_{out} reflection coefficient chart is mapped onto the Γ_{source} plane, then either the inside of the resultant circular locus or the outside of the resultant circular locus will be representative of the inside of the Γ_{out} reflection coefficient chart. Which case is described will be determined shortly. Likewise, if the unit circle of the Γ_{in} reflection coefficient chart is mapped onto the Γ_{load} plane, then either the inside of the resultant circular locus or the outside of the resultant circular locus will be representative of the inside of the Γ_{in} reflection coefficient chart.

What does the locus given by the following equation look like?

$$\Gamma_{\text{source}}\,(\Gamma_{\text{out}})|_{|\Gamma_{\text{out}}|=1}$$

In other words, what is the locus in the source plane that gives an output reflection coefficient locus that has a unity magnitude? Likewise, what does the locus given by the following equation look like?

$$\Gamma_{\text{load}}(\Gamma_{\text{in}})|_{|\Gamma_{\text{in}}|=1}$$

In other words, what is the locus in the load plane that gives an input reflection coefficient locus that has a unity magnitude? This can be determined by starting with:

$$\Gamma_{\text{in}} = \frac{S_{11} - \Delta_S\Gamma_{\text{load}}}{1 - S_{22}\Gamma_{\text{load}}}$$

Positive real input immittances occur when:

$$|\Gamma_{\text{in}}| < 1$$

which is equivalent to:

$$\Gamma_{\text{in}}\Gamma_{\text{in}}^* < 1$$

Setting:

$$\Gamma_{\text{in}}\Gamma_{\text{in}}^* = 1$$

gives:

$$\left(\frac{S_{11} - \Delta_S\Gamma_{\text{load}}}{1 - S_{22}\Gamma_{\text{load}}}\right)\left(\frac{S_{11} - \Delta_S\Gamma_{\text{load}}}{1 - S_{22}\Gamma_{\text{load}}}\right)^* = 1$$

Multiplying this out gives:

$$S_{11}S_{11}^* + \Delta_S\Delta_S^*\Gamma_{\text{load}}\Gamma_{\text{load}}^* - \Delta_S\Gamma_{\text{load}}S_{11}^* - \Delta_S^*\Gamma_{\text{load}}^*S_{11}$$
$$= 1 - S_{22}\Gamma_{\text{load}} - S_{22}^*\Gamma_{\text{load}}^* + S_{22}S_{22}^*\Gamma_{\text{load}}\Gamma_{\text{load}}^*$$

or:

$$\Gamma_{\text{load}}\Gamma_{\text{load}}^* - \Gamma_{\text{load}}^*\left(\frac{S_{22}^* - \Delta_S^*S_{11}}{|S_{22}|^2 - |\Delta_S|^2}\right) - \Gamma_{\text{load}}\left(\frac{S_{22} - \Delta_SS_{11}^*}{|S_{22}|^2 - |\Delta_S|^2}\right) = \frac{|S_{11}|^2 - 1}{|S_{22}|^2 - |\Delta_S|^2}$$

   This is not quite in the form of a circle. The magnitude squared of the coefficient of the Γ_{load} term needs to be added to both sides resulting in:

$$\Gamma_{\text{load}}\Gamma_{\text{load}}^* - \Gamma_{\text{load}}^*\left(\frac{S_{22}^* - \Delta_S^* S_{11}}{|S_{22}|^2 - |\Delta_S|^2}\right) - \Gamma_{\text{load}}\left(\frac{S_{22} - \Delta_S S_{11}^*}{|S_{22}|^2 - |\Delta_S|^2}\right) + \left|\frac{S_{22} - \Delta_S S_{11}^*}{|S_{22}|^2 - |\Delta_S|^2}\right|^2$$

$$= \frac{|S_{11}|^2 - 1}{|S_{22}|^2 - |\Delta_S|^2} + \frac{\left(S_{22} - \Delta_S S_{11}^*\right)\left(S_{22}^* - \Delta_S^* S_{11}\right)}{\left||S_{22}|^2 - |\Delta_S|^2\right|^2}$$

The numerator of the right-hand side can be rearranged like:

$$|S_{11}|^2|S_{22}|^2 - |\Delta_S|^2|S_{11}|^2 - |S_{22}|^2 + |\Delta_S|^2 + |S_{22}|^2 - \Delta_S^* S_{11} S_{22} - \Delta_S S_{11}^* S_{22}^* + |\Delta_S|^2|S_{11}|^2$$

$$= |S_{11}|^2|S_{22}|^2 - \Delta_S^* S_{11} S_{22} - \Delta_S S_{11}^* S_{22}^* + |\Delta_S|^2$$

$$= (S_{11} S_{22} - \Delta_S)(S_{11} S_{22} - \Delta_S)^*$$

$$= S_{12} S_{21} S_{12}^* S_{21}^*$$

Substituting for the right side:

$$\Gamma_{\text{load}}\Gamma_{\text{load}}^* - \Gamma_{\text{load}}^*\left(\frac{S_{22}^* - \Delta_S^* S_{11}}{|S_{22}|^2 - |\Delta_S|^2}\right) - \Gamma_{\text{load}}\left(\frac{S_{22} - \Delta_S S_{11}^*}{|S_{22}|^2 - |\Delta_S|^2}\right) + \left|\frac{S_{22} - \Delta_S S_{11}^*}{|S_{22}|^2 - |\Delta_S|^2}\right|^2$$

$$= \frac{|S_{12} S_{21}|^2}{\left||S_{22}|^2 - |\Delta_S|^2\right|^2}$$

   This implies that the center of the circle in the load reflection coefficient plane that maps to a unit circle in the input reflection coefficient plane is:

$$\Gamma_{\text{load\_c}} = \frac{S_{22}^* - \Delta_S^* S_{11}}{|S_{22}|^2 - |\Delta_S|^2}$$

The circle has a radius of:

$$\rho_{\text{load}} = \frac{|S_{12} S_{21}|}{\left||S_{22}|^2 - |\Delta_S|^2\right|}$$

This circle is called the *load stability circle* and it maps into the unit circle in the input reflection coefficient plane.

   A similar derivation can be done for the source plane using the output reflection coefficient. This results in the center of the source stability circle $\Gamma_{\text{source\_c}}$ given by:

$$\Gamma_{\text{source\_c}} = \frac{S_{11}^* - \Delta_S^* S_{22}}{|S_{11}|^2 - |\Delta_S|^2}$$

with radius of:

$$\rho_{\text{source}} = \frac{|S_{12} S_{21}|}{\left||S_{11}|^2 - |\Delta_S|^2\right|}$$

   This circle maps into the unit circle in the output reflection coefficient plane. Keep in mind that the source stability circle implies that the output reflection coefficient Γ_{out} is unity in magnitude and that the load stability circle implies that the input reflection coefficient Γ_{in} is unity in magnitude. It needs to be determined whether immittances lying inside or outside of these circles result in positive real port immittances looking into the opposite port. The stable region may lie inside a circle or outside a circle. The circle to consider is the stability region circle and not the unit reflection coefficient circle! The region is determined by considering whether S_{ii} is greater or less than 1 in magnitude and whether the point S_{ii} maps to the inside or outside of the stability region circle.

   Consider the input reflection coefficient for different loads. Which region of the load plane maps back into positive real input immittances? If the load reflection coefficient

$\Gamma_{\text{load}} = 0$, then $\Gamma_{\text{in}} = S_{11}$. If $|S_{11}| < 1$, then the region that the point $\Gamma_{\text{load}} = 0$ is contained in gives positive real input immittances. Similarly, which region of the source plane results in positive real output immittances? If the source reflection coefficient $\Gamma_{\text{source}} = 0$, then $\Gamma_{\text{out}} = S_{22}$. If $|S_{22}| < 1$, then the region that the point $\Gamma_{\text{source}} = 0$ is contained in gives positive real output immittances. The opposite arguments hold whenever $S_{ii} > 1$. Since the equations are cross coupled, if a region of the positive real load plane results in a negative real input immittance, then some region of the positive real source immittance plane will result in a negative real output immittance. The opposite is also true.

Now consider the reverse mapping case shown in the top row of charts in Figure 5.2. Do positive real load immittances map into positive real input immittances? The right half of the immittance planes lies within the unity reflection coefficient circle. The largest locus for a positive real immittance load reflection coefficient is the unit circle. Find the input reflection coefficient circle that represents the load unit circle. The mapping is as follows.

$$\Gamma_L \Gamma_L^* = \left(\frac{S_{11} - \Gamma_{\text{in}}}{\Delta_S - S_{22}\Gamma_{\text{in}}} \right) \left(\frac{S_{11} - \Gamma_{\text{in}}}{\Delta_S - S_{22}\Gamma_{\text{in}}} \right)^* = 1$$

This gives the circle in the input reflection coefficient plane as follows.

$$\Gamma_{\text{in}} \Gamma_{\text{in}}^* - \Gamma_{\text{in}}^* \frac{S_{11} - \Delta_S S_{22}^*}{1 - |S_{22}|^2} - \Gamma_{\text{in}} \frac{S_{11}^* - \Delta_S^* S_{22}}{1 - |S_{22}|^2} + \frac{\left| S_{11} - \Delta_S S_{22}^* \right|^2}{\left(1 - |S_{22}|^2 \right)^2}$$

$$= \frac{|\Delta_S|^2 - |S_{11}|^2}{1 - |S_{22}|^2} + \frac{\left| S_{11} - \Delta_S S_{22}^* \right|^2}{\left(1 - |S_{22}|^2 \right)^2}$$

Consider Figure 5.4. The magnitude of the vector to the center of the circle plus the magnitude of the radius must be less than 1 if the input reflection coefficients are to be less than 1 in magnitude. Since:

$$\left| S_{11} - \Delta_S S_{22}^* \right|^2 = \left(1 - |S_{22}|^2 \right) \left(|S_{11}|^2 - |\Delta_S|^2 \right) + |S_{12} S_{21}|^2$$

the right side of the equation is:

$$\rho^2 = \frac{|S_{12} S_{21}|^2}{\left(1 - |S_{22}|^2 \right)^2}$$

the sum of the magnitude of the center of the circle plus the radius is:

$$\sqrt{\frac{\left(1 - |S_{22}|^2 \right) \left(|S_{11}|^2 - |\Delta_S|^2 \right) + |S_{12} S_{21}|^2}{\left(1 - |S_{22}|^2 \right)^2}} + \frac{|S_{12} S_{21}|}{\left| 1 - |S_{22}|^2 \right|} < 1$$

Moving the right-hand term on the left side to the right side of the inequality and squaring the result introduces two conditions. In order to keep from squaring a negative number and introducing an extraneous root, the conditions are:

$$|S_{12} S_{21}| < 1 - |S_{22}|^2 \qquad |S_{22}|^2 < 1$$

or:

$$|S_{22}|^2 < 1 - |S_{12} S_{21}| < 1$$

Since the second part of the above equation is always true, the left side of the equation is the true constraint. Reducing the equation after carrying out the squaring operation gives the following inequality.

$$K = \frac{1 - |S_{22}|^2 - |S_{11}|^2 + |\Delta_S|^2}{2 |S_{12} S_{21}|} > 1$$

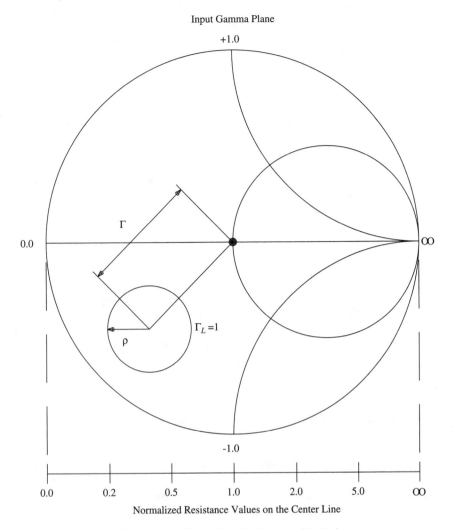

Figure 5.4 Stability circle within the unit radius circle.

where the left-hand side has been given the designation of K. This is the inverse of the *Linville C number* [1] and the condition is called the *Rollett condition* [2]. Starting from the source plane and looking at the output reflection coefficient gives the same result for K. It introduces another condition:

$$|S_{12}S_{21}| < 1 - |S_{11}|^2$$

Applying these conditions to:

$$\left|S_{11} - \Delta_S S_{22}^*\right|^2 = \left(1 - |S_{22}|^2\right)\left(|S_{11}|^2 - |\Delta_S|^2\right) + |S_{12}S_{21}|^2 > 0$$

gives:

$$|\Delta_S|^2 < \frac{|S_{12}S_{21}|}{\left(1 - |S_{22}|^2\right)}\,|S_{12}S_{21}| + |S_{11}|^2 < |S_{12}S_{21}| + |S_{11}|^2 < 1$$

Multiplying through the formula for K gives:

$$1 + |\Delta_S|^2 > 2\,|S_{12}S_{21}| + |S_{11}|^2 + |S_{22}|^2$$

$$|\Delta_S|^2 > 2\,|S_{12}S_{21}| + |S_{11}|^2 + |S_{22}|^2 - 1$$

Combining the results of the two sets of constraints gives the following result.

$$2\left|S_{12}S_{21}\right| + \left|S_{11}\right|^2 + \left|S_{22}\right|^2 - 1 < \left|\Delta_S\right|^2 < 1$$

Define a term L such that [3,4]:

$$L = \left|S_{12}S_{21}\right| + \frac{\left|S_{11}\right|^2 + \left|S_{22}\right|^2}{2}$$

Then:

$$\left|\left|\Delta_S\right|^2 - L\right| + L < 1$$

When this inequality is satisfied, the device is unconditionally stable in the even mode. This is a symmetrical and single condition equation for unconditional even-mode stability of linear two-port circuits. If one multiplies the last equation by -1 and then adds 2 to both sides of the resulting equation one gets:

$$2 - L - \left|\left|\Delta_S\right|^2 - L\right| > 1$$

$$2 - \left|S_{12}S_{21}\right| - \frac{\left|S_{11}\right|^2 + \left|S_{22}\right|^2}{2} - \left|\left|\Delta_S\right|^2 - \left|S_{12}S_{21}\right| - \frac{\left|S_{11}\right|^2 + \left|S_{22}\right|^2}{2}\right| > 1$$

This equation can be used like the K factor except that this equation is both necessary and sufficient for even-mode stability. The quantity L is expanded in the second part of the above equation to show the symmetry in the above condition for even-mode stability.

It has been shown [5] that the following equation is another single necessary and sufficient equation for stability in the even mode.

$$\mu = \frac{1 - \left|S_{11}\right|^2}{\left|S_{22} - S_{11}^*\Delta_S\right| + \left|S_{12}S_{21}\right|} > 1$$

Notice that when the two-port is matched, all three of the stability conditions indicate that:

$$\left|S_{12}S_{21}\right| < 1$$

The quantity K is used later in a maximum gain equation for devices that have a simultaneous conjugate match. The simultaneous conjugate match condition will be discussed after odd-mode stability is considered. It is shown in Section 5.5.4 that K does not vary with reactive port matching. However, the μ and L stability factors do vary as a function of matching.

5.3.3 Resistances Needed for Even-Mode Stability

How does one stabilize a circuit in the even mode? If a reactive element is placed in series with a port, the K factor will not change. If a reactive element is placed in series on a port, another series reactance of the opposite sign can be added to the new network and the original network is obtained and the K factor has not changed. The same can be said for a parallel susceptance. If one puts a transmission line on the port, it only shifts the reference plane on the network and the stability circles will be rotated but K will not change. However, if a series resistance or parallel conductance is added to the network, the K factor will change. Consider the network shown in the bottom of Figure 5.5. The resistance R_1 that is added in series with the input is shown as 0.2 normalized. This resistance plus the impedance Z_1 limits the impedance to that part of the right-half impedance plane shown by the shaded area in the lower-left graph in Figure 5.5. That region transformed into the reflection coefficient plane is shown by the shaded area in the reflection coefficient chart shown in the lower-right graph in Figure 5.5. The series resistance limits the impedance presented to the network to

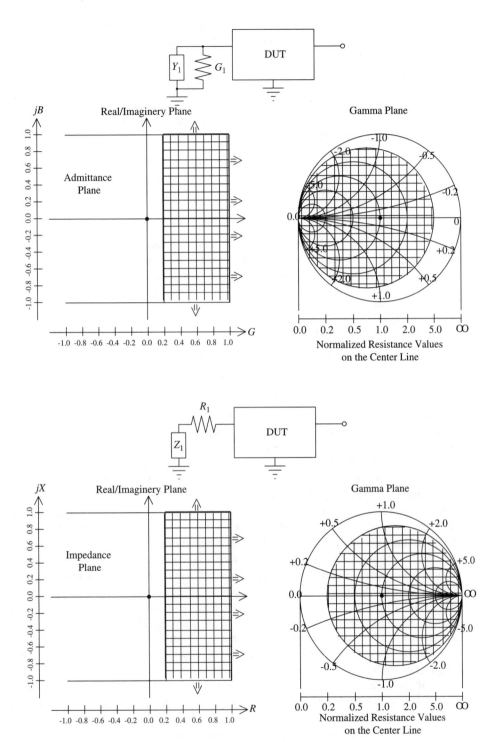

Figure 5.5 Excluded regions when using stabilizing resistors.

that subregion of the chart. A dual circuit using a normalized conductance of 0.2 is shown in the upper part of Figure 5.5 for a conductance placed on the network. The region presented to the network is then the region within that conductance circle.

Now consider the stability circle shown in the center chart in Figure 5.6. The circle has its center at two at 116.565 degrees and has a radius of four-thirds. What are the values for the impedance and conductance circles that are tangent to that circle? If (x, jy) are the coordinates of the center of the stability circle, ρ is the radius of the circle, and r is the normalized value of the resistance circle tangent to the stability circle, the following equation holds.

$$\left(x - \frac{r}{r+1}\right)^2 + y^2 = \left(\rho \pm \frac{1}{|r+1|}\right)^2$$

$$r^2\left(x^2 + y^2 - \rho^2 - 2x + 1\right) + r\left(2x^2 + 2y^2 - 2\rho^2 - 2x \mp 2\rho\right)$$
$$+ \left(x^2 + y^2 - \rho^2 \mp 2\rho - 1\right) = 0$$

The second line of the equation is valid only for positive real immittances. The top sign is used when the resistance circle lies outside the stability circle and the bottom sign is used when it lies inside the stability circle. Likewise, the normalized conductance value g of the conductance circle that is tangent to the stability circle is given as follows.

$$\left(x + \frac{g}{g+1}\right)^2 + y^2 = \left(\rho \pm \frac{1}{|g+1|}\right)^2$$

$$g^2\left(x^2 + y^2 - \rho^2 + 2x + 1\right) + g\left(2x^2 + 2y^2 - 2\rho^2 + 2x \mp 2\rho\right)$$
$$+ \left(x^2 + y^2 - \rho^2 \mp 2\rho - 1\right) = 0$$

Again, the second line is valid only for positive real immittances and the top sign is used when the conductance circle lies outside the stability circle and the bottom sign is used when it lies inside the stability circle. One solution to these equations is r equal to minus 1. The other solution is the desired solution. For the stability circle shown, the resistance circle $r = 0.28825$ and the conductance circle $g = 1.0077$ are solutions. Would these be the optimum values of resistances or conductances to put on the network? They would not give the minimum loss of power. They would, however, stabilize the network. A solution with less loss but a few more components would be the solution shown in the left chart or right chart of Figure 5.6. The left chart shows the stability circle rotated 63.4 degrees counterclockwise so that the short-circuit point lies within the circle. This is accomplished by inserting a 31.7-degree-long line on the input of the network. Notice that a series resistance is required to stabilize the resulting network. A shunt resistance would not prevent a short from being applied to the network but a series resistance would. The value of the series resistor is $r = 0.2$ normalized. This value is almost 50% less than the value required without rotation and would represent significantly less insertion loss of the source to the network. The right chart shows the stability circle with a 121.7°-long transmission line inserted on the port. Now a shunt resistance is required to stabilize the resulting network. In this case, the normalized resistance of the shunt conductance would be $r = 5$.

If the unstable region of the stability circle contains the short-circuit point, a series resistance equal to or greater than the r value of the tangent constant resistance circle is necessary to stabilize the network. If the unstable region of the stability circle contains the open-circuit point, a shunt resistance equal to or less than the r value of the tangent constant conductance circle is necessary to stabilize the network. What happens if the stability circles lies within the unit circle? Assume the stable immittances lie within the stability circle. This situation is shown in Figure 5.7. Either network of the two ell networks shown in Figure 5.7 will stabilize the network. Two resistors are required. The values of the resistors are related

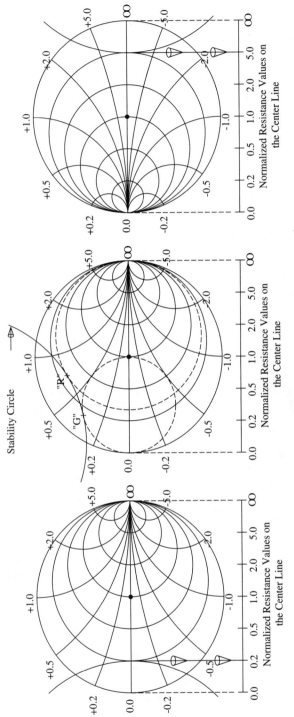

Figure 5.6 Rotating stability circles for optimum power transfer.

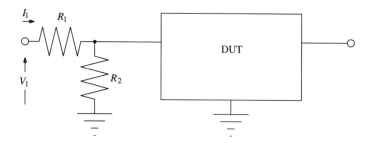

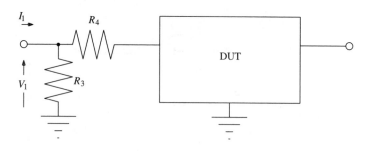

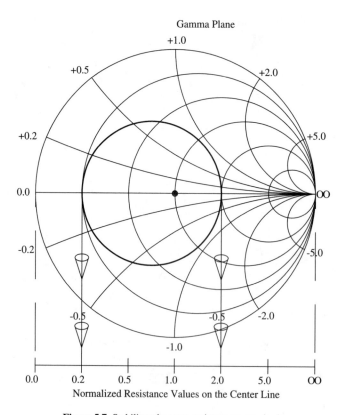

Figure 5.7 Stability when two resistors are required.

to the minimum and maximum resistances contained on the stability circle as follows.

$$R_1 = \frac{R_{\max} R_{\min}}{R_{\max} - R_{\min}}$$

$$R_2 = R_{\max}$$

$$R_3 = R_{\min}$$

$$R_4 = R_{\max} - R_{\min}$$

Figure 5.7 assumes that the stability circle has first been rotated with a transmission line to be symmetrically located across the real axis. What can be said about the cascade of unconditionally stable stages? If one stage is unconditionally stable, then there is no positive real immittance that can be placed on the device to produce a negative real immittance on the opposite port. Therefore, in a cascade of two networks, the next or previous network will also have a positive real immittance presented to it assuming the first stage is terminated in a positive real immittance. Since the second network is also unconditionally stable, it will have a positive real immittance on the next port because the adjacent network presents it with a positive real immittance. The argument can be continued for any number of networks in cascade. The inverse to the argument is not necessarily valid. It is possible to cascade some potentially unstable networks and have the cascade be unconditionally stable.

5.3.4 Two-Port Odd-Mode Stability Considerations [6]

For the odd mode of operation for two networks in parallel, construct a Y matrix for the overall network as shown in Figure 5.8. Note that there are no terminations shown on ports 3 through n. The Y matrix is partitioned so that the first row and column is port 1 and the second row and column is port 2. The rest of the internal nodes are contained in nodes 3 through n. The matrix then would look something like:

$$\begin{pmatrix} I_1 \\ I_2 \\ I_3 \\ \cdot \\ \cdot \\ \cdot \\ I_n \end{pmatrix} = (I) = (Y)(V) = \begin{pmatrix} Y_{11} & Y_{12} & Y_{13} & \cdot & \cdot & \cdot & Y_{1n} \\ Y_{21} & Y_{22} & Y_{23} & \cdot & \cdot & \cdot & Y_{2n} \\ Y_{31} & Y_{32} & Y_{33} & \cdot & \cdot & \cdot & Y_{3n} \\ & & & \cdot & & & \\ & & & & \cdot & & \\ & & & & & \cdot & \\ Y_{n1} & Y_{n2} & Y_{n3} & \cdot & \cdot & \cdot & Y_{nn} \end{pmatrix} \begin{pmatrix} V_1 \\ V_2 \\ V_3 \\ \cdot \\ \cdot \\ \cdot \\ V_n \end{pmatrix}$$

where rows and columns 3 through n contain admittances or zeroes depending on active or passive connections between any port pairs. For the odd mode, if it exists, the matrix would look like:

$$\begin{pmatrix} 0 \\ 0 \\ 0 \\ \cdot \\ \cdot \\ \cdot \\ 0 \end{pmatrix} = (I) = (Y)(V) = \begin{pmatrix} Y_{11} & Y_{12} & Y_{13} & \cdot & \cdot & \cdot & Y_{1n} \\ Y_{21} & Y_{22} & Y_{23} & \cdot & \cdot & \cdot & Y_{2n} \\ Y_{31} & Y_{32} & Y_{33} & \cdot & \cdot & \cdot & Y_{3n} \\ & & & \cdot & & & \\ & & & & \cdot & & \\ & & & & & \cdot & \\ Y_{n1} & Y_{n2} & Y_{n3} & \cdot & \cdot & \cdot & Y_{nn} \end{pmatrix} \begin{pmatrix} 0 \\ 0 \\ V_{3\text{odd}} \\ \cdot \\ \cdot \\ \cdot \\ V_{n\text{odd}} \end{pmatrix}$$

The currents at nodes 3 through n are zero since there are no connections to those nodes. However, for voltages to exist in the steady state at some or all of nodes 3 through n would

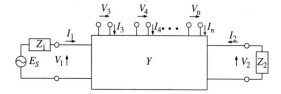

Figure 5.8 Admittance circuit for odd-mode stability of parallel connected networks. [*Figure 2, Robert J. Weber, "Even Mode versus Odd Mode Stability,"* 40th Midwest Symposium on Circuits and Systems, 1997, pp. 607–610, August 1997.]

require that:

$$
\begin{vmatrix}
Y_{33} & \cdot & \cdot & \cdot & Y_{3n} \\
& & \cdot & & \\
& & \cdot & & \\
& & \cdot & & \\
Y_{n3} & \cdot & \cdot & \cdot & Y_{nn}
\end{vmatrix} = 0
$$

This would be the condition for steady-state oscillations to occur internal to the network. What is of more interest than steady-state oscillations is whether the internal nodes are potentially unstable. In other words, does negative resistance exist at any of these nodes? Invert the Y matrix and get the Z matrix. The sub-Z matrix of the total Z matrix is:

$$
\begin{pmatrix} V_3 \\ \cdot \\ \cdot \\ \cdot \\ V_n \end{pmatrix} =
\begin{pmatrix}
Z_{33} & \cdot & \cdot & \cdot & Z_{3n} \\
& & \cdot & & \\
& & \cdot & & \\
& & \cdot & & \\
Z_{n3} & \cdot & \cdot & \cdot & Z_{nn}
\end{pmatrix}
\begin{pmatrix} I_3 \\ \cdot \\ \cdot \\ \cdot \\ I_n \end{pmatrix}
$$

where $Z = Y^{-1}$. The current and the voltage at nodes 1 and 2 for the odd mode are zero. Let currents at all other nodes remain at zero except at a node under consideration. Let the current into a node under consideration be a very small test current. The voltage and current at that node are related by:

$$
V_i = Z_{ii} I_i
$$

Therefore, if any of the diagonal terms of the Z submatrix are negative real, negative resistance exists and a potential oscillation exists. If all the diagonal terms of the Z submatrix are positive real, then no odd-mode oscillation will build up. Note that if the Z matrix does not exist because the determinant of the Y matrix is zero, then a steady-state oscillation condition already exists. One could reduce the rank of the submatrix of the Y matrix (i.e., consider a subset of nodes and check for stability with that submatrix. However, in order to stabilize the oscillation condition, the circuit changes necessary will likely change the rest of the matrix. Therefore, the suggested procedure is to fix the oscillation, form a new Y matrix, and then invert the new matrix and perform the test. In a symmetrical circuit, sometimes a fix is to put a resistance between two symmetrical nodes showing negative resistance. That procedure does not affect the common-mode voltage of a symmetrical network since the common-mode voltage between symmetrical nodes is the same. However, in the odd mode, the voltages in a symmetrical circuit are equal and opposite and the appropriate value of resistance will often stabilize the circuit. Just as in even-mode stability, sometimes a parallel resistance is needed and sometimes a series resistance is needed. It is more difficult to insert resistance in series in the odd mode. It introduces another node. Often, a resistor placed between two symmetrical nodes that do not show negative resistance will cause other nodes that show negative resistance to go into the positive real area.

Figure 5.9 shows a circuit similar to that used in [6]. The Y matrix for the circuit is:

$$(Y) = \begin{pmatrix} 0.2 & 0. & -0.1 & 0. & -0.1 & 0. \\ 0. & 0. & 0. & j0.2 & 0. & j0.2 \\ -0.1 & 0. & 0.1032 & -0.002581 & 0. & 0. \\ 0. & j0.2 & 0.0258 & -.00064 & 0. & 0. \\ -0.1 & 0. & 0. & 0. & 0.1032 & -0.002581 \\ 0. & j0.2 & 0. & 0. & 0.0258 & -.00064 \end{pmatrix}$$

The 33, 34, 55, and 56 terms are rounded off. The Z matrix for the circuit is:

$$(Z) = \begin{pmatrix} 160. & -j2.0 & 155. & 0. & 155. & 0 \\ j19.995 & 0.25 & j19.995 & -j2.5 & 19.995 & -j2.5 \\ 155. & -j2.0 & -465 & 2500 & 775 & -2500 \\ 0.0284 & -j2.5 & -25000 & 100000 & 25000 & -100000 \\ 155. & -j2.0 & 775 & -2500 & -465 & 2500 \\ -0.0284 & -j2.5 & 25000 & -100000 & -25000 & 100000 \end{pmatrix}$$

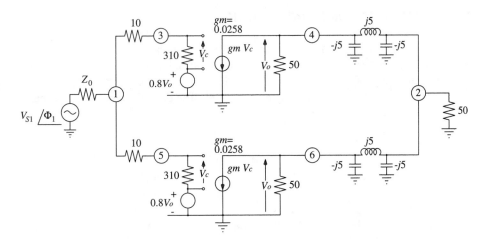

Figure 5.9 Device circuit for odd-mode stability example.

Putting a resistor between nodes 3 and 5 puts that resistance across stabilization resistors between these ports and the input, reducing their effectiveness. Putting a low-value resistor from ports 3 to 5 causes the new impedance matrix to show negative resistance between nodes 4 and 6. An analysis of the circuit with nodes 3 and 5 as a differential input and nodes 4 and 6 as a differential output shows that resistance will stabilize that "new" odd-mode circuit while the resistance between ports 3 and 5 needs to be more than 2480 ohms. This value in parallel with 20 ohms gives twice 9.92 ohms, the value of the negative resistance of the staring circuit. In [6], a resistor of 10 ohms between nodes 3 and 5 and a resistance of 250 ohms between nodes 4 and 6 was used to stabilize the circuit. The resistor between nodes 3 and 5 is less than the recommended value but that is compensated for by the resistor between nodes 4 and 6.

It must be kept in mind that the even- and odd-mode stability conditions have been considered in the frequency domain. These tests must be done at all frequencies where potential gain exists (i.e., up to the maximum frequency of oscillation of the ensemble of devices).

5.3.5 Mixed-Mode Scattering Parameter Stability

In Chapter 4, mixed-mode (differential/common-mode) scattering parameters were discussed. We discussed odd-mode stability earlier in this chapter. The differential- and common-mode scattering parameters discussed in Chapter 4 are related to even- and odd-mode stability. However, the mixed-mode scattering parameters are specifically for a four-port. When there are no cross-matrix terms in the mixed-mode matrix (i.e., when the differential to common-mode terms or common-mode to differential-mode terms do not exist), the differential and common-mode currents and voltages in a linear circuit can exist without interaction. When that case exists, then stability for the common mode is determined by the lower-right two-by-two matrix of the mixed-mode scattering matrix as if it were the only solution that exists. The stability in the common mode then depends only on that part of the matrix. The same thing holds for stability for the differential mode. When the cross terms exist, a differential/common-mode Y matrix can be formed as discussed in Chapter 4. For differential-mode circuits, the input and output ports are the differential-mode terms. If the even-mode terminations are fixed, then they can be included in the Y matrix. The Y matrix can then be inverted to find a Z matrix and the diagonal terms can be considered for stability in the same manner that the analysis was done for even- and odd-mode stability. Stabilization depends on putting positive real components into the circuit. When new parts are added to the circuit, then a new mixed-mode scattering matrix or immittance matrix needs to be formed and checked again for stability.

5.4 SIMULTANEOUS CONJUGATE MATCH

Simultaneous conjugate match is closely linked to stability of a network. Simultaneous conjugate match occurs when the output load results in an input impedance to the device that is a conjugate match of the source at the same time the source on the device results in an output impedance that is a conjugate match of the load. The conditions for simultaneous conjugate match are:

$$\Gamma_{in}(\Gamma_{load}) = \Gamma_{source}^*$$

$$\Gamma_{out}(\Gamma_{source}) = \Gamma_{load}^*$$

Not every network has a simultaneous conjugate match. A simple circuit that does not have a simultaneous conjugate match is shown in Figure 5.10 if Z is positive real and nonzero. An input match requires $Z_1 = Z^* + Z_2^*$. It is also required that $Z_2^* = Z + Z_1$. However, then Z_1 would want to be $Z_1 = Z^* + Z + Z_1$. This can be true only for $Re(Z) = 0$. Therefore, it is possible to find a simultaneous conjugate match for the circuit in Figure 5.10 only when Z is purely imaginary.

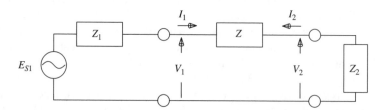

Figure 5.10 A simple circuit that does not have a simultaneous conjugate match.

What are the conditions for simultaneous conjugate match? Using the conditions for simultaneous conjugate match:

$$\Gamma_{\text{in}}(\Gamma_{\text{load}}) = \Gamma^*_{\text{source}} = \frac{S_{11} - \Delta_S \Gamma_{\text{load}}}{1 - S_{22} \Gamma_{\text{load}}}$$

$$\Gamma_{\text{out}}(\Gamma_{\text{source}} = \Gamma^*_{\text{load}} = \frac{S_{22} - \Delta_S \Gamma_{\text{source}}}{1 - S_{11} \Gamma_{\text{source}}}$$

Substituting the first equation into the second equation yields:

$$\Gamma_{\text{out}}(\Gamma_{\text{source}}) = \Gamma^*_{\text{load}} = \frac{S_{22} - \Delta_S \left(\dfrac{S_{11} - \Delta_S \Gamma_{\text{load}}}{1 - S_{22} \Gamma_{\text{load}}} \right)^*}{1 - S_{11} \left(\dfrac{S_{11} - \Delta_S \Gamma_{\text{load}}}{1 - S_{22} \Gamma_{\text{load}}} \right)^*}$$

$$\Gamma_{\text{load}} = \frac{S^*_{22} (1 - S_{22} \Gamma_{\text{load}}) - \Delta^*_S (S_{11} - \Delta_S \Gamma_{\text{load}})}{1 - S_{22} \Gamma_{\text{load}} - S^*_{11} (S_{11} - \Delta_S \Gamma_{\text{load}})}$$

This is a quadratic equation in Γ_{load}. The quadratic equation is as follows.

$$(S_{22} - \Delta_S S^*_{11}) \Gamma^2_{\text{load}} - (1 - |S_{11}|^2 + |S_{22}|^2 - |\Delta_S|^2) \Gamma_{\text{load}} + (S^*_{22} - \Delta^*_S S_{11}) = 0$$

Using the same procedure, the quadratic equation for the source reflection coefficient is as follows.

$$(S_{11} - \Delta_S S^*_{22}) \Gamma^2_{\text{source}} - (1 + |S_{11}|^2 - |S_{22}|^2 - |\Delta_S|^2) \Gamma_{\text{source}} + (S^*_{11} - \Delta^*_S S_{22}) = 0$$

The literature has identified the coefficients of this equation and assigned names to them [7,8]. These terms are as follows.

$$(S_{22} - \Delta_S S^*_{11}) \Gamma^2_{\text{load}} - \left(1 - |S_{11}|^2 + |S_{22}|^2 - |\Delta_S|^2\right) \Gamma_{\text{load}} + \left(S^*_{22} - \Delta^*_S S_{11}\right) = 0$$

$$\left(S_{11} - \Delta_S S^*_{22}\right) \Gamma^2_{\text{source}} - \left(1 + |S_{11}|^2 - |S_{22}|^2 - |\Delta_S|^2\right) \Gamma_{\text{source}} + \left(S^*_{11} - \Delta^*_S S_{22}\right) = 0$$

$$B_1 = \left(1 - |S_{11}|^2 + |S_{22}|^2 - |\Delta_S|^2\right)$$

$$B_2 = \left(1 + |S_{11}|^2 - |S_{22}|^2 - |\Delta_S|^2\right)$$

$$C_1 = \left(S_{11} - \Delta_S S^*_{22}\right)$$

$$C_2 = \left(S_{22} - \Delta_S S^*_{11}\right)$$

The solutions to these equations are as given here.

$$\Gamma_{\text{source}} = \frac{B_1 \pm \sqrt{B_1^2 - 4|C_1|^2}}{2C_1}$$

$$\Gamma_{\text{load}} = \frac{B_2 \pm \sqrt{B_2^2 - 4|C_2|^2}}{2C_2}$$

Using:

$$|C_1|^2 = \left| S_{11} - \Delta_S S^*_{22} \right|^2 = \left(1 - |S_{22}|^2\right) \left(|S_{11}|^2 - |\Delta_S|^2\right) + |S_{12} S_{21}|^2$$

$$|C_2|^2 = \left| S_{22} - \Delta_S S^*_{11} \right|^2 = \left(1 - |S_{11}|^2\right) \left(|S_{22}|^2 - |\Delta_S|^2\right) + |S_{12} S_{21}|^2$$

$$4|S_{12} S_{21}|^2 K^2 - 4|S_{12} S_{21}|^2 = 4|S_{12} S_{21}|^2 \left(K^2 - 1\right) = B_2^2 - 4|C_2|^2 = B_1^2 - 4|C_1|^2$$

and substituting into the simultaneous conjugate match equation the result is:

$$\Gamma_{\text{source}} = \frac{B_1 \pm \sqrt{B_1^2 - 4|C_1|^2}}{2C_1} = \frac{B_1 \pm 2|S_{12} S_{21}| \sqrt{K^2 - 1}}{2|C_1|^2} C_1^*$$

$$\Gamma_{\text{load}} = \frac{B_2 \pm \sqrt{B_2^2 - 4|C_2|^2}}{2C_2} = \frac{B_2 \pm 2|S_{12} S_{21}| \sqrt{K^2 - 1}}{2|C_2|^2} C_2^*$$

Notice that when K is greater than 1, the conjugate match points are on the same ray from the origin that the stability circles are. Recall that the centers of the stability circles are also on the rays of C_1 and C_2 conjugate. Note also that when the magnitude of $K = 1$:

$$|B_1| = 2|C_1|_{|K|=1}$$

$$|B_2| = 2|C_2|_{|K|=1}$$

Since the quadratic goes to zero when $K = 1$, the source and load conjugate matches have a magnitude of 1 (purely imaginary immittances) in the reflection coefficient planes. The conjugate match point get closer and closer to the edge of the chart as K gets closer and closer to 1. Since the conjugate match points lie on the same ray as the stability circles, when a part is matched at the simultaneous conjugate match point, it is approaching the stability circle, especially as K gets close to 1. One has to ensure that with the parameter variations of a part, the circuit does not become unstable for a given match point as the parameters of the device change and the stability circles move. Notice that when the magnitude of K is less than 1, the equations for conjugate match can be rewritten as:

$$\Gamma_{\text{source}} = \frac{B_1 \pm j2\,|S_{12}S_{21}|\,\sqrt{1 - K^2}}{2\,|C_1|^2}C_1^*$$

$$\Gamma_{\text{load}} = \frac{B_2 \pm j2\,|S_{12}S_{21}|\,\sqrt{1 - K^2}}{2\,|C_2|^2}C_2^*$$

$$|\Gamma_{\text{source}}| = \left.\sqrt{\frac{\left||B_1|^2 - 4\,|S_{12}S_{21}|^2\left(1 - K^2\right)\right|}{4\,|C_1|^2}}\right|_{|K|<1} = 1$$

$$|\Gamma_{\text{load}}| = \left.\sqrt{\frac{\left||B_2|^2 - 4\,|S_{12}S_{21}|^2\left(1 - K^2\right)\right|}{4\,|C_2|^2}}\right|_{|K|<1} = 1$$

When the magnitude of K is greater than 1, positive real immittance simultaneous match occurs (the correct choice of roots in front of the square root sign needs to be made). An unconditionally stable device has K greater than 1. Therefore, all unconditionally stable devices have simultaneous conjugate match. However, some conditionally stable devices can also have positive real immittance simultaneous conjugate match. Some devices have the magnitude of K greater than 1. Those devices that have K less than -1 or those devices that have K greater than 1 and the magnitude of the determinant greater than 1 are potentially unstable but have positive real immittance simultaneous conjugate match.

5.5 GAIN DEFINITIONS

In Figure 5.1, there are four reflection coefficients shown. With each of the reflection coefficients, a power quantity can be identified. Just as there was maximum available power from a one-port source, there is maximum available power from the output of a two-port network. The amount of available power from the output of the two-port will depend on the type of source connected to the input of the two-port. Figure 5.11 identifies the four types of power under consideration. Several types of gain can be described as ratios of these four powers. The four power quantities represent:

$$P_{\text{av\_source}} = \text{Power available from the source}$$

$$P_{\text{in}} = \text{Power actually delivered into the device}$$

$$P_{\text{av\_out}} = \text{Power available from the output of the device}$$

$$P_{\text{load}} = \text{Power actually delivered to the load}$$

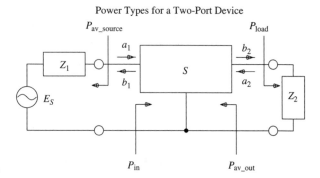

Power Types for a Two-Port Device

Figure 5.11 Four separate power directions.

The ratio of reflected power to incident power is called return loss, R.L., when expressed in negative dBs. Return gain, R.G., would be without the minus sign. R.G. $= -$R.L.

$$\text{R.L.} = -10 \log_{10}\left(\left|\frac{b}{a}\right|^2\right) = -20 \log_{10}\left(\left|\frac{b}{a}\right|\right)$$

Return loss can be given at both the input and the output ports of a two-port network.
The following transmission gains are defined as:

$$G_p = \frac{P_{\text{load}}}{P_{\text{in}}}$$

$$G_a = \frac{P_{\text{av\_out}}}{P_{\text{av\_source}}}$$

$$G_t = \frac{P_{\text{load}}}{P_{\text{av\_source}}}$$

where G_p is power gain, G_a is available gain, and G_t is transducer gain. G_p uses the two arrows to the right, G_a uses the two arrows to the left, and G_t uses the two outer arrows on the drawing shown above. G_p is useful for power amplifiers, G_a is useful in noise analysis, and G_t is useful for cascading circuits. An insertion gain can also be defined. Let P_{ins} be the power delivered to a load on a two-port when the two-port is driven by a source. Let P_{uns} be the power delivered by the same source coupled directly to that same load. Notice that impedance mismatch is not considered in either of these quantities. The insertion gain G_{ins} is defined as:

$$G_{\text{ins}} = \frac{P_{\text{ins}}}{P_{\text{uns}}}$$

Insertion gain is a measured gain and displays no knowledge of the two-port parameters of the device. Maximum power gains cannot be derived from it. We will not consider insertion gain further in this text but the reader should be aware of its definition. Sometimes authors use the term insertion gain or insertion loss to mean transducer gain or loss. Insertion gain is the same as transducer gain when the source and load impedances are equal and positive real.

5.5.1 Transducer Gain

Consider the circuit shown in Figure 5.11. The power into the load and the power available from the source need to be found. Specifically, the power available from the source needs to be expressed in terms of port one's normalization impedance.

The power into the load is:

$$P_{\text{load}} = |b_2|^2 - |a_2|^2$$
$$= |b_2|^2 - |\Gamma_{\text{load}}|^2 |b_2|^2$$
$$= |b_2|^2 \left(1 - |\Gamma_{\text{load}}|^2\right)$$

but:

$$b_2 = S_{21}a_1 + S_{22}a_2 = S_{21}a_1 + S_{22}\Gamma_{\text{load}}b_2$$

$$|b_2|^2 = \frac{|S_{21}|^2 |a_1|^2}{|1 - S_{22}\Gamma_{\text{load}}|^2}$$

Therefore:

$$P_{\text{load}} = \frac{|S_{21}|^2 |a_1|^2 \left(1 - |\Gamma_{\text{load}}|^2\right)}{|1 - S_{22}\Gamma_{\text{load}}|^2}$$

In order to describe transducer gain, the power into the load needs to be divided by the power available from the source. The power available from the source depends on its source impedance. If the source impedance is different from the normalization impedance, then that amount of power needs to be expressed in terms of the normalization impedance. Using input voltage and current V_1 and I_1:

$$I_1 = \frac{E_S}{Z_{\text{source}} + Z_{\text{in}}}$$

$$V_1 = \frac{E_S Z_{\text{in}}}{Z_{\text{source}} + Z_{\text{in}}}$$

$$a_1 = \frac{V_1 + I_1 Z_0}{2\sqrt{Z_0}} = \frac{\left(\dfrac{E_S Z_{\text{in}} + E_S Z_0}{Z_{\text{source}} + Z_{\text{in}}}\right)}{2\sqrt{Z_0}}$$

$$P_{\text{av\_source}} = \frac{|E_S|^2}{2\left(Z_{\text{source}} + Z_{\text{source}}^*\right)}$$

This gives:

$$\frac{|a_1|^2}{P_{\text{av\_source}}} = \frac{(Z_{\text{in}} + Z_0)(Z_{\text{in}} + Z_0)^*}{(Z_{\text{source}} + Z_{\text{in}})(Z_{\text{source}} + Z_{\text{in}})^*} \frac{\left(Z_{\text{source}} + Z_{\text{source}}^*\right)}{2Z_0}$$

Using:

$$\frac{\left(Z_{\text{source}} + Z_{\text{source}}^*\right)}{2Z_0} = \frac{1}{2}\left(\frac{1 + \Gamma_{\text{source}}}{1 - \Gamma_{\text{source}}} + \frac{1 + \Gamma_{\text{source}}^*}{1 - \Gamma_{\text{source}}^*}\right)$$

$$= \frac{1 - |\Gamma_{\text{source}}|^2}{(1 - \Gamma_{\text{source}})\left(1 - \Gamma_{\text{source}}^*\right)}$$

and:

$$\frac{(Z_{\text{in}} + Z_0)}{(Z_{\text{source}} + Z_{\text{in}})} = \frac{\dfrac{1 + \Gamma_{\text{in}}}{1 - \Gamma_{\text{in}}} + 1}{\dfrac{1 + \Gamma_{\text{source}}}{1 - \Gamma_{\text{source}}} + \dfrac{1 + \Gamma_{\text{in}}}{1 - \Gamma_{\text{in}}}} = \frac{1 - \Gamma_{\text{source}}}{1 - \Gamma_{\text{source}}\Gamma_{\text{in}}}$$

the ratio of incident power to power available is:

$$\frac{|a_1|^2}{P_{av\_source}} = \frac{1 - |\Gamma_{source}|^2}{(1 - \Gamma_{source})\left(1 - \Gamma_{source}^*\right)} \frac{1 - \Gamma_{source}}{1 - \Gamma_{source}\Gamma_{in}} \frac{1 - \Gamma_{source}^*}{1 - \Gamma_{source}^*\Gamma_{in}^*}$$

$$= \frac{1 - |\Gamma_{source}|^2}{|1 - \Gamma_{source}\Gamma_{in}|^2}$$

The following transducer gain equation results.

$$G_t = \frac{1 - |\Gamma_{source}|^2}{|1 - \Gamma_{source}\Gamma_{in}|^2} |S_{21}|^2 \frac{1 - |\Gamma_{load}|^2}{|1 - S_{22}\Gamma_{load}|^2}$$

The denominator can be rearranged as follows.

$$(1 - \Gamma_{source}\Gamma_{in})(1 - S_{22}\Gamma_{load}) \qquad\qquad \Leftarrow$$

$$= 1 - S_{22}\Gamma_{load} - (1 - S_{22}\Gamma_{load})\Gamma_{in}\Gamma_{source}$$

$$= 1 - S_{22}\Gamma_{load} - (1 - S_{22}\Gamma_{load})\left(\frac{S_{11} - \Delta_S\Gamma_{load}}{1 - S_{22}\Gamma_{load}}\right)\Gamma_{source}$$

$$= 1 - S_{11}\Gamma_{source} - S_{22}\Gamma_{load} + \Delta_S\Gamma_{source}\Gamma_{load} \qquad\qquad \Leftarrow$$

$$= 1 - S_{11}\Gamma_{source} - \Gamma_{load}(S_{22} - \Delta_S\Gamma_{source})$$

$$= 1 - S_{11}\Gamma_{source} - \Gamma_{load}\Gamma_{in}(1 - S_{11}\Gamma_{source})$$

$$= (1 - S_{11}\Gamma_{source})(1 - \Gamma_{out}\Gamma_{load}) \qquad\qquad \Leftarrow$$

Using the above equations, the following three formulas are identical and can be used to calculate transducer gain. It depends on whether one is interested in load properties, source properties, or transfer properties as to which one is used.

$$G_t = \frac{\left(1 - |\Gamma_{source}|^2\right)|S_{21}|^2\left(1 - |\Gamma_{load}|^2\right)}{|1 - \Gamma_{source}\Gamma_{in}|^2\,|1 - S_{22}\Gamma_{load}|^2}$$

$$G_t = \frac{\left(1 - |\Gamma_{source}|^2\right)|S_{21}|^2\left(1 - |\Gamma_{load}|^2\right)}{|1 - S_{11}\Gamma_{source}|^2\,|1 - \Gamma_{out}\Gamma_{load}|^2}$$

$$G_t = \frac{\left(1 - |\Gamma_{source}|^2\right)|S_{21}|^2\left(1 - |\Gamma_{load}|^2\right)}{|1 - S_{11}\Gamma_{source} - S_{22}\Gamma_{load} + \Delta_S\Gamma_{source}\Gamma_{load}|^2}$$

Notice that the first equation is concerned primarily with the input immittance, the second equation is concerned primarily with the output immittance, and the third equation is concerned with both the source and the load.

5.5.2 Power Gain

Power gain is the ratio of the power delivered to the load divided by the actual power into a circuit. It is important to understand the definition because many specifications are expressed in terms of power gain rather than transducer gain. If a part has a very large input reflection coefficient (i.e., one approaching a magnitude of 1), then the transformation or matching circuit will likely be complex. The measurement necessary to quantify the gain will also be more difficult because of losses in the matching circuit. Therefore, many specifications are in terms of power gain rather than transducer gain. The actual power into a device may be very small and the power gain then will be large. However, it may be difficult to achieve that gain in practice.

The power into a device is:

$$P_{in} = a_1^2 - b_1^2 = a_1^2\left(1 - |\Gamma_{in}|^2\right)$$

The power into the load was given in the transducer gain derivation. Therefore the power gain is as follows.

$$\frac{P_{\text{load}}}{P_{\text{in}}} = \frac{|S_{21}|^2 |a_1|^2 \left(1 - |\Gamma_{\text{load}}|^2\right)}{a_1^2 \left(1 - |\Gamma_{\text{in}}|^2\right) |1 - S_{22}\Gamma_{\text{load}}|^2} = \frac{|S_{21}|^2 \left(1 - |\Gamma_{\text{load}}|^2\right)}{\left(1 - |\Gamma_{\text{in}}|^2\right) |1 - S_{22}\Gamma_{\text{load}}|^2}$$

Notice that this equation depends only on the load immittance. It might be observed that the input reflection coefficient is in the denominator. Remember that the input reflection coefficient is a function only of the load reflection coefficient and not of the source. Power gain assumes that one is looking only to the right in Figure 5.11.

5.5.3 Available Gain

Available gain is the ratio of the power available at the output of a device divided by the power available from the source. The power available at the output of a circuit depends on the input terminations. This will be evident in the available gain equation. The power available at the output does not depend on the value of the output load. The actual power delivered depends on the value of the load immittance but the power available does not. Available gain is used in noise figure and signal-to-noise ratio calculations since the definition of noise figure references available power.

The power available from a source expressed in terms of an input normalization impedance was derived in the transducer gain discussion. The power available at the output would be the power delivered to a load if that load were a conjugate of the output immittance. The power available at the output is determined as follows.

$$b_2 = S_{21}a_1 + S_{22}a_2 = S_{21}a_1 + S_{22}\Gamma_{\text{out}}^* b_2$$

$$|b_2|^2 = \frac{|S_{21}|^2 |a_1|^2}{\left|1 - S_{22}\Gamma_{\text{out}}^*\right|^2}$$

$$P_{\text{out}} = |b_2|^2 \left(1 - \left|\Gamma_{\text{out}}^*\right|^2\right) = \frac{|S_{21}|^2 |a_1|^2 \left(1 - \left|\Gamma_{\text{out}}^*\right|^2\right)}{\left|1 - S_{22}\Gamma_{\text{out}}^*\right|^2}$$

The available gain is this power divided by the available power from the source.

$$G_{\text{av}} = \frac{P_{\text{out}}}{P_{\text{av\_source}}} = \frac{\left(1 - |\Gamma_{\text{source}}|^2\right) |S_{21}|^2 \left(1 - \left|\Gamma_{\text{out}}^*\right|^2\right)}{|1 - \Gamma_{\text{source}}\Gamma_{\text{in}}|^2 \left|1 - S_{22}\Gamma_{\text{out}}^*\right|^2}$$

This equation can be rearranged in the same manner that the transducer gain equation was. This results in the following formula for available gain.

$$G_{\text{av}} = \frac{\left(1 - |\Gamma_{\text{source}}|^2\right) |S_{21}|^2}{|1 - \Gamma_{\text{source}}S_{11}|^2 \left(1 - |\Gamma_{\text{out}}|^2\right)}$$

Notice that the only variable in the above equation is the source reflection coefficient. It might be observed that the output reflection coefficient is in the equation but again remember that the output reflection coefficient is a function only of the source reflection coefficient.

5.5.4 Gain at Simultaneous Match

Substituting the simultaneous conjugate match reflection coefficients given at the end of Section 5.4 into the transducer gain equations given in Section 5.5.1, it can be shown that the value of transducer gain at simultaneous conjugate match is:

$$G_{\text{tmax}} = \left|\frac{S_{21}}{S_{12}}\right| \left(K - \sqrt{K^2 - 1}\right)$$

where the magnitude of the ratio of S_{21} to S_{12} is called the *figure of merit* for a device. That value is called the *maximum stable gain*. When K is equal to 1, then G_{tmax} is equal to the maximum stable gain. When K is very large, a large margin of stability, the maximum gain tends toward zero. The designer makes a tradeoff between stability and gain in a particular circuit implementation. Notice that when the part is terminated with the simultaneous conjugate match, there are no reflections at either of the ports. Then the value of power gain, available gain, and transducer gain are all equal to each other and equal to the G_{tmax} value.

The quotient of S_{21} over S_{12} does not change by putting series impedances or shunt admittances on either of the ports of the device.

$$S_{12} = \frac{2(AD - BC)}{A + BY_0 + CZ_0 + D}$$

$$S_{21} = \frac{2}{A + BY_0 + CZ_0 + D}$$

$$\frac{S_{21}}{S_{12}} = AD - BC$$

Adding a series impedance or shunt admittance at the input port gives the following value for $AD - BC$.

$$\begin{pmatrix} 1 & Z_S \\ 0 & 1 \end{pmatrix} \begin{pmatrix} A & B \\ C & D \end{pmatrix} = \begin{pmatrix} A + Z_S C & B + Z_S D \\ C & D \end{pmatrix} = \begin{pmatrix} A' & B' \\ C' & D' \end{pmatrix}$$

$$A'D' - B'C' = AD + Z_S CD - BC - CZ_S D = AD - BC$$

$$\begin{pmatrix} 1 & 0 \\ Y_P & 1 \end{pmatrix} \begin{pmatrix} A & B \\ C & D \end{pmatrix} = \begin{pmatrix} A & B \\ Y_P A + C & Y_P B + D \end{pmatrix} = \begin{pmatrix} A' & B' \\ C' & D' \end{pmatrix}$$

$$A'D' - B'C' = AD + AY_P B - Y_P AB - BC = AD - BC$$

The determinant of the $ABCD$ matrix is $AD - BC$. The determinant of the matrix multiplying the $ABCD$ matrix is equal to 1. Therefore the determinant of the product would also be $AD - BC$. If a transmission line is used at one of the ports, the answer is the same since the determinant of the $ABCD$ matrix of a transmission line is equal to 1. This shows that the quotient is invariant to adding a series impedance or shunt admittance at either port and indicates as well, therefore, that K does not change as a function of reactive port matching. This allows one to calculate the value of K needed to get a desired gain, G_{des}.

$$G_{\text{des}} = \frac{|S_{21}|}{|S_{12}|} \left(K - \sqrt{K^2 - 1} \right)$$

$$K = \frac{|S_{21}|^2 + G_{\text{des}}^2 |S_{12}|^2}{2 G_{\text{des}} |S_{12} S_{21}|}$$

This formula can be used in an engineering trade-off study between desired gain and stability margin.

5.5.5 Unilateral Transducer Gain

The formulas given in the transducer gain discussion show an interaction between the source and the load immittances. The load immittance is a conjugate of the output immittance that is a function of the source immittance. The source immittance is a conjugate of the input immittance that is a function of the load immittance. The equation is coupled. In order to make some estimate of gain without the coupling, the feedback term S_{12} can be set equal to zero. This eliminates the coupling in the equation. When this is done, the gain is called the *unilateral gain.* Notice that the input reflection coefficient for the unilateral case is just S_{11} and the output reflection coefficient for the unilateral case is just S_{22}. It is also of interest to see how much the approximate calculation for unilateral gain differs from the actual calculation for transducer gain.

$$\Gamma_{\text{in}} = S_{11} + \frac{S_{12} S_{21} \Gamma_{\text{load}}}{1 - S_{22} \Gamma_{\text{load}}}\bigg|_{S_{12}=0} = S_{11}$$

$$\Gamma_{\text{out}} = S_{22} + \frac{S_{12} S_{21} \Gamma_{\text{source}}}{1 - S_{11} \Gamma_{\text{source}}}\bigg|_{S_{12}=0} = S_{22}$$

The transducer gain under this condition is:

$$G_t = \frac{1 - |\Gamma_{\text{source}}|^2}{|1 - \Gamma_{\text{source}} S_{11}|^2} |S_{21}|^2 \frac{1 - |\Gamma_{\text{load}}|^2}{|1 - S_{22} \Gamma_{\text{load}}|^2}\bigg|_{S_{12}=0}$$

It is easily shown that the above equation has its maximum when:

$$\Gamma_{\text{source}} = S_{11}^*$$

$$\Gamma_{\text{load}} = S_{22}^*$$

The maximum value of unilateral gain is:

$$G_t = \frac{1}{1 - |S_{11}|^2} |S_{21}|^2 \frac{1}{1 - |S_{22}|^2}\bigg|_{\substack{S_{12}=0 \\ \Gamma_{\text{source}}=S_{11}^* \\ \Gamma_{\text{load}}=S_{22}^*}}$$

The true value of transducer gain evaluated under the assigned source and load immittances is:

$$G_t = \frac{\left(1 - |S_{11}|^2\right) |S_{21}|^2 \left(1 - |S_{22}|^2\right)}{\left|\left(1 - |S_{11}|^2\right)\left(1 - |S_{22}|^2\right) - S_{12} S_{21} S_{11}^* S_{22}^*\right|^2}$$

$$= \frac{\left(1 - |S_{11}|^2\right) |S_{21}|^2 \left(1 - |S_{22}|^2\right)}{\left(1 - |S_{11}|^2\right)\left(1 - |S_{22}|^2\right)^2} \frac{1}{\left|1 - \dfrac{S_{12} S_{21} S_{11}^* S_{22}^*}{\left(1 - |S_{11}|^2\right)\left(1 - |S_{22}|^2\right)}\right|}$$

$$= \frac{G_{\text{tumax}}}{\left|1 - \dfrac{S_{12} S_{21} S_{11}^* S_{22}^*}{\left(1 - |S_{11}|^2\right)\left(1 - |S_{22}|^2\right)}\right|}$$

Let:

$$U = \left|\frac{S_{11} S_{12} S_{21} S_{22}}{\left(1 - |S_{11}|^2\right)\left(1 - |S_{22}|^2\right)}\right|$$

Then:

$$G_t\big|_{\substack{\Gamma_{\text{source}}=S_{11}^* \\ \Gamma_{\text{load}}=S_{22}^*}} = \tilde{G}_t = \frac{G_{\text{tumax}}}{|1 - U e^{j\Phi}|}$$

Notice that this is not maximum transducer gain but transducer gain using S_{11}^* and S_{22}^* as the termination reflection coefficients. This gain should be close to the maximum available gain if S_{12} is small. This gain has a maximum when the phase angle Φ is 0 degrees and this gain has a minimum when the phase angle Φ is 180 degrees. Thus:

$$\frac{1}{|1 + U|^2} < \frac{\tilde{G}_t}{G_{\text{tumax}}} < \frac{1}{|1 - U|^2}$$

If the quantity U is small, then the calculation of unilateral gain is a good approximation. However, if U is an appreciable part of 1, the approximation is not good. The unilateral maximum transducer gain is often used for a first calculation since it is easy to calculate.

5.5.6 Unilateral Gain Circles

Notice that the unilateral gain expressions contain the term:

$$G_i = \frac{1 - |\Gamma_i|^2}{|1 - S_{ii}\Gamma_i|^2}$$

Whenever:

$$\Gamma_i = S_{ii}^*$$

$$G_{i\_max} = \frac{1}{1 - |S_{ii}|^2}\bigg|_{|S_{ii}|<1}$$

Let:

$$g_i = \frac{G_i}{G_{i\_max}} = G_i \left(1 - |S_{ii}|^2\right)$$

Then:

$$0 < g_i < 1$$

$$g_i = \frac{\left(1 - |\Gamma_i|^2\right)\left(1 - |S_{ii}|^2\right)}{|1 - S_{ii}\Gamma_i|^2}$$

The equation can be arranged in the form of an equation of a circle as:

$$\Gamma_i\Gamma_i^* - \frac{S_{ii}^* g_i}{\left[1 + |S_{ii}|^2 (g_i - 1)\right]}\Gamma_i^* - \frac{S_{ii} g_i}{\left[1 + |S_{ii}|^2 (g_i - 1)\right]}\Gamma_i + \frac{|S_{ii}|^2 g_i^2}{\left[1 + |S_{ii}|^2 (g_i - 1)\right]^2}$$

$$= \frac{\left[\left(1 - |S_{ii}|^2\right)\sqrt{1 - g_i}\right]^2}{\left[1 + |S_{ii}|^2 (g_i - 1)\right]^2}$$

From this the center and radius of a circle with the running parameter of g_i is:

$$\Gamma_{cent} = \frac{S_{ii}^* g_i}{\left[1 + |S_{ii}|^2 (g_i - 1)\right]}$$

$$\rho = \frac{\left(1 - |S_{ii}|^2\right)\sqrt{1 - g_i}}{\left|1 + |S_{ii}|^2 (g_i - 1)\right|}$$

Therefore, if one chooses a normalized gain g_i, the gain generated by any reflection coefficient on that circle is down $-10\log_{10}(g_i)$ from $G_{i\,max}$. Plots of these gain circles are useful for plotting expected gains of circuits over different frequency bandwidths. Computer-based optimizers will likely be used to generate the final circuit topology; however, the computer will generally find an optimum solution faster if it is prompted with a near-optimum solution generated by hand. These circles can be used to give an approximate value of gain for various load and source terminations. This is also useful in conjunction with noise figure calculations and noise figure circles when noise is considered in Chapter 10.

5.6 MAXIMUM GAIN VERSUS COMMON-MODE INDUCTANCE

Often matched amplifier modules are purchased and used in microwave assemblies. These have been very successful. Because of this success, there is often a desire to quickly test the devices for some use. The user will quickly hook up the circuits with pieces of wire. When the circuit is tested, the results are not what are expected. The following analysis will attempt to try to identify some of the difficulties with testing these modules without using good microwave techniques. Just hooking up the devices with wires on the top of a printed circuit board is not a good microwave technique. Many matched modules are similar to the circuit shown in Figure 5.12. The current source shown cannot necessarily be identified with a single transistor since the modules are often multistage units. However, the analysis of the device shown in Figure 5.12 will give some design guidelines. The capacitor C_{12} is a feedback capacitor from the output lead to the input lead. The inductor L_3 is a common-mode inductor. Notice that the currents I_1 and I_2 have to go through the inductor unless $I_1 = -I_2$, in which case there is no current gain and, if the terminal impedances are the same value, no power gain. In cascaded amplifiers, R_1 and R_2 are often equal to the characteristic impedance of the system and often 50 ohms is used, although 75- and 93-ohm units can be fabricated. The analysis will start by assuming the unit is a very-high-frequency unit for which C_{12} is negligible and L_3 is negligible. It is easily shown that the Z matrix for the system with $L_3 = 0$ and $C_{12} = 0$ is:

$$\begin{pmatrix} V_1 \\ V_2 \end{pmatrix} = \begin{pmatrix} Z_1 & 0 \\ -G_m Z_1 Z_2 & Z_2 \end{pmatrix} \begin{pmatrix} I_1 \\ I_2 \end{pmatrix}$$

where Z_1 is used for R_1 and Z_2 is used for R_2. The common-mode inductance, jX_3, appears in series with the ground terminal. It can be shown that the series combination of the above matrix and the inductance is:

$$\begin{pmatrix} V_1 \\ I_1 \end{pmatrix} = \left(\begin{pmatrix} Z_1 & 0 \\ -G_m Z_1 Z_2 & Z_2 \end{pmatrix} + \begin{pmatrix} jX_3 & jX_3 \\ jX_3 & jX_3 \end{pmatrix} \right) \begin{pmatrix} I_1 \\ I_2 \end{pmatrix}$$

$$= \begin{pmatrix} Z_1 + jX_3 & jX_3 \\ -G_m Z_1 Z_2 + jX_3 & Z_2 + jX_3 \end{pmatrix} \begin{pmatrix} I_1 \\ I_2 \end{pmatrix}$$

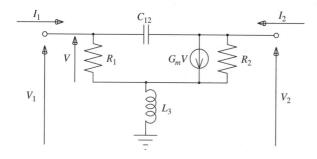

Figure 5.12 A matched module equivalent circuit.

The impedance into a device described by a Z matrix and loaded by an impedance Z_2 is:

$$Z_{in} = Z_{11} - \frac{Z_{12}Z_{21}}{Z_L + Z_{22}} = \frac{Z_{11}Z_L + \Delta_Z}{Z_L + Z_{22}}$$

If $Z_1 = Z_2 = R$, $Z_L = R + jX_L$, and $G_m = 2A/R$, where A is the voltage gain of the impedance-matched cascaded stage, then Z_{in} is:

$$Z_{in} = \frac{2R^2 - X_3X_L + j\left[(3R + 2AR) + RX_L\right]}{2R + j\left(X_3 + X_L\right)}$$

If the real part of the impedance is less than zero (negative real) for any loads put on the device (or vice-versa if the output impedance is negative real for any source impedance), the part can potentially oscillate. The real part of the input impedance is:

$$RE\left[Z_{in}\right] = \frac{2R\left(2R^2 - X_3X_L\right) + (X_3 + X_L)\left[(3R + 2AR)X_3 + RX_L\right]}{4R^2 + (X_3 + X_L)^2}$$

The denominator is always positive. If the numerator is negative, the input impedance will be negative. The numerator can be rearranged into:

$$\text{NUM} = \left[4R^2 + (X_3 + X_L)^2\right] RE\,[Z_{in}]$$
$$= R\left[(2A + 3)\,X_3^2 + (2 + 2A)\,X_L X_3 + \left(4R^2 + X_L^2\right)\right]$$

which is a quadratic equation. If the load reactance is varied, the minimum value of the numerator will occur where:

$$R\frac{d(\text{NUM})}{dX_L} = R\left[2X_L + (2 + 2A)\,X_3\right] = 0$$
$$\Rightarrow X_L = (1 + A)\,X_3$$

Substituting this back into the Z_{in} equation results in:

$$A \geq \sqrt{2 + \frac{4R^2}{\omega^2 L^2}}$$

If the gain A is greater than that given by the above equation, there is a very high likelihood that the part will oscillate. A detailed analysis with the actual scattering parameters will give a more accurate stability analysis, but the above equation will give a rough idea of how much gain one can get on a PCB or MMIC with a given amount of common-mode inductance. For instance, if R is 50 ohms, L is 1 nH, and f is 1 GHz, the maximum gain is $A = 16 = 24$ dB.

EXERCISES

5-1 A device has the following S parameters.
$S_{11} = -0.8$, $S_{12} = 0.05$, $S_{21} = -4$, $S_{22} = 0.8$
($A = 0.02$, $B = -3.5$, $C = -0.0014$, $D = -0.38$)
(a) Stabilize the device at the input.
(b) Stabilize the device at the output.
(c) First calculate the stability circles, then determine the value of resistance or conductance to be used. Determine the maximum available gain for the stabilized network in each case.

Hint: For manual calculations, *ABCD* parameters of the device are given to facilitate finding the result of a cascade of the device with the stabilizing resistor. Find the *ABCD* parameters of the cascade of the resulting two networks and convert the net *ABCD* parameters back to a scattering matrix. Use the maximum available gain equation on the resultant scattering matrix.

5-2 For the device in Exercise 5-1, put one-half the resistor value necessary on the input for stability. Determine the resistor that needs to be added to the output for stability. Determine the maximum available gain for the stabilized network.

5-3 Determine the odd-mode and even-mode stability for a four-port differential amplifier formed from the following two devices.
$h_{11} = 100$, $h_{12} = 0.0001$, $h_{21} = 0.010$, $h_{22} = 0.01$
$h_{11} = 100$, $h_{12} = 0.0001$, $h_{21} = 0.008$, $h_{22} = 0.01$

5-4 Using the device from Exercise 5-1, stabilize the device for a maximum available gain of 15 dB. Determine the simultaneous conjugate match for the stabilized amplifier. Assume the frequency is 2 GHz. Calculate the approximate amount of common-mode inductance that would cause the stabilized amplifier to go unstable at 2 GHz. In actual use, the common-mode inductance should be significantly less than this.

5-5 Plot the unilateral gain circles and stability circles for the device given in Exercise 5-1. Plot the two circles each for the input and the output. Use an increment of 2 dB down for each circle. Determine whether the unilateral approximation is a good idea to use for the part. Plot the unilateral gain circles and stability circles when the amplifier is stabilized at the input for 15 dB of gain. Determine whether the unilateral gain approximation is a good idea to use for the stabilized part.

6

Matching Networks, Attenuators, and Phase Shifters

6.1 MATCHING NETWORKS

This chapter considers different types of circuits to perform impedance matching. Impedance-matched attenuator and phase shifter circuits are also included. A consideration of the broad band matching [1] in general and the limit on the bandwidth of matching possible [2] is beyond the scope of this book. However, some general comments will be made with regard to bandwidth as impedance matching is considered.

Power amplifiers require a given mismatch to perform with high efficiency. The degree of mismatch is discussed in Chapter 7. Low-noise amplifiers (discussed in Chapter 10) usually require some mismatch in order to obtain optimum noise performance. Therefore, when match is considered, the circuit designer is not always interested in low-VSWR circuits but is interested in presenting specific impedances to a device. That is what is intended when impedance match is concerned. When circuits are designed for maximum gain, then a low VSWR is desired. A low VSWR is measured with respect to the resistance of the source. If the actual transmission lines used in the matching network do not have that value of characteristic impedance, the actual VSWR on the transmission line may not be low.

For conjugate match, the desired input immittance looking into the matching network that is terminated in the given load immittance is the conjugate of the source immittance. The input immittance looking into the matching network when a particular degree of mismatch is desired may be quite different but that value would be known from other considerations besides maximum gain. In the immittance or reflection coefficient plane, one would like to take the shortest route from the load immittance to the desired input immittance into the network. However, the paths that can be taken in the immittance planes are straight lines or circles. Consider the match from $5 + j10$ to 50 ohms shown in Figure 6.1. There is no single element to take one from point A to point D. The first element used is a shunt capacitor. This takes the immittance from point A to point B. One procedure used in matching is to resonate the load first. This takes the locus to the real axis. It would have been possible to put a series capacitor on the load at point A. This would have resulted in the locus passing through the point $5 + j0$. This point is further from the desired input impedance than point B. It depends on the circuit and the topology of the load whether one should series resonate the load or parallel resonate

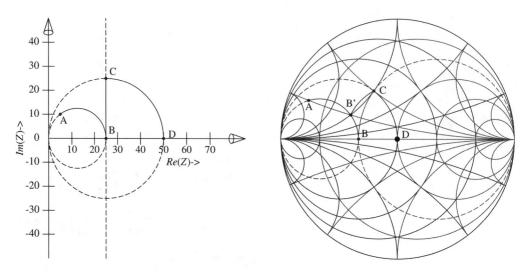

Figure 6.1 Loci followed to match $5 + j10$ to $50 + j0$.

the load. As discussed later in this chapter, if the load appears to be a series circuit, it should be parallel resonated, and vice versa.

In Figure 6.1, the next step is to put a series inductance at point B. This takes the immittance locus to point C. At that point, a shunt capacitance is used to bring the locus to point D, the desired end point. The reader can follow the sequence on the reflection coefficient chart shown on the right side of Figure 6.1. Each step in impedance matching should take the locus closer to the desired point. For lumped-constant reactive elements, the only options are to constrain the movements along constant resistance or constant conductance lines. The movement initially takes the locus farther away from the desired point but the next movement brings the locus much closer to the desired point. When movement along a transmission line is allowed, then movement along a constant reflection coefficient magnitude is permitted. This is movement along a constant radius circle in the reflection coefficient plane. That movement would also be a circle in the immittance planes but not along something easily defined as constant in those planes. In Figure 6.1, a smaller shunt capacitor could have been used to move the locus from point A to point B'. From point B', a length of transmission line would then move the locus from point B' to point C along a constant reflection coefficient magnitude locus. A shunt capacitor then would move the locus from point C to point D. Single component matching, including matching networks with resistance for circuit stabilization, moves the locus along constant circles in the reflection coefficient plane. For shunt reactive elements, the locus moves along a constant conductance circle. For series reactive elements, the locus moves along a constant resistance circle. For adding a shunt stabilization resistance, the locus moves along a constant susceptance circle. For adding a series stabilization resistance, the locus moves along a constant reactance circle. Finally, inserting a transmission line that has a characteristic impedance equal to the chart's normalization resistance between two points is a movement along a constant reflection coefficient magnitude locus.

6.2 FINDING THE OPTIMUM SUSCEPTANCE
WHEN THE LOAD CONDUCTANCE IS GIVEN [3]

Devices often have reactive feedback, for instance, the collector to base capacitance in a bipolar or the drain to gate capacitance in a FET transistor. When power amplifiers are considered in Chapter 7, it will be shown that a transistor should be loaded with a given resistance in parallel with an inductance that is in resonance with the output reactance of the device. However, the output capacity of the device or two-port depends on the input source match and matching the input immittance to the source depends on the output load. What is the correct value of the parallel reactance to put on the device? This problem has a closed form solution and is given here. Let the value of the optimum load admittance be Y_L consisting of a known G_L in parallel with some unknown jB_L. Consider the system described in terms of $ABCD$ matrices as shown in Figure 6.2. The $ABCD$ matrix for the complete network will be identified using carets over the $ABCD$ entries. The $ABCD$ matrix for the active device or two-port will be identified as simply $ABCD$. Then:

$$\begin{pmatrix} \hat{A} & \hat{B} \\ \hat{C} & \hat{D} \end{pmatrix} = \begin{pmatrix} 1 & Z_S \\ 0 & 1 \end{pmatrix} \begin{pmatrix} A & B \\ C & D \end{pmatrix} \begin{pmatrix} 1 & 0 \\ G_L & 1 \end{pmatrix} \begin{pmatrix} 1 & 0 \\ jB_L & 1 \end{pmatrix}$$

$$= \begin{pmatrix} A + BY_L + CZ_S + DZ_SY_L & B + DZ_S \\ C + DY_L & D \end{pmatrix}$$

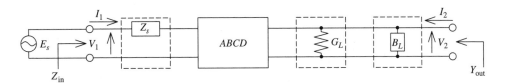

Figure 6.2 Circuit used to determine the optimum load susceptance. [*R. Weber, Microwave Engineering Course Notes, ©1987–1997, August 1997 edition, p. 80.*]

In terms of $ABCD$ matrices, the ratio A divided by C is the input impedance of a network that is terminated in an open circuit. The output admittance of a network terminated with a short circuit at its input is the ratio A divided by B. In terms of the $ABCD$ matrix of the device or two-port, the input impedance to the network shown in Figure 6.2 is:

$$Z_{in} = \frac{A + BY_L + CZ_S + DZ_SY_L}{C + DY_L}$$

and the output admittance is:

$$Y_{out} = \frac{A + BY_L + CZ_S + DZ_SY_L}{B + DZ_S}$$

The product of Z_{in} and Z_{out} is:

$$\frac{Z_{in}}{Y_{out}} = \frac{B + DZ_S}{C + DY_L}$$

Whenever the part is matched at the input, the source is equal to the conjugate of the input impedance of the device. The input impedance of the network shown in Figure 6.2 is equal to the input impedance of the device Z_s^* plus the source impedance Z_s, or Z_{in} is the sum of Z_s and Z_s^*. For the output immittance to be purely real, Y_{out} minus Y_{out}^* is equal to zero. Using these two constraints, the following relationship results.

$$Y_{out} = Z_{in}\frac{C + DY_L}{B + DZ_S} = (Z_S + Z_S^*)\frac{C + DY_L}{B + DZ_S} = Y_{out}^* = (Z_S + Z_S^*)\frac{C^* + D^*Y_L^*}{B^* + D^*Z_S^*}$$

$$(B^* + D^*Z_S^*)(C + DY_L) = (B + DZ_S)(C^* + D^*Y_L^*)$$

Notice that Z_s^* is equal to the impedance looking to the right at the junction between Z_s and the device. This can be determined by letting Z_s be zero in the input impedance equation for the network or:

$$Z_{in} = \frac{A + BY_L}{C + DY_L}$$

Substituting in for Z_s and Z_s^* in the above equation and rearranging gives:

$$(C + DY_L)B^* + D^*(A + BY_L) = B(C^* + D^*Y_L^*) + D(A^* + B^*Y_L^*)$$

or:

$$Y_L - Y_L^* = 2jB_L = \frac{(BC^* - B^*C) - (AD^* - A^*D)}{B^*D + BD^*}$$

$$B_L = j\frac{(AD^* - A^*D) - (BC^* - B^*C)}{2(B^*D + BD^*)}$$

This value of B will give the following input impedance.

$$Z_{in} = \frac{A + B(G_L + jB_L)}{C + D(G_L + jB_L)}$$

When the conjugate of this impedance is used for a source and the output terminals have the optimum B_L placed across it, the output admittance of the network will be purely real.

$$Z_{out} = R_{out} + j0 = \frac{1}{\dfrac{A + CZ_S^*}{B + DZ_S^*} + B_L}$$

The output conductance will not usually be equal to the G_L. There will be a mismatch at the output but that is to be expected when the load is given independent of the scattering parameters. However, this value of B_L will allow the load line on the device to be purely real and allow the maximum excursions of the voltage and current without additional reactive currents flowing on the output. The output susceptance is parallel resonated out by the calculated load susceptance.

For example, let the optimum load line impedance of a device be 400 ohms and let the scattering parameters of the device be:

$$S_{11} = 0.50\angle135°$$
$$S_{12} = 0.05\angle45°$$
$$S_{11} = 2.50\angle60°$$
$$S_{11} = 0.80\angle -30°$$

The *ABCD* parameters of the device are:

$$A = (0.087, 0.0445)$$
$$B = (8.24, -9.89)$$
$$C = (0.00226, -0.00142)$$
$$D = (0.035, -0.469)$$

The optimum reactance in parallel with the output and the input and output impedance of the device under this operating condition is:

$$X_p = j185.8 \text{ ohms}$$
$$Y_L = 0.0025 - j0.00538 \text{ Siemans}$$
$$Z_{in} = 7.58 + j20.0 \text{ ohms}$$
$$R_{pout} = 2088 \text{ ohms} \qquad X_{pout} = -j185.8$$

Notice that the parallel output resistance is not 400 ohms. However, the parallel output reactance is the negative of the parallel reactance to put on the part, resulting in a net zero reactance when the reactance is added.

6.3 SIMPLE MATCHING NETWORKS FOR MATCHING BETWEEN TWO VALUES OF RESISTANCE

6.3.1 $Q^2 + 1$ Method [4]

It is often convenient to use one or more ell sections to match to the input or output impedances of devices under consideration. In order to use the formulas in this section, Q of series and parallel R-C and R-L circuits need to be correctly identified. The formula for Q of a series circuit is the inverse of the formula for a parallel circuit. The formulas for series and parallel R-C and R-L circuits are as follows.

<table>
<tr><td></td><td align="center">Series Circuits</td><td align="center">Parallel Circuits</td></tr>
<tr><td>R-C</td><td>$Q_S = \dfrac{1}{\omega C_S R_S}$</td><td>$Q_P = \dfrac{\omega C_P}{G_P} = \omega C_P R_P$</td></tr>
<tr><td>R-L</td><td>$Q_S = \dfrac{\omega L_S}{R_S}$</td><td>$Q_P = \dfrac{1}{\omega L_P G_P} = \dfrac{R_P}{\omega L_P}$</td></tr>
</table>

The first step in using this matching method is to add a reactance or susceptance to the load to series or parallel resonant the impedance under consideration so that the resulting impedance is purely real. Then the matching sections shown in Figure 6.3 can be used to match the remaining resistance to some Z_0. Whether one chooses to use a series capacitor or series

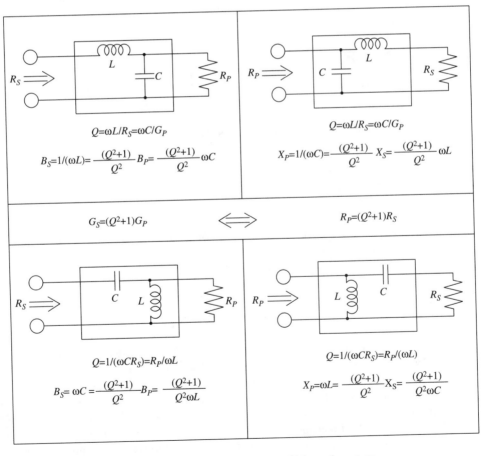

Figure 6.3 Simple matching networks with lumped constants.

inductor depends on the application and the bandwidth of the match desired. The matching network phase shift is different when series capacitors are used than when series inductors are used. In addition, there are times when series capacitors can serve as the dc blocking capacitor and the shunt inductor can be used as a bias inductor.

A short example will illustrate the procedure. Suppose one wishes to present a 400-ohm load line resistance to a small signal device at one gigahertz. Assume the parallel output reactance of the device is $-j200$ ohms. The required network presented to the device needs to look like 400 ohms in parallel with $j200$ ohms. First place $j200$ ohms across the device as shown in Figure 6.4. This configuration might be chosen to match the output of a transistor to 50 ohms. The series capacitor would act as a blocking capacitor and tuning capacitor and the shunt inductors would be combined in parallel and would be used as a bias inductor. Keep in mind that the inductors are shown connected to ground. A capacitor that has a net zero reactance (capacitance plus parasitic inductance) would be placed on the lower end of the inductors to ground for dc blocking. One would use the equivalent circuit models as discussed in Chapter 2. The series capacitor is chosen as follows.

$$Q^2 + 1 = \frac{400}{50} = 8$$

$$Q = \sqrt{7} = \frac{X_{Cm}}{50} = \frac{1}{50\omega Cm}$$

$$Cm = \frac{1}{50\omega\sqrt{7}} = 1.203\,pF\,|_{\omega=2\pi\,10^9}$$

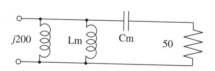

Figure 6.4 Matching network example using Q squared plus 1.

The value of the shunt inductor is chosen to provide the same Q for the parallel circuit as exists for the series circuit.

$$Q = \sqrt{7} = \frac{B_{Lm}}{G} = \frac{400}{\omega Lm}$$

$$Lm = \frac{400}{\omega\sqrt{7}}$$

This inductor is in parallel with the resonating inductor. Therefore a single inductor can be used in parallel for the network and has the following value.

$$Lr = \frac{200}{\omega} \qquad Lm = \frac{400}{\omega\sqrt{7}}$$

$$Lpar = \frac{Lm\,Lr}{Lm + Lr} = \frac{86.1}{\omega} = 13.7\,nH\,|_{\omega=2\pi\,10^9}$$

The Q-squared-plus-one ($Q^2 + 1$) technique will always work. When the impedance is series or parallel resonated, the remaining impedance is purely real. That resistance is transformed up by using a series reactance followed by a shunt reactance. Likewise, that resistance is transformed down by using a shunt reactance followed by a series reactance according to the design equations. The resulting network will then be a network of three components. The first component is the reactance or susceptance used to resonate the load. The next two reactances are in the impedance transformation circuit. The three components can often be reduced to two components. When the component used to resonate the load is a series reactance and the resulting resistance is lower than the desired resistance, the transformation network will require a series reactance followed by a shunt reactance. The two series reactances can be combined into one reactance. The net reactance may be of a different type (i.e., capacitor rather than an inductor) from the reactance first used to resonate the load. However, the transformation reactance can be chosen to be of the same type or opposite type. In order to minimize Q and therefore obtain the widest bandwidth, the transformation reactance can be chosen of the opposite type used for the resonating reactance.

If the circuit one uses to do the transformation consists, for example, of a parallel resonating reactance, followed by a series reactance, followed by a shunt reactance, the resulting three-reactance circuit is a pi circuit. That pi circuit can be transformed into a tee circuit using delta-wye transformations. It might turn out that the resulting tee circuit can be combined with other circuitry to reduce the number of components.

One should check for the lowest Q of the circuit topologies. A shunt resonant circuit can be tried and a series resonant circuit can be tried. The value of the reactance is different in each case and will result in different Q values for the transformation circuits.

6.3.1.1 Multiple-step Q^2+1 Transformations.

If the transformation Q is much higher than the load Q, then the transformation can be broken into several steps. One approach is to use n geometric means to transform the networks. The impedance is raised by a given ratio r at each step. Starting from a resistance R, for n steps, the resultant resistance and transformation Q for each step is:

$$R_t = Rr^n$$

$$r = \left(\frac{R_t}{R}\right)^{\frac{1}{n}}$$

$$Q_t = \sqrt{r-1}$$

For example, if one desires to transform between 50 ohms and 500 ohms, the single-step transformation Q would be 3. One could perform the transformation in n steps. Suppose three steps are chosen. Then the first transformation would be from 50 ohms to the cube root (three steps) of 10 times 50. This would be a transformation from 50 ohms to 107.7 ohms for a transformation Q of 1.074. The next step would be a transformation from 107.7 ohms to 232.1 ohms and the final transformation would be from 232.1 ohms to 500 ohms. Each of the transformation steps would be with a transformation Q of 1.074. The two different transformations are shown on Figure 6.5. The Q circles have a center and locus defined by the following circle:

$$\Gamma\Gamma^* \pm j\frac{\Gamma}{Q} \mp j\frac{\Gamma^*}{Q} + \frac{1}{Q^2} = \frac{Q^2+1}{Q^2}$$

$$\Gamma_C = \mp\frac{1}{Q}$$

$$\rho = \sqrt{\frac{Q^2+1}{Q^2}}$$

where the top part of the plus or minus sign is used when the reactance is positive. The match shown on Figure 6.5 uses series inductances and parallel capacitances. The passband of the match may be enhanced if the series component is alternated between a series inductor and a series capacitor and the shunt component is alternated between a shunt capacitor and a shunt inductor. This is given as one of the exercises. When a series capacitor is used, the locus goes into the bottom part of the chart and is brought back to the real axis with a shunt inductor. The effective bandwidth of the three low Q transformations would be larger than the bandwidth

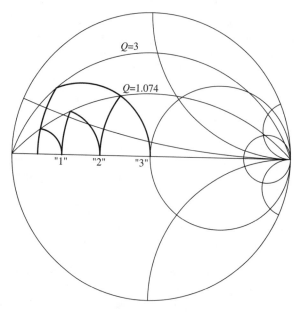

Figure 6.5 Reflection coefficient chart for a single-step and three-step match.

of one transformation with a Q of 3. Loss in the components would have to be considered as well and one needs to minimize the number of components to minimize loss. Therefore, an engineering trade-off can be performed.

If the circuit topology of the load is a parallel combination of a resistance and single reactance, then it is often a broader band solution to series resonate the load. In like manner, if the circuit topology of the load is a series combination of a resistance and single reactance, then it is often a broader band solution to parallel resonate the load. Resonating the load with a shunt reactance when the load appears as a series circuit or using a series reactance when the load appears as a parallel circuit in effect performs the first step of an impedance transformation. The next transformation then completes a two-step rather than a single-step transformation with the resulting lower transformation Q. Plotting R and X versus frequency and G and B versus frequency often reveals the topology of the load over the band of interest.

Matthaei [5] has given a set of tables for short-step impedance transformers that use transmission lines. The transmission lines in that approach alternate between low-impedance and high-impedance transmission lines. When the lines are short, that structure appears at low frequencies to be series L shunt C structures. The locus of match on the reflection coefficient plane will be similar to the multiple-step $Q^2 + 1$ one match. However, those tables predict the band shape of the transformation network as well.

6.3.2 Matching with a Single Transmission Line

For narrowband circuits, a simple match might be just using a single transmission line of the correct characteristic impedance. It is not always possible to match a network with just a single transmission line. The following analysis will show what the characteristic impedance should be if it is possible. If one plots the source and load impedance in a reflection coefficient plane that is normalized to the desired characteristic transmission line impedance, the source and load points would each have to be at the same radius. This allows one to rotate the source to the conjugate of the load (or vice versa) along a constant radius path and obtain conjugate match. The length of the line is found from the rotation required. The magnitudes squared of the reflection coefficients and the required characteristic impedance are given in the next equation.

$$\left(\frac{Z_S - Z_0}{Z_S + Z_0}\right)\left(\frac{Z_S^* - Z_0}{Z_S^* + Z_0}\right) = \left(\frac{Z_L - Z_0}{Z_L + Z_0}\right)\left(\frac{Z_L^* - Z_0}{Z_L^* + Z_0}\right)$$

$$Z_0 = \sqrt{\frac{R_S|Z_L|^2 - R_L|Z_S|^2}{R_L - R_S}}$$

If the quantity within the square root sign is negative, a single transmission line solution does not exist. Since Z_0^2 is a positive number the following conditions must be met.

$$\frac{R_S}{1 + |Z_S|^2} > \frac{R_L}{1 + |Z_L|^2}\bigg|_{R_L > R_S}$$

$$\frac{R_S}{1 + |Z_S|^2} < \frac{R_L}{1 + |Z_L|^2}\bigg|_{R_L < R_S}$$

When these conditions are met, a single transmission line can be used to match the two impedances. The length of the transmission line can easily be calculated once the characteristic impedance is known.

$$e^{-j2\beta D} = \left(\frac{Z_L^* - Z_0}{Z_L^* + Z_0}\right)\left(\frac{Z_S + Z_0}{Z_S - Z_0}\right)$$

As an example, let:

$$Z_S = 2 - j5$$
$$Z_L = 30 - j15$$
$$\Rightarrow$$
$$Z_0 = 7.0204$$
$$\beta D = 301.4°$$

When the two impedances are real, a match always exists. The formulas for the characteristic impedance and line length reduce to:

$$Z_0^2 = R_S R_L \Rightarrow R_S = \frac{Z_0^2}{R_L}$$

$$e^{-j2\beta D} = \left(\frac{R_L - Z_0}{R_L + Z_0}\right)\left(\frac{\frac{Z_0^2}{R_L} + Z_0}{\frac{Z_0^2}{R_L} - Z_0}\right) = \left(\frac{R_L - Z_0}{R_L + Z_0}\right)\left(\frac{Z_0 + R_L}{Z_0 - R_L}\right) = -1$$

This is the well-known quarter-wavelength transformer. Just as in the $Q^2 + 1$ procedure, one can use multiple quarter-wavelength transformers to match the lines. One procedure is to use the same nth-order geometric means for the intermediate impedance level.

6.3.3 Matching Using Two Transmission Lines of Different Characteristic Impedance

As an alternate approach to the lumped-constant match using an L and a C or a single transmission line, impedance matches using two transmission lines with different impedances can be tried. The value of the two line impedances might be given or their values might be dictated by other considerations. The question is twofold. Can the two impedances be matched using two different transmission lines and, if so, what are the lengths of the lines? The circuit configuration appears as shown in Figure 6.6. This configuration can be plotted

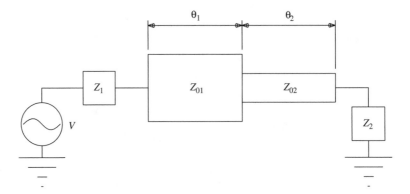

Figure 6.6 Two-section transmission line match.

on different reflection coefficient planes using different normalization impedances. If one chooses to normalize the impedance to either Z_{01} or Z_{02}, the charts appear quite simple as shown in Figure 6.7. The left chart is normalized to Z_{01} and the right chart is normalized to Z_{02}. The circles on the charts are bilinear transformations of each other. The circles resulting from rotating the terminal impedances along their respective transmission lines must intersect for a solution to exist. There are two intersection points. One of the immittances must be rotated to one of the two intersections and then the other immittance to the opposite one. Note that the two points of intersection represent immittances that are conjugates of each other giving conjugate match. Since the left impedance could be rotated to either the top or bottom intersection point and vice versa, there are two solutions for the right impedance and thus the match.

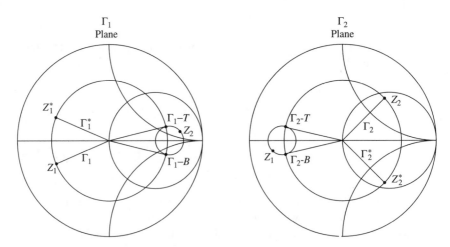

Figure 6.7 Reflection coefficient representation of two-section match.

When the immittances are rotated down the transmission lines, the immittance will become real at two different points. The low value and high values of resistance are:

$$R_{i\_low} = Z_{0i} \frac{1 - |\Gamma_i|}{1 + |\Gamma_i|}$$

$$R_{i\_high} = Z_{0i} \frac{1 + |\Gamma_i|}{1 - |\Gamma_i|}$$

Now using a chart normalized to Z_{01}, the minimum and maximum values of the Γ_2 circle plotted in the Z_{01} plane are:

$$\Gamma_{2\_low} = \frac{Z_{02}\dfrac{1 - \Gamma_{20}}{1 + \Gamma_{20}} - Z_{01}}{Z_{02}\dfrac{1 - \Gamma_{20}}{1 + \Gamma_{20}} + Z_{01}}$$

$$\Gamma_{2\_high} = \frac{Z_{02}\dfrac{1 + \Gamma_{20}}{1 - \Gamma_{20}} - Z_{01}}{Z_{02}\dfrac{1 + \Gamma_{20}}{1 - \Gamma_{20}} + Z_{01}}$$

where Γ_{20} is the radius of the reflection coefficient circle of Z_2 in the gamma plane normalized to Z_{02}. The difference between these reflection coefficient values is the diameter of the Z_2 reflection coefficient circle plotted in the gamma plane normalized to Z_{01}. This difference is:

$$\Gamma_{2\_high} - \Gamma_{2\_low} = \frac{Z_{02}(1 + \Gamma_{20}) - Z_{01}(1 - \Gamma_{20})}{Z_{02}(1 + \Gamma_{20}) + Z_{01}(1 - \Gamma_{20})} - \frac{Z_{02}(1 - \Gamma_{20}) - Z_{01}(1 + \Gamma_{20})}{Z_{02}(1 - \Gamma_{20}) + Z_{01}(1 + \Gamma_{20})}$$

$$= \frac{2\Gamma_{20}(Z_{02} + Z_{01})^2 - 2\Gamma_{20}(Z_{02} - Z_{01})^2}{(Z_{02} + Z_{01})^2 - \Gamma_{20}^2(Z_{02} - Z_{01})^2}$$

giving a radius of:

$$\rho_1 = \frac{4\Gamma_{20}Z_{01}Z_{02}}{(Z_{02} + Z_{01})^2 - \Gamma_{20}^2(Z_{02} - Z_{01})^2}$$

Likewise, the center of the circle is:

$$\Gamma_{C1} = \frac{\Gamma_{2\_high} + \Gamma_{2\_low}}{2} = \frac{(1 - \Gamma_{20}^2)(Z_{02}^2 - Z_{01}^2)}{(Z_{02} + Z_{01})^2 - \Gamma_{20}^2(Z_{02} - Z_{01})^2}$$

The points of intersection of the circle due to Z_1 and the circle due to Z_2 in the reflection coefficient plane normalized to Z_{01} are:

$$\Gamma_{10}e^{j\theta_1} = \Gamma_{C1} + \rho e^{j\phi_1}$$

Using real and imaginary parts:

$$\Gamma_{10}\cos(\theta) = \Gamma_{C1} + \rho_1\cos(\phi)$$
$$\Gamma_{10}\sin(\theta) = \rho_1\sin(\phi)$$

and squaring both sides gives the following result.

$$\Gamma_{10}^2[\cos^2(\theta_1) + \sin^2(\theta_1)] = \Gamma_{C1}^2 + 2\Gamma_{C1}\rho_1\cos(\phi_1) + \rho_1^2[\cos^2(\phi_1) + \sin^2(\phi_1)]$$
$$\Gamma_{10}^2 = \Gamma_{C1}^2 + 2\Gamma_{C1}\rho_1\cos(\phi_1) + \rho_1^2$$
$$\cos(\phi_1) = \frac{\Gamma_{10}^2 - \Gamma_{C1}^2 - \rho_1^2}{2\Gamma_{C1}\rho_1}$$
$$\cos(\theta_1) = \frac{\Gamma_{10}^2 + \Gamma_{C1}^2 - \rho_1^2}{2\Gamma_{C1}\Gamma_{10}}$$

Likewise, for the circles in the second transmission line plane:

$$\rho_2 = \frac{4\Gamma_{10}Z_{01}Z_{02}}{(Z_{02} + Z_{01})^2 - \Gamma_{10}^2(Z_{02} - Z_{01})^2}$$

and:

$$\Gamma_{C2} = \frac{(1 - \Gamma_{10}^2)(Z_{02}^2 - Z_{01}^2)}{(Z_{02} + Z_{01})^2 - \Gamma_{10}^2(Z_{02} - Z_{01})^2}$$

The angles in the second reflection coefficient plane are:

$$\cos(\phi_2) = \frac{\Gamma_{20}^2 - \Gamma_{C2}^2 - \rho_2^2}{2\Gamma_{C2}\rho_2}$$

$$\cos(\theta_1) = \frac{\Gamma_{20}^2 + \Gamma_{C2}^2 - \rho_2^2}{2\Gamma_{C2}\Gamma_{20}}$$

There are two values for each cosine. If the right sides of these equations are greater than 1, then no solution exists.

As an example, use the same impedances that were used in the single transmission line example. For this example, use a 5-ohm and a 20-ohm line.

$$Z_S = 2 - j5$$

$$Z_L = 30 - j15$$

$$\Gamma_1 = 0.6778\angle - 85.43°$$

$$\Gamma_2 = 0.3453\angle - 39.61°$$

$$\Gamma_1^* = 0.6778\angle + 85.43°$$

$$\Rightarrow$$

$$\theta_1 = 18.22°$$

$$\theta_2 = 90.02°$$

One solution is to rotate from a θ_1 of 18.22 degrees to the conjugate of the source that is at 85.43 degrees. That is a rotation of 256 degrees, giving a line length of 128 degrees for the 5-ohm line. The line impedance choice is definitely not optimum but this demonstrates the use of two transmission lines for the match.

6.4 KURODA'S IDENTITIES

When using the $Q^2 + 1$ matching procedure, sometimes a series inductive reactance is obtained. If one wishes to synthesize the impedances in the circuit with transmission line stubs, a series-shorted stub can be used to get the inductive reactance. However, the physical head height required for a series-shorted stub might not be available. How can this circuit be synthesized without the use of a series-shorted stub?

Kuroda [6] showed a set of transformations for two section circuit structures consisting of either a shunt stub and a transmission line or series stub and a transmission line. The electrical length of the stub is equal in length to transmission line. These identities are often used in filter design [7] when the stub and transmission lines are an eighth of a wavelength long at the center frequency of the passband of interest or as modified impedance inverters when their lengths are a quarter of a wavelength long. Impedance inverters will be discussed in Chapter 9. However, these transformations have a broader application and can be used to change the topology of a circuit. The transformations are shown in Figure 6.8. The reader can show their equivalence with the use of ABCD matrices for each of the elements. The identities

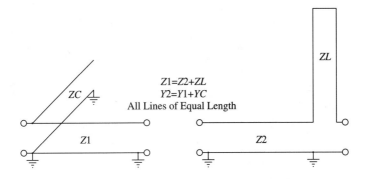

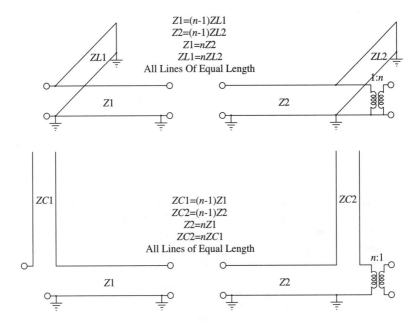

Figure 6.8 Kuroda's identities.

are often shown with unit elements and either inductor or capacitor symbols. They are shown here in their transmission line form.

As an example, consider matching from 50 ohms to 5 ohms with a $Q^2 + 1$ circuit. A low-pass type of structure will be used. The series inductance and shunt capacitance are:

$$50 = (Q^2 + 1)5 \Rightarrow Q = 3$$

$$jX_S = j5 * 3 = j15$$

$$jB_P = j3/50 = j0.06$$

If this match were at one gigahertz, the series inductance would be 2.4 nH and the shunt capacitance would be 9.5 pF. The circuit can be built with stubs as well. The circuit shown in the top of Figure 6.9 shows the lumped-constant version and the version using stubs. Short-length stubs 30 degrees long are chosen for this example. At one gigahertz, that would be 25

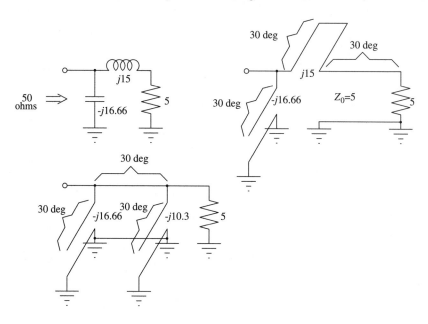

Figure 6.9 Fifty-ohm to 5-ohm match using a Kuroda identity.

mm long in air. The impedances of the stubs are:

$$j15 = jZ_0 \tan(30°) \Rightarrow Z_{0S} = 25.98 \text{ ohms}$$

$$j0.06 = jY_0 \tan(30°) \Rightarrow Z_{O0} = 9.62 \text{ ohms}$$

However, if a 5-ohm transmission line is inserted between the series stub and the load and it is also 30 degrees long, then the Kuroda identify shown in the bottom part of Figure 6.9 can be used to replace the series stub with an open shunt stub as shown in the bottom part of Figure 6.9. Using the identities, the transmission line impedance that is 30 degrees long between the stubs and the second stub characteristic impedance is:

$$Z_{TL} = Z_{0S} + 5 = 25.98 + 5 = 30.98 \text{ ohms}$$

$$Y_{OO2} = \frac{1}{5} - \frac{1}{30.98} \Rightarrow Z_{0O2} = 5.96 \text{ ohms}$$

This match could have been done using different-length stubs. The reader may want to try different values for the electrical length of the stubs. Figure 6.10 shows the insertion loss for the lumped-constant match and the match performed using Kuroda's identity. Since the transmission lines were short in length, there is not much difference in the match at low frequencies. At higher frequencies, the transmission line and lumped-constant match deviate from each other as expected.

It might actually be easier to design the series-shorted stub in actual practice. The low-impedance shunted stub on the 5-ohm load would be quite wide (about 7.5 mm wide by 14 mm long on a .25-mm thick 3.38-dielectric constant material). The 5-ohm line on the same material would be close to 10 mm wide. The stub width and length are about the same as the 5-ohm line width and the design procedures for these structures are often off since there is no well-defined start and stop for the stub. The series stub might be fashioned with ribbon and placed in series with the load. The reverse procedure used in this example might be used to transform a two-step shunt open stub match to a single shunt open stub and a single series-shorted stub in applications that permit the head height.

Kuroda's identities find additional applications in filter design and syntheses.

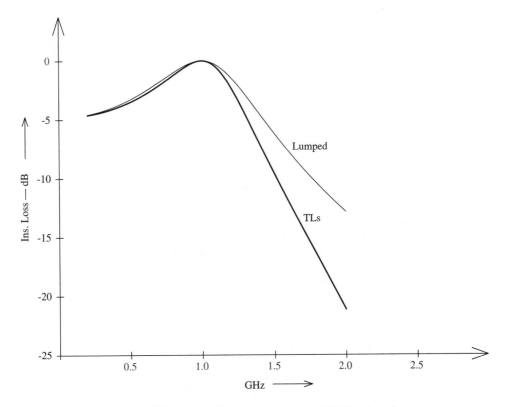

Figure 6.10 Response of the lumped-constant and distributed match.

6.5 DE-EMBEDDING PROCEDURES

In order to determine the value of a component that is embedded inside a network, a scheme called *de-embedding* is often used. If the components that exist between the unknown immittance and the input port are known, then the techniques shown for a specific example in Figure 6.11 will work. Notice that for a transmission line, either the length or the characteristic impedance can be made the negative of the original values. However, they cannot both be made the negatives of the original values.

6.5.1 Using De-embedding to Determine Device Impedances

Using *ABCD* parameters, it is quite easy to show that this method works. The *ABCD* matrix sequence for the network in the upper left of Figure 6.11 follows. This sequence assumes that there is a set of unused terminals to the right of $Z_1 = 1/Y_1$.

$$\begin{pmatrix} A & B \\ C & D \end{pmatrix} = \begin{pmatrix} 1 & 0 \\ \dfrac{1}{-j\omega L} & 1 \end{pmatrix} \begin{pmatrix} 1 & 0 \\ \dfrac{1}{j\omega L} & 1 \end{pmatrix} \begin{pmatrix} 1 & 0 \\ Y_1 & 1 \end{pmatrix}$$

$$= \begin{pmatrix} 1 & 0 \\ 0 & 1 \end{pmatrix} \begin{pmatrix} 1 & 0 \\ Y_1 & 1 \end{pmatrix} = \begin{pmatrix} 1 & 0 \\ Y_1 & 1 \end{pmatrix}$$

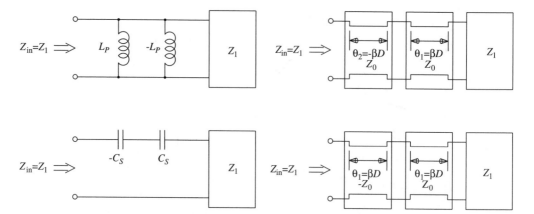

Figure 6.11 De-embedding reactive components and transmission lines. [*R. Weber, Microwave Engineering Course Notes, ©1987–1997, August 1997 edition, p. 83.*]

Similarly, for the circuit on the lower left of Figure 6.11:

$$\begin{pmatrix} A & B \\ C & D \end{pmatrix} = \begin{pmatrix} 1 & \dfrac{1}{-j\omega C} \\ 0 & 1 \end{pmatrix} \begin{pmatrix} 1 & \dfrac{1}{j\omega C} \\ 0 & 1 \end{pmatrix} \begin{pmatrix} 1 & 0 \\ Y_1 & 1 \end{pmatrix}$$

$$= \begin{pmatrix} 1 & 0 \\ 0 & 1 \end{pmatrix} \begin{pmatrix} 1 & 0 \\ Y_1 & 1 \end{pmatrix} = \begin{pmatrix} 1 & 0 \\ Y_1 & 1 \end{pmatrix}$$

For the transmission line network in the upper right of Figure 6.11 where de-embedding is used for a negative length of line, the sequence is as follows.

$$\begin{pmatrix} A & B \\ C & D \end{pmatrix} = \begin{pmatrix} \cos(-\beta D) & jZ_0\sin(-\beta D) \\ jY_0\sin(-\beta D) & \cos(-\beta D) \end{pmatrix} \begin{pmatrix} \cos(\beta D) & jZ_0\sin(\beta D) \\ jY_0\sin(\beta D) & \cos(\beta D) \end{pmatrix} \begin{pmatrix} 1 & 0 \\ Y_1 & 1 \end{pmatrix}$$

$$= \begin{pmatrix} 1 & 0 \\ 0 & 1 \end{pmatrix} \begin{pmatrix} 1 & 0 \\ Y_1 & 1 \end{pmatrix} = \begin{pmatrix} 1 & 0 \\ Y_1 & 1 \end{pmatrix}$$

For the transmission line network in the lower right of Figure 6.11 where the de-embedding uses a negative characteristic impedance of the transmission line, the sequence is as follows.

$$\begin{pmatrix} A & B \\ C & D \end{pmatrix} = \begin{pmatrix} \cos(\beta D) & -jZ_0\sin(\beta D) \\ -jY_0\sin(\beta D) & \cos(\beta D) \end{pmatrix} \begin{pmatrix} \cos(\beta D) & jZ_0\sin(\beta D) \\ jY_0\sin(\beta D) & \cos(\beta D) \end{pmatrix} \begin{pmatrix} 1 & 0 \\ Y_1 & 1 \end{pmatrix}$$

$$= \begin{pmatrix} 1 & 0 \\ 0 & 1 \end{pmatrix} \begin{pmatrix} 1 & 0 \\ Y_1 & 1 \end{pmatrix} = \begin{pmatrix} 1 & 0 \\ Y_1 & 1 \end{pmatrix}$$

This technique is shown in Figure 6.12 for a more complicated example. Notice that the order of the component types is interchanged and that the component values are made the negative of the component they mirror. This technique is very useful in determining the characteristics of different devices when the parasitic embedding component values are known. These embedding component values are often determined by bonding up packages without active device elements in them in various configurations and measuring the resultant configurations. It is important that the bonded-up packages use the same length of bond wires in them as used in the packages containing actual devices.

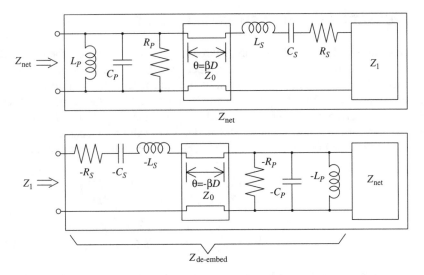

Figure 6.12 A de-embedding example. [*R. Weber, Microwave Engineering Course Notes,* ©*1987–1997, August 1997 edition, p. 83.*]

6.5.2 Using De-embedding to Compensate for Short Transmission Lines

When a component such as a transistor is inserted onto a PCB, it needs to be attached to a small pad or short length of line. De-embedding can be used to calculate the value of an impedance to put on the end of a short transmission line when the impedance level is known only at the device. In Figure 6.13, Z_1 represents the desired impedance to be presented to a device. If one uses in the calculation a negative length of transmission line and a positive length of transmission line between the load and Z_{in}, then one still sees Z_1. However, at the junction between the positive and negative length of line, one sees Z_{TL} looking to the right. If one synthesizes Z_{TL} instead of Z_1 and presents that at the end of the positive-length line, one gets Z_1 at the input. This procedure is quite helpful in synthesis when parasitic transmission lines or bonding pads have to be used.

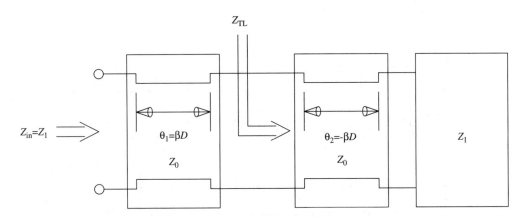

Figure 6.13 De-embedding for parasitic solder pads or bond pads.

6.6 COMPENSATION FOR TRANSMISSION LINE PADS

In the section on de-embedding, a procedure was given to compensate for a short transmission line. If a matching network has been developed for a device without transmission lines there is another technique that can be used to compensate the matching capacitors and inductors when a transmission line is added to allow them to be mounted. From the Z and Y matrices for transmission lines given in Chapter 2, one can derive a tee and pi model of short transmission lines that is valid for this compensation. The model is not valid for reactance slope compensation as will be discussed in Chapter 9 on filters. The Z and Y matrices are given here.

$$(Y) = \begin{pmatrix} -jY_0\cot(\beta D) & jY_0\csc(\beta D) \\ jY_0\csc(\beta D) & -jY_0\cot(\beta D) \end{pmatrix}$$

$$(Z) = \begin{pmatrix} -jZ_0\cot(\beta D) & -jZ_0\csc(\beta D) \\ -jZ_0\csc(\beta D) & -jZ_0\cot(\beta D) \end{pmatrix}$$

Assuming the lines are lossless, only reactances appear in the circuit. For a tee circuit, as shown in Figure 6.14, the following reactances result assuming series L's for Z_1 and Z_2 and a shunt C for Z_3.

$$X_1 = X_2 = Z_0 \frac{1 - \cos(\beta D)}{\sin(\beta D)} = \omega L_1$$

$$X_3 = \frac{-Z_0}{\sin(\beta D)} = \frac{-1}{\omega C_3}$$

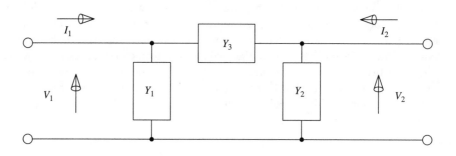

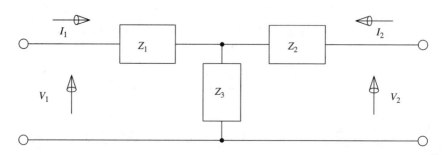

Figure 6.14 Tee and Pi models of a transmission line.

For a pi circuit, as shown in Figure 6.14, the following susceptances result assuming shunt C's for Y_1 and Y_2 and a series L for Y_3.

$$B_1 = B_2 = Y_0 \frac{1 - \cos(\beta D)}{\sin(\beta D)} = \omega C_1$$

$$B_3 = \frac{-Y_0}{\sin(\beta D)} = \frac{-1}{\omega L_3}$$

After determining the value of the capacitors and inductors in the equivalent circuit, the value of the component mounted on the pad can be reduced or increased to compensate for the effect of the mounting or bonding pad. If the impedance levels are low at the point the pad is inserted, the shunt capacitance of the pad may be able to be ignored but the inductance of the pad would have to be included. If the impedance level is high where the pad is inserted, then it may be possible to ignore the series inductance of the pad but compensate for its shunt capacitance. If the impedance level is on the order of the impedance of the pad, then all of the components need to be accounted for.

6.7 SYMMETRICAL ATTENUATORS

When cascading stages, sometimes it is necessary to provide some loss between stages. One might want to introduce some loss if potentially unstable stages are cascaded. Often because of part size availability, attenuation is needed on the output of one stage to prevent overdrive on the next stage. Attenuation placed between stages should not introduce any mismatch or VSWR. This section gives the values of matched symmetrical attenuators that can be used for this purpose.

6.7.1 Matched Attenuators

Attenuators are circuit elements that attenuate the signal through them while maintaining impedance match. The attenuators that are being considered here are symmetrical attenuators. These attenuators have the same match impedance on both sides.

The S matrix of these attenuators is:

$$(S) = \begin{pmatrix} 0 & M \\ M & 0 \end{pmatrix}$$

When gain was discussed in Chapter 5, insertion gain was only briefly discussed and it was stated that its use depended on source and load match. Typically attenuators are specified in terms of insertion loss and match is assumed. Insertion loss of the attenuator is usually expressed in dBs. The insertion gain of a network usually means dB transducer gain. Insertion loss (IL) in dBs is the negative of insertion gain in dBs. The insertion loss of the attenuator is:

$$IL = -20 \log_{10}(S_{21}) = -20 \log_{10}(M)$$

Figure 6.15 shows both a tee type and a pi type of symmetrical attenuator. When the attenuators are terminated in a Z_0 impedance, the ratio V_2/V_1 is M. Consider the tee attenuator first. In order for the attenuators to be matched when the attenuators are terminated in their characteristic impedance Z_0, the input impedance must also be equal to Z_0. The two constraints, (1) $M = V_2/V_1$ and (2) $Z_{in} = Z_0$, allow the resistances of R_1 and R_2 to be determined. The voltage

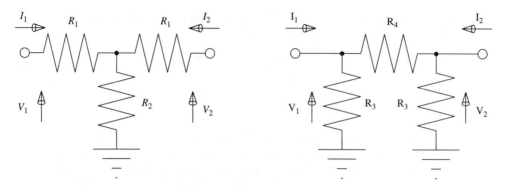

Figure 6.15 Tee and Pi attenuator models.

ratio M is given by:

$$M = \frac{V_{\text{out}}}{V_{\text{in}}} = \left[\frac{\dfrac{(R_1 + Z_0)R_2}{(R_1 + R_2 + Z_0)}}{R_1 + \dfrac{(R_1 + Z_0)R_2}{(R_1 + R_2 + Z_0)}} \right] \left[\frac{Z_0}{(R_1 + Z_0)} \right]$$

$$= \frac{R_2 Z_0}{R_1(R_1 + 2R_2) + Z_0(R_1 + R_2)}$$

The input impedance for the tee attenuator terminated in its characteristic impedance is:

$$Z_{\text{in}} = R_1 + \frac{(R_1 + Z_0)R_2}{(R_1 + R_2 + Z_0)} = Z_0$$

where the third part of the equation is the constraint that the input impedance must be equal to the characteristic impedance. These equations give:

$$Z_0^2 = R_1(R_1 + 2R_2) \qquad r_1(r_1 + 2r_2) = 1$$

$$M = \frac{R_2}{Z_0 + R_1 + R_2} = \frac{r_2}{1 + r_1 + r_2}$$

$$r_1 = \frac{1 - M}{1 + M} \qquad R_1 = \frac{1 - M}{1 + M} Z_0$$

$$r_2 = \frac{2M}{1 - M^2} \qquad R_2 = \frac{2M}{1 - M^2} Z_0$$

This gives the solution for a tee attenuator. The pi attenuator can be determined by doing a (Z) to (Y) matrix conversion. The (Z) matrix of the tee attenuator is:

$$(Z) = \begin{pmatrix} R_1 + R_2 & R_2 \\ R_2 & R_1 + R_2 \end{pmatrix}$$

The determinant of this is:

$$(Z) = \begin{pmatrix} R_1 + R_2 & R_2 \\ R_2 & R_1 + R_2 \end{pmatrix}$$

$$\Delta_Z = (R_1 + R_2)^2 - R_2^2 = R_1^2 + 2R_1 R_2$$

$$= \left[\left(\frac{1 - M}{1 + M} \right)^2 + 2 \left(\frac{1 - M}{1 + M} \right) \left(\frac{2M}{1 - M^2} \right) \right] Z_0^2 = Z_0^2$$

This gives:

$$y_{11} = \left(\frac{1-M}{1+M} + \frac{2M}{1-M^2}\right) Y_0 = \left(\frac{1+M^2}{1-M^2}\right) Y_0$$

$$y_{12} = \left(\frac{-2M}{1-M^2}\right) Y_0$$

Therefore:

$$R_3 = (y_{11} + y_{12})^{-1} = \left(\frac{1+M}{1-M}\right) Z_0$$

$$R_4 = \frac{1}{-y_{12}} = \left(\frac{1-M^2}{2M}\right) Z_0$$

As an example, a 3-dB attenuator would have:

$$M = 10^{\left(-\frac{3}{20}\right)} = \frac{\sqrt{2}}{2}$$

For a 50-ohm, 3-dB attenuator, this gives $R_1 = 8.550$ ohms, $R_2 = 141.9$ ohms, $R_3 = 292.4$ ohms, and $R_4 = 17.61$ ohms. The first two values are used for the tee attenuator and the last two values are used for a pi attenuator.

6.8 LUMPED-CONSTANT PHASE SHIFTERS

Sometimes it is necessary to put in a matched phase shift between stages. This can be done to rotate the stable and unstable regions of a part around so that when two circuits are cascaded the overall circuit is stable. Matched symmetrical phase shifters that can be used for this purpose are discussed in this section.

Lumped-constant, lossless, and matched phase shifter values can be calculated for the circuits in Figure 6.16. Note that by convention the propagation constant for a positive-length transmission line is specified as positive. Therefore, the phase constant $\theta = \beta D$ for a transmission line results in circuit transmission phase shift ϕ of $-\theta$! The scattering matrix of the circuits is therefore:

$$(S) = \begin{pmatrix} 0 & e^{-j\theta} \\ e^{-j\theta} & 0 \end{pmatrix} = \begin{pmatrix} 0 & e^{j\phi} \\ e^{j\phi} & 0 \end{pmatrix}$$

With respect to this scattering matrix, the susceptance of the shunt L's and C's in Figure 6.16 are as follows. The circuits on the left side have a positive phase shift ϕ but a negative value of θ. The circuits on the right side have a negative phase shift ϕ but a positive value of θ. For the pi circuits the following formulas hold.

$$\omega C_{\text{sh\_pi}} = -Y_0 \cot(\theta) + Y_0 \csc(\theta)|_{\theta>0}$$

$$\omega L_{\text{se\_pi}} = \frac{1}{Y_0 \csc(\theta)}\bigg|_{\theta>0}$$

$$\omega L_{\text{sh\_pi}} = \frac{1}{Y_0 \cot(\theta) - Y_0 \csc(\theta)}\bigg|_{\theta<0}$$

$$\omega C_{\text{se\_pi}} = -Y_0 \csc(\theta)|_{\theta<0}$$

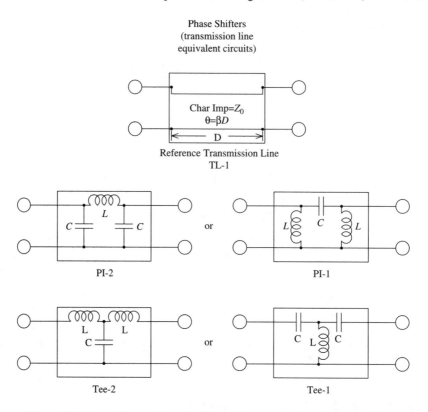

Figure 6.16 Phase shifter models. [R. Weber, Microwave Engineering Course Notes, ©1987–1997, August 1997 edition, p. 116.]

For the tee circuits, the following formulas hold.

$$\omega L_{\text{se\_tee}} = -Z_0 \cot(\theta) + Z_0 \csc(\theta)|_{\theta>0}$$

$$\omega C_{\text{sh\_tee}} = \frac{1}{Z_0 \csc(\theta)}\bigg|_{\theta>0}$$

$$\omega C_{\text{se\_tee}} = \frac{1}{Z_0 \cot(\theta) - Z_0 \csc(\theta)}\bigg|_{\theta<0}$$

$$\omega L_{\text{sh\_tee}} = -Z_0 \csc(\theta)|_{\theta<0}$$

If the phase shift desired is small, the ell sections shown in Figure 6.17 can be used to provide small amounts of phase shift. The sections are similar to the Q^2+1 matching sections. Note that the magnitude of S_{21} is not exactly equal to 1 and thus the magnitude of S_{11} is not zero but very close to 1 and zero respectively if the normalized x or b's are very small. The

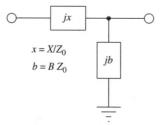

$x = X/Z_0$
$b = B Z_0$

Figure 6.17 An approximate phase shifter. [R. Weber, Microwave Engineering Course Notes, ©1987–1997, August 1997 edition, p. 117.]

ABCD matrix of the product is used to determine the transducer gain.

$$(ABCD) = \begin{pmatrix} 1 & jx \\ 0 & 1 \end{pmatrix} \begin{pmatrix} 1 & 0 \\ jb & 1 \end{pmatrix} = \begin{pmatrix} 1-bx & jx \\ jb & 1 \end{pmatrix}$$

$$S_{21} = \left.\frac{2}{2 - bx + j(b+x)}\right|_{b=x} = \frac{2}{2 - x^2 + 2jx} = \frac{1}{\sqrt{1 + \dfrac{x^4}{4}}} e^{\phi_{21}}$$

$$\phi_{21} = \left.\mathrm{atan}\left(\frac{-2x}{2-x^2}\right)\right|_{x<\sqrt{2}}$$

$$x = \frac{1}{\tan(\phi_{21})} - \sqrt{\frac{1}{\tan^2(\phi_{21})} + 2}$$

EXERCISES

6-1 Using a CAD program, calculate the match passband response for the single-step and three-step match shown in Figure 6.5. Compare the response with a match that alternates the series component between a series inductor and a series capacitor.

6-2 Determine the optimum load susceptance for a part that requires a net resistive load with the following parameters at 2 GHz. The part is to be loaded with a load line resistance of 200 ohms. Determine the power gain when the part is loaded with 200 ohms in parallel with the optimum load susceptance.

$$S_{11} = -0.8, S_{12} = 0.07 \text{ at } 45 \text{ deg.}, S_{21} = 3.2 \text{ at } 135 \text{ deg.}, S_{22} = 0.8 \text{ at } -90 \text{ deg.}$$

6-3 Find a $Q^2 + 1$ match from 10 ohms to 100 ohms at 2 GHz using a series capacitor.

6-4 Change the match in Exercise 6-3 to a series inductor and then perform a transformation using Kuroda's identities and derive a match using only shunt stubs. Use a frequency of 2 GHz.

6-5 Find a two-transmission-line match for matching between 10 ohms and 100 ohms.

6-6 Find a shunt stub and two-transmission-line match to match between $5 - j5$ ohms and 100 ohms.

6-7 Find a one-transmission-line match for matching between 10 ohms and 100 ohms.

6-8 Find a lumped-constant 50-ohm phase shifter for a phase shift of plus 75 degrees at 2 GHz. Determine the component values. Choose either a pi or a tee model.

6-9 A power source puts out 100 mW. A power amplifier requires 30 mW of input power. The source and the amplifier are designed for 200-ohm-level operation. Design a matched attenuator to be placed between the source and the amplifier.

6-10 A device requires an impedance of $20 + j30$ to be presented to it at 1.5 GHz. The device requires that a 40-mm-long 80-ohm transmission line with an effective dielectric constant is placed between it and the load. Determine the load to be presented to the transmission line so that the device is presented with the proper impedance. Design a $Q^2 + 1$ match for the load. The load transforms 50 ohms to the required impedance. Capacitors have a parasitic inductance of 0.5 nH and inductors have a parasitic capacity of 0.2 pF.

6-11 A source has an output impedance of 1 ohm in series with 0.32 nH. A load has an impedance of 100 ohms in series with 2 pF. The frequency is 1.5 GHz. Design a circuit to match the source to the load. Use a three-section $Q^2 + 1$ match. Parallel resonate the source and load before determining the match. Use series inductor, shunt capacitor sections. Plot the impedance seen in the matching network for each component on a reflection coefficient chart. Calculate the passband from 800 MHz to 2500 MHz. Determine what is limiting the bandwidth. Redo the match using four sections of a $Q^2 + 1$ match and recalculate the passband. Use a series L shunt C Series L shunt C for one match and a Series L shunt C Series C shunt L for a second match. Compare the bandwidths of the matches.

7

RF/Microwave Power Generation Considerations

7.1 POWER DEVICE CONSIDERATIONS

There is a major difference between designing amplifiers for maximum gain and designing amplifiers for maximum power conversion efficiency. Designing for maximum gain means getting the largest amount of rf power out of the amplifier for a given amount of rf input power. Designing for maximum dc to rf efficiency means getting the most rf power out for a given amount of dc input power. Maximum power out of an amplifier without regard for either gain, harmonic generation, or efficiency can often be accomplished with higher input drive levels and load lines that allow the device waveforms to be nonlinear over part of their cycle. For example, the load line and bias combination can be positioned for linear operation over most of the region of operation but allow the instantaneous voltage on the device to swing to a low-voltage region where the gain of the part drops. This results in a larger voltage swing and a larger power output at some expense in harmonic generation and nonlinear effects. Often when a part is matched for maximum efficiency or maximum power, the power gain is low compared to the maximum available gain. When a part is matched for maximum gain, the dc-to-rf efficiency is often quite low. These conditions result in design trade-offs.

For a maximum gain design with a device that has a simultaneous conjugate match, the input and output ports of the device have simultaneous conjugate match. For a maximum power conversion efficiency amplifier, one is not interested in maximum power gain but in maximum power conversion from dc input power to output rf power and the load is considerably different than the load that would be used to match the output impedance of the part. This results in a lower gain for the high power conversion efficiency amplifier when compared to the impedance-matched amplifier.

In a circuit where the device is matched for power gain, the load impedance required to achieve a specified power gain is dictated by the device's scattering parameters. In a power amplifier, where the device is matched to achieve maximum dc-to-rf power conversion the output load required is dictated by the device's V-I curves.

There are many modes of operation that can be used with a device to convert dc power into rf power. These modes or classes of operation are called Class A, B, C, and so on [1, 2]. The different modes of operation have to do with the percentage of time the active device is conducting over an rf cycle and the shape of the device's voltage and current waveforms.

It will be obvious when the different amplifier configurations are discussed that the output power of an amplifier cannot continue to increase beyond a given amount as the input level is increased. As the maximum power point is reached, the gain of the amplifier starts to decrease and the amplifier starts to become nonlinear in its input/output characteristics. This nonlinearity manifests itself in several ways. The decrease in gain is one way: however, the nonlinear behavior of the amplifier causes mixing to occur in the amplifier. The noise that exists at all frequencies in the amplifier can be converted up to the passband frequencies. Mixing and frequency conversion is discussed in Chapter 11. When two separate input frequencies are introduced into the input of the amplifier, when the amplifier goes into nonlinear behavior, these two signals will interact and create new frequencies. This is called *intermodulation* and that is also discussed in Section 11.12. As the gain of the amplifier decreases as its maximum power is reached, a point is reached for which the gain is decreased by one dB. This is called the *one-dB compression point* and this is discussed further in Section 11.12.

7.2 BIAS CONDITIONS

Figure 7.1 shows some of the considerations to be taken for a power amplifier circuit when the device is operating in a linear manner over all time. A bipolar device is shown but the same design considerations hold for FET devices as well. The components are shown as lumped-constant components. However, some of these components could be replaced with their equivalent distributed components in microwave circuits. The considerations are the same. Since the average voltage across the inductor L is zero, the average value of the device voltage is V_o. The rf voltage across R_L in this circuit is the same as the rf voltage across the active device since the output coupling capacitor is assumed to be infinite in value.

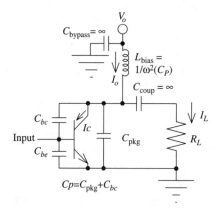

Figure 7.1 Circuit to consider for power for generation. [*R. Weber, Microwave Engineering Course Notes, ©1987–1997, August 1997 edition, p. 64.*]

Notice that an inductor is shown between V_o and the device rather than a resistor. A resistor is often used in small signal amplifiers. However, in a power circuit, the rf and dc power lost in the bias resistor is considered wasted power. Such a bias resistor would be in parallel with R_L as far as the rf voltage is concerned. The goal in a power amplifier is to produce as much power as possible in the load resistor for as little power as possible from the bias source V_o.

7.3 CONDUCTION DUTY CYCLES

In Class A circuits, the active device is on (conducting current) 100% of the time (i.e., over a complete rf cycle). In Class B circuits, the active device is on for 50% of the rf cycle. In Class C circuits, the active device is on less than 50% of an rf cycle. There are push-pull versions of

Class A, B, or C circuits in which the current in the two devices in the circuits are 180 degrees out of phase with each other. In Class B and Class C circuits, it is the device current that is zero over part of the rf cycle. In Class D and Class E circuits the device voltage is held at zero or a constant voltage over part of the cycle. In voltage mode Class D circuits, the part is saturated during part (usually 50%) of the rf cycle. In Class E circuits, the device voltage is held at zero volts primarily by the circuit that is put on the output of the device. Class D and E circuits will be discussed briefly after a more detailed analysis is given for Class A, B, and C circuits.

7.4 LOAD LINE CONSIDERATIONS

Figure 7.2 shows a typical set of load lines for either a FET or bipolar transistor. The circuit shown in Figure 7.1 supports the load line shown as "1" or "2" as long as the part is not allowed to go to zero current or the voltage permitted to reach the minimum value allowed. These load lines are dc load lines or load lines at rf when parasitic reactances have been removed from the V-I relationships. If maximum gain is desired from the part, then the load line slope must be approximately the negative of the slope of the V-I lines—Curve "1". This would give maximum gain but not maximum power conversion from the source V_o. Notice that for this load line, for linear operation the voltage swing on the part can vary from V_{sat} to approximately twice V_{oa} but that the rf current would not vary very far from I_o. Curve "2" is the curve which allows the rf current to swing a maximum amount limited by the load line intersection with V_{sat} to a minimum amount limited by the device turn-off point. Simultaneously the voltage on the part changes from V_{sat} to approximately twice V_{oa}. This load line allows the largest linear voltage and current swing in the part without permitting the device to go into nonlinear behavior. Linear operation is assumed for load line "2". If the part is allowed to go nonlinear over part of the cycle and have an excursion to lower voltage values for higher power, then load line "2a" would be used. For those load lines that allow the part to experience nonlinear operation, the circuit needs to be modified from the one shown in Figure 7.1. The load line resistance for load line "2" is as follows.

$$R_L = \frac{V_o - V_{sat}}{I_o - I_{min}}$$

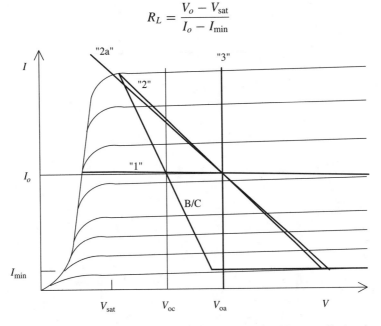

Figure 7.2 Power load lines for V-I curves of a device. [*R. Weber, Microwave Engineering Course Notes, ©1987–1997, August 1997 edition, p. 64.*]

The power delivered to a load of this value is:

$$P_{\mathrm{RF}} = \frac{(V_o - V_{\mathrm{sat}})^2}{2R_L} = \frac{(I_o - I_{\min})^2}{2} R_L$$

    Both curves "1" and "2" allow Class A operation of the part. In fact, load lines with slopes between the slopes of curves "2" and "3" would also give Class A behavior if the part is not driven into saturation or cutoff. Some designers design for a Class A type of load line but allow the device to go harder into saturation. More power is obtained at the expense of harmonic generation in the device. There is an engineering trade-off in those choices. This analysis considers power generation for Class A as limited by the saturation line and linear operation. Curve B/C is the load line of the device-driven Class A-B, Class B or Class C (i.e., conducting less than 100% of the rf cycle). Class A-B load lines have conduction cycles between 50% and 100%. Class B load lines are similar except that the percentage of an rf cycle is fixed at 50%. Class C load lines are similar also but the percentage of time the part is conducting is less than 50%. As the duty cycle changes from 100% to some lower value, the portion of the load line that is not at a low value of current get steeper. This results because a lower impedance is needed to reach the same current in a smaller period of time. Class A-B load lines are used to reduce the amount of nonlinearity that results from the gain nonlinearity that occurs at low current levels. Figure 7.3 shows time waveforms of the device current and voltage as they relate to Figure 7.2. Only the Class A time waveforms apply to Figure 7.1.

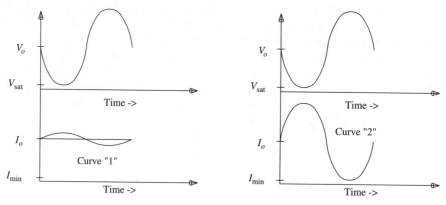

Collector Voltage and Current Waveforms

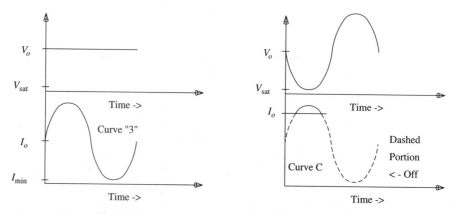

Figure 7.3 Time waveforms for Figure 7.2. [*R. Weber, Microwave Engineering Course Notes, ©1987–1997, August 1997 edition, p. 65.*]

For bipolar parts, a large variation of the collector to base voltage exists in power amplifiers and produces a large variation of the collector to base capacity. This results in potential nonlinear instabilities. Nonlinear instabilities are discussed in a section on stability in nonlinear circuits in Chapter 14. A consideration of Figure 7.4 shows that the primary nonlinear effect for voltage variation in a bipolar part is the collector to base capacity since the base to emitter capacitor value is usually at least an order of magnitude higher than the collector to base capacity. Later in this chapter, it is shown that if curve "2" is used to bias a bipolar transistor operating in the Class A mode, the collector to base capacity does not vary even though the collector to base voltage is varying. This is a result of the displacement current and conduction current balancing effects in the collector to base depletion region. However, due to mismatches that exist in the output circuit, the load line may deviate from an ideal curve "2". This then results in some deviation of the collector to base capacity over an rf cycle.

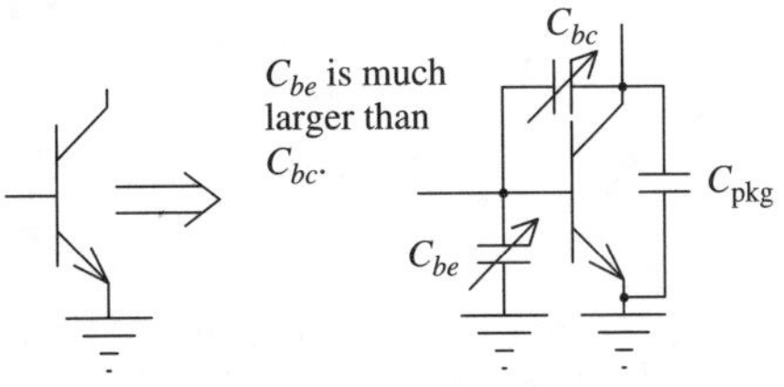

Figure 7.4 Collector to base capacity is important in bipolar designs. [*R. Weber, Microwave Engineering Course Notes, ©1987–1997, August 1997 edition, p. 67.*]

7.5 PUSH-PULL CLASS B

Figure 7.5 shows a typical push-pull Class B circuit. The transformer is shown having a $1:1:1$ ratio. However, the ratio on the load side could be some other value. Note from the circuit waveforms in Figure 7.6 that there is current in the circuit from the dc supply at all times, but that each individual part is only conducting for 50% of the time. The value of $I_{\min}$ is shown being above zero to demonstrate that the device might have some leakage current, not that it is purposely biased above zero current. However, as indicated in Section 7.4, a Class A-B circuit might have a nonzero bias current flowing in the "off" part to minimize the amount of

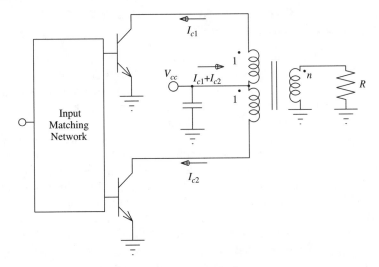

Figure 7.5 Typical Class B output stage. [*Adapted from* Solid State Radio Engineering, *H. L. Krauss, C. W. Bostian, and F. H. Raab, Copyright ©1980, John Wiley & Sons, Inc. Reprinted by permission of John Wiley & Sons, Inc.*]

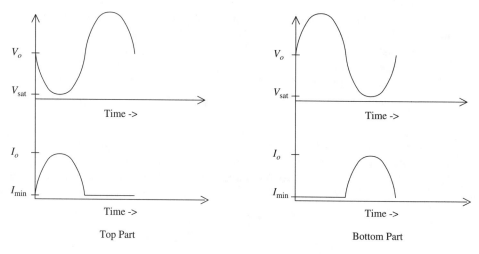

Figure 7.6 Typical Class B voltage and current waveforms

harmonic generation at low device currents. One often misses the fact that in push-pull Class B circuits a device that is in the cutoff state has voltage across it. In a transformer coupled circuit as shown in Figure 7.5, the output coupling transformer produces that voltage at zero current in that part of the winding. The active device therefore needs to be able to withstand a voltage stress of at least twice the bias voltage V_{cc}. Notice that the current that exists in the center tap of the primary side of the transformer is the sum of the two device currents and therefore consists of two half-sine waves or the equivalent of a rectified sine wave of current. It is important that a very-high-quality capacitor is used to bypasses the center tap point. The capacitor must be able to provide a large second harmonic current at a low impedance level in order to allow the center tap to stay at a constant voltage of V_{cc}. The parasitic series resistance of the capacitor needs to be very low as well.

7.6 LESS THAN ONE-HUNDRED-PERCENT CONDUCTION DUTY FACTOR—CLASS A, B, AND C

Figure 7.7 shows a circuit similar to that shown in Figure 7.1 except that a parallel resonator (tank) has been added to the output of the circuit. This is done to allow the device current to be only a part of a sinusoid or a pulse but yet allow the device voltage to remain sinusoidal. If the Q of the tank circuit is large, the tank circuit will provide the rf current to the load during the time that the active device is not conducting.

Approximate voltage and current waveforms for the device in Figure 7.7 are shown in Figure 7.8. The voltage and current waveforms are the device waveforms that are the in-phase voltage and current waveforms (i.e., waveforms of the device active area). Thus as shown in Figure 7.7, the current I_c is the current into the device and not the current idling through the collector to base depletion capacitance. When the current through the collector capacity is parallel resonated by an inductor and the parallel resonator is at resonance, the load current is the same as the fundamental component of the device rf current. It is also assumed that the Q of the parallel tank is high enough to prevent any substantial harmonic components in the voltage waveform. Under these conditions, the device current and voltage can then be

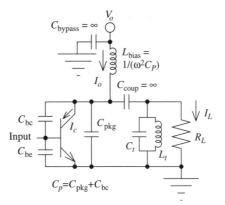

Figure 7.7 Power circuit with shunt tank on the output. [*Adapted from* Solid State Radio Engineering, *H. L. Krauss, C. W. Bostian, and F. H. Raab, Copyright ©1980, John Wiley & Sons, Inc. Reprinted by permission of John Wiley & Sons, Inc.*]

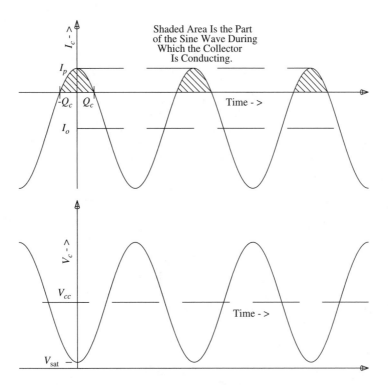

Figure 7.8 Voltage and current waveforms for Figure 7.7. [*R. Weber, Microwave Engineering Course Notes, ©1987–1997, August 1997 edition, p. 68.*]

described as:

$$I_c = I_o + (I_p - I_o)\cos(\theta); \qquad \theta_c < \theta < \theta_c$$

$$V_c = V_{cc} - (V_{cc} - V_{sat})\cos(\theta)$$

$$= V_{cc} + (I_o - I_p)R_{LL}\cos(\theta)$$

where R_{LL} is the load line resistance.

Looking at the current when it is equal to zero allows the relationship between I_o, I_p, and θ_c, to be found. This relationship can then be used in the voltage waveform equation.

$$\cos\left(\frac{\theta_c}{2}\right) = \frac{I_o}{I_o - I_p}$$

$$V_c = V_{cc} + \frac{I_o R_{LL}}{\cos\left(\dfrac{\theta_c}{2}\right)} \cos(\theta)$$

$$V_c = V_{cc} + \frac{-I_p R_{LL}}{1 - \cos\left(\dfrac{\theta_c}{2}\right)} \cos(\theta)$$

The average value of I_c is given by:

$$I_{dc} = \frac{1}{2\pi} \int_{-\frac{\theta_c}{2}}^{\frac{\theta_c}{2}} [I_o + (I_p - I_o)\cos(\theta)]\, d\theta$$

$$= -\frac{I_p}{2\pi} \int_{-\frac{\theta_c}{2}}^{\frac{\theta_c}{2}} \left[\frac{\cos\left(\dfrac{\theta_c}{2}\right)}{1 - \cos\left(\dfrac{\theta_c}{2}\right)} - \frac{\cos(\theta)}{1 - \cos\left(\dfrac{\theta_c}{2}\right)} \right] d\theta$$

$$= \frac{I_p \theta_c}{2\pi \left[1 - \cos\left(\dfrac{\theta_c}{2}\right) \right]} \left[\frac{\sin\left(\dfrac{\theta_c}{2}\right)}{\dfrac{\theta_c}{2}} - \cos\left(\dfrac{\theta_c}{2}\right) \right]$$

Therefore the dc power input is:

$$P_{dc} = \frac{V_{cc} I_p}{\pi} \frac{\left[\sin\left(\dfrac{\theta_c}{2}\right) - \dfrac{\theta_c}{2}\cos\left(\dfrac{\theta_c}{2}\right) \right]}{\left[1 - \cos\left(\dfrac{\theta_c}{2}\right) \right]}$$

The fundamental term for the Fourier series is:

$$I_{1st\ harmonic} = \frac{1}{\pi} \int_{-\frac{\theta_c}{2}}^{\frac{\theta_c}{2}} [I_o + (I_p - I_o)\cos(\theta)]\cos(\theta)\, d\theta$$

$$= \frac{I_p}{2\pi} \frac{\theta_c - \sin(\theta_c)}{\left[1 - \cos\left(\dfrac{\theta_c}{2}\right) \right]}$$

Assuming that only the fundamental current flows in the load, the voltage across the load is:

$$V_{R_L} = \frac{-I_p R_L}{2\pi} \frac{\theta_c - \sin(\theta_c)}{\left[1 - \cos\left(\dfrac{\theta_c}{2}\right) \right]} \cos(\theta) = -(V_{cc} - V_{sat})\cos(\theta)$$

Since the load is purely resistive, the ac power across the load is the product of the peak voltage times the peak current divided by two. The output power is therefore:

$$P_{RF} = \frac{(V_{cc} - V_{sat}) I_p [\theta_c - \sin(\theta_c)]}{4\pi \left[1 - \cos\left(\dfrac{\theta_c}{2}\right) \right]} \cos(\theta)$$

This gives a device efficiency of:

$$\eta = \frac{(V_{cc} - V_{sat})}{2V_{cc}} \frac{[\theta_c - \sin(\theta_c)]}{\left[2\sin\left(\dfrac{\theta_c}{2}\right) - \theta_c \cos\left(\dfrac{\theta_c}{2}\right)\right]}$$

The load resistor from the ratio of the fundamental current to the voltage is:

$$R_L = \frac{2\pi(V_{cc} - V_{sat})\left[1 - \cos\left(\dfrac{\theta_c}{2}\right)\right]}{I_p[\theta_c - \sin(\theta_c)]}$$

These equations hold for the Class A, Class AB, Class B, and Class C classes of operation when the output resonator Q is high enough to allow substantially only fundamental power in the load. The circuit component loss has been ignored. For small conduction angles, circuit component losses can significantly reduce the efficiency and also alter the required value of the load resistor. The efficiency is plotted in Figure 7.9 and the device load value is plotted in Figure 7.10 for the same value of I_p and V_{cc}. In using these formulas and curves, keep in mind that as defined, I_p is not the magnitude of the sine wave but the peak magnitude of the truncated sine wave.

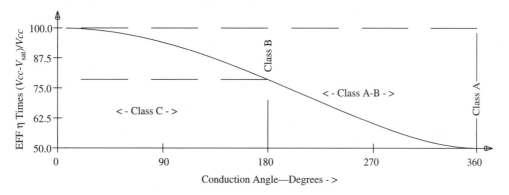

Figure 7.9 Efficiency vs. device conduction angle. [*Adapted from* Solid State Radio Engineering, *H. L. Krauss, C. W. Bostian, and F. H. Raab, Copyright ©1980, John Wiley & Sons, Inc. Reprinted by permission of John Wiley & Sons, Inc.*]

The slope of the load line in the region where I_c is not equal to zero is:

$$R_{LL} = \frac{(V_{cc} - V_{sat})}{(I_p - I_o)} = \frac{(V_{cc} - V_{sat})\left[1 - \cos\left(\dfrac{\theta_c}{2}\right)\right]}{I_p}$$

This gives the ratio between the load line slope and the value of the load resistor as follows.

$$\frac{R_{LL}}{R_L} = \frac{[\theta_c - \sin(\theta_c)]}{2\pi}$$

Letting the peak current for Class A operation equal I_{pA} and setting the power output for an arbitrary θ_c equal to the power output for Class A operation gives:

$$P_{RF} = \frac{(V_{cc} - V_{sat})I_p[\theta_c - \sin(\theta_c)]}{4\pi\left[1 - \cos\left(\dfrac{\theta_c}{2}\right)\right]} = \frac{(V_{cc} - V_{sat})I_{pA}}{4}$$

This results in the following relationship between the peak current for Class A and the peak

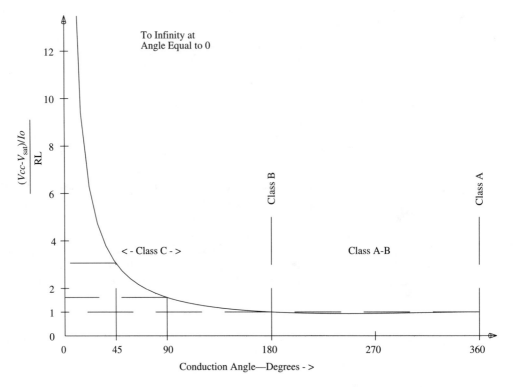

Figure 7.10 Load resistance RL vs. conduction angle. [*R. Weber, Microwave Engineering Course Notes, ©1987–1997, August 1997 edition, p. 71.*]

current for other conduction angles.

$$I_p = I_{pA} \frac{\pi \left[1 - \cos\left(\dfrac{\theta_c}{2}\right) \right]}{[\theta_c - \sin(\theta_c)]}$$

Plots of the resistance ratio and current ratio are shown in Figure 7.11 and Figure 7.12 respectively.

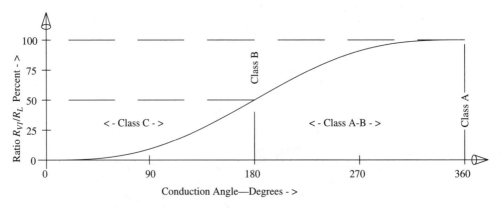

Figure 7.11 Ratio of load resistor to Class A load resistance. [*R. Weber, Microwave Engineering Course Notes, ©1987–1997, August 1997 edition, p. 72.*]

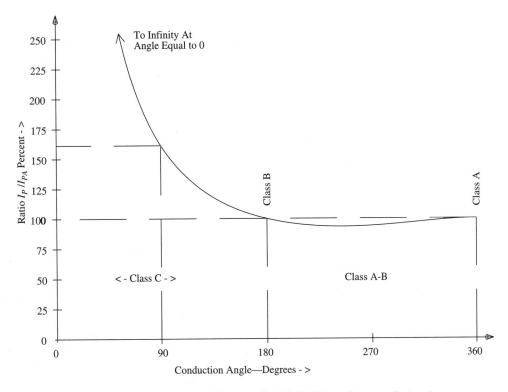

Figure 7.12 Peak current ratio—Class A to Class C. [*R. Weber, Microwave Engineering Course Notes, ©1987–1997, August 1997 edition, p. 72.*]

7.7 CLASS D CIRCUITS

Figure 7.13 shows the schematic of a typical single-ended Class D circuit. The devices in a Class D circuit are driven with a large enough voltage to act as switches. The charge storage that exists in a bipolar part will prevent the device from turning off rapidly. The bipolar part is often replaced with different types of FET devices to allow the devices to turn off more rapidly. If the devices can be approximated as switches, then the circuit shown in the lower half of Figure 7.13 approximates the behavior of a Class D circuit. The series R-L-C circuit on the output of the switch allows only fundamental current to flow in the load if its Q is high enough. Due to charge storage effects, there are losses in the Class D circuit that come from the period of time current flows through both transistors simultaneously. Other losses include losses from parasitic resistance in the transistors and finite Q's of the output inductor and capacitor. If the parasitic losses were not present, the dc-to-rf conversion efficiency from the V_{cc} power source to the load resistor R would be 100%.

The voltage and current waveforms in the Class D circuit are easily described. Since the part either switches between V_{cc} or ground, the voltage across the device is a square wave. Since the rf current in the load is the same as the rf current in the transistors, each transistor has a one-half cycle of current through it. This current flows when the part is turned on. If the voltage drop across the device were zero, there would be no power dissipation in the device at that time. The circuit can be implemented on an integrated circuit. The device that is on the ground side will have its parasitic capacity from the body of the device to substrate discharged during each cycle. That is lost energy and therefore lost efficiency. Careful layout of the part will minimize this loss.

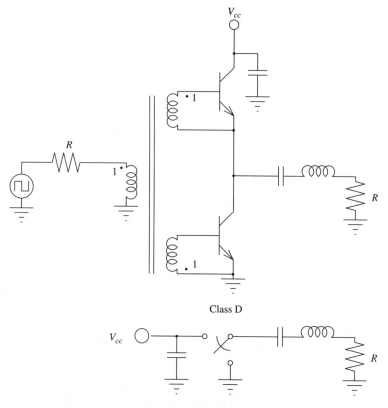

Class D

Figure 7.13 Single-ended Class D amplifier. [*Adapted from* Solid State Radio Engineering,
*H. L. Krauss, C. W. Bostian, and F. H. Raab, Copyright ©1980, John Wiley &
Sons, Inc. Reprinted by permission of John Wiley & Sons, Inc.*]

A transformer version of a voltage mode Class D circuit exists similar to a Class B circuit
except for the load circuit. This circuit is shown in Figure 7.14. The same relationship holds for
the voltage and current through each device. The output transformer applies nominally 2 V_{cc}
to the device that is off since the center tap voltage is V_{cc} and the opposite device is at ground.
The Class D circuits already discussed have square waves of voltage across each device. A
dual circuit would have square waves of current through each device. A Class D circuit that
has square waves of current through each device is shown in Figure 7.15. The voltage across
each device is now one-half of a sine wave since the load current and voltage are sine waves.
When one part is on, the transformer forces the voltage from the center tap to ground to be the
same as that from the other transistor to the center tap. The center tap is now not at a virtual
ground but the voltage of the center tap has two half sine waves on it of the same polarity. The
average value of that waveform is equal to the dc power supply value. Therefore, the peak
value of the voltage is just over one-and-one-half times the power supply voltage. The input
inductor forces a dc current only into the network so each transistor conducts that value of
current for each half cycle. The theoretical efficiency of Class D transformer circuits is 100%.
However, losses in the core of the transformers, the parasitic capacity of the transformers, and
the finite Q of the output resonators will limit the overall efficiency obtained from the Class
D circuit. The bandwidth of the circuit is limited. In order to have high efficiency, the Q of
the output circuit needs to be high. If the Q is high, the bandwidth will be low.

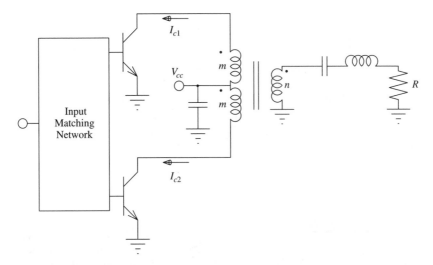

Figure 7.14 Push-pull Class D voltage mode amplifier. [*Adapted from* Solid State Radio Engineering, *H. L. Krauss, C. W. Bostian, and F. H. Raab, Copyright ©1980, John Wiley & Sons, Inc. Reprinted by permission of John Wiley & Sons, Inc.*]

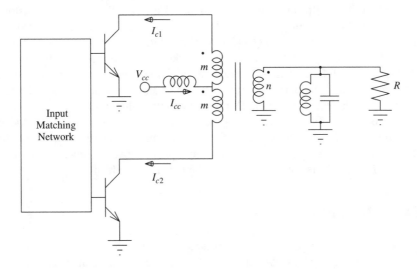

Figure 7.15 Push-pull Class D current mode amplifier. [*Adapted from* Solid State Radio Engineering, *H. L. Krauss, C. W. Bostian, and F. H. Raab, Copyright ©1980, John Wiley & Sons, Inc. Reprinted by permission of John Wiley & Sons, Inc.*].

7.8 CLASS E CIRCUITS

Figure 7.16 shows a typical Class E circuit [3,4]. There are dual versions of this circuit as well. The capacitor is added in parallel with the active device's capacity. This capacity is used along with the power supply bias inductor and the series R-L-C circuit on the output of the device to allow the voltage across the device to go to zero while the part is turned off. Preferably it will reach zero just before the part is turned on by the input rf drive. The energy that exists in the device capacity would be lost if the device turned on before the device voltage went to zero.

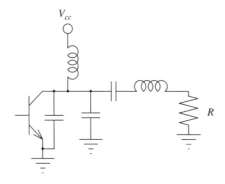

Class E

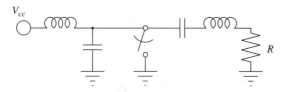

Figure 7.16 Typical Class E amplifier. [*Adapted from* Solid State Radio Engineering, *H. L. Krauss, C. W. Bostian, and F. H. Raab, Copyright ©1980, John Wiley & Sons, Inc. Reprinted by permission of John Wiley & Sons, Inc.*]

The current in the circuit external to the device preferably also goes to zero just before the part is turned off. This helps limit charge injection and storage effects in the transistor. The active device in a Class E circuit acts as a switch. The efficiency of the circuit approaches 100% if the components have very high Q and the device's parasitic resistances are minimized. During each half-cycle of operation, the circuit acts like a damped resonant circuit. The damping coefficient in each half cycle is different because when the transistor is on, the bias inductor is not connected to the output circuit. When the transistor is on, the output circuit acts like a damped single resonant circuit. During the on cycle, the bias inductor is charging up with energy from the bias supply. This energy is used when the transistor is off to produce rf energy in the output. When the transistor is off, the circuit responds like two coupled filter sections. The bias inductor and parallel capacity form a parallel L-C resonant circuit and the output series R-L-C circuit is coupled to it when the transistor is off.

7.9 CLASS F CIRCUITS

One goal of a power amplifier circuit is to attempt to have either a square wave of voltage across the device or a square wave of current through the device. This allows the power dissipated in the device to approach zero if the voltage is high when the current is zero or the current is high when the voltage is zero. Square waves require high harmonic content. High harmonic content requires that the active device respond to voltages and currents at frequencies much higher than the fundamental output frequency. A high harmonic content also requires that the circuit be able to support very high frequencies.

One method to shape the device current and voltage waveforms while limiting the number of harmonics in the circuit is to put resonant circuits in the output matching network at a few harmonics. These resonant circuits will shape the current and voltage waveforms and allow them to become more "square." Figure 7.17 shows a modification of Figure 7.7. A third harmonic parallel resonant circuit is put in series with the output circuit. This gives some flattening to the device voltage waveform if the device is driven into saturation. Driving the device a little into saturation forces the phase of the third harmonic with respect to the phase of the fundamental. If the part is biased similar to the Class B stage, this phase relationship is established.

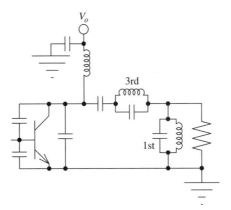

Figure 7.17 Class F—third harmonic peaked amplifier circuit. [*Adapted from* Solid State Radio Engineering, *H. L. Krauss, C. W. Bostian, and F. H. Raab, Copyright ©1980, John Wiley & Sons, Inc. Reprinted by permission of John Wiley & Sons, Inc.*]

Figure 7.18 shows another modification of Figure 7.7. In this circuit, a quarter-wavelength line is put between the output and the device. At the fundamental, the quarter-wavelength line is just an impedance transformer. At other frequencies, the output resonator acts as a short or open circuit. The even harmonics at the input of the quarter-wavelength line are shorted out since the short is transformed through a multiple of two pi around the reflection coefficient plane. This transforms a short into a short. However, the odd harmonics voltage waveforms see an open circuit looking into the quarter-wavelength line. This allows a square wave of voltage to form across the device. Recall that a square wave has only odd harmonics. The phase of the odd harmonics needs to be controlled but if the part is driven somewhat into saturation, that phase relationship is established. The current through the device has fundamental and only even harmonic components since the quarter-wave line plus output resonator in conjunction with a high-impedance bias inductor places a high impedance across the transistor. Since the voltage has only fundamental and odd harmonics and the current has fundamental and even harmonic content, no power is generated at any harmonic in the ideal case.

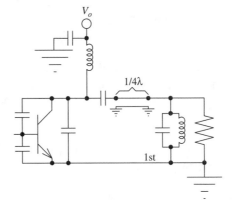

Figure 7.18 Class F—quarter-wave resonator matched amplifier circuit. [*Adapted from* Solid State Radio Engineering, *H. L. Krauss, C. W. Bostian, and F. H. Raab, Copyright ©1980, John Wiley & Sons, Inc. Reprinted by permission of John Wiley & Sons, Inc.*]

7.10 CLASS S OPERATION

Figure 7.19 shows still another type of switching amplifier circuit. However, in this circuit the device is not switched at the rf output frequency but at many times the rf output frequency. The part can be pulse-width modulated or the part can be turned on at various times with a constant width pulse. The short-term duty factor of the input voltage into the low-pass filter roughly determines the amplitude of the output of the low-pass filter. This circuit can be used to generate amplitude-modulated signals from a constant-amplitude power supply. The

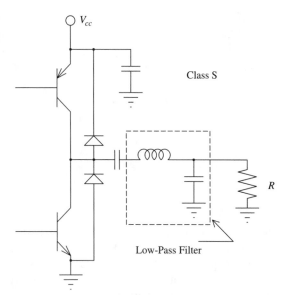

Class S

Low-Pass Filter

R

Figure 7.19 Class S—switching amplifier circuit. [*Adapted from* Solid State Radio Engineering, *H. L. Krauss, C. W. Bostian, and F. H. Raab, Copyright ©1980, John Wiley & Sons, Inc. Reprinted by permission of John Wiley & Sons, Inc.*]

efficiency of the circuit also approaches 100% since the losses in these circuits are similar to the losses in the other switching amplifier circuits. The loss depends on component Q, parasitic resistances, and discharge of parasitic capacity in the active devices and circuit.

7.11 GENERAL COMMENTS ABOUT CLASSES OF OPERATION

One should be able to recognize the various classes of circuits given a few details of their operation. Many variations of the circuits exist in the industry. Switching amplifier circuits are used at lower frequencies (usually below UHF); however, with the advances in active device technology such as PHEMTs and HBTs (high electron mobility transistors and heterojunction bipolar transistors), it is now possible to fabricate switching amplifiers well up into the GHz regime.

Class G and Class H circuits exist and are sometimes found in audio amplifier circuits. Class G and Class H circuits often use multiple devices in unique arrangements.

A more detailed analysis of the various modes of amplification can be found in the literature [1,2].

7.12 LOAD MISMATCH EFFECTS ON POWER AMPLIFIER CIRCUITS

Load mismatch affects the amount of power delivered from a power amplifier to a much greater extent than the effect on a linear source. Load mismatch means deviation from the optimum load value on the amplifier and not minimum VSWR of the load on the amplifier. Recall that a power amplifier is loaded with a resistance that is usually quite different from that required for a low load VSWR when compared with the output conductance of the power amplifier. A power amplifier that uses a bipolar or FET part operates only in a finite area of the V-I plane— only positive currents and voltages for *npn* or *n*-channel devices or only negative currents and voltages for a *pnp* or *p*-channel device. A linear source can operate with current and voltage swings in both the positive and negative areas of the V-I plane. Figure 7.20 shows four different load lines for a device. Load line "A" is the load line for conjugate match, load line "B" would be the load line for maximum power while remaining linear, load line "C" is for

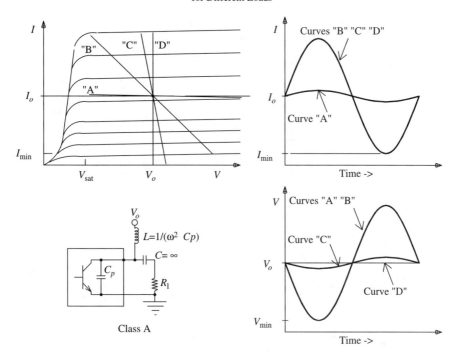

Figure 7.20 Amplifier bias conditions for different loads. [*R. Weber, Microwave Engineering Course Notes,* ©1987–1997, August 1997 edition, p. 76.]

a low impedance, and load line "D" is for a short circuit. An open-circuit line (horizontal line) could have been shown also. The maximum Class A power output for a load corresponding to curve "B" is:

$$P_{max} = \frac{(V_o - V_{sat})(I_o - I_{min})}{2} = \frac{(V_o - V_{sat})^2}{2R_{opt}} = \frac{(I_o - I_{min})^2}{2} R_{opt}$$

where R_{opt} is the load resistance for optimum power generation. Notice that the device capacitance is in parallel resonance with the shunt inductor so that the load current and transistor current are equal and opposite when the inductor and capacitor are in parallel resonance. When the load resistance R_l is smaller than R_{opt}, then the maximum Class A power is limited by current swing. Then the maximum Class A power generated is:

$$P_{max} = \frac{(I_o - I_{min})^2}{2} R_l$$

When the load resistance R_l is larger than R_{opt}, then the maximum Class A power is limited by voltage swing. Then the maximum Class A power generated is:

$$P_{max} = \frac{(V_o - V_{sat})^2}{2R_{opt}}$$

The maximum efficiency for Class A operation is then:

$$\eta_{opt} = \frac{P_{ac}}{P_{dc}} = \frac{(V_o - V_{sat})(I_o - I_{min})}{2V_o I_o} = \frac{(I_o - I_{min})^2 R_{opt}}{2V_o I_o} = \frac{(V_o - V_{sat})^2}{2V_o I_o R_{opt}}$$

These equations assume that the part is correctly biased for Class A operation. When the part is incorrectly biased as shown in Figure 7.21, the efficiency is reduced. If the part is

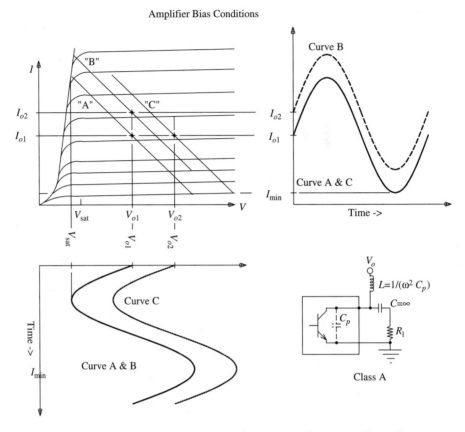

Figure 7.21 Amplifier bias conditions. [*R. Weber, Microwave Engineering Course Notes,* ©*1987–1997, August 1997 edition, p. 76.*]

biased with too much current, then the new current is:

$$\hat{I}_o = I_o + I_{ex}$$

giving an efficiency of:

$$\eta = \eta_{opt} \frac{I_o}{I_o + I_{ex}}$$

Likewise, if the voltage is too high, the efficiency is:

$$\eta = \eta_{opt} \frac{V_o}{V_o + V_{ex}}$$

where V_{ex} is the excess bias voltage used.

The above analysis assumed that the load mismatches were purely resistive. When reactive loads are put on the transistor, then the power reduction is even greater. The power reduction will be reduced since the effective voltage swing across the output resistance is limited by the elliptical locus that the load line follows for a reactive load. Figure 7.22 portrays the locus for purely reactive loads and for partially reactive loads. Keep in mind that the power conversion analyses were done with the output capacitance resonated by a shunt inductor. When the load is partially reactive, the output capacitance has to be de-embedded from the device to see what the true V-I locus is on the active portion of the device. These effects are shown in Figure 7.22. De-embedding techniques were discussed in Section 6.5.

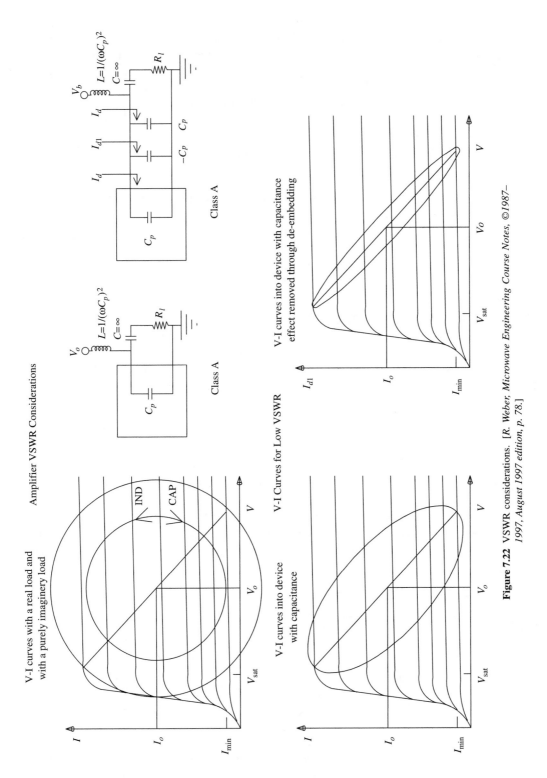

Figure 7.22 VSWR considerations. [*R. Weber, Microwave Engineering Course Notes, ©1987–1997, August 1997 edition, p. 78.*]

Notice that as the reactive portion of the load impedance rises, in effect the voltage swing across the resistive load is reduced. Whenever a power part is mismatched, the power lost is much greater than would be expected from a mismatch on a linear source. This is due to the limitation placed on the voltage and/or current swing having to remain in the linear region of operation on the part. The effect of resistive mismatch on a power device versus the effect on a linear source can be quantified by noting that the power output from a power part varies either directly or inversely with the value of the load resistance. Let the load resistance be normalized to the optimum load resistance as follows.

$$r_l = \frac{R_l}{R_{\text{opt}}}$$

When R_l is less than R_{opt}, r_l is less than 1 and vice versa. When the ratio is less than 1, the power output is:

$$P_{\max} = \frac{(I_o - I_{\min})^2}{2} R_{\text{opt}} \qquad P_{\text{out}} = \frac{(I_o - I_{\min})^2}{2} R_l$$

$$\frac{P_{\text{out}}}{P_{\max}} = \frac{R_l}{R_{\text{opt}}} = r_l$$

and when the ratio is greater than 1, the power output is:

$$P_{\max} = \frac{(V_o - V_{\text{sat}})^2}{2 R_{\text{opt}}} \qquad P_{\text{out}} = \frac{(V_o - V_{\text{sat}})^2}{2 R_l}$$

$$\frac{P_{\text{out}}}{P_{\max}} = \frac{R_{\text{opt}}}{R_l} = \frac{1}{r_l}$$

Power from a linear source under mismatch is:

$$\frac{P_{\text{out}}}{P_{\text{avail}}} = \frac{4 r_l}{(1 + r_l)^2} = \frac{\dfrac{4}{r_l}}{\left(1 + \dfrac{1}{r_l}\right)^2}$$

The results of these equations are plotted in dB loss in Figure 7.23. Notice that the amount of maximum generated power decreases much faster for a power device than for a

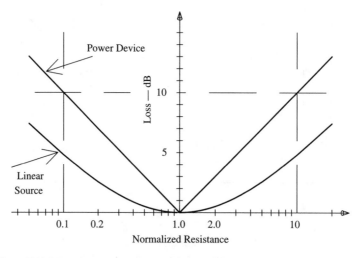

Figure 7.23 RF power converted from dc vs. linear power under VSWR. [*R. Weber, Microwave Engineering Course Notes, ©1987–1997, August 1997 edition, p. 79.*]

linear source. This is extremely important in system design because the effect of a $2:1$ VSWR on a power amplifier is a 3-dB loss but it is only 0.5 dB for a linear source.

7.13 FINDING THE OPTIMUM REACTANCE

The previous section indicates that the control of VSWR is extremely important when regionally linear power devices are used. It is necessary to determine the optimum parallel inductance on the output of a power amplifier device in order to ensure that the load line on the active portion of the device is purely resistive to minimize the load line elliptical swing. However, the output capacity of the device depends on the input source and matching the input immittance to the source depends on the output load. The value of the load line resistance is known from the V-I curves. The conductance that should be placed on the part is the reciprocal of the load line resistance. What is the value of the parallel reactance? The solution to this was given in Section 6.2.

7.14 COLLECTOR TO BASE HARMONIC GENERATION IN POWER BIPOLAR DEVICES

Load impedance variations affect the collector delay and output capacity variation for a bipolar transistor. This section relies somewhat on basic device physics material; however, this section is not necessary for understanding the remainder of the text. It is included to guide those who wish to provide minimal harmonic generation from the varactor action of the collector to base depletion region movement.

   This section discusses some general considerations for time delay in the collector-base drift region of a bipolar transistor. This is in relation to rf power generation in a bipolar transistor. As the collector voltage varies over a period of the rf cycle, one would expect that a variation of collector to base capacity would generate harmonics. This section shows that if the rf load line on the part is the load line for optimum power then the effective capacity does not change as the collector to base voltage changes over a sine wave cycle. The time delay through the collector to base region, and thus phase shift due to the collector to base junction, does not change, giving minimum am-to-pm conversion in the collector to base region. The analysis considers the drift region from a quasi-static charge control approach and further assumes that carriers drift across the collector to base region at their saturated velocity. It shows that the delay due to the collector depletion region varies from $\tau_{tr}/2$ to infinity where τ_{tr} is equal to the total drift time for a carrier across the width of the collector depletion region. The time delay, usually considered to be $\tau_{tr}/2$, is shown to apply only for the case for which no ac power can be delivered by the part.

   For a fixed voltage across a bipolar part, Early [5] gives the collector depletion region time delay as:

$$\tau_c = \frac{L}{2v_{\text{sat}}}$$

where L is the collector depletion width and v_{sat} is the saturated electron drift velocity in the depletion region. This is the time delay under the condition of a constant collector voltage drop across the collector-base depletion region. A constant voltage is used for that calculation because the common emitter current gain h_{21} is used to define f_t where f_t is defined as the frequency for which the current gain drops to 1. The small signal, common emitter, current gain, h_{21} is defined as:

$$h_{21} = \frac{I_2}{I_1}\bigg|_{V_2=0}$$

The condition $V_2 = 0$ means that the ac voltage from the input to the output is held at zero. When there is no ac voltage, there is no power generation. How does the time delay, and thus the phase, vary as the voltage is allowed to change? Does the device stay linear? The total delay time τ_{ec} from the emitter to the collector is related to f_t by:

$$f_t = \frac{1}{2\pi\tau_{ec}}$$

where τ_{ec} is the sum of all the time delays of the transistor.

$$\tau_{ec} = \sum_i \tau_i = \tau_{rest} + \tau_c$$

This analysis considers only the time delay, τ_c, associated with the collector drift region.

A depletion region as shown in Figure 7.24 is considered. Uniform doping across the depletion region is considered although many of the results are independent of the shape of the collector diffusion profile. The depletion widths and the collector region delay times are derived for a uniform current across the depletion region as shown in Figure 7.25. This condition arises from using a saturated velocity for the carriers. In this analysis an npn transistor is assumed where N_a is the acceptor charge density in the base region, N_d is the donor charge density in the collector region, x_b is the depletion width in the base, x_c is the depletion width in the collector, q is the electronic charge value, and ε is the permittivity of the material. The peak electric field is:

$$E_{pk} = \frac{q}{\varepsilon}(N_a + n)|x_b| = \frac{q}{\varepsilon}(N_d - n)x_c$$

The voltage across the depletion region is then:

$$\phi = \frac{q}{\varepsilon}\frac{(N_a + n)x_b^2}{2} + \frac{q}{\varepsilon}\frac{(N_d - n)x_c^2}{2}$$

$$= V_{ext} + \phi = V_{ext} + V_{built\text{-}in\ potential}$$

Substituting the electric field equation into the potential equation ϕ becomes:

$$\phi = \frac{q}{\varepsilon}\frac{(N_d + N_a)}{2}\frac{(N_d - n)}{(N_a + n)}x_c^2$$

The device is considered to be operating under Class A amplifier conditions as shown schematically in Figure 7.26. Three device load lines will be considered as shown in Figure 7.27. The time waveform for these three load lines is shown in Figure 7.28. The load line conditions for curve B of Figure 7.27 are given by:

$$R_L = \frac{(V_o - V_{sat})}{(I_o - I_{min})}$$

$$\eta = \frac{(V_o - V_{sat})(I_o - I_{min})}{2V_oI_o} < 50\%$$

From the electric field equation, the change of the base depletion width and collector depletion width are related as:

$$x_c(N_d - n) = -x_b(N_a + n)$$

$$\frac{dx_c}{dn} = \frac{1}{(N_d - n)}\left[x_c - x_b - (N_a + n)\frac{dx_b}{dn}\right]$$

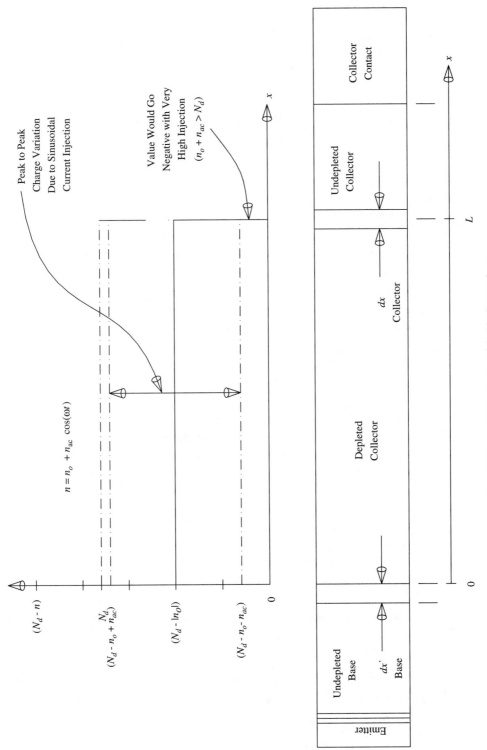

Peak to Peak Charge Variation Due to Sinusoidal Current Injection

Value Would Go Negative with Very High Injection $(n_o + n_{ac} > N_d)$

$n = n_o + n_{ac} \cos(\omega t)$

$(N_d - n)$

$(N_d - n_o + n_{ac})$ N_d

$(N_d - |n_o|)$

$(N_d - n_o - n_{ac})$

Undepleted Base dx' Base

Depleted Collector dx Collector

Undepleted Collector   Collector Contact

Emitter

L

Figure 7.24 Assumed structure of the bipolar transistor.

181

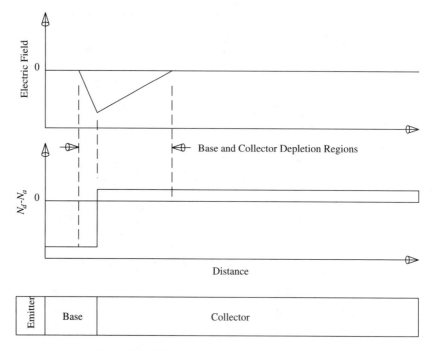

Figure 7.25 Diffusion profile and electric field intensity.

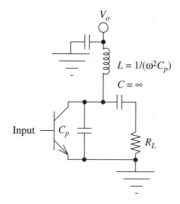

Figure 7.26 Amplifier used for optimum load line analysis.

Differentiating the potential equation for x_c versus voltage with respect to carrier density gives:

$$\frac{d\phi}{dn} = \frac{dV_{\text{ext}}}{dn} = \frac{q}{\varepsilon}(N_d + N_a)\left(\frac{N_d - n}{N_a + n}\right)x_c\frac{dx_c}{dn} - \frac{q}{2\varepsilon}\left(\frac{N_d + N_a}{N_a + n}\right)^2 x_c^2$$

For the case of the power match load line shown in Figure 7.27:

$$V_{\text{ext}} = V_c = V_o - (I_c - I_o)R_L$$

$$\frac{dV_{\text{ext}}}{dn} = -R_L\frac{dI_c}{dn} = -R_L\frac{d(qnv_{\text{sat}})}{dn} = -R_Lqv_{\text{sat}}$$

If the condition $dx_c/dn = 0$ (no depletion region movement and thus the depletion layer capacity remaining constant) is imposed on $d\phi/dn$, comparing this to the previous equation

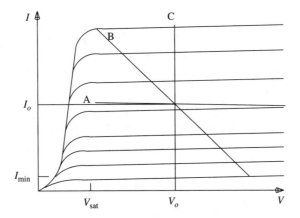

Figure 7.27 Resistive load line analysis for a bipolar.

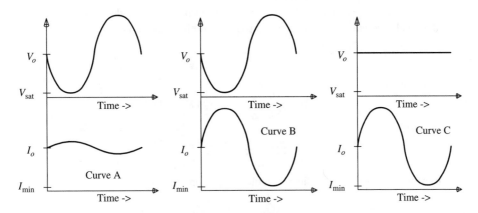

Figure 7.28 Voltage and current waveforms for the bipolar part.

gives the result:

$$R_L = \left(\frac{N_d + N_a}{N_a + n} \right)^2 \frac{x_c^2}{2\varepsilon v_{\text{sat}}} = R_{\text{opt}}$$

$$\approx \frac{x_c^2}{2\varepsilon v_{\text{sat}}} \bigg|_{N_a \gg N_d; \ N_a \gg n}$$

Let the general load resistance be equal to K_v times R_{opt} or:

$$R = K_v R_{\text{opt}}$$

$$= \frac{K_v x_c^2}{2\varepsilon v_{\text{sat}}}$$

Equating the potential derived from $d\phi/dn$ and from the circuit and assuming $N_a \gg Nd$ and $N_a \gg n$:

$$-q v_{\text{sat}} K_v R_{\text{opt}} = \frac{q}{\varepsilon} (N_d + N_a) \left(\frac{N_d - n}{N_a + n} \right) x_c \frac{dx_c}{dn} - \frac{q}{2\varepsilon} \left(\frac{N_d + N_a}{N_a + n} \right)^2 x_c^2$$

$$\approx \frac{q}{\varepsilon} (N_d - n) x_c \frac{dx_c}{dn} - q v_{\text{sat}} R_{\text{opt}}$$

or:

$$\frac{dx_c}{dn} \approx \frac{x_c}{2(N_d - n)}(1 - K_v)$$

Using this charge relationship across the base collector junction and assuming $N_n \gg n$ and $x_c \gg x_b$:

$$-\frac{N_a}{(N_d - n)}\frac{dx_b}{dn} + \frac{x_c}{(N_d - n)} \approx \frac{x_c}{2(N_d - n)}(1 - K_v)$$

$$\frac{dx_b}{dn} \approx \frac{x_c}{N_a}\frac{(1 + K_v)}{2}$$

Since charge divided by current is time delay:

$$\tau = \frac{\Delta Q_b}{\Delta I_c} = \frac{q N_a}{q v_{sat}}\frac{\Delta x_b}{\Delta n} \approx \frac{N_a}{v_{sat}}\frac{dx_b}{dn} = \frac{N_a}{v_{sat}}\frac{x_c}{N_a}\frac{(1 + K_v)}{2} = \frac{x_c}{v_{sat}}\frac{(1 + K_v)}{2} = \tau_{tr}\frac{(1 + K_v)}{2}$$

Considering the collector side of the junction, the total collector charge is the total stored charge variation plus the induced charge variation due to the depletion width change. The induced charge variation due to the depletion width variation is:

$$\Delta Q_{induced} = -q N_d \Delta x_c = \frac{x_c}{2(N_d - n)}(K_v - 1)q N_d \Delta n$$

Assuming $N_d \gg n$:

$$\Delta Q_{induced} = \frac{x_c}{2}\frac{(K_v - 1)}{v_{sat}}\Delta I_c$$

$$\frac{\Delta Q_{induced}}{\Delta I_c} = \frac{x_c}{2}\frac{(K_v - 1)}{v_{sat}}$$

The stored charge in the collector depletion region is:

$$\Delta Q_{stored} = q x_c \Delta n = \frac{x_c}{v_{sat}}\Delta I_c$$

$$\frac{\Delta Q_{stored}}{\Delta I_c} = \frac{x_c}{v_{sat}}$$

Therefore, the time delay expression is:

$$\tau = \frac{\Delta Q_{induced} + \Delta Q_{stored}}{\Delta I_c} = \frac{\Delta Q_{total}}{\Delta I_c} = \left[\frac{x_c(K_v - 1)}{2v_{sat}} + \frac{x_c}{v_{sat}}\right]$$

$$= \frac{x_c}{2v_{sat}}(1 + K_v) = \tau_{tr}\frac{(1 + K_v)}{2}$$

giving the same time delay as derived based on base charge.

From the variation of collector width versus carrier concentration, when the voltage across the depletion region is held constant, the length of the depletion region in the collector will vary as:

$$\Delta x \approx \frac{x_c \Delta n}{2(N_d - n)} \approx \frac{x_c \Delta n}{2N_d}\bigg|_{N_d \gg n; \ \frac{d\phi}{dn}=0}$$

Substituting the potential versus x_c in this equation, if the length of the depletion region is held constant and the voltage is allowed to vary the depletion width is related to the voltage as:

$$\Delta \phi = \Delta V_{ext} = \frac{-q x_c^2}{2\varepsilon}\left(\frac{N_d + N_a}{N_a + n}\right)\Delta n\bigg|_{\frac{dx_c}{dn}=0; \ K_v=1}$$

The external voltage is then linearly related to the carrier density for all levels of n for which $N_a \gg n$. Let the external voltage be related to the load current via the resistive load lines of Figure 7.27. Substituting a dc bias voltage plus an ac voltage into the equation for potential versus collector depletion width with $N_a \gg N_d$ and $N_a \gg n$:

$$V_{\text{ext}} = V_o - V_{\text{ac}} \cos(\omega t + \theta_1)$$

$$\approx -\phi + \frac{q x_c^2 (N_d - n_o)}{2\varepsilon} - \frac{q x_c^2 n_{\text{ac}} \cos(\omega t + \theta_2)}{2\varepsilon}$$

Letting $I_c = q n v_{\text{sat}}$:

$$|V_{\text{ac}}| = \frac{q x_c^2 n_{\text{ac}}}{2\varepsilon} = \frac{q x_c^2 n_{\text{ac}} v_{\text{sat}}}{2\varepsilon v_{\text{sat}}} = |I_{\text{ac}}| \frac{q x_c^2}{2\varepsilon v_{\text{sat}}} = |I_{\text{ac}}| R_{\text{opt}}$$

where $I_{\text{ac}} = $ ac current per unit area and R_{opt} is the normalized optimum load resistance for maximum class A efficiency, which is shown as Curve B of Figure 7.27 and given by the equation for R_{opt}.

The three curves shown in Figure 7.27 give curves for three different bias terminations on the bipolar part. Curve A represents conjugate match, maximum transducer gain, and requires a large K_v since the load is much greater than R_{opt}. Even though the transducer gain is at a maximum, the dc-to-rf conversion efficiency is poor. The large value of K_v results in a large time delay. Note that the slope of curve A is approximately the negative of the slope of the $V I$ curves. Harmonic generation is high since according to the collector depletion width versus K_v relation, the depletion region movement is large when K_v is large. The depletion width continues to decrease as the carriers injected at the base boundary increase. The depletion area on the collector side captures some of the carriers as they come across the depletion area. A significant fraction of them do not make it to the collector contact area, causing a delay in collector current. The opposite happens as the current decreases.

Curve B is for power match (maximum dc-to-ac power conversion) and results in no depletion width movement (and thus low harmonic generation) and no induced charge on the collector contact according to the induced charge equation. There are carriers captured or emitted by the depletion boundary on the collector side. This results in a collector delay time that is equal to the transit delay τ_{tr} across the device collector base depletion region.

Curve C is for an ac short on the output of the device and $K_v = 0$. No real power is transferred for this load line. The delay time is the shortest and equal to $\tau_{tr}/2$. This is due to the largest induced charge in the collector but no power is generated in R_L. The depletion width increases as the carriers injected at the base boundary increase. The depletion area on the collector side gives up the carriers as other carriers come across the depletion area. A significant fraction of carriers are added to the carriers coming across the depletion region, creating a larger collector current. The opposite happens as the current decreases.

Time delay can be plotted versus K_v as shown in Figure 7.29. Note that as $K_v > 1$, the induced charge is negative and the delay time increases significantly. The derivative of the depletion width versus n changes sign at $K_v = 1$. For low impedances, x_c increases as n increases. For high impedances relative to R_{opt}, the depletion region decreases in width as n increases.

These results indicate that minimum harmonic generation should be possible for the optimum power load line. Minimum harmonic generation leads to minimum intermodulation distortion. Maximum efficiency would be produced for high harmonic generation as shown in the discussions on Class B, C, D, E, F, and S. However, when low-harmonic-distortion Class A operation is required and bipolar transistors are used, the suggestion is to try the optimum load line.

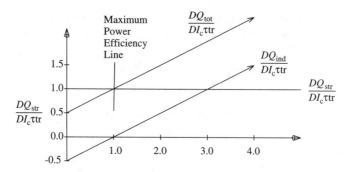

Figure 7.29 Induced and constant charge vs. K.

EXERCISES

7-1 Determine the load line required for a power FET operating in the Class A mode. The part has an operating voltage of 20 volts, a minimum drain voltage of 1 volt at the peak value of current, a minimum drain current of 100 mA, and an operating current of 2 amps.

7-2 Design a transformer to use on the output of a Class B stage. The dc supply voltage is 12.6 volts, and each part is operating at 1 amp. Determine the maximum Class B power available from the part. Assume the minimum voltage on each part is 0.5 volt and the minimum current is negligible. The load resistance is 50 ohms.

7-3 Determine the load resistance value that is required for a Class D voltage mode amplifier that delivers 5 watts and operates from 28 volts dc. Assume a conversion efficiency of 90 percent. Determine the amount of device current that will be drawn when each device is on.

7-4 A Class A part is being operated at 10 volts and 100 mA. The part has a minimum voltage of 2 volts. The minimum device current is 1 mA. The load VSWR can be as high as 1.8:1. Determine how much the power drops for the maximum VSWR if the part is designed for maximum power at 1:1 VSWR. Determine the change in power with a 1.8:1 VSWR if the part is matched with a resistance that is 80% of the optimum resistance value.

7-5 Determine the amount of voltage and current that a Class A part needs to be biased at to support three sinusoidal signals. Each signal is at a different frequency. The part is to stay linear in operation at all times. The dc voltage available is 10 volts. Each signal needs to deliver 0.25 watts. The part has negligible minimum current and a minimum voltage of 1 volt. Determine the efficiency of power generation when only one signal is present. Determine the efficiency of power generation when all three signals are present. Assume that the part is operating in the linear mode.

7-6 Design a load match for a part operating in the Class F mode using a quarter-wave transmission line in the match and a high Q load resonator. The power is delivered to a 50-ohm load. Assume the load and part can support many harmonics. The part is operating at 24 volts and delivers 1 watt. If the power conversion efficiency is 85% (including component losses), determine the approximate device operating current.

8

Resonators and Oscillators

8.1 OSCILLATORS

An oscillator consists of a network that has some amplification resulting in power gain, some circuit for feedback, and some circuit that is resonant at a frequency to limit the bandwidth over which the circuit has gain in order to constrain the oscillator output spectrum to be at a specified frequency. Various analyses that deal with oscillators as if the amplifying device had zero output impedance and infinite input impedance do not consider power transfer. Those analyses only consider satisfying some feedback criteria based on linear variables. There is no constraint on the amount of power available from a source with zero output impedance and there is no power transfer to a load of infinite resistance. Those types of networks do not exist and for rf and microwave applications, the effect of finite source and load resistances must be considered. The resonant frequency circuit can be synthesized in several ways. A network called a *resonator* can be used to provide the frequency control. Three kinds of oscillation circuits are considered in this chapter. The three types are the one-port oscillator, the two-port or three-terminal device oscillator that uses series feedback, and the loop oscillator. Resonator analysis will be considered before oscillator analysis. Resonator analysis will also be used in filter design when filters are considered in Chapter 9. The effect of load VSWR and the impedance loci that present stable loads for an oscillator will be considered at the end of this chapter.

8.2 RESONATORS

In low-frequency circuits, a resonator consists of an inductor and a capacitor connected in series or parallel. The lossless parallel resonator has infinite impedance across it at the resonant frequency. The lossless series resonator has zero impedance across it at the resonant frequency. For our purposes, an ideal parallel resonator still has infinite impedance across it at the resonant frequency and an ideal series resonator still has zero impedance across it at the resonant frequency. However, in most resonators used for storing magnetic energy or electric energy at microwave frequencies there is no single inductor or capacitor that can be identified . A resonator consists of a volume of material capable of storing large amounts of electric and magnetic energy in comparison to the energy dissipated per cycle of that energy. At the resonant frequency, the peak value of the magnetic energy during a cycle of the signal is equal to the peak value of the electric energy during a cycle of the signal. The magnetic and electric energies are summed over the complete volume contained in the resonator. If a port exists for sampling the energy of the resonator, the input immittance into that port is very close to zero at the resonant frequency. From a terminal (circuit) standpoint, a resonator looks like either a series R-L-C circuit or a parallel R-L-C circuit. As expected, the series R-L-C circuit is called a *series resonator* and the shunt R-L-C circuit is called a *shunt resonator*. Since a resonator can store energy for a short period of time, it acts as a reservoir of energy. A resonator is sometimes referred to as a *tank circuit* [1]. The input port of a resonator often has parasitic capacitance, inductance, or resistance associated with it.

8.2.1 Characterizing Resonators

Figure 8.1 shows conceptual resonators that do not have input parasitics. A series resonator that has parasitics is shown in Figure 8.2. Note that there may be low Q parasitic parallel components in parallel with the series resonator. All, part, or none of the parasitics may exist in a specific resonator circuit. The reactance slope of the series portion of the resonator appears as shown in Figure 8.3. For infinite Q of the series part of the resonator, the parasitic elements do not change the series resonance of the resonator, since at series resonant frequency, the shunt components have no voltage across them and thus no energy in them. However, for a finite Q, the resonant frequency is somewhat perturbed. In addition, the parasitic elements introduce a parallel resonance just below (for net inductive parasitics) or just above (for net capacitive parasitics) the series resonant frequency.

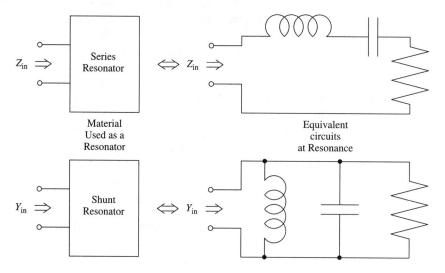

Figure 8.1 Series and shunt resonators. [*R. Weber, Microwave Engineering Course Notes, ©1987–1997, August 1997 edition, p. 88.*]

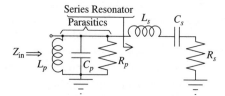

Figure 8.2 Series resonator with parasitics. [*R. Weber, Microwave Engineering Course Notes, ©1987–1997, August 1997 edition, p. 89.*]

Since Q is defined for an isolated single resonator, a single parallel resonance can be defined for a series resonator only if either L_p is negligible or C_p is negligible. If both of

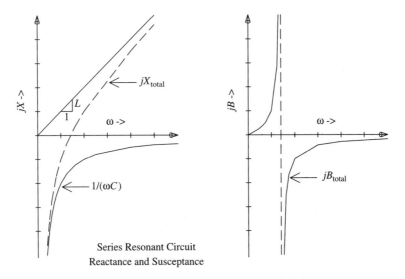

Series Resonant Circuit
Reactance and Susceptance

Figure 8.3 Series resonator reactance and susceptance slopes. [*R. Weber, Microwave Engineering Course Notes, ©1987–1997, August 1997 edition, p. 89.*]

these parasitics are zero there is no parallel resonance. When there is no load on the Z_{in} port, the parasitic elements are in series with the series resonator components. When the impedance of the parallel circuit is added in series with the impedance of the series cicuit, the new net resonant frequency is the new parallel resonant frequency. If both C_p and L_p contain appreciable energy at the resonant frequency, Q must be defined for each portion of the circuit individually and coupled resonator theory needs to be used to describe the circuit's behavior.

A shunt or parallel tank that has parasitics is shown in Figure 8.4. As in the series tank, all, part, or none of the parasitic series components may exist. The susceptance slope of a shunt resonator without parasitics is shown in Figure 8.5. The dual to the discussion of the series resonator applies to the shunt resonator. The parasitic series portion of the circuit can resonate with the off resonance reactance of the shunt resonator to create a series resonance slightly lower or higher than the desired parallel resonance. The parasitic resonances can create difficulties when applying resonators in specific applications.

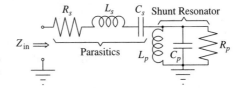

Figure 8.4 Shunt resonator with parasitics. [*R. Weber, Microwave Engineering Course Notes, ©1987–1997, August 1997 edition, p. 90.*]

A crystal equivalent circuit, given in Chapter 11, has a series resonant circuit with a parasitic parallel capacitor. The YIG resonator's equivalent circuit, given in Chapter 11, has a parallel resonant circuit with a parasitic series inductance. These parasitic components modify the circuits such that there are additional resonances at the input ports of the devices. For instance, in a crystal, series resonance occurs whenever the series tank is resonant. However, there is a parallel resonance (high impedance) at the input port whenever its parallel capacitance resonates with the net inductive reactance of the series circuit. This occurs whenever the series inductance of the crystal resonates with the series combination of the two capacitors in the crystal circuit. The parallel resonant frequency is higher than the series resonant frequency. Two resonances also occur in the YIG circuit. In the YIG circuit, the parallel resonant frequency is lower than the series resonant frequency.

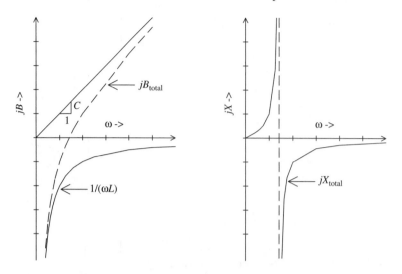

Figure 8.5 Shunt resonator reactance and susceptance slopes. [*R. Weber, Microwave Engineering Course Notes, ©1987–1997, August 1997 edition, p. 91.*]

Q for a series resonant circuit is given by:

$$Q_s = \frac{\omega L_s}{R_s} = \frac{1}{\omega C_s R_s}$$

Q for a parallel resonant circuit is given by:

$$Q_p = \frac{\omega C_p}{G_p} = \frac{1}{\omega L_p G_p}$$

$$= \omega C_p R_p = \frac{R_p}{\omega L_p}$$

Q can be measured by using reactance or susceptance slope parameters. For a series circuit, the reactance slope parameter x is defined as [2]:

$$x = \frac{\omega_0}{2} \frac{dX_s(\omega)}{d\omega}\bigg|_{\omega=\omega_0}$$

where ω_o is the frequency at which the reactance is zero. This procedure works for resonators with parasitics also. In fact, it includes the effects of the parasitics around the resonant frequency and allows one to determine an equivalent resonant circuit for the resonator. A general resonator circuit might have a complex impedance expression but the reactance slope method yields the correct equivalent inductance and capacitance to generate the same change in reactance versus frequency curve at resonance that the original circuit does. The equivalent inductor and capacitor are valid only at and near resonance. Consider a series LC circuit. The reactance slope parameter x for a series LC gives the equivalent series inductance as L as expected.

$$X_s = \frac{\omega^2 LC - 1}{\omega C} \qquad \omega_0^2 = \frac{1}{LC}$$

$$x = \omega_0 L_{eq-ser} = \frac{\omega_0}{2} \frac{dX_s(\omega)}{d\omega}\bigg|_{\omega=\omega_0} = \frac{\omega_0}{2} \left[L \frac{\omega^2 LC + 1}{\omega^2 LC} \right]\bigg|_{\omega=\omega_0} = \omega_0 L$$

A dual expression exists for the susceptance slope parameter b for parallel resonant circuits.

$$b = \frac{\omega_0}{2} \left. \frac{d B_p(\omega)}{d\omega} \right|_{\omega=\omega_0}$$

The susceptance slope parameter b for a parallel LC gives the equivalent parallel capacitance as C as expected.

$$B_p = \frac{\omega^2 LC - 1}{\omega L} \qquad \omega_0^2 = \frac{1}{LC}$$

$$b = \omega_0 C_{\text{eq-par}} = \frac{\omega_0}{2} \left. \frac{d B_p(\omega)}{d\omega} \right|_{\omega=\omega_0} = \frac{\omega_0}{2} \left[C \frac{\omega^2 LC + 1}{\omega^2 LC} \right]\Bigg|_{\omega=\omega_0} = \omega_0 C$$

At resonance, $d R_s/d\omega = 0$ for a series resonant circuit and $d G_p/d\omega = 0$ for a parallel resonant circuit. The Q of a series resonator is determined by dividing x by the series resistance at resonance. The Q of a parallel resonator is determined by dividing b by the parallel conductance (multiplying b by the parallel resistance) at resonance.

$$Q_s = \frac{x}{R_s(\omega_s)} = \frac{\omega_0 L_{\text{eq-ser}}}{R_s(\omega_0)}$$

$$Q_p = \frac{b}{G_p(\omega_0)} = \frac{\omega_0 C_{\text{eq-par}}}{G_p(\omega_0)} = \omega_0 C_{\text{eq-par}} R_p(\omega_0)$$

8.2.2 Reactance Slopes and Q of Immittances and Reflection Coefficients

Q or quality factor has just been related to the reactance and susceptance slope parameters. The reactance and susceptance slope parameters can be related to the phase slope of immittance at resonance. The phase slope of the reflection coefficient can be related to Q at resonance but care must be exercised when doing this or invalid Q values will result. It might appear that Q approaches infinity when in reality the true value of Q is quite small. In order to relate the reactance slope to the phase slope of S_{11}, consider the curve shown in Figure 8.6. The curves are shown for a series R-L-C but the dual equivalent circuit will yield the same results as a shunt R-L-C except that the reflection coefficient chart locus will be the mirror image on the left side of the chart. The reflection coefficient chart locus shown in Figure 8.6 is defined by:

$$\Gamma = \Gamma_c + \Gamma_r e^{j\Psi} = \Gamma_o e^{j\theta}$$

$$\left. \frac{d\Gamma}{d\omega} \right|_{\omega=\omega_o} = \left. j\Gamma_r e^{j\Psi} \frac{d\Psi}{d\omega} \right|_{\omega=\omega_o} = \left. \left(\frac{d\Gamma_o}{d\omega} e^{j\theta} + j\Gamma_o e^{j\theta} \frac{d\theta}{d\omega} \right) \right|_{\omega=\omega_o}$$

$$\left. \frac{d\Gamma}{d\omega} \right|_{\omega=\omega_o} = \left. j\Gamma_r \frac{d\Psi}{d\omega} \right|_{\omega=\omega_o} = \left. \left(\frac{d\Gamma_o}{d\omega} + j\Gamma_o \frac{d\theta}{d\omega} \right) \right|_{\omega=\omega_o}$$

$$\left(\left. \frac{d\Gamma_o}{d\omega} \right|_{\omega=\omega_o} = 0 \ \text{ if } \ \Gamma_o|_{\omega=\omega_0} \neq 0 \right)$$

Using the impedance locus shown in the right side of Figure 8.6, the quality factor of the

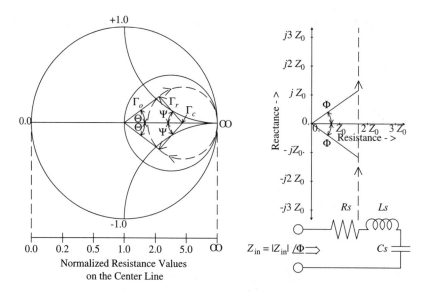

Figure 8.6 Reflection coefficient plane and impedance plane for a series resonator. [*R. Weber, Microwave Engineering Course Notes, ©1987–1997, August 1997 edition, p. 92.*]

resonator can be found. The Q for the series R-L-C is:

$$Q = \frac{\omega L_s}{R_s} = \frac{\omega_0}{2R_s}\frac{dX_s}{d\omega}\bigg|_{\omega=\omega_0} = \frac{f_0}{2R_s}\frac{dX_s}{df}\bigg|_{f=f_0} = \frac{\omega_0}{2}\frac{d(X_s/R_s)}{d\omega}\bigg|_{\omega=\omega_0}$$

$$= \frac{\omega_0}{2}\frac{d\tan(\phi)}{d\omega}\bigg|_{\omega=\omega_0} = \frac{\omega_0}{2\cos^2(\phi)}\frac{d\phi}{d\omega}\bigg|_{\omega=\omega_0} = \frac{\omega_0}{2}\frac{d\phi}{d\omega}\bigg|_{\omega=\omega_0} \approx \frac{f_0}{2\Delta f}\frac{2\pi}{360}\Delta\phi\bigg|_{\substack{\omega=\omega_0 \\ \phi=0}}$$

The relationship between the input reflection coefficient and impedance allows us to establish a relationship between the rate of change of the angles ψ, ϕ, and θ.

$$\frac{d\Gamma}{d\omega}\bigg|_{\omega=\omega_0} = \frac{(Z+Z_0)\dfrac{dZ}{d\omega} - (Z-Z_0)\dfrac{dZ}{d\omega}}{(Z+Z_0)^2}\bigg|_{\omega=\omega_0} = \frac{2Z_0\dfrac{dZ}{d\omega}}{(Z+Z_0)^2}\bigg|_{\omega=\omega_0} = \frac{j2Z_0\dfrac{dX_s}{d\omega}}{(R_s+Z_0)^2}\bigg|_{\omega=\omega_0}$$

$$= \frac{j4Z_0R_s}{\omega_0(R_s+Z_0)^2}\frac{\omega_0}{2R_s}\frac{dX_s}{d\omega}\bigg|_{\omega=\omega_0} = \frac{j4Z_0R_s}{\omega_0(R_s+Z_0)^2}Q$$

The fact that the resistance slope does not change at resonance (i.e., that $dZ/d\omega = jdX_s/d\omega$ or $dR_s/d\omega = 0$ at resonance) is used in the above derivation. Equating the two expressions for the derivatives of $d\Gamma/d\omega$ evaluated at $\omega = \omega_0$ gives the following result.

$$Q = \frac{\omega_0\Gamma_0(Z_0+R_s)^2}{4Z_0R_s}\frac{d\theta}{d\omega}\bigg|_{\omega=\omega_0} = \frac{\omega_0\Gamma_0(Z_0+R_s)^2}{4Z_0R_s}\tau$$

$$\approx \frac{|Z_0-R_s|(Z_0+R_s)}{4Z_0R_s}\frac{f_0}{\Delta f}\frac{2\pi}{360}\Delta\theta° \approx \frac{|Z_0^2-R_s^2|}{Z_0R_s}\frac{f_0}{\Delta f}\frac{\pi}{720}\Delta\theta° \approx \frac{|Z_0^2-R_s^2|}{Z_0R_s}\frac{\pi}{2}f_0\tau$$

$$= \frac{\omega_0}{2}\frac{d\phi}{d\omega}\bigg|_{\omega=\omega_0} \approx \frac{\pi}{360}f_0\frac{\Delta\phi°}{\Delta f}$$

The differential time delay, τ, is related to the phase slope of the input reflection coefficient.

In order to relate θ and ψ, there are two cases to be considered. If the locus is entirely to the right side of the reflection coefficient chart origin, then the angle θ is less than 90 degrees. If the locus crosses the real axis to the left of the origin, then the angle θ is greater than 90 degrees. If Γ_r is less than $\frac{1}{2}$, then the locus crosses to the right of the origin. If Γ_r is greater than $\frac{1}{2}$, then the locus crosses to the left of the origin. In the following equation, the top sign of a plus-minus sign signifies that the locus crosses to the right of the reflection coefficient chart origin and the bottom sign signifies that the locus crosses to the left side of the reflection coefficient chart origin. Then:

$$\text{since } \Gamma_r = \frac{1 \mp \Gamma_o}{2}$$

$$\frac{d\theta}{d\omega} = \frac{1 \mp \Gamma_o}{2\Gamma_o} \frac{d\Psi}{d\omega}$$

An alternate way of measuring Q using reflection coefficient magnitudes will now be given [3]. The left side of Figure 8.7 shows a series resonant circuit locus plotted on an admittance reflection coefficient chart with parts of two other circles. These circles are the circles for which $R = X$ and $R = -X$. The loci for the circles for $R = |X|$ are given by:

$$z + z^* = 2r = \pm j(z - z^*) = 2|x|$$

$$\frac{1 + \Gamma}{1 - \Gamma} + \frac{1 + \Gamma^*}{1 - \Gamma^*} = \pm j\left(\frac{1 + \Gamma}{1 - \Gamma} - \frac{1 + \Gamma^*}{1 - \Gamma^*}\right)$$

$$1 - \Gamma\Gamma^* = \pm j(\Gamma - \Gamma^*)$$

$$\Gamma\Gamma^* \pm j\Gamma \mp j\Gamma^* + 1 = 2$$

$$\Gamma_{\text{cent}} = \pm j$$

$$\rho = \sqrt{2}$$

The top curve can be described by:

$$\Gamma = -j + \sqrt{2}e^{j\phi}$$

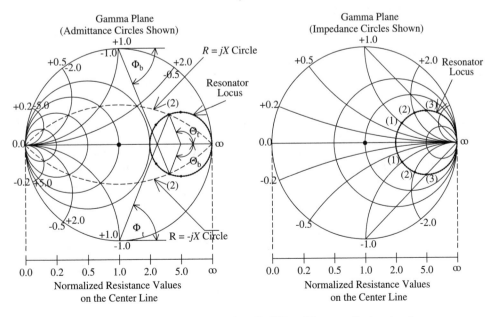

Figure 8.7 Q reflection coefficient circles. [*R. Weber, Microwave Engineering Course Notes, ©1987–1997, August 1997 edition, p. 95.*]

the bottom curve by:

$$\Gamma = +j + \sqrt{2}e^{j\phi}$$

and the resonator curve by:

$$\Gamma_{\text{res}} = \Gamma_c + \rho e^{j\theta}$$

The intersection of the resonator curve with the $R = |X|$ curves is given by:

$$1 - \rho + \rho e^{+j\theta_t} = -j + \sqrt{2}e^{+j\phi_t}$$
$$1 - \rho + \rho e^{-j\theta_t} = +j + \sqrt{2}e^{-j\phi_t}$$
$$1 + j - \rho + \rho e^{+j\theta_t} = \sqrt{2}e^{+j\phi_t}$$
$$1 - j - \rho + \rho e^{-j\theta_t} = \sqrt{2}e^{-j\phi_t}$$

Multiplying both sides of these equations together gives:

$$(1 - \rho + j)(1 - \rho - j) + (1 - \rho - j)\rho e^{+j\theta_t} + (1 - \rho + j)\rho e^{-j\theta_t} + \rho^2 = 2$$
$$(1 - \rho)\cos(\theta_t) + \sin(\theta_t) = 1 - \rho$$
$$(1 - \rho) = \frac{\sin(\theta_t)}{[1 - \cos(\theta_t)]} = \cot\left(\frac{\theta_t}{2}\right)$$
$$\theta_t = 2\arctan\left(\frac{1}{1-\rho}\right)$$

Substituting this θ_t back into:

$$\Gamma_{\text{res}} = \Gamma_c + \rho e^{j\theta} = (1 - \rho) + \rho e^{j\theta}$$

gives the reflection coefficient at the intersection. The return loss, R.L., at that point is a given here.

$$\text{R.L.} = -20\log_{10}(\Gamma_{\text{res}})|_{\theta=\theta_t}$$

At resonance the reflection coefficient of the resonator is:

$$\Gamma_{\text{res}} = \Gamma_c + \rho e^{j\pi}$$
$$= \Gamma_c - \rho$$
$$= 1 - \rho - \rho$$
$$= 1 - 2\rho$$
$$\rho = \frac{1 - \Gamma_{\text{res}}}{2}$$

It will now be shown that the Q of the circuit can be derived easily from the value of the return loss at the intersection points. Let ω_o be the center radian frequency and $\Delta\omega$ be the radian frequency difference between the upper and the lower intersections. Then for a series resonator:

$$Z_{\text{top}} = R + j\left(\omega + \frac{\Delta\omega}{2}\right)L - \frac{j}{\left(\omega + \dfrac{\Delta\omega}{2}\right)C}$$

$$Z_{\text{bot}} = R + j\left(\omega - \frac{\Delta\omega}{2}\right)L - \frac{j}{\left(\omega - \dfrac{\Delta\omega}{2}\right)C}$$

Equating the real part of the impedances to the absolute value of the reactances gives:

$$R = \left| \left(\omega + \frac{\Delta\omega}{2} \right) L - \frac{1}{\left(\omega + \frac{\Delta\omega}{2} \right) C} \right| = \left| \left(\omega - \frac{\Delta\omega}{2} \right) L - \frac{1}{\left(\omega - \frac{\Delta\omega}{2} \right) C} \right|$$

$$\left(\omega + \frac{\Delta\omega}{2} \right)^2 L - \left(\omega + \frac{\Delta\omega}{2} \right) R - \frac{1}{C} = 0 = \left(\omega - \frac{\Delta\omega}{2} \right)^2 L - \left(\omega - \frac{\Delta\omega}{2} \right) R - \frac{1}{C}$$

$$\Delta\omega\omega_0 L = \omega_0 R$$

$$Q = \frac{\omega_0 L}{R} = \frac{\omega_0}{\Delta\omega} = \frac{f_0}{\Delta f}$$

If a network analyzer with a polar display of the reflection coefficient is available, the Q of a resonator is easily determined. If an overlay is available that has the $R = |X|$ lines etched on it, then the two frequencies for which the resonator curve and the $R = |X|$ lines intersect are used. The center frequency is divided by difference in frequencies to determine Q. Usually there are parasitics involved with the resonator. To a first-order approximation, the Q can be determined by rotating the resonance locus on the chart until it is symmetrically located about the real axis. Then the frequencies are read and Q can be determined. If the resonator has extra parasitic R, then the Q calculation is not quite right since the resonator locus does not start and end on the unit reflection coefficient circle. De-embedding techniques can be used to determine the Q of those resonators. This $\Delta\omega$ procedure works for series resonant and shunt resonant circuits and also works to the first-order approximation when the locus is shifted up or down from the real immittance axis by a single small parasitic reactive element.

The same procedure can be used to measure external Q and loaded Q of the resonator. This assumes that the resonator is loaded externally with Z_0 of the measurement system. For external Q, the difference frequency for which the reactance is equal to Z_0 is found. When the reactance is equal to Z_0, that value is found when the reactance crosses the normalized plus or minus $X = 1$ circle on the chart. That frequency is shown as the frequency points labeled (1) on Figure 8.7. The frequency points labeled (2) on the chart are for the unloaded Q just discussed. Using the difference frequency for which the reactance is equal to the sum of the resonator resistance plus Z_0 results in a loaded Q measurement. It can be shown that that locus is the line in the reflection coefficient plane given as follows.

$$|X| = R + Z_0$$

$$\Gamma(1 \mp j) + \Gamma^*(1 \pm j) - 2 = 0$$

This is the equation of a straight line with imaginary axis intercepts at plus or minus one and real axis intercepts at plus one. With one measurement, external, unloaded, and loaded Q can be measured. The same procedure can be followed for shunt resonators except that the arcs and straight line segments are on the left-hand side of the chart and the resonator locus starts at the short-circuit point.

When a series resonator has a parallel parasitic resonator, the following procedure will give the value of the series resonator Q easily and accurately even when the shunt reactance is appreciable. The procedure also works for the dual circuit with admittances replaced by impedances and vice versa. The accuracy of the Q measurement depends on the accuracy to which the data points can be read and the limitation caused by some of the assumptions. It is assumed that the parasitic parallel resonator susceptance slope parameter is much lower than the series resonator reactance slope parameter such that the susceptance of the parallel resonator does not change appreciably over the frequency range for which the series resonator goes through its reactance excursion. Then B_{par} is essentially a constant over the frequency for which the impedance varies as a result of the series resonance. Since the impedance locus for the series resonator is a straight line in the impedance plane (a circle of infinite radius), the impedance locus in the admittance plane will be a circle. This is a result of the admittance locus being a bilinear transformation of the impedance locus. Some network analyzers plot the circular locus [4] in the admittance plane for resonators. A circular locus in either the impedance plane or admittance plane results in a circular locus in the reflection coefficient plane. The parasitic elements of a series R-L-C shifts the resonance circle vertically (susceptance parasitic) and to the right (conductance parasitic) in the admittance plane. This is shown for a specific example in Figure 8.8.

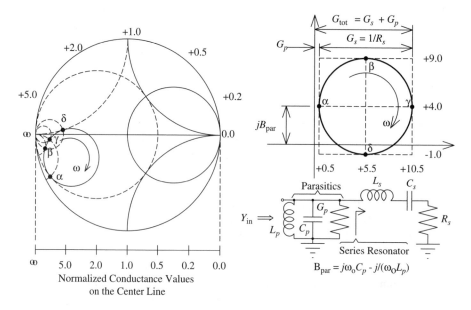

Figure 8.8 Reflection coefficient locus for series resonator with parasitics.

Because the circular locus may be perturbed by higher-order resonances, care needs to be given to reading the data from a reflection coefficient chart. The admittance is obtained from the reflection coefficient chart as shown in Figure 8.8. As discussed above, Q can be calculated by using the frequencies for which $|X| = R$ or $G = |B|$. In the admittance chart shown in Figure 8.8, that is point β and point δ. Notice that it would be difficult to determine the frequency exactly around point β or point δ from the minimum or maximum B. This is because maximum or minimum B is at the tangent points to the circle. If the measurement instrument does not have a maximum or minimum search for B, then an increase in accuracy can be generated by first determining the maximum and minimum B. Then calculating the average B, find the minimum and maximum G. Taking the average G (the vertical line going through the center of the locus in the admittance plane), find the frequency for which the locus passes through this average G. The G value will change more rapidly than the B values at those points. Q is then determined by dividing the frequency at point γ by the difference in frequencies from point δ to point β. It is important to recognize that the series resonant points are not the points for which the locus crosses the real axis in the reflection coefficient plane.

8.2.3 Resonator Q Measured as a One-Port and as a Two-Port

Insertion measurements or reflection measurements can be used to measure resonator Q. The previous section gave Q in term of the phase slope of S_{11} of the resonator measured as a one-port. Given a series R-L-C resonator, there are five possible ways to measure it. These are shown in Figure 8.9. If series resonators are measured as two-ports, one can make S_{11} and S_{21} measurements on them. They can be measured either in series or in shunt. The resonator can also be measured as a one-port as discussed in the previous section. The resonator has an impedance $Z = R + jX$. The S parameters for a series impedance in a two-port configuration are:

$$S_{11} = \frac{Z}{2Z_0 + Z} \qquad S_{21} = \frac{2Z_0}{2Z_0 + Z}$$

Differentiating the S_{21} equation with respect to frequency gives:

$$\frac{dS_{21}}{d\omega} = \frac{-2Z_0}{(2Z_0 + Z)^2}\frac{dZ}{d\omega} = \frac{d(|S_{21}|e^{j\phi_{21}})}{d\omega} = jS_{21}\frac{d\phi_{21}}{d\omega} + e^{j\phi_{21}}\frac{d|S_{21}|}{d\omega}$$

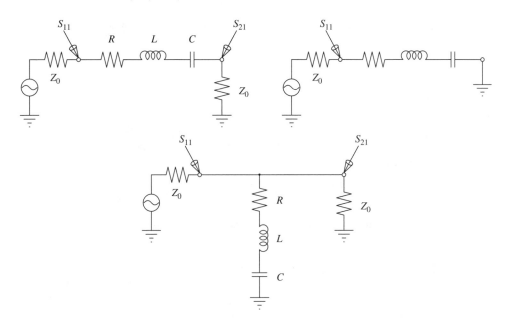

Figure 8.9 Five different ways to measure a resonator.

The first portion of the equation is a differentiation of the S_{21} formula. The second part of the equation is an implicit differentiation of S_{21}. The derivative of the magnitude of S_{21} is equal to zero at ω_o since that is the point of maximum transmission gain. Note that $dZ/d\omega$ evaluated at ω_o is $jdX/d\omega$ and $Z(\omega_o) = R$ where R is the series resistance in the resonator. Evaluating both sides of the equation of this derivative at $\omega = \omega_0$ gives:

$$\frac{-2Z_0}{(2Z_0 + R)^2}\frac{dX}{d\omega}\bigg|_{\omega=\omega_o} = \frac{2Z_0}{2Z_0 + R}\frac{d\phi_{21}}{d\omega}\bigg|_{\omega=\omega_o}$$

$$\frac{dX}{d\omega}\bigg|_{\omega=\omega_o} = -(2Z_0 + R)\frac{d\phi_{21}}{d\omega}\bigg|_{\omega=\omega_o}$$

The unloaded Q derived from reactance parameters for a series resonator is found from:

$$Q = \frac{\omega_o}{2R} \frac{dX}{d\omega}\bigg|_{\omega=\omega_o}$$

The Q derived from S_{21} measurements of a series two-port consisting of a series resonator is:

$$Q = -\frac{2Z_0 + R}{R} \frac{\omega_o}{2} \frac{d\phi_{21}}{d\omega}\bigg|_{\omega=\omega_o} = -\frac{\omega_o}{2}\left[\frac{1}{S_{11}} \frac{d\phi_{21}}{d\omega}\right]\bigg|_{\omega=\omega_o}$$

Likewise, using S_{11} of the series resonator measured as a two-port:

$$\frac{dS_{11}}{d\omega} = \frac{2Z_0}{(2Z_0 + Z)^2} \frac{dZ}{d\omega} = \frac{d(|S_{11}|e^{j\phi_{11}})}{d\omega} = jS_{11}\frac{d\phi_{11}}{d\omega} + e^{j\phi_{11}}\frac{d|S_{11}|}{d\omega}$$

The derivative of the magnitude of S_{11} is equal to zero at ω_o since that is the point where the locus of S_{11} crosses the real axis at right angles. Evaluating both sides of this derivative equation gives the following result at the center frequency.

$$Q = \frac{2Z_0 + R}{2Z_0} \frac{\omega_o}{2} \frac{d\phi_{11}}{d\omega}\bigg|_{\omega=\omega_o} = \frac{\omega_o}{2}\left[\frac{1}{S_{21}} \frac{d\phi_{11}}{d\omega}\right]\bigg|_{\omega=\omega_o}$$

The S parameters for a shunt impedance in a two-port configuration are:

$$S_{11} = \frac{-Z_0}{Z_0 + 2Z} \qquad S_{21} = \frac{2Z}{Z_0 + 2Z}$$

Using the S_{21} measurements of the series resonator measured as a two-port with the resonator in shunt with the line gives:

$$\frac{dS_{21}}{d\omega} = \frac{2Z_0}{(Z_0 + 2Z)^2} \frac{dZ}{d\omega} = \frac{d(|S_{21}|e^{j\phi_{21}})}{d\omega} = jS_{21}\frac{d\phi_{21}}{d\omega} + e^{j\phi_{21}}\frac{d|S_{21}|}{d\omega}$$

This results in a Q of:

$$Q = \frac{Z_0 + 2R}{Z_0} \frac{\omega_o}{2} \frac{d\phi_{21}}{d\omega}\bigg|_{\omega=\omega_o} = -\frac{\omega_o}{2}\left[\frac{1}{S_{11}} \frac{d\phi_{21}}{d\omega}\right]\bigg|_{\omega=\omega_o}$$

Likewise, using two-port measurements of the series resonator measured in shunt with the transmission line gives:

$$\frac{dS_{11}}{d\omega} = \frac{2Z_0}{(Z_0 + 2Z)^2} \frac{dZ}{d\omega} = \frac{d(|S_{11}|e^{j\phi_{11}})}{d\omega} = jS_{11}\frac{d\phi_{11}}{d\omega} + e^{j\phi_{11}}\frac{d|S_{11}|}{d\omega}$$

This results in a Q of:

$$Q = -\frac{Z_0 + 2R}{2R} \frac{\omega_o}{2} \frac{d\phi_{11}}{d\omega}\bigg|_{\omega=\omega_o} = -\frac{\omega_o}{2}\left[\frac{1}{S_{21}} \frac{d\phi_{11}}{d\omega}\right]\bigg|_{\omega=\omega_o}$$

Finally, measuring the series resonator as a one-port gives:

$$S_{11} = \frac{Z - Z_0}{Z + Z_0}$$

Differentiating this equation results in:

$$\frac{dS_{11}}{d\omega} = \frac{2Z_0}{(Z_0 + Z)^2} \frac{dZ}{d\omega} = \frac{d(|S_{11}|e^{j\phi_{11}})}{d\omega} = jS_{11}\frac{d\phi_{11}}{d\omega} + e^{j\phi_{11}}\frac{d|S_{11}|}{d\omega}$$

This results in a Q of:

$$Q = \frac{R^2 - Z_0^2}{2RZ_0} \frac{\omega_o}{2} \frac{d\phi_{11}}{d\omega}\bigg|_{\omega=\omega_o} = \frac{\omega_o}{2}\left[\frac{2S_{11}}{1 - S_{11}^2} \frac{d\phi_{11}}{d\omega}\right]\bigg|_{\omega=\omega_o}$$

8.3 EQUATION OF A CIRCLE GIVEN THREE POINTS

In order to do data processing to find the Q of a resonator, it is often helpful to find the equation of a circle given three points on that circle. The procedure to be discussed works best if the points are spaced somewhat evenly around the circle. However, it will work if the precision of the data is good for points that are not spaced evenly around the circle.

Starting from the equation of a circle in the complex plane with three points Z_1, Z_2, and Z_3 located on it we then have the following three equations.

$$Z_1 Z_1^* - Z_c Z_1^* - Z_c^* Z_1 + Z_c Z_c^* = \rho^2$$

$$Z_2 Z_2^* - Z_c Z_2^* - Z_c^* Z_2 + Z_c Z_c^* = \rho^2$$

$$Z_3 Z_3^* - Z_c Z_3^* - Z_c^* Z_3 + Z_c Z_c^* = \rho^2$$

Subtracting the first from the second and the third from the second equations and eliminating Z_c^* from the result gives Z_c.

$$Z_c = \frac{(Z_3 - Z_2)(Z_1 Z_1^* - Z_2 Z_2^*) - (Z_2 - Z_1)(Z_2 Z_2^* - Z_3 Z_3^*)}{(Z_2 - Z_1)(Z_3^* - Z_2^*) - (Z_2^* - Z_1^*)(Z_3 - Z_2)}$$

Knowing Z_c, the radius is found from any of the first three equations. For example:

$$\rho = \sqrt{Z_1 Z_1^* - Z_1 Z_c^* - Z_1^* Z_c + Z_c Z_c^*}$$

If necessary, more accuracy may possibly be found by using more points and cycling through these points (three at a time) to find an "average" radius and an "average" center from a series of points.

8.4 OSCILLATOR DESIGN

The theory presented in this book covers two types of feedback oscillator designs. The first type to be discussed is an oscillator configuration that uses feedback directly in series with the common terminal of a device. This is sometimes referred to as a *series oscillator configuration*. The second type of device is a loop feedback design. The circuitry external to the device in a loop feedback design can be characterized as a two-port in parallel with the device and thus the loop oscillator appears as a shunt configuration oscillator.

8.4.1 Oscillator Conditions

An oscillator is a circuit that puts out power at a given frequency without any incident wave. Sometimes this type of circuit is called a self-excited oscillator. The following questions are often asked. What types of circuit generate power without any incident energy? How can an unconditionally stable circuit be made to oscillate? What are the optimum load impedances? Can potentially unstable devices be used? Appropriate feedback needs to be put around some devices to get them to oscillate. There are an infinite number of feedback circuits that will make a device or circuit oscillate! The circuit designer needs to find the appropriate value for the feedback immittance for the desired conditions of operation.

8.4.2 One-Port Oscillator

Consider the one-port series circuit shown in Figure 8.10. When the switch in the circuit is closed as shown at time $t = 0$, the current I will grow exponentially if $R_s + R_l < 0$ and will decay exponentially if $R_s + R_l > 0$. The voltage source V_o and the current source I_o are initial conditions. If $R_s + R_l = 0$, the circuit will "ring" or have signals in it with a constant amplitude and at a frequency equal to the series resonant frequency of the inductor and capacitor.

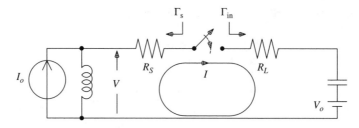

Figure 8.10 One-port oscillation conditions. [*Robert J. Weber, "Oscillator Design Techniques Using Calculated and Measured S Parameters," Copyright ©1991, Short Course, 45th Annual Symposium on Frequency Control, May 1991, Los Angeles, CA.*]

Oscillation at the series resonant frequency of the total loop implies that $Z_s = -Z_l$. Therefore:

$$\Gamma_S = \frac{Z_S - Z_0}{Z_S + Z_0} = \frac{-Z_L - Z_0}{-Z_L + Z_0} = \frac{Z_L + Z_0}{Z_L - Z_0} = \frac{1}{\Gamma_L}$$

$$\Rightarrow \Gamma_S \Gamma_L = 1$$

This condition will exist at any port of a circuit that is oscillating and generating power!

In order for the loop current to grow, it must have a starting value other than zero. This starting value can be provided as shown in Figure 8.10 by initial conditions or it can be provided by noise. In almost all of the oscillators the author has built, the oscillation was started by initial conditions. Turning the circuit on causes a set of initial conditions to be placed on the loop. The loop must have a net negative resistance for the loop current to grow. As the magnitude of the loop signals grows, at some time, due to nonlinear effects, either the value of the magnitude of the negative resistance drops or the value of the positive resistance rises. This causes the net value of resistance to reach zero and the loop signals then remain constant at the value needed to cause net-zero resistance. If the magnitude of the negative resistance rises with signal level or the value of the positive resistance drops with signal level, an unstable oscillation is produced. The series circuit should be replaced with a parallel circuit under those conditions. At resonance, a series circuit and a parallel circuit both consist of two resistors tied together. Therefore the oscillation condition at resonance is the same for both configurations. In the multiport oscillators that will be discussed, loop stability is usually reached by having a forward gain parameter reduce to cause a constant value of some signal. Choosing which parameter varies with signal level can be left to the device, can be determined by load lines, or can be purposely dictated by the introduction of limiters (circuits that clip the peaks from sine waves or reduce their magnitudes in a nonlinear manner). The parameters used to characterize the device need to be the parameters valid for some level of saturation. The small signal values can be used to determine whether a circuit will start to oscillate. The saturated values are used to determine the steady-state operating conditions of the circuit. In the discussions to follow, the values of the scattering parameters used in oscillator design are assumed to be those for the device under a small amount of limiting or saturation. These parameters can be measured or calculated using the load pull technique discussed in Section 3.1.6.

8.4.3 Some *n*-Port Oscillator Design Considerations

A two-port circuit will be considered next. Suppose power is being generated by oscillation at one of the ports of a two-port circuit. What are the conditions on the other port of the two-port under the conditions of oscillation? Consider the two-port circuit as shown in

Figure 8.11. The input reflection coefficient of this circuit is:

$$\Gamma_{in} = \frac{S_{11} - \Delta_S \Gamma_2}{1 - S_{22}\Gamma_2}$$

If the oscillation condition is met at port one, then:

$$\Gamma_1 = \frac{1}{\Gamma_{in}} = \frac{1 - S_{22}\Gamma_2}{S_{11} - \Delta_S\Gamma_2}$$

The output reflection coefficient is then:

$$\Gamma_{out} = \frac{S_{22} - \Delta_S\Gamma_1}{1 - S_{11}\Gamma_1}$$

Substituting for Γ_1 gives:

$$\Gamma_1 = \frac{1}{\Gamma_{in}} = \frac{1 - S_{22}\Gamma_2}{S_{11} - \Delta_S\Gamma_2}$$

$$\Gamma_{out} = \frac{S_{22} - \Delta_S\dfrac{1 - S_{22}\Gamma_2}{S_{11} - \Delta_S\Gamma_2}}{1 - S_{11}\dfrac{1 - S_{22}\Gamma_2}{S_{11} - \Delta_S\Gamma_2}} = \frac{S_{11}S_{22} - \Delta_S}{S_{11}S_{22}\Gamma_2 - \Delta_S\Gamma_2} = \frac{1}{\Gamma_2}$$

$$\Rightarrow \Gamma_{out}\Gamma_2 = 1$$

This implies that the oscillation condition is met at port two if the condition is met at port one [5].

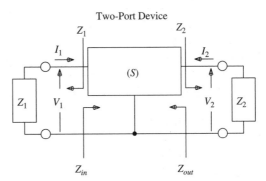

Figure 8.11 Two-port circuit for oscillator analysis.

Consider next the n-port network shown in Figure 8.12. If the oscillation conditions are met at port one, then any two-port constructed from port one and any other port i will meet the above two-port criteria. Since port i can be any port of the circuit, the oscillation conditions are met at all ports of an n-port network as expected from the one-port analysis. This significant result implies that negative resistance exists at any terminal pair or port of a circuit containing an oscillator. It is particularly significant that this condition exists in the bias terminals of the network and care must be given to ensure that this negative resistance does not cause some other circuit connected to the power supply to oscillate.

8.4.4 Three-Terminal Device Oscillator Design

Consider a two-port circuit that is potentially unstable as shown by the stability circles in Figure 8.13. There are an infinite number of ways to form an oscillator. For instance, choose any point inside the source reflection coefficient chart for which the output impedance of the device has a negative real impedance and place the negative of that impedance on the output. You will produce on oscillator. If a further constraint is placed on the oscillator that all the

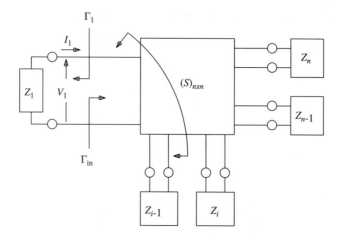

Figure 8.12 *N*-port circuit for oscillator analysis. [*Robert J. Weber, "Oscillator Design Techniques Using Calculated and Measured S Parameters," Copyright ©1991, Short Course*, 45th Annual Symposium on Frequency Control, *May 1991, Los Angeles, CA.*]

Four Different Reflection Coeffiecient Planes

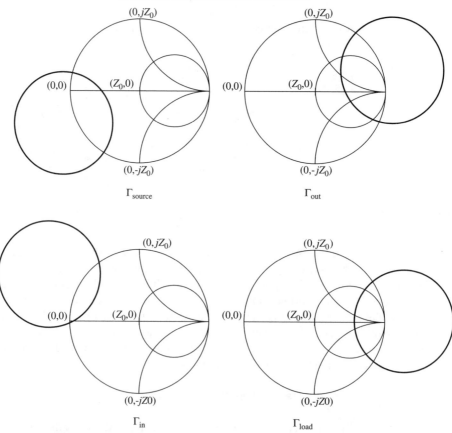

Figure 8.13 Four different reflection coefficients for an oscillator. [*Robert J. Weber, "Oscillator Design Techniques Using Calculated and Measured S Parameters," Copyright ©1991, Short Course*, 45th Annual Symposium on Frequency Control, *May 1991, Los Angeles, CA.*]

power generated by the oscillator is delivered at the output, then source immittances that are purely reactive would be chosen. Any point on the unit circle of the source reflection coefficient chart within the unstable region will allow all of the power that is generated to be dissipated at the output and none at the input. There are again an infinite number of possibilities. What is the most optimum point for source terminations to give the most power at the output? Does simply using some source reflection coefficient form the optimum oscillator or does some additional feedback need to be supplied to the circuit? The answer to these questions will be given shortly.

The infinite set of terminations that cause a circuit to oscillate can also be demonstrated by considering the network shown in Figure 8.14. The oscillation conditions can be demonstrated by considering the total three-port circuit matrix of the circuit. Assuming real Z_0 normalization impedances for each of the three ports, the port reflection coefficients and circuit network equations are expressed as:

$$\Gamma_i = \frac{a_i}{b_i}$$

$$(b) = (S)(a) = (S)(\Gamma_i)(b)$$

$$\Rightarrow |(S)(\Gamma_i) - (I)| = 0$$

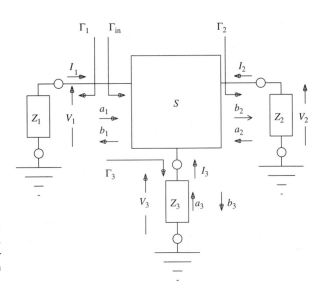

Figure 8.14 Series oscillator configuration. [*R. J. Weber, "Oscillator Design Using S-Parameters and a Predetermined Source or Load,"* Proceedings 45th Annual Symposium on Frequency Control, *pp. 364–367, May 1991.*]

This matrix equation has an infinite number of solutions if (S) is of rank 3. Another constraint needs to be applied to the system to identify the optimum load. It will be shown that a single constraint added to the oscillator design will result in an equation that has no physically realizable solution, one physically realizable solution, or two realizable solutions. A solution that is not physically realizable is possible since any specific constraint may require that an active device have more gain than is available.

Several constraints could be applied to the two-port active device. Two important constraints would be maximum dc-to-rf power conversion or minimum in band noise figure for the device. In the rest of this section, the constraint of maximum dc-to-rf power conversion will be applied. Maximum dc-to-rf power conversion will take place when the active device sees an optimum load line that is the same as it sees in an amplifier circuit biased for maximum dc-to-rf power conversion. This is the condition that will be imposed on three-terminal oscillators.

When the active device is loaded with the optimum power load as described in Chapter 7, the output voltage waveform will be symmetrically clipped if the circuit is driven to just beyond the peak to peak limit of the voltage swing. The phase of the fundamental frequency component of a symmetrically clipped sine wave is the same as the phase of the original sine wave. However, the amplitude of the fundamental frequency component of the clipped sine wave will be somewhat less than the amplitude of the original sine wave. To a first-order approximation, the input and output impedances and often the feedback transimpedances do not change very much when an amplifier is driven into a small amount of saturation. Therefore, the primary change in parameters is a small decrease in magnitude only of S_{21}. There are some other frequency components that are generated but they are usually filtered out from passing around the loop due to the effects of the loop filter. The designer can measure the forward scattering parameters of the device at a level producing a small amount of saturation or use the load pull technique to measure the scattering parameters of the device under a small amount of saturation. However, good results have been obtained by using small signal scattering parameters except with the magnitude of S_{21} reduced to 80% of its measured value. This corresponds to about two dBs of loop saturation and accounts for some level of parameter variation, lot-to-lot variation in the transistors, and some temperature gain variations.

8.4.5 Three-Terminal Device Oscillator with a Load Impedance Constraint [6]

If one obtains an optimum impedance Z_{opt} for the load of the device shown in the Figure 8.14 as an amplifier (i.e., with terminal three grounded), then:

$$Z_{opt} = -\frac{V_{23}}{I_2} = \frac{V_3 - V_2}{I_2} = \frac{a_3 + b_3 - a_2 - b_2}{a_2 - b_2} Z_0$$

$$\Gamma_{opt} = \frac{Z_{opt} - Z_0}{Z_{opt} + Z_0} = \frac{a_3 + b_3 - 2a_2}{a_3 + b_3 - 2b_2}$$

The input reflection coefficient equation for the three-port network is derived from the three scattering parameter equations for the three-port plus two constraints. The first constraint is the constraint given by the above relationship between a_2, b_2, a_3, and b_3 in terms of the optimum impedance. The second constraint is:

$$\Gamma_3 = \frac{a_3}{b_3}$$

The scattering parameter equations can be written in a straightforward way for the second constraint. The optimum impedance constraint can also be rewritten.

$$b_1 = S_{11}a_1 + S_{12}a_2 + S_{13}a_3 = S_{11}a_1 + S_{12}a_2 + S_{13}\Gamma_3 b_3$$

$$b_2 = S_{21}a_1 + S_{22}a_2 + S_{23}a_3 = S_{21}a_1 + S_{22}a_2 + S_{23}\Gamma_3 b_3$$

$$b_3 = S_{31}a_1 + S_{32}a_2 + S_{33}a_3 = S_{31}a_1 + S_{32}a_2 + S_{33}\Gamma_3 b_3$$

$$\Gamma_{opt} = \frac{a_3 + b_3 - 2a_2}{a_3 + b_3 - 2b_2} = \frac{b_3(\Gamma_3 + 1) - 2a_2}{b_3(\Gamma_3 + 1) - 2b_2}$$

The equations can be solved in terms of a_1 and a_2.

$$b_3 = S_{31}a_1 + S_{32}a_2 + S_{33}\Gamma_3 b_3$$

$$b_3 = \frac{S_{31}}{(1 - S_{33}\Gamma_3)}a_1 + \frac{S_{32}}{(1 - S_{33}\Gamma_3)}a_2$$

$$b_2 = S_{21}a_1 + S_{22}a_2 + S_{23}\Gamma_3 \frac{S_{31}}{(1 - S_{33}\Gamma_3)}a_1 + \frac{S_{32}}{(1 - S_{33}\Gamma_3)}a_2$$

$$b_1 = S_{11}a_1 + S_{12}a_2 + S_{13}\Gamma_3 \frac{S_{31}}{(1 - S_{33}\Gamma_3)}a_1 + \frac{S_{32}}{(1 - S_{33}\Gamma_3)}a_2$$

$$\left[\frac{S_{31}(\Gamma_3 + 1)}{(1 - S_{33}\Gamma_3)}a_1 + \frac{S_{32}(\Gamma_3 + 1)}{(1 - S_{33}\Gamma_3)}a_2 - 2b_2\right]\Gamma_{opt} = \left[\frac{S_{31}(\Gamma_3 + 1)}{(1 - S_{33}\Gamma_3)}a_1 + \frac{S_{32}(\Gamma_3 + 1)}{(1 - S_{33}\Gamma_3)}a_2 - 2a_2\right]$$

The last equation above is an equation relating b_2 to a_1 and a_2. The b_2 equation and the last equation can be used to relate a_1 to a_2. This relationship can be substituted into the b_1 equation resulting in b_1 in terms of a_1, Γ_3, Γ_{opt}, and the scattering parameters [6].

The equation is long and is presented here in parts. The numerator and denominator of Γ_{in}, under the constraint that the voltage between terminals 2 and 3 divided by the output current 2 is fixed, are:

$$\Gamma_{in} = \frac{Num[\Gamma_{in}]}{Den[\Gamma_{in}]}$$

where:

$$
\begin{aligned}
Num[\Gamma_{in}] = &+\Gamma_{opt}(2S_{12}S_{21}S_{31} - S_{12}S_{31}^2 + S_{11}S_{31}S_{32} - 2S_{11}S_{22}S_{31}) \\
&+\Gamma_{opt}\Gamma_3(2S_{12}S_{23}S_{31}^2 - 2S_{12}S_{21}S_{31}S_{33} - 2S_{13}S_{22}S_{31}^2 + 2S_{11}S_{22}S_{33}S_{31}) \\
&+\Gamma_{opt}\Gamma_3(-S_{12}S_{31}^2 - 2S_{11}S_{23}S_{31}S_{32} + S_{11}S_{31}S_{32} + 2S_{13}S_{31}S_{21}S_{32}) \\
&+\Gamma_3(S_{12}S_{31}^2 - S_{11}S_{31}S_{32} + 2S_{13}S_{31}^2 - 2S_{11}S_{33}S_{31}) \\
&+(2S_{11}S_{31} + S_{12}S_{31}^2 - S_{11}S_{31}S_{32})
\end{aligned}
$$

and:

$$
\begin{aligned}
Den[\Gamma_{in}] = &+\Gamma_{opt}(S_{31}S_{32} - 2S_{22}S_{31}) \\
&+\Gamma_{opt}\Gamma_3(2S_{22}S_{31}S_{33} - 2S_{23}S_{31}S_{32} + S_{31}S_{32}) \\
&-\Gamma_3(S_{31}S_{32} + 2S_{31}S_{33}) \\
&+(2S_{31} - S_{31}S_{32})
\end{aligned}
$$

This results in an input reflection coefficient written as:

$$\Gamma_{in} = \frac{A\Gamma_{opt}\Gamma_3 + B\Gamma_{opt} + C\Gamma_3 + D}{E\Gamma_{opt}\Gamma_3 + F\Gamma_{opt} + G\Gamma_3 + H}$$

where:

$$A = (2S_{12}S_{23}S_{31}^2 - 2S_{12}S_{21}S_{31}S_{33} - 2S_{13}S_{22}S_{31}^2 + 2S_{11}S_{22}S_{31}S_{33})$$
$$+ (-S_{12}S_{31}^2 - 2S_{11}S_{23}S_{31}S_{32} + S_{11}S_{31}S_{32} + 2S_{21}S_{13}S_{31}S_{32})$$

$$B = (2S_{12}S_{21}S_{31} - S_{12}S_{31}^2 + S_{11}S_{31}S_{32} - 2S_{11}S_{22}S_{31})$$

$$C = (S_{12}S_{31}^2 - S_{11}S_{31}S_{32} + 2S_{13}S_{31}^2 - 2S_{11}S_{31}S_{33})$$

$$D = (2S_{11}S_{31} + S_{12}S_{31}^2 - S_{11}S_{31}S_{32})$$

$$E = (2S_{22}S_{31}S_{33} - 2S_{23}S_{31}S_{32} + S_{31}S_{32})$$

$$F = (S_{31}S_{32} - 2S_{22}S_{31})$$

$$G = -(S_{31}S_{32} + 2S_{31}S_{33})$$

$$H = (2S_{31} - S_{31}S_{32})$$

Assigning a value to Γ_{opt} in the Γ_{in} equation results in:

$$\Gamma_{\text{in}} = \frac{\hat{A}\Gamma_3 + \hat{B}}{\hat{C}\Gamma_3 + \hat{D}}$$

where

$$\hat{A} = A\Gamma_{\text{opt}} + C$$
$$\hat{B} = B\Gamma_{\text{opt}} + D$$
$$\hat{C} = E\Gamma_{\text{opt}} + G$$
$$\hat{D} = F\Gamma_{\text{opt}} + H$$

If the magnitudes of the input reflection coefficients on ports one and three are equal to 1, then no power will be dissipated on the loads on these ports. If some loss is associated with the terminations on ports one and three, then the magnitudes of the reflection coefficients are less than 1.

Often the loci for these reflection coefficients can be described as circles. Let circles describing the reflection coefficients looking into port one and out to load three be defined as:

$$\Gamma_{\text{in}} = \Gamma_{10} + \Gamma_{1\rho}e^{j\phi_{1\rho}}$$
$$\Gamma_3 = \Gamma_{30} + \Gamma_{3\rho}e^{j\phi_{3\rho}}$$

Substituting these equations into the formula for Γ_{in} allows two values of Γ_3 to be solved for.

$$\Gamma_{\text{in}} = \Gamma_{10} + \Gamma_{1p}e^{j\phi_{1\rho}} = \frac{\hat{A}(\Gamma_{30} + \Gamma_{3p}e^{j\phi_{3\rho}}) + \hat{B}}{\hat{C}(\Gamma_{30} + \Gamma_{3p}e^{j\phi_{3\rho}}) + \hat{D}}$$

$$\Gamma_{1p}e^{j\phi_{1\rho}} = \frac{(\hat{A}\Gamma_{3p} - \hat{C}\Gamma_{3p}\Gamma_{10})e^{j\phi_{3\rho}} + \hat{B} + \hat{A}\Gamma_{30} - \hat{C}\Gamma_{30}\Gamma_{10} - \hat{D}\Gamma_{10}}{\hat{C}(\Gamma_{30} + \Gamma_{3p}e^{j\phi_{3\rho}}) + \hat{D}} = \frac{\bar{A}e^{j\phi_{3\rho}} + \bar{B}}{\bar{C}e^{j\phi_{3\rho}} + \bar{D}}$$

$$|\Gamma_{1p}|^2 = \left(\frac{\bar{A}e^{j\phi_{3\rho}} + \bar{B}}{\bar{C}e^{j\phi_{3\rho}} + \bar{D}}\right)\left(\frac{\bar{A}e^{j\phi_{3\rho}} + \bar{B}}{\bar{C}e^{j\phi_{3\rho}} + \bar{D}}\right)^* = \frac{\bar{A}\bar{A}^* + \bar{A}^*\bar{B}e^{-j\phi_{3\rho}} + \bar{A}\bar{B}^*e^{j\phi_{3\rho}} + \bar{B}\bar{B}^*}{\bar{C}\bar{C}^* + \bar{C}^*\bar{D}e^{-j\phi_{3\rho}} + \bar{C}\bar{D}^*e^{j\phi_{3\rho}} + \bar{D}\bar{D}^*}$$

Rearranging this equation with the left side containing terms with ϕ and the right side containing only constant terms gives:

$$[\bar{A}\bar{B}^* - \bar{C}\bar{D}^*|\Gamma_{1p}|^2]e^{j\phi_{3\rho}} + [\bar{A}^*\bar{B} - \bar{C}^*\bar{D}|\Gamma_{1p}|^2]e^{-j\phi_{3\rho}} = (\bar{C}\bar{C}^* + \bar{D}\bar{D}^*)|\Gamma_{1p}|^2 - \bar{A}\bar{A}^* - \bar{B}\bar{B}^*$$

Define K as a real constant and w as a complex constant as follows.

$$K = (\bar{C}\bar{C}^* + \bar{D}\bar{D}^*)|\Gamma_{1\rho}|^2 - \bar{A}\bar{A}^* - \bar{B}\bar{B}^*$$

$$w = (\bar{A}\bar{B}^* - \bar{C}\bar{D}^*|\Gamma_{1\rho}|^2)$$

Then:

$$K = we^{j\phi_{3\rho}} + w^* e^{-j\phi_{3\rho}}$$

$$= (u + jv)[\cos(\phi_{3\rho}) + j\sin(\phi_{3\rho})] + (u - jv)[\cos(\phi_{3\rho}) - j\sin(\phi_{3\rho})]$$

$$= 2u\cos(\phi_{3\rho}) - 2v\sin(\phi_{3\rho})$$

$$= 2\sqrt{u^2 + v^2}\cos(\phi_{3\rho} + \theta)$$

$$u = \sqrt{u^2 + v^2}\cos(\theta)$$

$$v = \sqrt{u^2 + v^2}\sin(\theta)$$

$$\theta = \arctan\left(\frac{v}{u}\right)$$

$$\phi_{3\rho} = -\theta + \arccos\left(\frac{K}{2\sqrt{u^2 + v^2}}\right)$$

This equation has two possible solutions for $\phi_{3\rho}$. If $\Gamma_{3\rho} = 1$, then one solution has $\Gamma_{1\rho} > 1$ and one solution has $\Gamma_{1\rho} < 1$.

If:

$$\frac{K}{2\sqrt{u^2 + v^2}} > 1$$

then no oscillator solution exists for the specified conditions. In other words, for the specified optimum load, terminal losses, and so on, there is not enough gain in the active device to support oscillation under the constraint imposed. The output reflection coefficient is determined by:

$$\begin{pmatrix} b_1 \\ b_2 \\ b_3 \end{pmatrix} = \begin{pmatrix} S_{11} & S_{12} & S_{13} \\ S_{21} & S_{22} & S_{23} \\ S_{31} & S_{32} & S_{33} \end{pmatrix} \begin{pmatrix} \Gamma_1 b_1 \\ a_2 \\ \Gamma_3 b_3 \end{pmatrix}$$

$$\Gamma_{\text{out}} = \frac{b_2}{a_2} = S_{22} + \frac{S_{21}\Gamma_1[S_{32}S_{13}\Gamma_3 + S_{12}(1 - S_{33}\Gamma_3)]}{(1 - S_{11}\Gamma_1)(1 - S_{33}\Gamma_3) - S_{13}S_{31}\Gamma_1\Gamma_3}$$

$$+ \frac{S_{23}\Gamma_3[S_{12}S_{31}\Gamma_1 + S_{32}(1 - S_{11}\Gamma_1)]}{(1 - S_{11}\Gamma_1)(1 - S_{33}\Gamma_3) - S_{13}S_{31}\Gamma_1\Gamma_3}$$

$$\Gamma_{\text{load}} = \frac{1}{\Gamma_{\text{out}}}$$

The circle equations for the common terminal and the input reflection coefficients can represent circles of constant loaded Q, unloaded Q, or external Q as well as circles of constant return loss. Even resistance or constant conductance circles can be used. If the circles are Q circles, then two sets of circles have the same Q and four possible solutions may exist. All four or none of the four may be realizable in any particular set of conditions for a device. Figure 8.15 shows the various Q circles in the reflection coefficient plane. The figure shows unloaded Q, loaded Q, and external Q circles. These circles appear different from the ones used to measure unloaded Q, and so on, of a resonator. The reason is that when measuring Q of a resonator, the reactance is set to a resistance. When plotting Q for a single reactance in series with a resistance, the value of Q is a variable. When Q is a running variable, a series of Q curves result in the reflection coefficient plane. Their position is found the same way that the Q circles for measuring Q were derived.

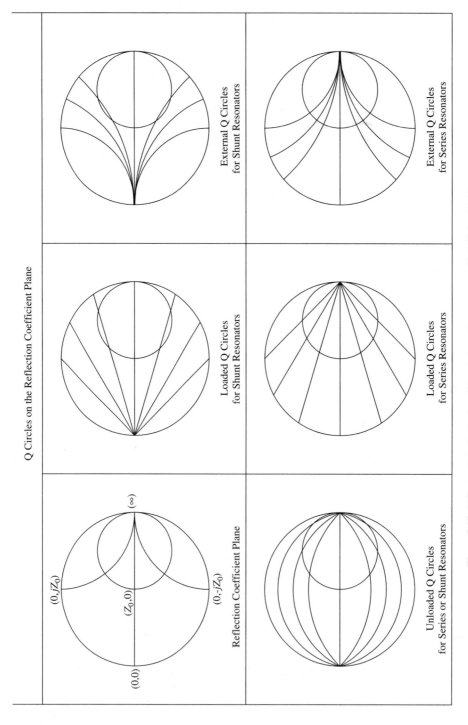

Figure 8.15 Different Q circles for describing the loss of components. [*R. J. Weber, "A Microwave Oscillator S-Parameter Design Technique and Computer Algorithm," Proceedings 34th Midwest Symposium on Circuits and Systems, pp. 303–306, May 1991.*]

So far the design has just been a solution for the basic device. The circuit must now be fabricated. The parasitic pads necessary for mounting the device must be accounted for. Figure 8.16 shows the components that need to be designed. Starting from the center of the figure, one works outward to complete the oscillator design accounting for the bonding or soldering pads necessary to fabricate the oscillator on a realizable mechanical circuit using the de-embedding techniques given in Chapter 6. A resonator to limit the bandwidth of the circuit is often placed in the network. In Figure 8.16 it is shown on the input side but it could be placed on the output as well. The loading effect of its loss needs to be included in the Q circle used in the oscillator solution. A phase shifter is shown on the input side to adjust the resonator position in the reflection coefficient plane. In some cases, it is possible to put the phase shifter on the output.

In order to ensure that the oscillation conditions are met at only one frequency, it might be necessary to rotate the output impedance around in the reflection coefficient plane. When plotting the negative of the output impedance, the optimum locus would be a locus that comes into the center of the reflection coefficient at right angles to the real axis and then on either side of the desired frequency goes to the left side of the chart. Consider the dashed-line locus in Figure 8.17. This locus does not come into the center of the chart at right angles to the real axis. At a frequency removed from the desired frequency, the real part of the output impedance has a larger negative value than the desired resistance. The oscillator will likely put out several frequencies at once. The locus is rotated using a phase shifter between the output and the load. The phase shifter rotates the locus so that away from resonance the reflection coefficient goes to the left side of the chart, meaning that the output resistance is less than the load resistance, and that prevents spurious oscillations at other frequencies. The oscillator stability conditions given later in this chapter confirm that the reflection coefficient locus should cross the real axis at right angles and proceed to the lower magnitudes of negative resistance as the frequency changes from the desired frequency.

For circuit analysis programs that do not plot the negative of an impedance, the circuit shown in Figure 8.18 will convert the negative output impedance into the negative of it, allowing one to plot the resulting positive real impedance on a standard reflection coefficient chart.

8.4.6 Loop Oscillator Design [7]

It is possible to design oscillators using shunt feedback also. Shunt feedback or loop feedback oscillator designs are often advantageous when discrete two-port resonators at a given impedance level are used as the frequency control elements. This section describes a design sequence using a closed-form solution to the loop oscillator equations.

If the devices used are matched to some given impedance (e.g., modular 50-ohm amplifiers), or if the resonator used is a two-port device, then a loop oscillator design might be the design choice. However, the resonator used in loop oscillators to control frequency could also be a shunt resonator. Figure 8.19 shows a drawing of a loop oscillator. The entire network is contained in the block called S. The lines from the input to the output are zero-length lines.

Notice that the output reflection coefficient of the device is also its source reflection coefficient and vice versa for the input and load reflection coefficients. Either $|\Gamma_{in}|$ or $|\Gamma_{out}|$ is greater than 1 but both cannot be. Which one is greater than 1 determines the direction of signal flow around the oscillator loop. The condition for oscillation is:

$$b_1 = S_{11}a_1 + S_{12}a_2 = a_2 \Rightarrow S_{11}a_1 = (1 - S_{12})a_2$$

$$b_2 = S_{21}a_1 + S_{22}a_2 = a_1 \Rightarrow (1 - S_{21})a_1 = S_{22}a_2$$

$$\frac{S_{11}}{1 - S_{21}} = \frac{1 - S_{12}}{S_{22}} \Rightarrow 1 - S_{12} - S_{21} - \Delta_S = 0$$

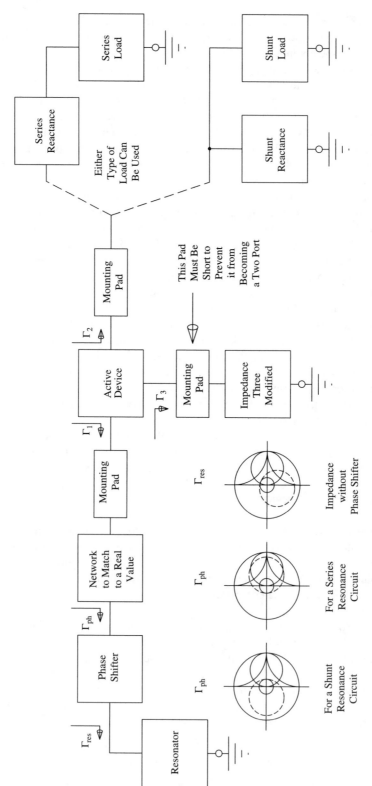

Figure 8.16 Design procedure for a series oscillator. [*R. J. Weber, "A Microwave Oscillator S-Parameter Design Technique and Computer Algorithm," Proceedings 34th Midwest Symposium on Circuits and Systems, pp. 303–306, May 1991.*]

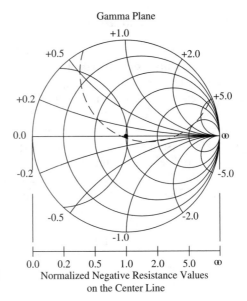

Figure 8.17 Typical oscillator output (plot of negative of the output impedance).

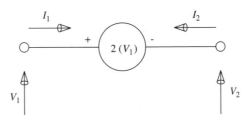

Figure 8.18 Circuit for a negative impedance converter.

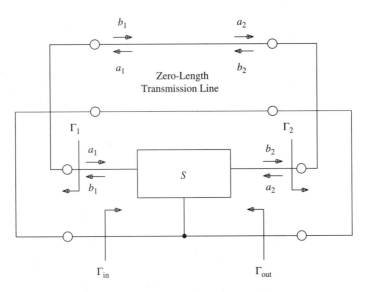

Figure 8.19 Loop oscillator (shunt oscillator). [*Robert J. Weber, "Oscillator Design Techniques Using Calculated and Measured S Parameters," Copyright ©1991, Short Course, 45th Annual Symposium on Frequency Control, May 1991, Los Angeles, CA.*]

If either S_{11} or S_{22} is equal to zero, then:

$$1 - S_{12} - S_{21} + S_{12}S_{21} = 0$$

This relationship is similar to the Barkhausen criteria. The Barkhausen criteria assumes that $S_{12} = 0$ and that the circuit block is matched. Notice that all of the circuit (device and matching, etc.) is included in the block designated S. A design technique that allows the external load to be separated out of the circuit will be described shortly. There are four possible (two redundant) solutions described by the above equations. These are identified in Figure 8.20. Assuming $|S_{12}| \ll 1$ and $|S_{21}| > 1$, the circuit has net gain only in the counterclockwise direction. Either power gain for optimum power conversion, transducer gain for a specified loop gain, or available gain for optimum in band noise figure can be used as design criteria. Design for power gain for optimum power conversion will now be described.

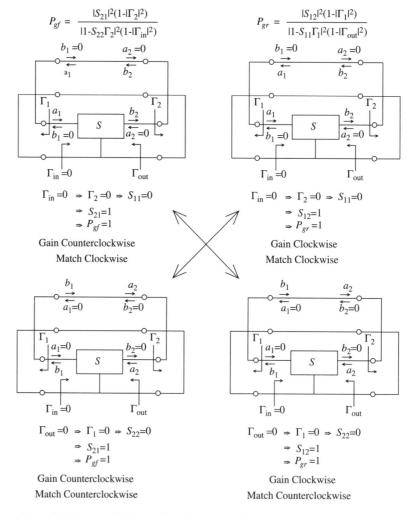

Figure 8.20 Four possible loop criteria for a loop oscillator. [R. J. Weber, R. Huisinga, and D. Ripley, "A Microwave Oscillator S-Parameter Design Procedure for Loop Oscillators," Proceedings 35th Midwest Symposium on Circuits and Systems, pp. 1341–4, August 1992.]

The loop oscillator is the shunt equivalent of the series oscillator described in the previous section. If one looks at the basic loop oscillator shown in Figure 8.21 as composed of two parts

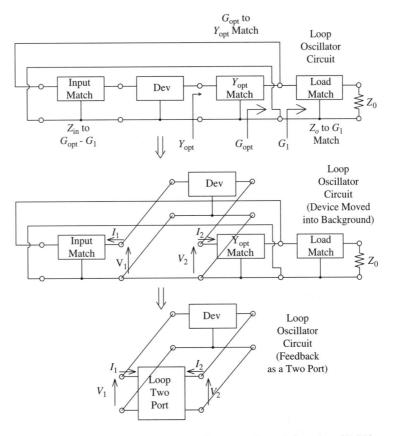

Figure 8.21 Procedure to move the loop oscillator to a shunt configuration. [*R. Weber, Microwave Engineering Course Notes, ©1987–1997, August 1997 Edition, p. 113.*]

(the active device and its parallel shunting admittances), then the shunt configuration is evident. Several parts of the loop feedback are determined based on a several-step design procedure to be described. However, when all of the external feedback components are considered as a two-port, the external loop is a two-port in parallel with the active device two-port as shown in Figure 8.21, where the actual load is included in "Loop Two-Port."

Figure 8.22 assumes that power gain is in the counterclockwise direction and that the device's optimum load admittance Y_{opt} is known. The analysis assumes that at each point in the loop except at the device terminals, the circuit is matched to some Z_0. The choice of Z_0 for the loop is a matter of convenience. When the circuit is fully designed, the part values are calculated as real L's and C's, and so forth, and the choice of Z_0 drops out of the calculations. For calculations, it is convenient to use a choice of Z_0 equal to the value of the measured S parameters for the active device or for the resonator. After the circuit is designed, impedance transformations can be made to change the impedance level if desired. When power is assumed to flow counterclockwise, the impedance transformations take place clockwise. The starting point is the device. The device has a Γ_{opt} load on it. The input reflection coefficient into the device is then:

$$\Gamma_{in} = \frac{S_{11} - \Delta_S \Gamma_{opt}}{1 - S_{22} \Gamma_{opt}}$$

The matching network from the device back to the point "B" matches the impedance corresponding to this reflection coefficient to Z_0. There is a phase shifter shown between points

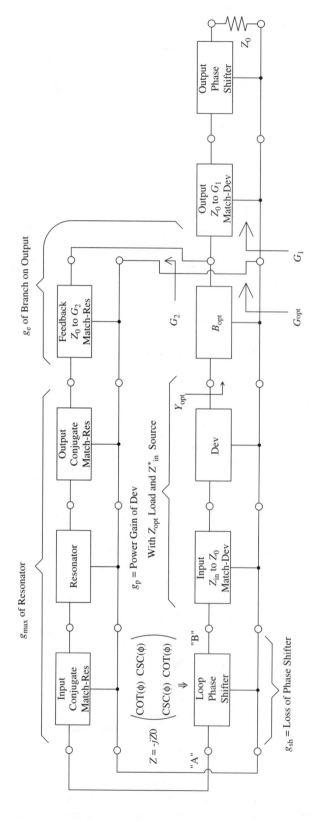

Figure 8.22 Design procedure for a loop (shunt) oscillator. [*R. J. Weber, R. Huisinga, and D. Rip-ley, "A Microwave Oscillator S-Parameter Design Procedure for Loop Oscillators," Proceedings 35th Midwest Symposium on Circuits and Systems, pp. 1341–4, August 1992.*]

"A" and "B". The phase shifter could be put at other points in the loop but this is a convenient point for it. It is assumed to be designed for a Z_0 impedance level. The input impedance into the phase shifter is also Z_0. The value of the phase shift will be determined later.

The next circuit block around the loop clockwise is the resonator. Because the resonator often exhibits a high reactance slope, the resonator scattering parameters usually have to be calculated from the resonator equivalent circuit around resonance. Measured values can be used at frequencies farther away from resonance. Depending on the resonator, it may or may not have a simultaneous conjugate. Recall that a series resistor used for a two-port does not have a simultaneous conjugate match. If the resonator does not have a simultaneous conjugate match, the input impedance of the resonator loaded with Z_0 is matched to Z_0. If the resonator has a simultaneous conjugate match, the resonator is matched with that match to Z_0 on both sides of the resonator. The external or loaded Q of the resonator can be adjusted at this point by impedance transforming the resonator to set the reactance slope of the resonator with respect to Z_0. The power gain for the resonator for power going counterclockwise with the output load (load shown into the phase shifter) of Z_0 is designated $g_{t\,max}$.

The next block in the feedback is the Z_0 to G_2 block. This block either can be a reactive lossless match or can be a resistive (e.g., a series resistor) match. The purpose of this block is to match Z_0 up to an impedance level of $1/G_2$. G_2, the value of the parallel conductance presented by the feedback loop, is determined by the loop gain requirements. If this block is constructed from reactive elements, it has no loss. If this block is constructed from a resistor, the value of the loss needs to be accounted for. The parallel optimum conductance placed on the device is G_1 in parallel with G_2 and is equal to G_{opt}. G_1 is the parallel conductance presented by the output match. The voltage across these two resistors is the same. The total input power into the node is the voltage magnitude squared times the total conductance. The power delivered into the feedback loop is the voltage magnitude squared times G_2. Therefore, the branch gain g_{br} for a reactive match is:

$$P_{dev} = |E|^2 G_{opt} = |E|^2 (G_1 + G_2)$$

$$P_{loop} = |E|^2 G_2$$

$$g_{br} = \frac{P_{loop}}{P_{dev}} = \frac{G_2}{(G_1 + G_2)}$$

When the feedback to G_2 match is made from a series resistor R_s, $Z_0 + R_s = R_2 = 1/G_2$. This series combination has an additional loss over what a reactive match has. That additional loss is given by a gain $g_e < 1$ as given here.

$$P_{loop} = |I|^2 R_2$$

$$\hat{P}_{loop} = |I|^2 Z_0$$

$$g_e = \frac{\hat{P}_{loop}}{P_{loop}} = \frac{Z_0}{R_2}$$

The total power gain g_t around the loop starting with the power gain of the device g_p is the product of the individual gains.

$$g_t = g_p g_{br} g_e g_{t\,max} g_{sh}$$

Assuming the phase shifter gain is equal to 1, the only unknown in the above equation is the value of the product $g_{br} g_e$. This product is a function of G_2 only. The value of G_2 is determined from this product allowing the total loop gain to be 1 in magnitude. The loop gain criteria, $g_t = |S_{21}| = 1$, is met by calculating the gain from point "B" in the circuit to point "A". An approximate value for the phase shifter gain is used. The net phase around the circuit is then calculated from point "B" to point "A". The total gain should be equal to 1 if the calculations are done properly. A phase shifter that has a characteristic impedance of Z_0 is then added to the loop to give a net loop phase of $2\pi n$ around the loop where n is an integer. This results in the total loop gain $S_{21} = 1$. Note that the loop is matched for power going in the counterclockwise direction. Loop gain for power existing in the clockwise direction would likely be very small since the magnitude of S_{12} is usually much smaller than the magnitude of S_{21}. If the circuit were designed for available gain, power would still be going in the counterclockwise direction but matching would be done in the counterclockwise direction (i.e., starting with a given source impedance and continuing around the loop clockwise).

After all the loop values are determined, the value of G_1 is found by subtracting G_2 from G_{opt}. The output match from Z_0 to G_1 is synthesized and the total circuit is analyzed. The output impedance out of the output match should be $-Z_0$. The locus of impedance variation versus frequency is determined, and just as in the series oscillator, a phase shifter can be added at the output to ensure that the negative of the oscillator impedance locus stays on the left-hand side of the reflection coefficient chart.

When the circuit design is complete, the components that make up the various blocks can be combined using delta-to-wye and wye-to-delta conversions to minimize the number of components in the loop. Typically components used to match the resonator to Z_0 are left untouched since they are usually very critical. Choosing Z_0 and the impedance level of the resonator sets the loaded Q or external Q of the resonator. Various constraints and considerations for minimizing phase noise are kept in mind when the resonator matches are obtained. Phase shifters can be made from transmission lines or lumped-constant phase shifters. Lumped-constant phase shifters were discussed in Chapter 6.

8.4.7 Oscillator Operating Point Stability [8]

The stability of the operating point is determined by considering the output impedance (or admittance) of the oscillator as a complex function of a complex number. The independent variable is a complex number consisting of a real part equal to the amplitude of voltage or current and the imaginary part equal to frequency:

$$Z_{\text{out}} = Z_{\text{out}} \, (\text{amp}, \omega)$$

where amp means either the magnitude of the voltage or current. For this output impedance to be an analytic function:

$$Z_{\text{out}} = R_{\text{out}} + j X_{\text{out}}$$

$$\frac{\partial R_{\text{out}}}{\partial \text{amp}} = +\frac{\partial X_{\text{out}}}{\partial \omega} \qquad \frac{\partial X_{\text{out}}}{\partial \text{amp}} = -\frac{\partial R_{\text{out}}}{\partial \omega}$$

$$\Rightarrow \frac{\partial R_{\text{out}}}{\partial \text{amp}} \frac{\partial X_{\text{out}}}{\partial \omega} > 0 \qquad \frac{\partial X_{\text{out}}}{\partial \text{amp}} \frac{\partial R_{\text{out}}}{\partial \omega} < 0$$

$$\frac{\partial R_{\text{out}}}{\partial \text{amp}} \frac{\partial X_{\text{out}}}{\partial \omega} - \frac{\partial X_{\text{out}}}{\partial \text{amp}} \frac{\partial R_{\text{out}}}{\partial \omega} > 0$$

The output immittance for some oscillators is expressed in terms of admittance. Then:

$$Y_{\text{out}} = G_{\text{out}} + j B_{\text{out}}$$

$$\frac{\partial G_{\text{out}}}{\partial \text{amp}} = +\frac{\partial B_{\text{out}}}{\partial \omega} \qquad \frac{\partial B_{\text{out}}}{\partial \text{amp}} = -\frac{\partial G_{\text{out}}}{\partial \omega}$$

$$\Rightarrow \frac{\partial G_{\text{out}}}{\partial \text{amp}} \frac{\partial B_{\text{out}}}{\partial \omega} > 0 \qquad \frac{\partial B_{\text{out}}}{\partial \text{amp}} \frac{\partial G_{\text{out}}}{\partial \omega} < 0$$

$$\frac{\partial G_{\text{out}}}{\partial \text{amp}} \frac{\partial B_{\text{out}}}{\partial \omega} - \frac{\partial B_{\text{out}}}{\partial \text{amp}} \frac{\partial G_{\text{out}}}{\partial \omega} > 0$$

Note that an analytic function must have each of the two individual equations satisfied. The sum of each of the individual conditions gives the last condition. When the locus of the output impedance crosses the real axis at right angles, the second term is zero. Reactance slopes and susceptance slopes are in the same direction. However, if R increases with amplitude, G decreases with amplitude. If the first term in the impedance expression is negative, then the resonator or matching of the oscillator to the load should be changed so that the oscillator looks like a parallel resonant circuit.

8.4.8 Oscillator Frequency versus Load Impedance Changes—Pulling

When the oscillator is built, it is often necessary to determine how much the oscillator frequency will change with load VSWR. This is called the *pulling* of an oscillator. The output impedance of the oscillator and thus the output reactance slope of the oscillator can be determined using various commercial microwave and rf circuit analysis programs. Consider the oscillator system shown in Figure 8.23. The scattering parameters of a matched attenuator are:

$$|S_{21}| = 10^{-\left(\frac{x_{dB}}{20}\right)}$$

$$(S_{att}) = \begin{pmatrix} 0 & |S_{21}|e^{j\phi} \\ |S_{21}|e^{j\phi} & 0 \end{pmatrix}$$

The input reflection coefficient looking into an unmatched attenuator of x dB that is terminated by a sliding short is:

$$|S_{21att}| = 10^{-\left(\frac{x_{dB}}{20}\right)}$$

$$\Gamma_{in} = |S_{21att}|^2 e^{j2\phi_{att}} e^{j\theta_{ss}}$$

The reflection coefficient looking into a matched isolation amplifier that is terminated in a sliding short is given here.

$$\Gamma_{in} = |S_{12amp}||S_{21amp}|e^{j(\phi_{21}+\phi_{12})} e^{j\theta_{ss}}$$

The isolation from an amplifier is the product of a gain term and an isolation term. The isolation from an attenuator is the product of two isolation terms. If an amplifier is used for isolation, it should be chosen because it has a lot of reverse isolation. Usually, an amplifier followed by an attenuator is needed to reach the desired isolation. If the attenuator is used first, then the signal is reduced in magnitude and noise could be a problem. If the amplifier is used first, the amplifier must be able to generate a large output signal. This is an engineering trade-off. So, typically an amplifier is added on the output of the oscillator followed by an attenuator in order to protect signal-to-noise ratio and limit compression in the following stages of any buffer amplifier string. When a matched amplifier and an attenuator are used to provide isolation, then the input reflection coefficient into the isolation network is:

$$\Gamma_{in} = |S_{12amp}||S_{21amp}||S_{21att}|^2 e^{j2\phi_{att}} e^{j(\phi_{21}+\phi_{12})} e^{j\theta_{ss}}$$

The following analysis assumes that the oscillator looks like a series resonant circuit at the output. A similar analysis holds when the oscillator looks like a parallel resonant circuit at its output. The external Q of a network is the ratio of the reactance of the equivalent inductance of the resonator divided by the external resistance. Normalize the load that is placed on the oscillator to the magnitude of the external resistance loading the oscillator. According to the oscillation requirements, the oscillator presents a negative resistance to the resonator equal to its internal resistance. When the resonator Q is the largest Q in the network, that equivalent Q will be reflected in the output impedance of the oscillator. Assume that the amount of VSWR placed on the resonator or oscillator is small. Any amount of reactance reflected back

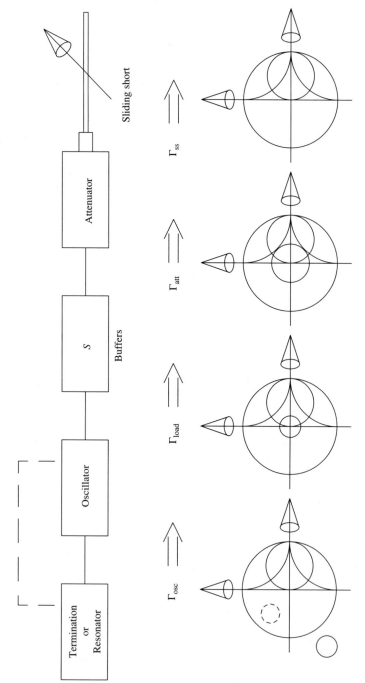

Figure 8.23 Configuration to determine the frequency pulling of an oscillator. [*Robert J. Weber, "Oscillator Design Techniques Using Calculated and Measured S Parameters," Copyright ©1991, Short Course, 45th Annual Symposium on Frequency Control, May 1991, Los Angeles, CA.*]

through the circuit and the oscillator network will appear across the resonator and change the frequency of the resonator. Let the locus of the reflection coefficient be a circle. The network between the resonator and the sliding short acts like a bilinear transformation. The maximum reactance presented by the load occurs when the circular locus presents a reflection coefficient angle near plus or minus 90 degrees. The resistance at those angles is very nearly the normal load resistance. Let the magnitude of the reflection coefficient circle be Γ_{load}. This can be calculated from the formulas given for looking into a network terminated by a sliding short, as has been just discussed. The initial value of X_{load} is zero (i.e., the oscillator is designed to be matched with a Z_0 load).

$$Z_{\text{load}} = Z_0 \frac{1 \pm j|\Gamma_{\text{load}}|}{1 \mp j|\Gamma_{\text{load}}|} \approx Z_0(1 \pm 2j|\Gamma_{\text{load}}|) = Z_0 \pm j2Z_0|\Gamma_{\text{load}}|$$

$$\left|\frac{X_{\text{load}}}{X_{\text{osc}}}\right| = \left|\frac{\Delta X_{\text{load}}}{X_{\text{osc}}}\right| = \left|\frac{\pm 2Z_0|\Gamma_{\text{load}}|}{X_{\text{osc}}}\right| = \frac{2|\Gamma_{\text{load}}|}{Q_{\text{ext}}} = \frac{\Delta X_{\text{load}}/Z_0}{Q_{\text{ext}}}$$

The amount of frequency shift due to a small variation of reactance in a series R-L-C is:

$$\omega = \frac{1}{\sqrt{LC}} \qquad \frac{d\omega}{dL} = \frac{d\omega}{dX}\frac{dX}{dL} = \omega\frac{d\omega}{dX}$$

$$\frac{d\omega}{dL} = \frac{d}{dL}\left(\frac{1}{\sqrt{LC}}\right) = -\frac{1}{\sqrt{LC}}\frac{1}{2L} = -\omega\frac{1}{2L} = \omega\frac{d\omega}{dX}$$

$$\left|\frac{\Delta f_{\text{BW}}}{f}\right| = \left|\frac{\Delta \omega_{\text{BW}}}{\omega}\right| = \left|\frac{2\Delta\omega}{\omega}\right| \approx \left|\frac{1}{\omega L}\Delta X\right| = \left|\frac{\Delta X}{X}\right| = \frac{2|\Gamma_{\text{load}}|}{Q_{\text{ext}}}$$

The total bandwidth $\Delta\omega_{\text{BW}}$ is the double-sided frequency bandwidth. In some technologies, that parameter divided by the operating frequency is called the *gamma* of the source. For a narrow gamma, the external Q needs to be high.

8.4.9 Calculating VSWR with Computer-Aided Design Circuits [8]

In order to determine whether an active device is loaded and biased for symmetrical limiting, the device's load line can be analyzed. When calculating the effect of the load line and load VSWR on the device's operating point in the oscillator circuit with a circuit analysis program, it is important that a true representation of the load line on the device is given. The device V-I curves assume that there are no reactive elements on the device. Therefore, the device parasitic capacities need to be de-embedded before looking at the V-I characteristics of the device. If they are not, a false picture of the load line results. Figure 7.22 shows the effect of a reactive impedance load on the device load line. The load line is elliptical in shape for reactive loads.

8.4.10 Power Supply Pushing

An important parameter in oscillator design is the change in frequency due to power supply variations. In oscillators that depend on voltage biases, the parameter is called *voltage pushing*. The amount of frequency change depends on the variation of saturation level in a part, or the variation of device parameters such as gain or input impedance as a function of bias changes in the part. If a high Q resonator is included in the oscillator loop and if the reactance slope at the point the resonator is included in the circuit is orthogonal to the resistance change, the voltage pushing can be minimized. In oscillators that include a limiter in the oscillator loop, the voltage pushing factor due to loop gain changes can be minimized. In oscillators that depend on a current bias such as an IMPATT diode oscillator, the appropriate determination would be current pushing or frequency change as a function of bias current.

EXERCISES

8-1 A series resonator has a Q of 1000 at 1 GHz. The series resistance is 5 ohms. Determine the series inductance and capacitance. A parasitic capacitance of 2 pF exists on the resonator. Determine the parallel resonant frequency and approximate Q at parallel resonance.

8-2 On a network analyzer, the frequencies for which the resistance is equal to the reactance for a resonator are 100.01 MHz and 100.05 MHz. The reflection coefficient at 100.03 MHz is .2 at 0 degrees. The resonator appears to have a reflection coefficient very nearly equal to minus one at 95 MHz. Determine the resonator equivalent circuit.

8-3 A device has an output resistance of −5 ohms in series with an inductive reactance of 25 ohms. Design a one-port oscillator for the part. The load on the oscillator should have a Q of 500. Assume that a shunt circuit gives a stable oscillation. The frequency is 50 MHz.

8-4 A part has the following scattering parameters at 4 GHz: $S_{11} = 0.5$ at −125 deg, $S_{12} = 0.1$ at 80 deg, $S_{21} = 1.8$ at 110 deg, $S_{22} = 0.5$ at −90 deg. Design a series oscillator for the part assuming that there are no parasitic elements from the common terminal to ground. The optimum load for maximum power transfer from the part is 300 ohms. Use series resistances of 1 ohm for the components to be placed on the part.

8-5 Using the same scattering parameters as the part in Exercise 8-4, design a loop oscillator at 4 GHz.

8-6 An oscillator has an output impedance of −30 ohms small signal and −50 ohms large signal. The reactance slope for the oscillator is positive. The output impedance locus crosses the real axis in the reflection coefficient plane at right angles. Is the oscillator circuit stable?

8-7 An oscillator operating at nominally 1 GHz has a change in frequency of plus and minus 100 kHz when the VSWR on the oscillator is 1 : 05 at all phases. Determine the external Q of the oscillator.

9

Microwave Filter Design

9.1 FILTER DESIGN

Filters that are used in the GHz range typically use component values that cannot be physically built if they are scaled from low-pass prototypes. Consider building a filter by frequency scaling that requires a 100 pF capacitor. At one gigahertz, the reactance of the capacitor is $-j1.59$ ohms. If the maximum parasitic inductive reactance allowable for this capacitor is 5% of the value, the inductive reactance needs to be less than $j0.08$ ohms. Using a rule of thumb that the inductance is approximately ten nanohenries per inch or four nanohenries per centimeter, the distance for the input to output for this capacitor needs to be less than eight-thousandths of an inch or less than 200 microns. It might be possible to build such a capacitor but it needs to be connected to the outside world with less than this amount of inductance also. If one uses 25-micron-diameter wire, the length-to-diameter ratio of the wire is only 8. Some other way of building filters needs to be found. When discussing filter parameters, insertion loss is often used instead of transducer gain. That terminology will be used in this chapter. However, if the source resistance is not the same as the load resistance, insertion loss is not the same as transducer gain. This was discussed in Chapter 5.

9.2 FILTER REVIEW

Filters are described by various transfer functions. Among the types of filters, there are Bessel, Butterworth, Cauer (elliptic), Chebyshev, Cohn (equal element), and Papoulis, each with their own distinct advantages. The procedure for the approximation problem that leads to each of these filter types is given in filter books [1,2]. The approximation problems lead to solutions for the inductors and capacitors to be used in low-pass filters. These low-pass filters are described in terms of circuit parameters by their g parameters. The g parameters in filter theory are not to be confused with g parameters in two-port circuit theory. The parameter g_0 is the input (source) termination immittance value and g_{n+1} is the output (load) termination immittance value. The g values are usually given as normalized values, normalized to an impedance of one ohm and with a low-pass cutoff frequency of one radian. The cutoff frequency for some filters is where the transmission loss is three dB and for other filters it takes a value related to another parameter such as passband ripple. It is important when using the g values from some tabular source that one knows what definition was used for the bandwidth of the passband. A low-pass filter described by g values normalized to a one-ohm level and a one-radian-per-second radian frequency value is called a low-pass prototype.

In a low-pass prototype such as shown in Figure 9.1, the successive g values, g_i, g_{i+1} alternate from an impedance to admittance or vice versa. For instance, if one uses g_0 as a normalized admittance, then g_1 specifies a normalized impedance parameter or a series inductance in a low-pass prototype. However, if the first g_0 is a normalized impedance, then g_1 specifies a normalized admittance parameter or a shunt capacitance in a low-pass prototype. A low-pass prototype and its dual will have the same transfer function (S_{21}) but may have different input and output impedances or reflection coefficients (S_{11} and S_{22}). Consider for instance the prototypes in Figure 9.1. The value g_0 in prototype (1) is considered an admittance while the g_0 in prototype (2) is considered an impedance. These prototypes are shown for $n = 4$. The value n stands for the number of poles in the low-pass prototype. The order n of the prototype gives the number of reactive elements in the prototype. The value g_5 in prototype (1) is an impedance since g_4 is an admittance or a shunt element. The value g_5 in prototype (2) is an admittance since g_4 is an impedance or a series element. These prototypes, (1) and (2), are even-order prototypes; thus the input and output terminations are of different types. If a prototype is of odd order, then the input and output terminations will be both of the same type, either both impedances or both admittances. It is necessary to make this distinction before one can properly scale the prototype.

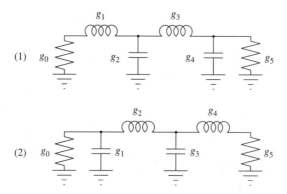

Figure 9.1 Two possible filter prototypes.

Equal-element filters have all g values equal to 1. Given a finite resonator Q, the equal element filter has been shown to have the largest reject band insertion loss (at a single frequency) for the smallest passband insertion loss (at the center frequency only) of all single-path filter types [3]. The derivation is based on low values of passband loss and equal-valued terminations. Sometimes it is called a *minimum loss filter*.

This book covers simple bandpass, high-pass, and bandstop transformations from low-pass prototypes and their implementations at microwave frequencies. There is also a set of microwave filter theory based on unit elements (e.g., one-eighth-wavelength lines) that is considered briefly at the end of this chapter. Sometimes filter design approaches similar to the unit element approach are called *commensurate line filters* since the line lengths are all the same and the trigonometric functions in the lines and stubs then have the same values for all lines and stubs.

9.3 SCALING A PROTOTYPE

Two types of scaling are possible with a prototype. The frequency at which the filter reaches a given insertion loss can be changed. This is called *frequency scaling*. The impedance level of the filter can be changed. This is called impedance scaling.

9.3.1 Impedance Scaling a Low-Pass Prototype

By convention g_0 is usually equal to 1 and g_{n+1} could be some other value depending on the filter type chosen. However, if one wishes to use the filter at some other impedance level, all the values can be scaled to that impedance level. If the impedance is scaled to a higher value, each inductor increases in value, each capacitor decreases in value, and each resistor increases in resistance value. Note that either g_0 or g_{n+1} may be given as an admittance depending on the low-pass prototype. Proper impedance scaling must change the admittance values to impedance values first to get the proper impedance scaling ratio. This is not a problem with g_0 since 1 ohm is 1 sieman (mho). However, g_{n+1} values may turn out to be an admittance value. The admittance value needs to be expressed as a resistance value before using the scaling factor. This would be the same as changing conductance values by the inverse of the scaling factor.

$$k_z = \frac{R_{zs}}{R_{\text{LPP}}} = \text{impedance scaling factor}$$

$$\frac{1}{k_z} = \frac{G_{zs}}{G_{\text{LPP}}} \Rightarrow k_z = \frac{G_{\text{LPP}}}{G_{zs}}$$

$$L_{zs} = k_z L_{\text{LPP}}$$

$$C_{zs} = \frac{C_{\text{LPP}}}{k_z}$$

$$R_{zs} = k_z R_{\text{LPP}}$$

where subscript zs denotes the impedance-scaled filter, subscript LPP denotes the low-pass prototype filter, and subscript z denotes impedance.

9.3.2 Frequency Scaling a Low-Pass Prototype

Frequency scaling scales the cutoff frequency of the prototype. Some prototypes are based on the filter having its minus-three-decibel point (or half-power point) at the cutoff frequency. Other prototypes (e.g., Chebyshev) are defined such that at the cutoff frequency, the insertion loss is at another insertion loss value—often the ripple magnitude. However, whatever way the cutoff frequency is expressed, frequency scaling will change the cutoff frequency as shown below.

$$k_f = \frac{\omega_{\text{fs}}}{\omega_{\text{LPP}}} = \text{frequency scaling factor}$$

$$L_{\text{fs}} = \frac{L_{\text{LPP}}}{k_f}$$

$$C_{\text{fs}} = \frac{C_{\text{LPP}}}{k_f}$$
$$R_{\text{fs}} = R_{\text{LPP}}$$

These relationships result from the need to maintain component impedance levels (e.g., $\omega_{\text{fs}} L_{\text{fs}} = \omega_{\text{Lpp}} L_{\text{LPP}}$) equal at each of their respective cutoff frequencies. The subscript fs denotes that the component is from the frequency scaled filter. For a combination of frequency and impedance scaling:

$$L_{cs} = L_{LPP}\left(\frac{k_z}{k_f}\right)$$

$$C_{cs} = C_{LPP}\left(\frac{1}{k_z k_f}\right)$$

$$R_{cs} = k_z R_{LPP}$$

where the subscript cs represents the combination-scaled filter parameters.

9.4 TRANSFORMING A LOW-PASS PROTOTYPE TO A BANDPASS, BANDSTOP, OR HIGH-PASS CONFIGURATION

The theory behind frequency transformations is not given here. However, a low-pass filter can be transformed to a high-pass, bandpass, or bandstop filter. For a low-pass-to-high-pass transformation, the cutoff frequency becomes the frequency where the passband starts rather than stops.

Transformations are not unique. For bandpass and bandstop filters, the centers of the pass or stop band and the width of the respective bands need to be defined. These are often defined as:

$$\omega_0 = \sqrt{\omega_1 \omega_2}$$

$$w = \frac{(\omega_2 - \omega_1)}{\omega_0}$$

where ω_1 = lower cutoff frequency, ω_2 = upper cutoff frequency, ω_0 = geometric cutoff frequency, and w = fractional frequency bandwidth. The frequency transformations for each of the elements of the low-pass filter from the low-pass prototype to bandpass filters, and so on, are shown in Figure 9.2. Impedance transformations are not included in this figure. Impedance

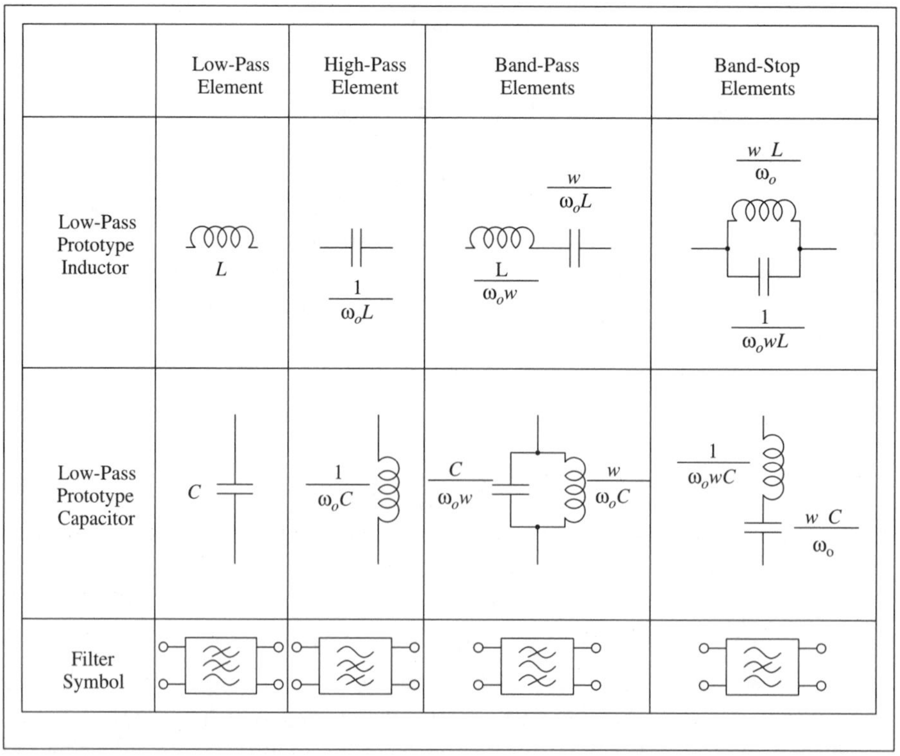

Figure 9.2 The transformations from the low-pass prototype to the other filter types.

transformations can be done after the low-pass-to-bandpass transformation has taken place. Note that each L, whether series or shunt, is replaced by the appropriate equivalent for an inductance. The same holds for each C. Therefore, a shunt C in a low pass becomes a shunt connection of a series LC circuit in a bandstop. Figure 9.3 shows an example for a low-pass-to-bandstop transformation while Figure 9.4 shows an example for a low-pass-to-bandpass transformation. For cases where there are transformations, impedance scaling, and frequency scaling, it is suggested that the transformation and frequency scaling be done in one step as given in the transformation equations above. After those transformations are done, then do the impedance scaling. With care, impedance scaling can be included in the transformation equations.

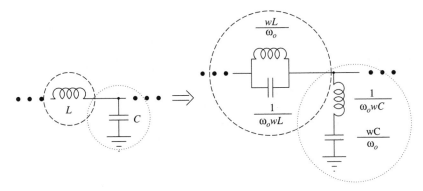

Figure 9.3 Low-pass to band-stop transformation of a series L and shunt C.

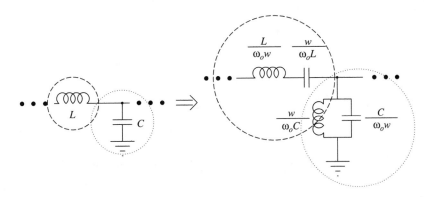

Figure 9.4 Low-pass to bandpass transformation of a series L and shunt C.

9.5 PRACTICAL FILTER CIRCUITS

Consider a low-pass-to-bandpass transformation. See Figure 9.4 for an example. Notice that two types of resonant circuits exist in the bandpass structure. Typically, one is not physically able to fabricate both types of resonators at high frequencies. It may not be possible to fabricate even one of the two types that result from using lumped-constant components. The parasitic capacitance or inductance of a physical inductor or capacitor may override the desired reactance value.

If more than one inductor value exists in the low-pass prototype, more than one set of LC values will exist in the transformed filter circuit. For manufacturing ease, cost savings, and so forth, it is often desirable to use the same resonators (L's and C's) everywhere. In addition, at many frequencies it is not possible to use discrete L's and C's. Distributed parameter circuits can be used as resonators. The various resonators available with distributed parameter circuits must somehow be described in terms of L's and C's if the low-pass prototype and transformation procedure described above is to be used.

9.6 REACTANCE SLOPE PARAMETERS

As discussed in Chapter 8, a resonator is a physical structure having energy stored alternately in the magnetic and electrical fields. In addition, the impedance or admittance at the terminals is purely real at resonance. Open-circuited and short-circuited transmission lines that are a multiple of a quarter-wavelength long satisfy this definition. What is the relationship between the values of the transmission line and the equivalent lumped-constant L and C for the resonator? It is desired that equivalent capacitances and inductances be found that have the same combined impedance variation with frequency at their terminals at resonance that the resonator has. It is the rate of impedance variation that is important rather than just the value of impedance. Consider the short-circuited quarter-wavelength line shown in Figure 9.5. Since the input looks like an open circuit at resonance, what are the effective L's and C's that describe this resonance? A parallel resonator has a high impedance at resonance. For parallel types of resonance, it was shown in Chapter 8 that the susceptance slope parameter,

$$b = \frac{\omega_o}{2} \frac{d B_p(\omega)}{d\omega} \bigg|_{\omega=\omega_o}$$

can be used to give the equivalent L's and C's of the parallel resonance valid around the resonant frequency.

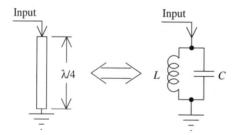

Figure 9.5 Circuit model of a quarter-wavelength T-line.

For the specific case of the short-circuited quarter-wavelength line of length d:

$$Y_{sc} = -jY_0 \cot(\beta d)$$

$$\Rightarrow B_{sc} = -Y_0 \cot(\beta d)$$

$$\frac{dB}{d\omega} = \frac{d B_{sc}}{d(\beta d)} \frac{d(\beta d)}{d\omega} = Y_0[\csc^2(\beta d)] \frac{d}{v} \bigg|_{\substack{\omega=\omega_o \\ \beta d = \frac{\pi}{2}}} = Y_0 \frac{d}{v}$$

$$\Rightarrow b = \frac{\omega_o Y_0 d}{2v} = \beta d \frac{Y_0}{2} = \frac{\pi}{2} \frac{Y_0}{2} = \frac{\pi}{4} Y_0 = \omega_o C_{eq}$$

$$C_{eq} = \frac{\pi Y_0}{4\omega_o}$$

Similarly, the discussion in Chapter 8 showed that the reactance slope parameter:

$$x = \frac{\omega_0}{2} \frac{d X_s(\omega)}{d\omega} \bigg|_{\omega=\omega_0}$$

gives the equivalent L's and C's of structures that exhibit series resonance.

9.6.1 Tapped *L-C* Resonators

Often parallel or series resonators are tapped (i.e., the resonator appears as shown in Figure 9.6). The tapped inductor resonant circuit has an input susceptance of:

$$B = \frac{\omega C}{1 - \omega^2 L_2 C} - \frac{1}{\omega L_1}$$

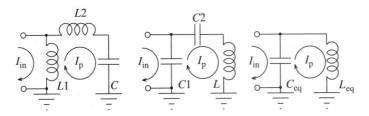

Figure 9.6 Tapped resonators for impedance transformation. [*R. Weber, Microwave Engineering Course Notes, ©1987–1997, August 1997 edition, p. 141.*]

Differentiating B with respect to ω and evaluating it at the parallel resonant frequency will give the equivalent C's and L's. Keep in mind that at the parallel resonant frequency, the total loop is series resonant since there is no current flow into the lossless network at parallel resonance. All the current is flowing around the loop. ($I_{in} = 0$, $I_{L1} = I_{L2} = I_p$ or $I_{C1} = I_{C2} = I_p$ at parallel resonance.) Therefore, the parallel resonant frequency is the frequency for which the internal loop is series resonant. The equivalent capacitance or capacitance at the tap point for the left circuit of Figure 9.6 is:

$$C_{tap} = \frac{1}{2} \frac{dB}{d\omega}\bigg|_{\omega^2 = \frac{1}{(L_1 + L_2)C}} = \frac{1}{2}\left[\frac{(1 - \omega^2 L_2 C)C + 2\omega^2 C^2 L_2}{(1 - \omega^2 L_2 C)^2} + \frac{1}{\omega^2 L_1}\right]\bigg|_{\omega^2 = \frac{1}{(L_1 + L_2)C}}$$

$$= \left(\frac{L_1 + L_2}{L_1}\right)^2 C$$

The susceptance for the center circuit is:

$$B = \omega C_1 + \frac{\omega C_2}{1 - \omega^2 L C_2}$$

The equivalent capacitance for the center circuit is:

$$C_{tap} = \frac{1}{2} \frac{dB}{d\omega}\bigg|_{\omega^2 = \frac{(C_1 + C_2)}{L C_1 C_2}} = \frac{1}{2}\left[C_1 + \frac{(1 - \omega^2 L C_2)C_2 + 2\omega^2 C_2^2 L}{(1 - \omega^2 L C_2)^2}\right]\bigg|_{\omega^2 = \frac{(C_1 + C_2)}{L C_1 C_2}}$$

$$= C_1\left(\frac{C_1 + C_2}{C_2}\right)$$

9.7 FILTER DESIGN USING INVERTERS

The low-pass-to-bandpass transformation gives a filter design that has both series and shunt resonators in it. In order to make a filter out of only one type of resonator, a technique must be used to transform series resonators into shunt resonators or shunt resonators into series resonators depending on the type desired. This is accomplished with the use of immittance inverters. This section follows filter design techniques given by Cohn [4] and by Matthaei [5].

9.7.1 Impedance Inverters

Figure 9.7 shows an impedance K inverter and an admittance J inverter. Note that an impedance inverter also can perform as an admittance inverter.

$$Z_{in} = \frac{K^2}{Z_L}$$

$$Y_{in} = \frac{J^2}{Y_L}$$

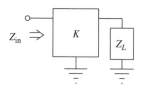
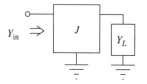

Figure 9.7 Symbols for a K inverter and a J inverter.

It will shortly be clear that K inverters are used with series circuits and J inverters are used with shunt circuits. Figures 9.8, 9.9, 9.10, and 9.11 show some inverters. The narrowband inverters shown in Figures 9.9 and 9.11 should not be used inside a filter design. They can be used at the input or the output of a filter with only a minor effect on the filter. A quarter-wavelength transmission line acts like either an impedance inverter or an admittance inverter.

$$Z_{in} Z_{load} = Z_0^2 \Rightarrow Z_{in} = \frac{Z_0^2}{Z_{load}}$$

$$Y_{in} Y_{load} = Y_0^2 \Rightarrow Y_{in} = \frac{Y_0^2}{Y_{load}}$$

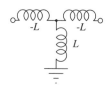
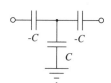

Figure 9.8 Two impedance or K inverters, $K = \omega L$ or $K = 1/\omega C$.

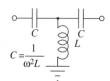
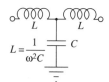

Figure 9.9 Two narrowband impedance or K inverters, $K = \omega L$ or $K = 1/\omega C$.

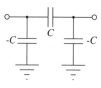
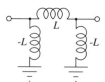

Figure 9.10 Two common admittance or J inverters, $J = \omega C$ or $J = 1/\omega L$.

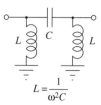
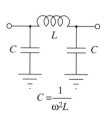

Figure 9.11 Two narrowband admittance or J inverters, $J = \omega C$, or $J = 1/\omega L$.

For the K inverter formed from reactances, the $ABCD$ matrix of a series jX, a shunt $-jX$, followed by a series jX and all terminated with a load is:

$$\begin{pmatrix} V_1 \\ I_1 \end{pmatrix} = \begin{pmatrix} 1 & jX \\ 0 & 1 \end{pmatrix} \begin{pmatrix} 1 & 0 \\ \dfrac{1}{-jX} & 1 \end{pmatrix} \begin{pmatrix} 1 & jX \\ 0 & 1 \end{pmatrix} \begin{pmatrix} 1 & 0 \\ Y_L & 1 \end{pmatrix} = \begin{pmatrix} jXY_L & jX \\ \dfrac{1}{-jX} & 0 \end{pmatrix}$$

$$Z_{\text{in}} = \frac{A}{C} = \frac{jXY_L}{\dfrac{1}{-jX}} = X^2 Y_L = \frac{X^2}{Z_L}$$

A similar analysis holds for the J inverter. The inverter circuits that have negative component values are used in filter designs. The negative values are absorbed into the adjacent positive-valued components forming a positive-valued component. Consider the case for a series inductor adjacent to a K inverter as shown in Figure 9.12. The resulting new inductors "L_1-L" and "L_2-L" have positive values whenever the magnitude of L coming from the K inverter is not greater than either L_1 or L_2. Keeping the net inductor positive is one of the design limitations when using inverters to generate filters. An inverter will change an impedance into an admittance and vice versa. Therefore, parallel resonators loading one side of an inverter appear as series resonators on the other side of an inverter and vice versa.

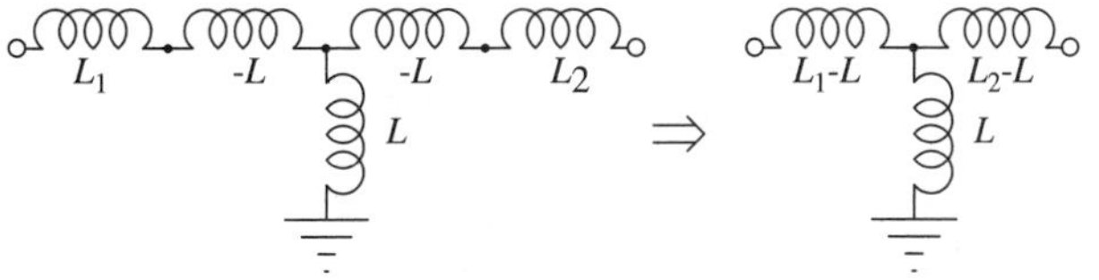

Figure 9.12 The negative-valued components are absorbed by adjacent positive-valued components.

The scattering matrix for an inverter normalized to the inverter value is:

$$(S) = \begin{pmatrix} 0 & -j \\ -j & 0 \end{pmatrix}$$

Substituting these values into the input reflection coefficient formula gives the result that the input reflection coefficient is the negative of the load reflection coefficient.

$$\Gamma_{\text{in}} = -\Gamma_L$$

$$Z_{\text{in}} = Z_0 \frac{1 + \Gamma_{\text{in}}}{1 - \Gamma_{\text{in}}} = Z_0 \frac{1 - \Gamma_L}{1 + \Gamma_L} = \frac{Z_0^2}{Z_L}$$

9.7.2 Filter Design Using Inverters—Center Sections

Consider the end section of a low-pass prototype as shown in Figure 9.13. The bottom part of the figure shows the bandpass circuit derived from the low-pass prototype. If a parallel resonator is required, then two J-inverters can be used to substitute for the series resonator as shown in Figure 9.14. Note that the g_{n+1} load in Figure 9.14 has also been simultaneously scaled to G_L. Applying this scale factor to the circuit in the bottom part of Figure 9.13 the impedance seen by the load G_L in Figure 9.14 should be:

$$Z'_n = Z_n R_L g_{n+1} = Z_n \frac{g_{n+1}}{G_L} = Z_n \text{ (scale factor)}$$

$$= \left(j\omega L + \frac{1}{j\omega C} \right) \frac{g_{n+1}}{G_L} + Z_{\text{rest}} \frac{g_{n+1}}{G_L}$$

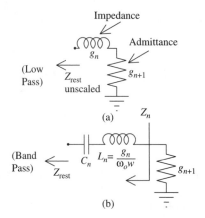

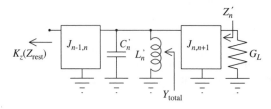

Figure 9.13 Low-pass prototype transformed to a bandpass structure.

Figure 9.14 Bandpass structure converted to a parallel resonator.

The impedance seen by the load G_L in Figure 9.14 is:

$$Y_n' = \frac{J_{n,n+1}^2}{Y_{total}} = \frac{J_{n,n+1}^2}{j\omega C_n' + \dfrac{1}{j\omega L_n'} + \cdots} = \frac{1}{Z_n'}$$

Equating the impedances from the above two equations results in:

$$j\omega L_n \frac{g_{n+1}}{G_L} = \frac{j\omega C_n'}{J_{n,n+1}^2} = j\omega \frac{g_n}{\omega_o w} \frac{g_{n+1}}{G_L}$$

$$\Rightarrow J_{n,n+1} = \sqrt{\frac{C_n' \omega_o w G_L}{g_n g_{n+1}}}$$

$$\frac{1}{j\omega C_n} \frac{g_{n+1}}{G_L} = \frac{1}{j\omega L_n' J_{n,n+1}^2}$$

A similar procedure can be used for the source-to-filter inverter. The inverter on the input of the filter is:

$$J_{0,1} = \sqrt{\frac{C_1' \omega_o w G_S}{g_0 g_1}}$$

Note that the value of $J_{0,1}$ or $J_{n,n+1}$ depends on C_n'. Therefore C_n' can be specified as a particular value.

If the low-pass-to-bandpass transformation requires an impedance rather than an admittance for the load, then an alternate termination as shown in Figure 9.15 is needed.

$$Y_n' = \frac{J_{n,n+1}^2}{G_{n+1}} = J_{n,n+1}^2 R_{n+1} = G_{n+1}'$$

$$\Rightarrow R_{n+1} = \frac{1}{J_{n,n+1}^2 R_{n+1}'}$$

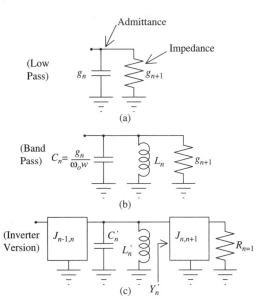

Figure 9.15 Low-pass prototype, bandpass transformation, and inverter versions of termination.

Constraining the Q's to be the same for the end sections of the low-pass-to-bandpass filter and the filter using J inverters we have:

$$C'_n R'_{n+1} = C_n g_{n+1} = \frac{g_n g_{n+1}}{\omega_0 w}$$

$$\Rightarrow C'_n = \frac{g_n g_{n+1}}{\omega_0 w R'_{n+1}} = \frac{g_n g_{n+1}}{\omega_0 w} J^2_{n,n+1}$$

$$\Rightarrow J_{n,n+1} = \sqrt{\frac{C'_n \omega_0 w G_L}{g_n g_{n+1}}}$$

This is the same J inverter equation as the J inverter equation found using Figure 9.14.

Now consider a center section as shown in Figure 9.16 for a low-pass-to-bandpass transformation somewhere in the middle of the filter. The component values and the impedance seen to the left of the middle of the filter are:

$$\hat{L}'_i = \frac{g_i}{\omega_0 w}$$

$$\hat{L}'_{i+1} = \frac{w}{\omega_0 g_{i+1}}$$

$$\hat{C}'_{i+1} = \frac{g_{i+1}}{\omega_0 w}$$

$$\hat{Y}_i^{-1} = \hat{Z}_i = j\omega \hat{L}'_i + \frac{1}{j\omega \hat{C}'_i} + \cdots$$

In the inverter version of the filter shown in Figure 9.17, the impedance is:

$$Z_i = \frac{j\omega C'_i + \dfrac{1}{j\omega L'_i} + \cdots}{J^2_{i,i+1}}$$

The ratio of impedances between the bandpass transformed filter and the inverter-based

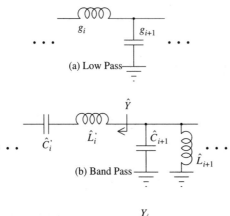

(a) Low Pass

(b) Band Pass

Figure 9.16 An example of a center section transformed to a bandpass via direct transformation.

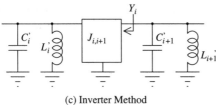

(c) Inverter Method

Figure 9.17 An example of a center section for the inverter method.

filter is:

$$\frac{Z_{\text{INV}}}{Z_{\text{BP}}} = \frac{Z_i}{\hat{Z}_i} = \frac{\dfrac{1}{\omega C'_{i+1}}}{\dfrac{1}{\omega \hat{C}'_{i+1}}} = \frac{\hat{C}'_{i+1}}{C'_{i+1}} = \frac{g_{i+1}}{\omega w C'_{i+1}}$$

Combining the equations for Figure 9.16 and Figure 9.17:

$$Z_i = \hat{Z}_i \frac{g_{i+1}}{\omega_0 w C'_{i+1}} = \frac{g_{i+1}}{\omega_0 w C'_{i+1}} \left(j\omega \hat{L}'_i + \frac{1}{j\omega \hat{C}'_i} + \cdots \right)$$

$$= \frac{g_{i+1}}{\omega_0 w C'_{i+1}} \left(j\omega \frac{g_i}{\omega_0 w} + \frac{1}{j\omega \hat{C}'_i} + \cdots \right)$$

$$= \frac{j\omega C'_i + \dfrac{1}{j\omega L'_i} + \cdots}{J^2_{i,i+1}}$$

$$\Rightarrow \frac{C'_i C'_{i+1}}{J^2_{i,i+1}} = \frac{g_i}{\omega_0 w} \frac{g_{i+1}}{\omega_0 w}$$

$$J_{i,i+1} = \omega_0 w \sqrt{\frac{C'_i C'_{i+1}}{g_i g_{i+1}}}$$

We now have the equations needed to design the J inverter needed between resonators of a filter. It transforms the immittance of adjacent parallel resonators to track the immittances seen in a filter formed from a low-pass-to-bandpass transformation. These J inverters allow a given or specified resonator to simulate the required circuit response corresponding to a resonator that would have resulted from a low-pass-to-bandpass transformation using these same g_i's.

The equations for three positions for J inverters, one for the source to first resonator, the inverters between resonators, and one for the last resonator to load, are summarized below.

$$J_{0,1} = \sqrt{\frac{C_1' \omega_0 w G_S}{g_0 g_1}}$$

$$J_{i,i+1} = \omega_0 w \sqrt{\frac{C_i' C_{i+1}'}{g_i g_{i+1}}}$$

$$J_{n,n+1} = \sqrt{\frac{C_n' \omega_0 w G_L}{g_n g_{n+1}}}$$

Similar results would be found if the resonators used were series resonators except that K inverters would be used. Therefore, by duality, the K values for an inverter-based filter using series resonators are:

$$K_{0,1} = \sqrt{\frac{L_1' \omega_0 w R_S}{g_0 g_1}}$$

$$K_{i,i+1} = \omega_0 w \sqrt{\frac{L_i' L_{i+1}'}{g_i g_{i+1}}}$$

$$K_{n,n+1} = \sqrt{\frac{L_n' \omega_0 w R_L}{g_n g_{n+1}}}$$

Figure 9.18 shows how each filter would look after it is built using J or K inverters. The end section inverters require some more discussion.

9.7.3 Filter Design Using Inverters—End Sections

Note that an end section inverter has no adjacent reactive component in the source or the load to absorb a negative component value. It may be necessary to use a narrowband version of the admittance inverter for the end section. This is accomplished by using an inductor in place of a negative capacitor or a capacitor in place of a negative inductor right at the source or load. One can also use a tapped resonator. The end section inverter transforms the load impedance to a value of $1/J^2 R_L$. Therefore, a $Q^2 + 1$ match (discussed in Chapter 8) can be also be used to perform this impedance transformation using all positive components. For example, if the transformed impedance is higher than the load impedance:

$$(Q^2 + 1) R_L = \frac{1}{J^2 R_L}$$

$$Q = \sqrt{\frac{1}{J^2 R_L^2} - 1}$$

The $Q^2 + 1$ match can be used in place of the last inverter. This eliminates the need for absorbing negative component values at the load. The $Q^2 + 1$ match should not be used inside the filter since the reactance slope parameters inside the filter will be altered. However, at the source or load end, the Q associated with the resistance of the source or load allows the $Q^2 + 1$ match to be used with minimal perturbation of the filter response. Similarly, if the impedance

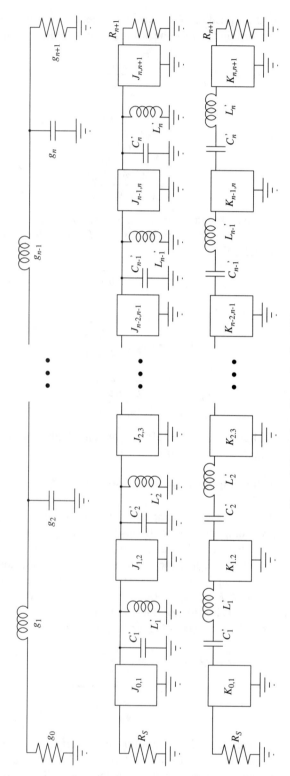

Figure 9.18 Bandpass filters using J and K inverters.

is lower than the load impedance:

$$R_L = \frac{1}{J^2 R_L}(Q^2 + 1)$$

$$Q = \sqrt{J^2 R_L^2 - 1}$$

If one is free to choose the value of the resonator reactances, then choosing $J_{01} = Y_S$ (or $K_{01} = Z_S$) and $J_{n,n+1} = Y_L$ (or $K_{n,n+1} = Z_L$) allows one to build a filter without the need for the first and last inverters.

$$Y_{\text{Filter}} = \frac{J^2}{Y_S} = \frac{Y_S^2}{Y_S} = Y_S$$

9.8 A TWO-POLE MICROSTRIP FILTER DESIGN EXAMPLE

9.8.1 Resonator Analysis—Comb Line Filter Example

Figure 9.19 shows a top view of a quarter-wavelength resonator. This resonator will be used to design a two-pole inverter coupled comb line filter. The filter resonators are all tied to ground at one end and several of them placed together to made a comb line filter look like the teeth of a comb. The resonator is shown as having the same width, and thus the same characteristic impedance over its whole length. Often filters are made out of resonators that have the top portion having a lower characteristic impedance than the lower portion, although then the total length of the resonator is no longer a quarter-wavelength. The correct resonance condition needs to be determined when the reactance slopes are calculated. The same design technique can be used with those resonators. Since the resonator that is being considered is a quarter-wavelength long, at any tap point along its length, it looks like an open circuit. The effective inductance and capacitance vary as a function of that tap point.

$$v_i = \frac{c}{\sqrt{\varepsilon_{\text{eff\_i}}}} \qquad \theta_1 = \frac{\omega D_1}{v_1} \qquad \theta_2 = \frac{\omega D_2}{v_2} \qquad D = D_1 + D_2$$

$$B = Y_{02} \tan(\theta_2) - Y_{01} \cot(\theta_1)$$

$$\frac{\omega \, dB}{2 \, d\omega}\bigg|_{\theta_2 = \frac{\pi}{2} - \theta_1} = \frac{\omega}{2}\left[\frac{Y_{02} D_2}{v_2} \sec^2(\theta_2) + \frac{Y_{01} D_1}{v_1} \csc^2(\theta_1)\right]\bigg|_{\theta_2 = \frac{\pi}{2} - \theta_1}$$

$$= \frac{\omega \csc^2(\theta_1)}{2}\left[\frac{Y_{02}(D - D_1)}{v_2} + \frac{Y_{01} D_1}{v_1}\right]$$

$$C_{\text{eff}} = \frac{\csc^2(\theta_1)}{2}\left[\frac{Y_{02}(D - D_1)}{v_2} + \frac{Y_{01} D_1}{v_1}\right]$$

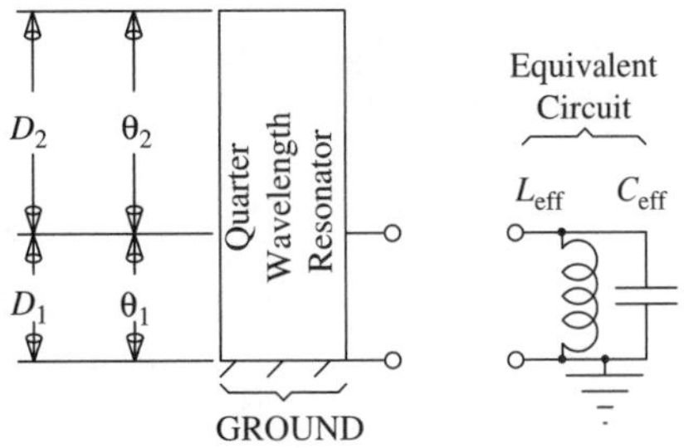

Figure 9.19 Quarter-wave resonator and equivalent circuit.

Y_{02} is the characteristic admittance of the top portion of the resonator and Y_{01} is the characteristic admittance of the bottom portion of the resonator. θ_2 is the radian length of the top portion of the resonator and θ_1 is the radian length of the bottom portion of the resonator.

When the characteristic impedance is the same at all points along the resonator, then the effective capacitance is given as:

$$C_{\text{eff}} = \frac{Y_0 D \csc^2(\theta_1)}{2v} = \frac{Y_0}{\omega} \frac{\pi}{4} \csc^2(\theta_1) = \frac{Y_0}{8f} \csc^2(\theta_1)$$

$$L_{\text{eff}} = \frac{1}{\omega^2 C_{\text{eff}}}$$

As expected, the effective capacity is an inverse function of frequency. The effective inductor is determined by the value needed to resonate the effective capacity at the given frequency.

9.8.2 Inverter Analysis—Comb Line Filter Example

The inverter used in the quarter-wavelength inverter coupled filter consists of two negative impedance transmission lines and one positive impedance transmission line as shown in Figure 9.20. Using ABCD matrices, the inverter can be described as:

$$\begin{pmatrix} A & B \\ C & D \end{pmatrix}\bigg|_{\text{inv}} = \begin{pmatrix} 1 & 0 \\ jY_0 \tan(\theta_1) & 1 \end{pmatrix} \begin{pmatrix} \cos(\theta_1) & jZ_0 \sin(\theta_1) \\ jY_0 \cos(\theta_1) & \cos(\theta_1) \end{pmatrix} \begin{pmatrix} 1 & 0 \\ jY_0 \tan(\theta_1) & 1 \end{pmatrix}$$

$$= \begin{pmatrix} 0 & jZ_0 \sin(\theta_1) \\ jY_0 \csc(\theta_1) & 0 \end{pmatrix}$$

If a load Y_L is added on the right side of the inverter and the input admittance is determined from C/A, the resulting input admittance and J inverter value are:

$$\begin{pmatrix} A & B \\ C & D \end{pmatrix} = \begin{pmatrix} 0 & jZ_0 \sin(\theta_1) \\ jY_0 \csc(\theta_1) & 0 \end{pmatrix} \begin{pmatrix} 1 & 0 \\ Y_L & 1 \end{pmatrix} = \begin{pmatrix} jZ_0 Y_L \sin(\theta_1) & jZ_0 \sin(\theta_1) \\ jY_0 \csc(\theta_1) & 0 \end{pmatrix}$$

$$Y_{\text{in}} = \frac{C}{A} = \frac{jY_0 \csc(\theta_1)}{jZ_0 Y_L \sin(\theta_1)} = \frac{Y_0^2 \csc^2(\theta_1)}{Y_L}$$

$$J_{\text{inv}} = Y_0 \csc(\theta_1)$$

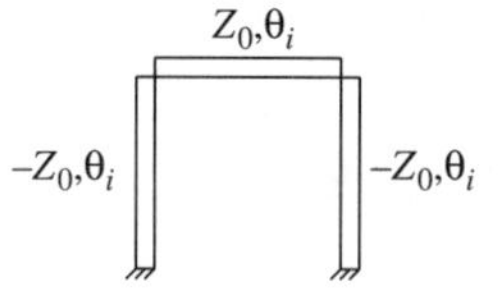

Figure 9.20 Inverter section for filter design example. [R. Weber, *Microwave Engineering Course Notes*, ©1987–1997, August 1997 edition, p. 138.]

9.8.3 End Section Analysis—Comb Line Filter Example

Figures 9.21 and 9.22 show series and shunt end sections respectively as they would be in a transformed bandpass. The external Q for the end section is:

$$Q_{\text{end section}} = \frac{\omega_0}{Z_S} L = \frac{\omega_0 Z_S g_1}{Z_S \omega_0 w} = \frac{g_1}{w}$$

For the dual shown in Figure 9.22, the external Q is the same.

$$Q_{\text{end section}} = \frac{\omega_0}{Y_S} C = \frac{\omega_0 Y_S g_1}{Y_S \omega_0 w} = \frac{g_1}{w}$$

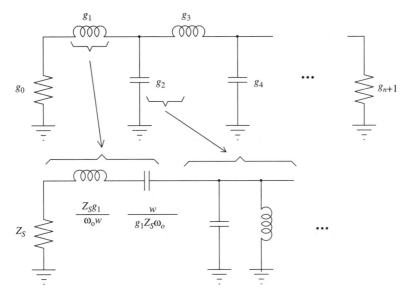

Figure 9.21 Series (impedance) end section for g_1.

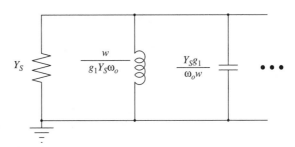

Figure 9.22 Shunt (admittance) end section for g_1.

The external Q of the end section sets the tap point of the end resonator. This external Q can often be set to the required value by varying the tap point of an L-C circuit by using the formulas for tapped L-C resonators. Tapping the resonator to perform impedance transformation changes the resonant frequency and thus the reactance slope of the resonator. The effect is small if the transformation ratio is small. Adding or subtracting a small amount of one resonator component usually compensates for this. Figure 9.23 shows an end resonator tap point for a quarter-wavelength resonator. For either a tapped L-C resonator or the distributed resonator the external Q is:

$$Q_{\text{ext}} = \frac{\omega C_{\text{tap}}}{Y_S}$$

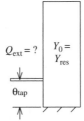

Figure 9.23 End resonator for a comb line quarter-wavelength filter. [*R. Weber, Microwave Engineering Course Notes, ©1987–1997, August 1997 edition, p. 142.*]

The external Q for the resonator shown in Figure 9.23 is:

$$C_{\text{tap}} = \frac{Y_0 \pi}{4\omega_0} \csc^2(\theta_{\text{tap}})$$

$$Q_{\text{ext}} = \frac{\omega C_{\text{tap}}}{Y_S} = \frac{\omega_0 Y_0 \pi}{4\omega_0 Y_S} \csc^2(\theta_{\text{tap}}) = \frac{Y_0 \pi}{4Y_S} \csc^2(\theta_{\text{tap}}) = \frac{g_1}{w}$$

The dual circuit and Q would be used for a series type of resonator.

9.8.4 Filter Example—Two-Pole Comb Line Butterworth Filter

Using the inverter formulas just derived, a microstrip line, two-pole, 1 GHz, 10% bandwidth Butterworth filter will be designed. The g parameters and design follow. Both parallel resonators will be made using the same characteristic impedance. This makes the resonator C's the same for all the resonators within the filter. The g values for a Butterworth response and the J inverter values are:

$$g_0 = 1 \quad g_1 = \sqrt{2} \quad g_2 = \sqrt{2} \quad g_3 = 1$$

$$J_{1,2} = Y_{\text{inv}} \csc(\theta_{\text{inv}}) = \frac{\omega_0 w}{g_1} \frac{Y_{\text{res}} \pi}{4\omega_0} \csc^2(\theta_{\text{inv}})$$

$$Y_{\text{inv}} = \frac{w Y_{\text{res}} \pi}{4 g_1} \csc(\theta_{\text{inv}})$$

$$\theta_{\text{inv}} = \arcsin\left(\frac{\pi w Z_{\text{inv}}}{4 g_1 Z_{\text{res}}}\right)$$

The input tap points are determined using the formula derived for external Q for the end inverter.

$$\csc^2(\theta_{\text{tap}}) = \frac{4 g_1 Y_S}{\pi w Y_{\text{res}}}$$

$$\theta_{\text{tap}} = \arcsin\left(\sqrt{\frac{\pi w Z_S}{4 g_1 Z_{\text{res}}}}\right)$$

In the following numerical example, an effective dielectric constant of 1 is used. In an actual filter built on a dielectric substrate the lengths need to be shortened by the square root of the effective dielectric constant of the substrate. The filter is shortened as appropriate when the effective dielectric constant of the substrate is known. Using a characteristic impedance of 20 ohms for the resonator, $Z_s = 50\Omega$, $\omega_0 = 2\pi \times 10^9/s$, $w = 0.1$, $Z_{\text{inv}} = 100\Omega$, and $\varepsilon_r = 1$, the tap points are:

$$\theta_{\text{tap}} = \arcsin\left(\sqrt{\frac{0.1\pi\, 50}{4\sqrt{2}\, 20}}\right) = \arcsin(0.3726)$$

$$= 21.9° = 0.3818\,\text{rad} = \frac{\omega D_{\text{tap}}}{v} = \frac{2\pi\, 10^9}{3 \times 10^8} D_{\text{tap}}$$

$$D_{\text{tap}} = 0.01823\,\text{m} = 0.7177\,\text{in}$$

$$\theta_{\text{inv}} = \arcsin\left(\frac{0.1\pi\, 100}{4\sqrt{2}\, 20}\right) = \arcsin(0.2777)$$

$$= 16.2° = 0.2814\,\text{rad} = \frac{\omega D_{\text{inv}}}{v} = \frac{2\pi\, 10^9}{3 \times 10^8} D_{\text{inv}}$$

$$D_{\text{inv}} = 0.01343\,\text{m} = 0.5289\,\text{in}$$

The characteristic admittance of the lower portion of the resonator is the sum of the resonator admittance value and the negative admittance value for the admittance inverter $(0.02S - 0.01S = 0.19S)$. This makes the lower portion of the filter resonator narrower than the upper portion. Figure 9.24 shows a top view of the filter.

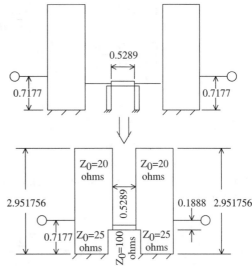

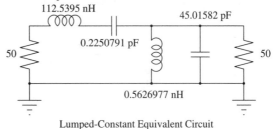

All lengths above need to be scaled by the square root of the effective dielectric constant. The effective dielectric constant depends on the material used for fabrication.

Lumped-Constant Equivalent Circuit

Figure 9.24 Final filter and scaled prototype. [*R. Weber, Microwave Engineering Course Notes, ©1987–1997, August 1997 edition, p. 145.*]

9.9 COUPLED LINE FILTERS

Two transmission lines that are in close proximity to each other will exchange energy from one line to the other. These transmission lines can be excited in phase. This is called even-mode excitation. There is a characteristic impedance Z_{0e} associated with this excitation. The transmission lines can also be excited out of phase. Then there is an odd-mode characteristic impedance Z_{0o} associated with the behavior of these transmission lines. Even- and odd-mode analysis of transmission lines was mentioned briefly in Chapter 4. Coupled lines will be discussed further in Chapter 12.

9.9.1 Coupled-Line Filters—Open-Circuit InterDigital Sections

When open-circuit or short-circuited half-wavelength lines are coupled together as will be discussed here, they look somewhat like the fingers of two hands pointing toward each other and then placed together with the fingers of one hand side by side the fingers of the other hand. Fingers are called *digits*. Filters formed this way are called *interdigital filters*. Two sections of open-circuited coupled sections are shown in Figure 9.25. Each coupled section has the following Z matrix (Section 11.11).

$$
\begin{pmatrix} V_1 \\ V_2 \end{pmatrix} = \begin{pmatrix} \dfrac{-j(Z_{0e} + Z_{0o})}{2\tan(\theta)} & \dfrac{-j(Z_{0e} - Z_{0o})}{2\sin(\theta)} \\ \dfrac{-j(Z_{0e} - Z_{0o})}{2\sin(\theta)} & \dfrac{-j(Z_{0e} + Z_{0o})}{2\tan(\theta)} \end{pmatrix} \begin{pmatrix} I_1 \\ I_2 \end{pmatrix}
$$

$$
= \left[\begin{pmatrix} \dfrac{-jZ_{0STUB}}{\tan(\theta)} & 0 \\ 0 & \dfrac{-jZ_{0STUB}}{\tan(\theta)} \end{pmatrix} + \begin{pmatrix} \dfrac{-jZ_{0INV}}{\tan(\theta)} & \dfrac{-jZ_{0INV}}{\sin(\theta)} \\ \dfrac{-jZ_{0INV}}{\sin(\theta)} & \dfrac{-jZ_{0INV}}{\tan(\theta)} \end{pmatrix} \right] \begin{pmatrix} I_1 \\ I_2 \end{pmatrix}
$$

$$
Z_{0STUB} = Z_{0o}
$$

$$
Z_{0INV} = \frac{Z_{0e} - Z_{0o}}{2}
$$

When the length of the open-circuited coupled section is $\theta = \pi/2$ ($\lambda/4$ line length), the input impedance is:

$$
Z_{in} = \frac{\left(\dfrac{Z_{0e} - Z_{0o}}{2}\right)^2}{Z_{load}} = \frac{K^2}{Z_{load}}
$$

$$
K = \frac{Z_{0e} - Z_{0o}}{2}
$$

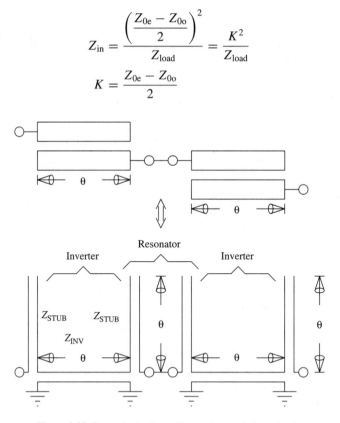

Figure 9.25 Open-circuited coupling section equivalent circuit.

Therefore, when the coupled section is a quarter-wavelength long, the section acts as an impedance inverter. Notice that the series open-circuited stubs on the input and the output of each equivalent circuit connect the inverter section to the input and output when the section is a quarter-wavelength long. The open-circuit stubs act as a resonator and the center section acts as an inverter.

A similar analysis (Section 11.11) for the shorted coupled section shown in Figure 9.26 gives the following Y matrix for part of the coupled section.

$$
\begin{pmatrix} V_1 \\ -V_2 \end{pmatrix} = \begin{pmatrix} \dfrac{-j(Y_{0o}+Y_{0e})}{2\tan(\theta)} & \dfrac{+j(Y_{0o}-Y_{0e})}{2\sin(\theta)} \\[2mm] \dfrac{+j(Y_{0o}-Y_{0e})}{2\sin(\theta)} & \dfrac{-j(Y_{0o}+Y_{0e})}{2\tan(\theta)} \end{pmatrix} \begin{pmatrix} I_1 \\ -I_2 \end{pmatrix}
$$

$$
= \left[\begin{pmatrix} \dfrac{-jY_{0STUB}}{\tan(\theta)} & 0 \\[2mm] 0 & \dfrac{-jY_{0STUB}}{\tan(\theta)} \end{pmatrix} + \begin{pmatrix} \dfrac{-jY_{0INV}}{\tan(\theta)} & \dfrac{+jY_{0INV}}{\sin(\theta)} \\[2mm] \dfrac{+jY_{0INV}}{\sin(\theta)} & \dfrac{-jY_{0INV}}{\tan(\theta)} \end{pmatrix} \right] \begin{pmatrix} I_1 \\ -I_2 \end{pmatrix}
$$

$$Y_{0STUB} = Y_{0e}$$

$$Y_{0INV} = \frac{Y_{0o}-Y_{0e}}{2}$$

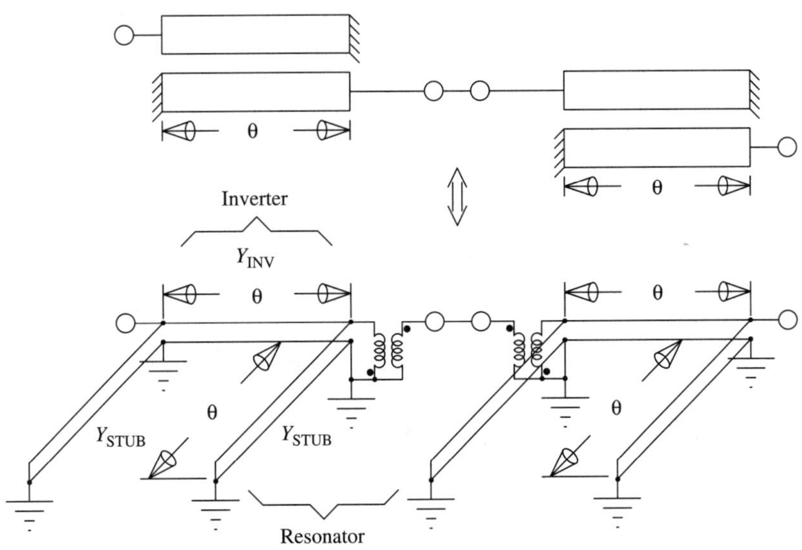

Figure 9.26 Short-circuited coupling section equivalent circuit.

Notice the minus signs on V_2 and I_2 because it is only part of the equivalent circuit. A transformer needs to be added to the equivalent circuit. The minus signs could have been put on V_1 and I_1 and not on V_2 and I_2 and the equation would be equally valid. The minus sign comes from the shorts giving a voltage reflection coefficient of -1. A unity-turns-ratio, inverting-transformer on either port of a two-port changes the sign of the off diagonal terms of the immittance matrix and leaves the diagonal terms unchanged. Adding a transformer to the equivalent circuit gives the following admittance matrix without the use of a minus sign on the second port voltage or current.

$$
\begin{pmatrix} V_1 \\ V_2 \end{pmatrix} = \begin{pmatrix} \dfrac{-j(Y_{0STUB}+Y_{0INV})}{\tan(\theta)} & \dfrac{-jY_{0INV}}{\sin(\theta)} \\[2mm] \dfrac{-jY_{0INV}}{\sin(\theta)} & \dfrac{-j(Y_{0STUB}+Y_{0INV})}{\tan(\theta)} \end{pmatrix} \begin{pmatrix} I_1 \\ I_2 \end{pmatrix}
$$

Notice that in a filter section, two inverting transformers in series in the model results in no net phase shift. The equivalent circuit would be equally valid if the transformer were inserted in the middle of the inverter section. This would yield a symmetrical circuit but the one shown in Figure 9.26 allows the two transformers to be placed in series for no net phase shift and thus not enter the analysis.

Two sections of either open or shorted-coupled sections can be cascaded together to provide a half-wavelength resonator and an impedance inverter. A schematic derivation of a two-pole filter using open-coupled sections is shown in Figure 9.27. In order to put these filter sections in the form in order that inverter theory can be used for its design, the lumped-constant equivalent of the filter, valid around the frequency for which the coupled sections are a quarter-wavelength long, needs to be found. One possible circuit is shown in the center of Figure 9.28. The inductances and capacitors L1, L2, L3, C1, C2, and C3 are determined from reactance slope parameters for the open-circuited stubs at resonance. The middle part of Figure 9.28 shows a narrowband version of the equivalent circuit. This equivalent circuit works well only at the center frequency since narrowband inverters are used. The bottom part of the figure shows a much broader-band version of the circuit that works over the passband of the filter and somewhat beyond. The inverter in the bottom circuit is much broader band since it uses only one type of reactive element. A tee equivalent circuit is used to model the behavior of the transmission line section that couples the open-circuited stubs together. The Z matrix of this transmission line is:

$$(Z) = \left(\frac{Z_{0e} - Z_{0o}}{2} \right) \begin{pmatrix} -j\cot(\theta) & -j\csc(\theta) \\ -j\csc(\theta) & -j\cot(\theta) \end{pmatrix}$$

Using the Z matrix of a transmission line, the impedances of L's and the C for the tee equivalent circuit of the transmission line are:

$$Z_L = \frac{j\left(\dfrac{Z_{0e} - Z_{0o}}{2} \right)[1 - \cos(\theta)]}{\sin(\theta)}$$

$$Z_C = \frac{-j\left(\dfrac{Z_{0e} - Z_{0o}}{2} \right)}{\sin(\theta)}$$

At the quarter-wavelength frequency, these impedances reduce to:

$$Z_L = +j\left(\frac{Z_{0e} - Z_{0o}}{2} \right)$$

$$Z_C = -j\left(\frac{Z_{0e} - Z_{0o}}{2} \right)$$

These values satisfy the conditions for an impedance converter. However, this inverter would be a narrowband frequency converter as shown in the middle of Figure 9.28. In the bottom part of Figure 9.28, a positive capacitor equal to the value of the inverter capacitor is inserted in series with the same value but negative capacitor. This series combination gives zero reactance and has not changed the circuit. However, by rearranging the components, a broadband frequency inverter is formed.

The two series open-circuit stubs along with the equivalent stubs from the coupled line inverter section form a series resonator. The lumped-constant equivalent of these stubs is shown Figure 9.29. The equivalent circuit shown in Figure 9.29 is valid only when the even- and odd-mode velocities of the coupled line section are not widely different. When the velocities

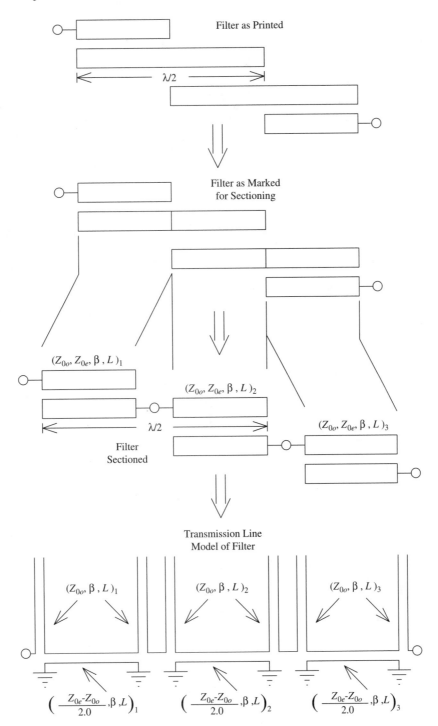

Figure 9.27 Schematic of a two-pole interdigital filter.

are widely different, an equivalent inductance is derived from the model shown in the center of Figure 9.29. When the sum of Z_{0e} plus Z_{0o} is set equal to some given value for all of the resonators, a simple design equation results for the synthesis of this filter. The series resonator

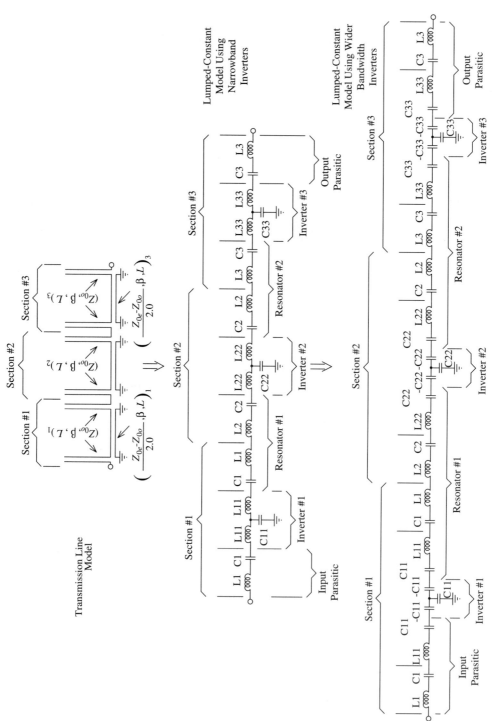

Figure 9.28 Procedure to go from coupled-line to K inverters.

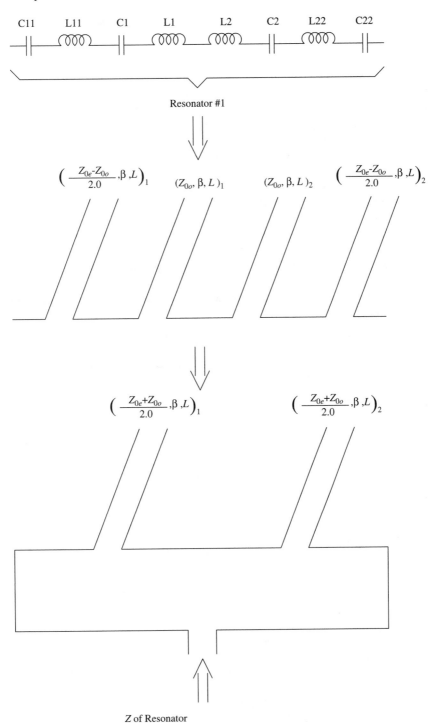

Figure 9.29 Procedure to form the coupled-line resonator.

has an equivalent inductance. The inductance can be derived from reactance slope parameters of a quarter-wavelength transmission line open-circuit stub of impedance $Z_{0e} + Z_{0o}$ and is:

$$L_{eq} = \frac{\pi(Z_{0e} + Z_{0o})}{4\omega}$$

The inverter value at the quarter-wavelength frequency as shown above is:

$$K = \frac{(Z_{0e} - Z_{0o})}{2}$$

The input series combination of L1, C1, L11, C11 is series resonant at the quarter-wavelength frequency. This input parasitic will not affect the performance of the coupled line filter over a narrowband around the center frequency and will have only a small effect over a broader bandwidth since the source resistance gives this resonator a small loaded Q. The reactance slope of these input (and output) parasitics is usually small when compared to the input and output source and load impedance magnitudes. A similar design technique can be used to design shorted-coupled sections using J inverters.

9.9.2 Coupled-Line Filter Design Example— Interdigital Sections

A two-pole Butterworth filter at one GHz and having a 2% bandwidth will be designed using open-circuited coupled-line sections. The input and output impedances are to be 50 ohms. The g values for a two-pole Butterworth filter were given in the design example for the comb line filter. K_{01} will be equal to K_{23} since this filter is symmetrical. Using a value of $Z_{0e} + Z_{0o} = 100$ ohms and $Z_0 = 50$ ohms, the values for K and the even- and odd-mode impedances for the end section inverters are:

$$K_{01} = K_{23} = \frac{(Z_{0e} - Z_{0o})|_1}{2} = \sqrt{\frac{L_{eq}\omega w R_S}{g_0 g_1}} = \sqrt{\frac{\pi w (Z_{0e} + Z_{0o})|_1 R_S}{4 g_0 g_1}}$$

$$= \sqrt{\frac{\pi 0.02(100)50}{4\sqrt{2}}} = 7.452$$

$$Z_{0e} - Z_{0o} = 14.9$$
$$Z_{0e} + Z_{0o} = 100$$
$$Z_{0e} = 57.45$$
$$Z_{0o} = 42.55$$

Also using $Z_{0e} + Z_{0o} = 100$ ohms, the center section K value and even- and odd-mode impedances are:

$$K_{12} = \frac{(Z_{0e} - Z_{0o})|_2}{2} = L_{eq}\omega w \sqrt{\frac{1}{g_1 g_2}} = \frac{\pi w (Z_{0e} + Z_{0o})|_2}{4} \sqrt{\frac{1}{g_1 g_2}}$$

$$= \frac{\pi 0.02(100)}{4} \sqrt{\frac{1}{\sqrt{2}\sqrt{2}}} = 1.111$$

$$Z_{0e} - Z_{0o} = 2.222$$
$$Z_{0e} + Z_{0o} = 100$$
$$Z_{0e} = 51.11$$
$$Z_{0o} = 48.89$$

An equivalent circuit for use in a CAD program is shown in Figure 9.30. Spice programs usually require that open-circuited nodes have a large value resistor placed on them. That resistor is not shown in the circuit but cannot be forgotten if the circuited is simulated in Spice. A response of the magnitude of S_{11} and S_{21} for the filter is shown in Figure 9.31. Notice that

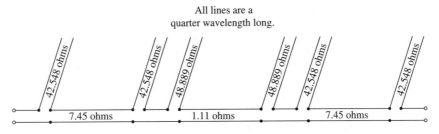

All lines are a
quarter wavelength long.

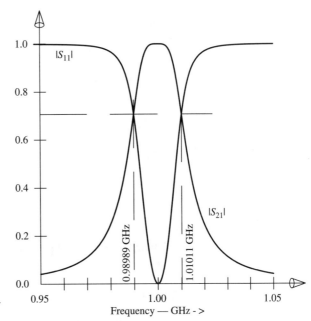

Figure 9.30 Equivalent circuit for a two-pole coupled-line filter.

Figure 9.31 Response of 2% bandwidth coupled-line filter.

the filter is quite close to 20 MHz wide at the 3-dB points and that the match at 1 GHz is almost perfect. Wider-bandwidth filters would tend to deviate from the expected response but this is a very good response.

9.10 SHUNT STUB FILTERS

Filters can be fabricated using shunt resonators made from open-circuit or short-circuit stubs. When quarter-wavelength lines separate these resonators, the design procedure is quite easy. Not all application requirements can be met with these filters but because they are easy to design, they can be tried. The shunt resonators can be either half-wavelength open-circuited resonators or quarter-wavelength short-circuited resonators. Each of the resonators can be made longer by multiples of a half-wavelength if a narrower band circuit is desired. The longer resonators have a higher susceptance slope and thus a narrower passband. However, the overmoded resonator will have spurious passbands whenever the resonator acts like an open circuit. Most transmission line filters exhibit spurious passbands whenever the lines forming the filters are excited with frequencies around a multiple of either a half-wavelength or quarter-wavelength.

Two versions of a three-pole filter using stubs and quarter-wavelength inverters are shown in Figure 9.32. For the resonators in the top part of this figure, the equivalent C of the resonator

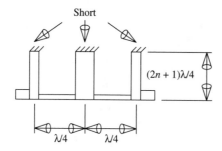

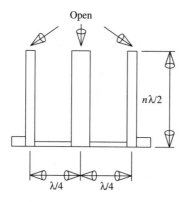

Figure 9.32 Drawing of open-circuit and short-circuit resonator filters.

is determined by:

$$Y_{\text{in}} = jY_0 \tan(\beta D) = jY_0 \tan\left(\frac{\omega}{v}D\right)$$

$$C_{\text{eq}} = \frac{1}{2}\frac{dB}{d\omega}\bigg|_{\omega=\omega_o} = \frac{1}{2}Y_0 \sec^2\left(\frac{\omega}{v}D\right)\frac{D}{v}\bigg|_{\frac{\omega}{v}D=n\pi} = \frac{n\pi Y_0}{2\omega} = \frac{DY_0}{2v}$$

For the resonators in the bottom of the figure, the equivalent C of the resonator is determined by:

$$Y_{\text{in}} = -jY_0 \cot(\beta D) = -jY_0 \cot\left(\frac{\omega}{v}D\right)$$

$$C_{\text{eq}} = \frac{1}{2}\frac{dB}{d\omega}\bigg|_{\omega=\omega_o} = \frac{1}{2}Y_0 \csc^2\left(\frac{\omega}{v}D\right)\frac{D}{v}\bigg|_{\frac{\omega}{v}D=\frac{2n+1}{2}\pi} = \frac{(2n+1)\pi Y_0}{4\omega} = \frac{DY_0}{2v}$$

The inverter value between each resonator is Y_0 of the transmission line used between the resonators. In order to have the input and output inverters eliminated as shown in these filters, those respective inverters are set equal to the inverse of the termination resistance.

$$Y_{\text{filter}} = \frac{J_{01}^2}{Y_S} = Y_S \Rightarrow J_{01} = Y_S$$

9.11 IMPEDANCE SCALING AND IMPEDANCE MATCHING WITH FILTERS

Notice that the formulas for intermediate inverters between filter resonators do not have the source or load immittance value in them. By changing the values of the resonator capacitors and inductors (their reactance or susceptance slopes), one can progressively raise or lower the impedance level as one progresses through the filter from the input to the output. The first and last inverter sections do have source and load immittances in them. There is a limit to the impedance transformation allowed by the last sections. The constraint is caused by the amount of negative reactance from these inverters that can be absorbed by the first or last resonator. The net value of the resonator inductor or capacitor must be no less than zero. The filter can be used to perform a matching network between a source and a reactive load. The reactive load can be treated as the last resonator if it resonated. A parallel or series resonator is determined by the configuration of the load: Does the load have a parallel or series structure? The Q of the last section sets a limit on the bandwidth of the filter and match [6]. The limit is often called the *Fano broadband limit*. The last section or resonator can sometimes be configured like a $Q^2 + 1$ inverter matching section. For a broadband match, a high number of poles are needed in the filter when the Q of the load is high.

9.12 UNIT ELEMENT FILTER DESIGN

Microwave filter design is often based on the concept of a unit element. The tangent or cotangent of pi over 4 is equal to 1. At the one-eighth-wavelength frequency, the impedance of a shorted stub is therefore $+jZ_0$ and the admittance of an open stub is $+jY_0$. An eight-wavelength section of line is considered to be a unit element. Since the immittance value of a transmission line unit element does not vary with frequency as does the immittance value of an inductor or capacitor, but varies as the tangent of frequency, a filter can be designed with the tangent function as the independent variable for plotting frequency response. This is called the *Richards transformation* [7]. For instance, if the low-pass prototype series g value is 1.5, the inductor used in a 50-ohm filter has a reactance of 50 times 1.5 at the cutoff frequency. With the Richards transformation, the inductor is replaced with a (50 times 1.5) ohm transmission line that is one-eighth of a wavelength long. Since each element has the same length, the relative rate of change of each element value will be the same. This type of filter is called a *commensurate line filter*. Note that the same reactance values will be produced by the tangent function whenever it is 1 in value. Therefore, the filter response repeats every four times the base frequency. The passband shape is different from the passband shape for lumped-element filters. However, when the passband is not too wide, the response of the filter is often acceptable.

9.13 MICROWAVE DIPLEXERS

The filters that have been discussed are doubly terminated circuits. Even though real sources have internal resistance, some circuit topologies use singly terminated filters. One of those circuits is a frequency diplexer. A frequency diplexer is a multiport network that takes input composed of several frequencies at one port and produces outputs at other ports with those outputs containing frequencies only in selected frequency bands. Other types of diplexers may route signals based on timing and this is often accomplished by switching circuits. These types of circuits are not considered in this section.

Frequency diplexers can be designed by combining two or more singly terminated networks in series or in parallel. Figures shown in this section contain four-pole low-pass and four-pole high-pass filters. Four-pole filters are shown for demonstration only. The number of poles can be greater than or less than four. In addition, the low-pass filters can be transformed to bandpass filters and the high-pass filters can be transformed to bandstop filters to form diplexers as well. In the bandpass or stop band case, the percentage bandwidth of each set of filters that will be discussed does not have to be the same. Consider the singly terminated Butterworth filters shown in Figure 9.33. This figure shows the top filter excited by a voltage source and the bottom filter excited by a current source. Generally, the g parameter tables for a single-terminated filter are given with g_{n+1} equal to infinity. However, for the purpose of

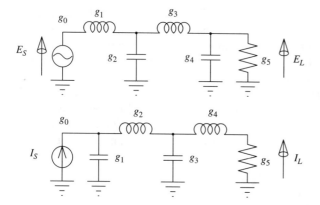

Figure 9.33 Low-pass prototypes using voltage or current sources.

using the singly terminated filter for a diplexer, the filter is shown flipped end for end with the infinite source immittance g_0 on the left-hand side and $g_{n+1} = 1$ (in the normalized case) on the right side. If the first component in the filter (starting from the left side) is a series element (an impedance), then the source immittance value is an admittance. An admittance equal to an infinite g_0 is a short circuit and thus a voltage source is used. Likewise, if the first component on the left side is a shunt element (an admittance), then the source immittance value is an impedance. An impedance equal to an infinite g_0 is an open circuit and thus a current source is used. The filters shown in Figure 9.33 are duals of each other. The dual of a voltage source is a current source and vice versa.

The transfer function of a Butterworth low-pass filter is [8]:

$$G_L^2 = \frac{1}{\omega^{2n} + 1}$$

In the low-pass-to-bandpass transformation, $1/\omega$ is replaced by ω. The transfer function for a Butterworth high-pass filter is then:

$$G_H^2 = \frac{1}{\omega^{-2n} + 1} = \frac{\omega^{2n}}{\omega^{2n} + 1}$$

G stands for voltage gain for voltage transfer functions and current gain for current transfer functions.

Consider singly terminated filters that have voltage sources. Connect a low-pass and a high-pass filter in parallel across the same voltage source as shown in Figure 9.34. Let each filter retain its own load. The load termination values are the same in a high-pass and a low-pass

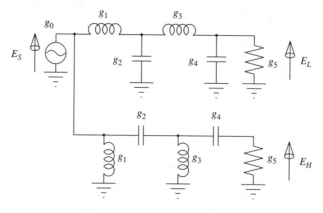

Figure 9.34 Two singly terminated filters connected in parallel.

filter. Then:

$$\frac{|V_L|^2}{R_L} = \frac{1}{\omega^{2n}+1}\frac{|V_S|^2}{R_L}$$

$$\frac{|V_H|^2}{R_L} = \frac{\omega^{2n}}{\omega^{2n}+1}\frac{|V_S|^2}{R_L}$$

Summing the two equations results in:

$$\frac{|V_S|^2}{R_L} = \frac{|V_L|^2}{R_L} + \frac{|V_H|^2}{R_L}$$

Let Y_{in1} be the input admittance of the first filter and Y_{in2} be the input admittance of the second filter. The power delivered to the two filters in parallel is:

$$P_{del} = |V_S|^2 Re(Y_{in1} + Y_{in2})$$

Since there is no power dissipated in the filter (the filters are assumed lossless in this derivation), the power delivered by the source to the two filters is the sum of the power dissipated in each of the loads. From the above equations, the sum of the real part of the input admittances is equal to the load conductance. It turns out that for Butterworth filters, the input admittances of a high-pass filter and a low-pass filter are complements of each other [9]. Therefore the sum of the total input admittance is purely real and equal to the load conductance.

Now consider singly terminated filters that use current sources. Connect a low-pass and a high-pass filter in series as shown in Figure 9.35 using current sources in series with this new input. Let each filter retain its own load. Then:

$$|I_L|^2 R_L = \frac{1}{\omega^{2n}+1}|I_S|^2 R_L$$

$$|I_H|^2 R_L = \frac{\omega^{2n}}{\omega^{2n}+1}|I_S|^2 R_L$$

Summing the two equations results in:

$$|I_S|^2 R_L = |I_L|^2 R_L + |I_H|^2 R_L$$

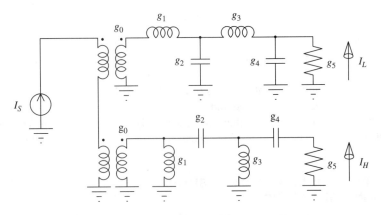

Figure 9.35 Two singly terminated filters connected in series.

Let Z_{in1} be the input impedance of the first filter and Z_{in2} be the input impedance of the second filter. The power delivered to the two filters in series is:

$$P_{del} = |I_S|^2 Re(Z_{in1} + Z_{in2}) = |I_S|^2 Re(Z_{in1}) + |I_S|^2 Re(Z_{in2})$$

Just as in the case with the filters that have a voltage source, the power delivered by the source to the two filters is the sum of the power dissipated in each of the loads. Now, however, the sum of the real part of the input impedances is equal to the load resistance. For Butterworth high-pass filters and low-pass filters, the input impedances are complements of each other. Therefore, the sum of the input impedances is purely real and equal to the load resistance.

The parallel input case is useful when using unbalanced transmission lines. The series input case is often useful when non-TEM waveguide structures are used. The series case is sometimes used in acoustical installations for putting two speaker assemblies in series [9]. The parallel case can be used to build transmission line diplexers as shown in Figure 9.36.

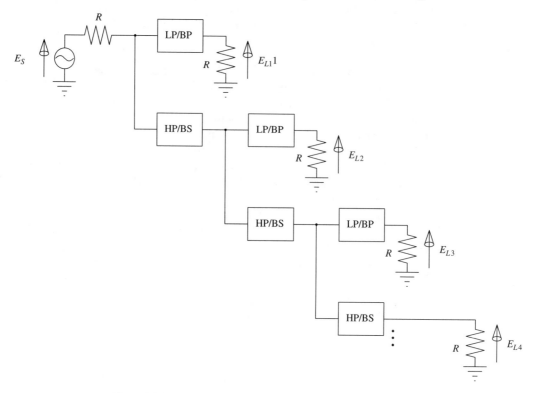

Figure 9.36 An example of a diplexer built from singly terminated filters.

At the input to each of the pairs of LP/BP-HP/BS, the impedance seen to the right is R. (LP/BP = Low Pass or Band Pass and HP/BS = High Pass or Band Stop.) The input impedance seen by the source on the left is therefore also R and the system is matched on the left side. Each of the loads is matched over the frequencies that are in the passband of the filter that they load.

9.14 MODELING OF LOSS IN DISTRIBUTED RESONATOR FILTERS

In the band pass filter that is transformed from the low pass, the loss of the filter resulting from finite Q's of the resonators can be easily calculated at the center frequency. If the Q of the resonator is known, the series LC sections are replaced with a resistor of value:

$$R_S = \frac{\omega L_S}{Q_S}$$

and the parallel LC sections are replaced with a resistor of value:

$$R_P = \frac{Q_P}{\omega C_P}$$

Recall that at resonance, the series reactance of the series resonator is zero and the shunt susceptance of the shunt resonator is zero, leaving only the resistors. The nose bandwidth shape depends on which one of the L or C really has the most loss, but at band center, the single resistor models the loss well. Note that when the resistors are added the filter may not be matched. The match may suffer and the passband shape may suffer.

Transmission line resonators can be modeled with lines that have attenuation in them. If a CAD package that is available to the designer does not have transmission lines with loss in them, a simple procedure can be used to introduce loss. It is not perfect but works very well for resonators with high Q's. One can cut the resonator at some point and insert an attenuator at that point that has the characteristic impedance of the resonator line. The attenuation value for the attenuator is the one path loss of the length of the resonator line. Since the attenuator, either a tee or pi circuit, can be put anywhere on the line, it can be put at the open circuit end where it can be reduced to a single resistor on the end of the line. It can be put on the shorted end of the resonator where it can be reduced to a single resistor as well. An easy way to approximate the value of the resistor is to calculate the equivalent LC circuit of the unloaded resonator at the open-circuit or short-circuit end. For instance, using the formula from Section 9.8.1 and looking at the open end of the quarter-wavelength line resonator, the equivalent parallel circuit components are:

$$C_{\text{eff}} = \frac{Y_0 D}{2v} = \frac{Y_0}{\omega} \frac{\pi}{4} = \frac{Y_0}{8f}$$

$$L_{\text{eff}} = \frac{1}{\omega^2 C_{\text{eff}}}$$

If one knows the unloaded Q of the resonator from measurements, then the value of a resistor to put across the open-circuit end of the resonator can be determined.

$$Q = \frac{\omega C_{\text{eff}}}{G_P} = \frac{\pi Y_0}{4} R_P$$

$$R_P = \frac{4Q Z_0}{\pi}$$

A similar derivation can be performed to determine the resistance that can be placed in series with the ground of a resonator to model the Q of the resonator. That derivation is left as an exercise.

9.15 TUNING OF MULTIPLE-POLE FILTERS

When multipole filters are fabricated, each of the resonators needs to be tuned to the center frequency. The question can be raised, how can that be done? Consider the resonators and inverters shown in Figure 9.18. There is an inverter on either side of each resonator. When the filter is constructed, part of the inverter is absorbed into the resonator. How does one tune the resonator properly? A technique described by Dishal [10] allows one to tune the resonators. When a parallel resonator is detuned, it tends toward a short circuit at the design frequency. When a series resonator is detuned, it tends toward an open circuit. For analysis, the shunt resonators will be described but the dual analysis works for series resonators. Consider the resonator shown in the center of Figure 9.37. In the top portion of the figure, the resonator is shown as having absorbed the negative capacitors (or inductors) from the adjacent J inverters. If the J inverters were removed, the resonate frequency of that circuit would not be at the center

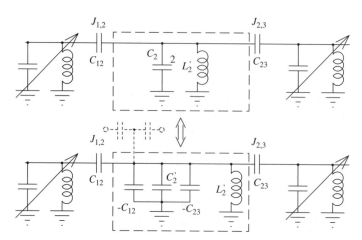

Figure 9.37 Example of tuning an internal resonator.

frequency. However, if the resonators on either side of the resonator under consideration are detuned, they effectively become a short circuit. That short circuit puts the series inverter capacitor in parallel with the *LC* circuit in the center of Figure 9.37. The bottom portion of Figure 9.37 shows the center resonator as it would be from the design phase with the resonator capacity on the adjacent negative inverter capacitors. Notice that when the adjacent resonators are detuned and act as short circuits the center resonator has the positive series inverter capacitors put in parallel with the negative inverter capacitors, leaving only the original designed resonator, which is at the center frequency. There is a dashed line connected to two other capacitors shown connected to the center resonator. These capacitors symbolize a very small coupling measurement that will be made. The capacitors are small enough not to measurably disturb the center resonator. When the outer two resonators are detuned and the center resonator is tuned, the insertion loss through the dashed capacitors will be at a minimum when the resonator is tuned to its center frequency.

The procedure for tuning the filter then consists of first detuning the outer resonators (usually with the temporary addition of a shorting stick without affecting the actual physical structure of the outer resonators). Then the resonator under consideration is tuned by peaking the response through the temporarily inserted lightly coupled set of probes. This is repeated for each resonator. The end resonator takes special care. If the end resonator has a full inverter on it, the input can be shorted for *J* inverters (or opened for *K* inverters). However, the end resonator may have had an impedance transforming network put on it or may have been tapped at a different point for impedance transformation. When this is the case, the adjacent resonator is detuned just as in the case of internal resonators but the circuit is driven from the input (or output as the case may be) and a probe is lightly coupled to that resonator. The resonator is then tuned for a peak response from the input (output) through the probe circuit.

Although the figure shows capacitive inverters, when the inverters are inductive inverters, this procedure works the same way. The adjacent inverters provide the correct reactance for the center resonator. It is observed that in the case of *K* inverters, when the adjacent resonators are detuned they become open circuit. The tee inverters (*K* inverters) between the resonators then include the positive reactance that was subtracted from the series inverters to bring the resonator on frequency. Dishal's paper also contains a procedure to set and check the coupling coefficient for each of the inverters.

EXERCISES

9-1 Design a 50-ohm bandpass filter using lumped-constant inductors and capacitors for a four-pole Butterworth filter. The g values for the filter are $g_0 = 1$, $g_1 = 0.7654$, $g_2 = 1.848$, $g_3 = 1.848$, $g_4 = 0.7654$, $g_5 = 1$. The frequency is 1.59155 GHz and the bandwidth is 2.5%. Comment on whether the components could be physically made using real-world components.

9-2 Design a three-pole, equal-element, 5% bandwidth bandpass filter at 500 MHz using combline construction. The characteristic impedance of the combline resonator is 15 ohms and the effective dielectric constant of the resonator is 36. Use whatever type of inverter section will work.

9-3 Determine the even- and odd-mode impedances for a stripline coupled-line band pass filter using open-circuit, half-wavelength resonators. The resonators have a characteristic impedance of 50 ohms. The center frequency is 4 GHz and the bandwidth is 2%. The dielectric constant for the stripline dielectric is 2.56.

9-4 Determine the value of the series resistance to put in series with the ground of a 20-ohm lossless quarter-wavelength line in order to have an unloaded Q of 100 at 1 GHz. The transmission line has an air dielectric.

9-5 Design a diplexer for selecting the harmonics of a 100 MHz signal. The filter load impedances are 50 ohms. The bandwidth of the spectrum of the 100 MHz signal is 4%. Provide separate loads for the fundamental, and the second through the fourth harmonics. The power in the fifth and higher harmonics should be terminated in 50 ohms also. Consider the number of poles required in the filter and whether the filter sections should be high-pass–low-pass or band pass-band stop sections.

9-6 Design a five-pole low-pass filter with a cutoff frequency of 2 GHz using one-eighth-wavelength unit elements. The filter is to be a 50 ohm filter. Use microstrip construction. Transform the filter to use only shunt stubs. The g values are $g_0 = 1$, $g_1 = 1.414$, $g_2 = 1.318$, $g_3 = 2.241$, $g_4 = 1.318$, $g_5 = 1.414$, and $g_6 = 1$.

9-7 Design a J inverter using capacitors at 1 GHz. The value of the inverter is 0.002. Redo the design using inductors. Which of the inverters could be built with real components if capacitors have parasitic inductances of 0.5 nH and inductors have parasitic capacitances of 0.2 pF? Could this inverter be built using a transmission line if the range of characteristic impedances available is 5 to 250 ohms?

9-8 Design a three-pole 5% bandwidth lumped-element 50-ohm bandpass filter at 50 MHz. Use capacitive J inverters with a 100-nH resonator inductors. Show that when the first pole resonator and the third pole resonator are detuned, the center resonator is at resonance.

9-9 For the filter given in Exercise 9-2, assume the resonator Q is 300. Use a circuit analysis program and calculate the passband insertion loss of the filter. You may use a shunt resistor on the open-circuit end of the resonator to model the Q of the resonator. Compare this loss with the loss calculated using a low-pass prototype to band pass transformation using lumped elements. Note that in the band pass filter formed with lumped elements, at the center frequency, the filter consists of a 50-ohm source followed by a tee or a pi circuit consisting of three resistors and then terminated in a 50-ohm load. The values of the resistors are determined by considering the element values in the resonant circuits and the Q of the circuits. For instance, when the first resonator is a series resonator, the inductance is 318.3 nH. The series resistance for a Q of 300 at 500 MHz is 3.33 ohms.

10

Noise Considerations
for Microwave Circuits

10.1 NOISE

There are often many questions asked about what noise is. Do noise voltages and noise currents act like linear variables? How do noise voltages and noise currents add? What are the characteristics of noise? What kinds of noise are there? This book addresses some of these questions as they apply to microwave circuits and amplifiers. There are several types of noise such as *shot noise*, *one-over-f* noise, *current noise*, and so on, which will not be discussed. Shot noise results from minute variations in current caused by discrete carrier conduction. The unit of charge is discrete. Part of charge cannot move. Although the charge can move with different velocities, it is the unit of charge that moves. When the charge stops moving, the current due to that charge stops abruptly. These small variations lead to shot noise. Current noise results from the different random paths that different carriers take through a region of resistance. Since each carrier takes a different path, the energy given up over a given distance is different. This minute difference of energy results in noise. Current noise depends on carrier flow (i.e., on current) and, like shot noise, is a function of the amount of current that flows. *Thermal noise*, often called *Johnson noise*, does not depend on current flow. All resistances generate thermal noise independent of whether current is flowing in the resistance. Thermal noise is a function of temperature rather than a function of current flow. A piece of chalk or a piece of wire generates the same amount of available noise power. The only difference is whether one can match the impedance necessary to extract that power.

10.2 NOISE VOLTAGES AND CURRENT AND SUPERPOSITION

Noise voltages and noise currents are linear variables just like the linear variables v and i that are used in circuit analysis. There are some important differences between the linear variables associated with the deterministic signals generally associated with circuit analysis and the linear variables associated with noise. These have to do with the random nature of amplitude, phase, and frequency of variables associated with noise as distinguished from the amplitude, phase, and frequency of variables associated with deterministic variables used in signal analysis of circuits. It is important to remember, however, that noise voltages and currents are linear variables and will add linearly to signal voltages and currents as well as to noise voltages and currents from different noise sources. When determining the range of amplitudes that an active device will encounter, the sum of the signal and noise variables will have to be determined or estimated. In order to gain some understanding of the difficulties encountered with summing voltages and currents to obtain power available from a source, consider a source as shown in Figure 10.1. (Remember that power is not a linear circuit variable but is a result of the product of voltage and current.) The voltage source shown in Figure 10.1 may contain discrete sine waves at several different frequencies. Let the voltage source be expressed as:

$$E(t) = E_1(t) = a_1 \sin(\omega_1 t + \phi_1) + a_2 \sin(\omega_2 t + \phi_2)$$
$$+ a_3 \sin(\omega_3 t + \phi_3) + a_4 \sin(\omega_4 t + \phi_4)$$

If the individual frequencies ω_i are all different, then the time average power dissipated in the load resistor R_L from this source is:

$$P_L = \left(\frac{a_1^2}{2} + \frac{a_2^2}{2} + \frac{a_3^2}{2} + \frac{a_4^2}{2} \right) \frac{R_L}{(R_L + R_S)^2}$$

If a different voltage source $E_2(t)$ is substituted for $E(t)$, for instance:

$$E(t) = E_2(t) = a_5 \sin(\omega_5 t + \phi_5) + a_6 \sin(\omega_6 t + \phi_6)$$
$$+ a_7 \sin(\omega_7 t + \phi_7)$$

Again, if all the ω_i are different, the power is also:

$$P_L = \left(\frac{a_5^2}{2} + \frac{a_6^2}{2} + \frac{a_7^2}{2} \right) \frac{R_L}{(R_L + R_S)^2}$$

However, if the two voltage sources are added such that:

$$E(t) = E_1(t) + E_2(t) = a_1 \sin(\omega_1 t + \phi_1) + a_2 \sin(\omega_2 t + \phi_2)$$
$$+ a_3 \sin(\omega_3 t + \phi_3) + a_4 \sin(\omega_4 t + \phi_4)$$
$$+ a_5 \sin(\omega_5 t + \phi_5) + a_6 \sin(\omega_6 t + \phi_6)$$
$$+ a_7 \sin(\omega_7 t + \phi_7)$$

what is the power dissipated in the load resistor? If all seven frequencies are different, the power is still:

$$P_L = \left(\frac{a_1^2}{2} + \frac{a_2^2}{2} + \frac{a_3^2}{2} + \frac{a_4^2}{2} + \frac{a_5^2}{2} + \frac{a_6^2}{2} + \frac{a_7^2}{2} \right) \frac{R_L}{(R_L + R_S)^2}$$

However, if for instance, ω_2 is equal to ω_7 and all the other ω_i are different from each other and from ω_2, then the voltages at the second and seventh frequencies add (the linear variable is voltage, not power) and the power is:

$$P_L = \left(\frac{a_1^2}{2} + \frac{a_3^2}{2} + \frac{a_4^2}{2} + \frac{a_5^2}{2} + \frac{a_6^2}{2} + \frac{c^2}{2} \right) \frac{R_L}{(R_L + R_S)^2}$$

where c is equal to:

$$c = \sqrt{[a_2 \cos(\phi_2) + a_7 \cos(\phi_7)]^2 + [a_2 \sin(\phi_2) + a_7 \sin(\phi_7)]^2}$$

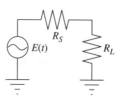

Figure 10.1 Thevenin voltage sources for determining noise power.

Note that the voltages at the individual frequencies have to be added first and then the power can be determined from the total voltages. When the amplitudes, frequencies, or phases of the individual signals are random variables, the power dissipated by the load resistor must be calculated by using the rms average of the signal variables similar to the procedure used above for discrete variables. If two separate random (noise) voltages have some frequency components related (correlated) to each other, that portion of those voltages have to be statistically added first and then the total power can be determined in the load resistor.

10.3 THERMAL NOISE

Any resistance R as shown in Figure 10.2, using either the Thevenin or Norton equivalent circuit and existing in nature, has an rms noise voltage v_n associated with it of magnitude:

$$v_n = \sqrt{\frac{4hfBR}{e^{\frac{hf}{kT}} - 1}}$$

and average value equal to zero. T is the temperature in degrees Kelvin (K), h is Planck's constant $= 6.626 \times 10^{-34}$ J-sec, k is Boltzmann's constant $= 1.380 \times 10^{-23}$, B is the bandwidth of the system in Hz, f is the center frequency of the bandwidth in Hz, and R is the resistance in ohms. This formula comes from Planck's black-body radiation law [1]. If the frequency is low enough (i.e., $hf \ll kT$), then the denominator inside the square root is approximately equal to hf/kT, and the noise voltage can be expressed by:

$$v_n = \sqrt{4kTBR}$$

$$i_n = \sqrt{4kTBG}$$

or the Rayleigh-Jeans approximation to Planck's radiation law. At room temperature, this approximation is valid up to approximately 1000 GHz and at 3 K the approximation is good up to approximately 10 GHz. The power spectral density (variation of noise power available versus frequency from the resistor) is flat over the frequencies for which this approximation is good. The available noise power from a resistor and source combination is the power available using a conjugate match. The resistor used for the load will also have a noise source. However, using superposition, the power available from the source is the power delivered to a load resistor of the same value. The voltage across the load resistor is then half of the value of the noise source voltage at each frequency component contained in the noise source. Therefore, the maximum power delivered to a load resistor from a resistor noise source voltage (power available from the source) is:

$$P_L = \frac{\left(\frac{v_n}{2}\right)^2}{R_L} = \frac{\left(\frac{v_n}{2}\right)^2}{R_S} = \frac{4kTBR_S}{4R_S} = kTB$$

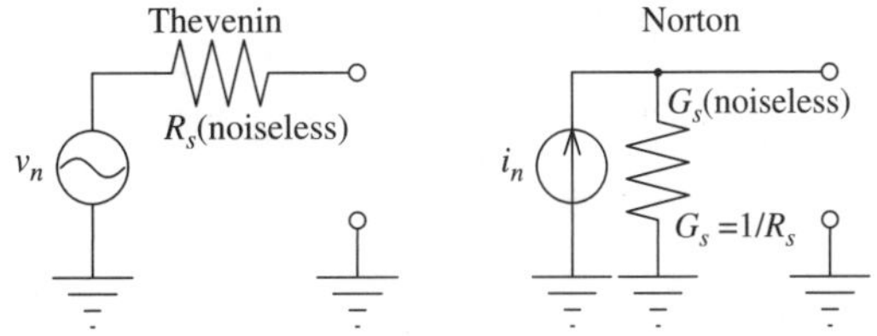

Figure 10.2 Thevenin or Norton equivalent noise sources.

If the temperature of both resistors is the same, then the power available from each resistance is the same independent of its resistance value. If two resistors are at the same temperature and are connected together electrically but are isolated thermally, one resistor transfers some of its power to the other resistor and vice-versa so that on the average the pair of resistors stay at the same temperature. However, when one resistor is hotter than the other, the hotter resistor transfers more power to the colder resistor than the colder resistor transfers to the hotter resistor. Therefore, on the average the hotter resistor will tend to become colder and vice-versa so that both resistors will reach the same temperature in the steady state.

10.4 THERMAL NOISE AND TWO-PORTS

Now consider the three-resistor network shown in Figure 10.3. Each of the resistors has a noise voltage associated with it. These voltages are independent from each other since the resistors are separate and distinct components. Each open-circuit voltage at port one and port two has a Thevenin voltage associated with it as shown in Figure 10.4. The voltage sources have been taken outside of the two-port and the two-port shown inside the dotted line is now modeled as a noiseless two-port. Notice, however, that the two voltage sources are not independent of each other. They each have a component of voltage associated with each of them from v_{n3}. They are correlated to each other. Each of the noise sources in the noisy three-resistor network in the original two-port were uncorrelated to each other since they come from separate isolated devices. In many system noise calculations, all of the noise sources are considered as existing at the input. Therefore, the effect of the noise source that exists on the output of the noiseless two-port is modeled by a Norton equivalent current source on the input of the noiseless two-port. The value of the input Norton current source depends on the values of R_1, R_2, and R_3 (or the Y matrix of the noiseless two-port). For an arbitrary two-port, the noise associated with the two-port can be modeled by an equivalent Thevenin voltage source and an equivalent Norton current source. Since these voltage and current sources are not necessarily independent of each other they might have a partial correlation to each other expressed by a relative magnitude and phase relationship to each other. There are four different parameters necessary to characterize the noise performance of a two-port. These are (1) the magnitude of the voltage source, (2) the magnitude of the current source, and (3 & 4) the complex number representing the partial correlation between the two sources. The diagram in Figure 10.5 shows this conceptually. If the Z_{ij} are the Z parameters of the noiseless two-port, and V_1 and V_2 are

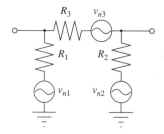

Figure 10.3 Resistor pi circuit with noise sources.

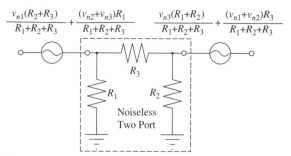

Figure 10.4 Thevenin equivalent of resistor pi circuit.

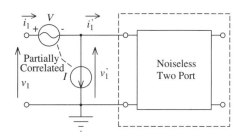

Figure 10.5 Two noise sources partially corre-lated. [*Adapted from H. Rothe and W. Dahlke, "Theory of Noisy Fourpoles,"* Proceedings of the IRE, *vol. 44, pp. 811–818, June 1959.*]

the Thevenin noise sources on ports one and two, then:

$$V_1 = V - I Z_{11}$$

$$V_2 = -I Z_{21}$$

$$V = V_1 - V_2 \frac{Z_{11}}{Z_{21}}$$

$$I = -\frac{V_2}{Z_{21}}$$

The two noise sources are partially correlated to each other [2]. If we let V consist of an uncorrelated part, V_u, plus a correlated part, V_c, then:

$$V = V_u + V_c = V_u + I Z_c$$

where Z_c is a correlation impedance. The currents and voltages in Figure 10.5 are related by the following equations.

$$V = V_u + V_c = V_u + I Z_c$$

$$i_i = I + i_i'$$

$$v_i = V + v_i' = V_u + I Z_c + v_i' = V_u + (i_i - i_i') Z_c + v'$$

These relationships are illustrated in the circuit shown in Figure 10.6. The noise sources v_n and i_n as given by:

$$\overline{v_n^2} = 4 k T_o R_u B$$

$$\overline{i_n^2} = 4 k T_o G B$$

are completely uncorrelated; the noise power can be calculated by using the principle of superposition using the circuit shown in Figure 10.6, where R_u is a resistance representing the value of v_n and G is a conductance representing the value of i_n. The impedances Z_c and $-Z_c$ are assumed to be at absolute zero in temperature so that they do not generate noise voltages. Note that four noise parameters are used in the above circuit: (1) v_n, (2) i_n, and (3 & 4) Z_c. The source v_g is the noise source from the generator resistance. Other simplifying circuits can be used with the above circuit. The Thevenin equivalent noise voltage v_t is:

$$\overline{v_t^2} = \overline{v_g^2} + \overline{v_n^2} + \overline{i_n^2} \left| Z_c + R_g \right|^2$$

Noise power and signal power available at the output of the two-port can therefore be calculated by superposition if Z_c is known since all of these sources are now uncorrelated. Notice that the available noise power out of the network depends on the value of the generator impedance.

A quantity F, defined as the noise figure, represents the ratio of available noise power out of the two-port divided by the product of available noise power at the input from the source

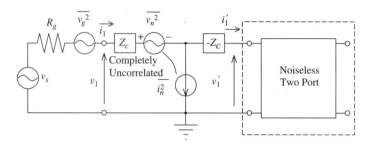

Figure 10.6 Circuit with noise sources and correlation separated. [*Adapted from H. Rothe and W. Dahlke, "Theory of Noisy Fourpoles,"* Proceedings of the IRE, *vol. 44, pp. 811–818, June 1959.*]

times available gain of the two-port:

$$F = \frac{P_{No}}{P_{Ni} G_a}$$

The available power at the input is kTB and, by convention, T is 290 degrees K. Since the available gain is the ratio of available power at the output divided by available power at the input, the ratio of signal powers is:

$$G_a = \frac{P_{So}}{P_{Si}}$$

then:

$$F = \frac{P_{No}}{P_{Ni}} \frac{P_{Si}}{P_{So}} = \frac{P_{Si}/P_{Ni}}{P_{So}/P_{No}}$$

or the ratio of the signal-to-noise ratio at the input divided by the signal-to-noise ratio at the output. Notice that available gains have been used in the above definitions and that the reference available noise at the input is from a generator resistance at 290 K. The noise figure is a numeric ratio of power ratios. Therefore, the noise figure expressed in dB is:

$$NF_{dB} = 10 \log_{10}(F)$$

and is always a positive number. The noise figure F is always a number greater than 1. It has been assumed in this analysis that the temperature of each component in the network is 290 K. When a source is not at this temperature, its available power is altered and alters the signal-to-noise ratio calculations.

10.5 NOISE FIGURE OF CASCADED STAGES

If the noise figure of two individual stages is known, what is the noise figure for two stages cascaded? Consider two stages as depicted in Figure 10.7. A capital N is used to designate the noise external to the two-port. A small n is used to designate the noise generated by the two-port. The noise out of the interstage consists of the noise available from the source resistor R_N equal to P_N multiplied by the available gain in the first two-port, G_a, added to the noise available from the two-port P_{n1}. This is the noise power available to the second two-port. The available power out of the second stage just from the input resistance is the available power from that resistance times the product of the available gains of the two two-ports. The first-stage noise figure is [3]:

$$F_1 = \frac{G_{a1} P_{N1} + P_{n1}}{G_{a1} P_{N1}} = 1 + \frac{P_{n1}}{G_{a1} P_{N1}}$$

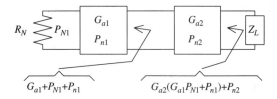

Figure 10.7 Diagram for cascaded noise networks.

Therefore, the noise figure for the two stages is:

$$F = \frac{G_{a2}(G_{a1}P_{N1} + P_{n1}) + P_{n2}}{G_{a1}G_{a2}P_{N1}} = 1 + \frac{P_{n1}}{G_{a1}P_{N1}} + \frac{P_{n2}}{G_{a1}G_{a2}P_{N1}}$$

$$= F_1 + \frac{1}{G_{a1}}\left(\frac{P_{n2}}{G_{a2}P_{N1}}\right) = F_1 + \frac{F_2 - 1}{G_{a1}}$$

where:

$$F_1 = 1 + \frac{P_{n1}}{G_{a1}P_{N1}}$$

$$F_2 = 1 + \frac{P_{n2}}{G_{a2}P_{N1}}$$

where for the second-stage noise figure, the noise from the input resistance is used since that is the source that would be used to determine its noise figure by itself.

10.6 NOISE FIGURE CIRCLES

It can be shown that the noise figure can be expressed by [4,5]:

$$F = F_{\min} + \frac{r_n}{g_S}|y_S - y_{\min}|^2$$

where $F_{\min}$ is the minimum noise figure obtainable from the two-port at a normalized source admittance of $y_{\min}$. The term r_n is a normalized noise resistance for the two-port. The normalized source admittance is $y_s = g_s + jb_s$. The normalized admittances can be expressed as:

$$y_S = \frac{1 - \Gamma_S}{1 + \Gamma_S}$$

$$y_{\min} = \frac{1 - \Gamma_{\min}}{1 + \Gamma_{\min}}$$

$$g_S = \frac{y_S + y_S^*}{2}$$

The source conductance can be rewritten as:

$$g_S = \frac{1}{2}\left(\frac{1 - \Gamma_S}{1 + \Gamma_S} + \frac{1 - \Gamma_S^*}{1 + \Gamma_S^*}\right) = \frac{1}{2}\left(\frac{1 - \Gamma_S}{1 + \Gamma_S}\frac{1 + \Gamma_S^*}{1 + \Gamma_S^*} + \frac{1 + \Gamma_S}{1 + \Gamma_S}\frac{1 - \Gamma_S^*}{1 + \Gamma_S^*}\right)$$

$$= \frac{1 - |\Gamma_S|^2}{(1 + \Gamma_S)(1 + \Gamma_S^*)}$$

The absolute value term in the noise figure equation can be rewritten as:

$$|y_S - y_{\min}|^2 = \left(\frac{1 - \Gamma_S}{1 + \Gamma_S} - \frac{1 - \Gamma_{\min}}{1 + \Gamma_{\min}}\right)\left(\frac{1 - \Gamma_S}{1 + \Gamma_S} - \frac{1 - \Gamma_{\min}}{1 + \Gamma_{\min}}\right)^*$$

$$= 4\frac{(\Gamma_{\min} - \Gamma_S)(\Gamma_{\min} - \Gamma_S)^*}{(1 + \Gamma_{\min})(1 + \Gamma_{\min})^*(1 + \Gamma_S)(1 + \Gamma_S)^*}$$

Therefore, the noise figure can be expressed as:

$$F = F_{min} + 4r_n \frac{|\Gamma_S - \Gamma_{min}|^2}{(1 - |\Gamma_S|^2)|1 + \Gamma_{min}|^2}$$

If the equation is rearranged and each side of the rearranged equation is set equal to a noise parameter N_i:

$$N_i = \frac{F - F_{min}}{4r_n}|1 + \Gamma_{min}|^2 = \frac{|\Gamma_S - \Gamma_{min}|^2}{1 - |\Gamma_S|^2}$$

Then:

$$(\Gamma_S - \Gamma_{min})(\Gamma_S - \Gamma_{min})^* = N_i(1 - |\Gamma_S|^2) = N_i - N_i\Gamma_S\Gamma_S^*$$

$$\Gamma_S\Gamma_S^*(1 + N_i) - \Gamma_{min}\Gamma_S^* - \Gamma_{min}^*\Gamma_S + \Gamma_{min}\Gamma_{min}^* = N_i$$

$$\Gamma_S\Gamma_S^* - \frac{\Gamma_{min}}{1 + N_i}\Gamma_S^* - \frac{\Gamma_{min}^*}{1 + N_i}\Gamma_S + \frac{\Gamma_{min}\Gamma_{min}^*}{(1 + N_i)^2} = \frac{N_i}{1 + N_i} - \frac{\Gamma_{min}\Gamma_{min}^*}{(1 + N_i)} + \frac{\Gamma_{min}\Gamma_{min}^*}{(1 + N_i)^2}$$

$$= \frac{(N_i - |\Gamma_{min}|^2)(1 + N_i) + |\Gamma_{min}|^2}{(1 + N_i)^2}$$

$$= \frac{N_i^2 + N_i(1 - |\Gamma_{min}|^2)}{(1 + N_i)^2}$$

This equation represents the equation of a series of circles in the source reflection coefficient plane. The center and radius of these circles depend on the parameter N_i, which depends on the difference between the minimum noise figure F_{min} and the given noise figure F. The center and radius are given by:

$$N_i \text{ (center)} = \frac{\Gamma_{min}}{1 + N_i}$$

$$N_i \text{ (radius)} = \frac{\sqrt{N_i^2 + N_i(1 - |\Gamma_{min}|^2)}}{1 + N_i}$$

Notice that the centers of these circles lie on a line from the origin of the reflection coefficient plane to Γ_{min}. These circles can be plotted on the source reflection coefficient plane along with the source unilateral gain circles.

10.7 GAIN AND NOISE FIGURE CIRCLES

The correspondence between minimum noise figure and maximum unilateral gain can be graphically depicted when both noise figure circles and gain circles are plotted on the same reflection coefficient chart. It is then quite apparent what amount of gain has to be sacrificed for minimum noise figure or what amount of noise figure degradation exists for maximum gain. The noise figure for Γ_s equal to zero is:

$$F(\Gamma_S = 0) = F_{min} + 4r_n \frac{|\Gamma_{min}|^2}{|1 + \Gamma_{min}|^2}$$

The reflection coefficient chart shown in Figure 10.8 has both unilateral gain circles for S_{11} and noise figure circles plotted on it. The noise figure circles overlap the unilateral gain circles only over a portion of the chart. This particular set of gain and noise figure circles is for illustration only. The two sets of circles could lie in quite different relationships to each other than that shown. Notice that the gain circle that goes through the origin has 0 dB of gain associated with it. The small dot inside the gain circles represents (for this case) 5.3 dB of possible gain that can be derived by matching the input. However, there would be a large noise

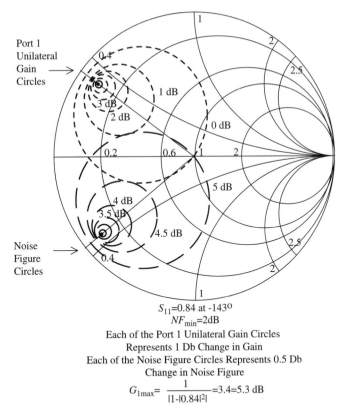

$S_{11}=0.84$ at -143°
$NF_{min}=2dB$
Each of the Port 1 Unilateral Gain Circles
Represents 1 Db Change in Gain
Each of the Noise Figure Circles Represents 0.5 Db
Change in Noise Figure

$$G_{1max}= \frac{1}{|1-|0.84|^2|}=3.4=5.3 \text{ dB}$$

Figure 10.8 Reflection coefficient chart with representative noise figure and input port unilateral gain circles.

figure penalty if maximum gain were chosen. Likewise, in order to get the minimum noise figure, there would be a large gain penalty. Some optimum compromise might be found, for instance, at the intersection of the 0-dB gain circle and the 4.5-dB noise figure circle. More gain circles and noise figure circles could have been plotted for other values of gain and noise figure. The gain and noise figure circles do not need to stop at the origin. The noise figure circles are not necessarily the conjugates of the unilateral gain circles, as it may appear from Figure 10.8.

EXERCISES

10-1 Determine the difference in noise voltage from a 150-ohm resistor using square root of $4KTB$ and the complete formula. The temperature is 50 K.

10-2 If a two-port consists of two resistors and one capacitor, how many independent noise parameters are associated with the two-port?

10-3 Determine the noise figure of a 15-dB attenuator. All parts of the attenuator are at 290 K.

10-4 A transistor has the following scattering parameters and noise parameters at 1 GHz. Draw the noise figure circles and the unilateral gain circles on a 50-ohm reflection coefficient chart. Determine the amount the noise figure is degraded to have input match and the amount the gain is degraded to have noise match.

Rn = 15, NFmin = 2.5 dB, Z for min noise figure 25 + j10 ohms.
$S_{11} = 0.7$ at 180 deg., $S_{12} = 0.075$ at 45 deg., $S_{21} = 5$ at 80 deg., $S_{22} = 0.25$ at -135 deg.

Is the part stable with a noise-matched source?

Design a stable noise matched stage using this part in a 50-ohm system. What is the circuit's gain?

10-5 What is the noise power available from a resistor at 1000 K at 1 GHz and a 30-MHz bandwidth? If a part is matched for 20-dB available gain, what is the available noise power from the part at 1 GHz over this bandwidth if the noise figure of the amplifier is 0.5 dB and the resistor of 1000 K is used as the source? The part remains at 290 K.

11

Detection and Mixing

11.1 MIXERS AND DETECTORS

This chapter is concerned with detecting microwave power and converting microwave energy from one frequency to another frequency. The detection of microwave power can be considered changing the amplitude information contained on a microwave signal to a base-band signal (i.e., changing the carrier frequency of the signal to dc). The process of changing the information from one carrier frequency to another carrier frequency is called *mixing*. The fundamentals of detection, mixing, and the accompanying frequency conversion are discussed in [1,2]. Although some of the devices have changed, the process is still the same. More recent publications include some newer circuitry and device types [3].

11.2 DIODE AND CRYSTAL DETECTORS

Historically, diode detectors were called *crystal detectors* because the first semiconductor detectors of the diode type were point-contact diodes. These diodes are formed by a sharp point contact of a metal on a semiconductor crystal. Later, Schottky diode devices were fabricated for detectors. These are formed by the deposition of a metal contact on a semiconductor crystal. The Schottky junction diode is a majority carrier device. This means that there is only a very small stored charge in the junction of the diode as contrasted with a *p-n* junction diode that is a minority carrier device. Most applications of diodes for power detection use the Schottky diode. Several applications of point contact diodes likely still exist. The *p-n* junction diode can be used as a power detector if it is used at frequencies for which the minority carrier lifetime is a small portion of a cycle of the alternating current waveform. The reactive element parasitics of each type of diode should be accounted for when the diode is applied to a specific power measurement problem.

Each of the types of diodes has a "dc" transfer function that can be expressed as:

$$\frac{I_d(t)}{I_o} = e^{\frac{qV(t)}{\eta kT}} - 1$$

where $I_d(t)$ is the time-dependent diode current, I_o is the saturation current constant, q is the electronic charge, $V(t)$ is the time-dependent voltage across the diode, η is the ideality constant, k is Boltzmann's constant, and T is absolute temperature. The diode is said to have a "dc" transfer function since the terminal relationship between voltage and current is only a steady-state relationship and not a transient relationship. This will be more evident when PIN diodes are studied. PIN diodes have a large charge storage time constant and can support large transient waveforms that do not satisfy the above terminal relationships. Since a Schottky diode does not have a large stored charge in it, its ac transfer curve is closer to its dc transfer curve.

The transfer function can be rewritten in terms of a Maclaurin series expansion around zero volts as:

$$I_d(t) = I_o \left(e^{\frac{qV(t)}{\eta KT}} - 1 \right) \approx \frac{qV(t)}{\eta KT} + \frac{1}{2!}\left[\frac{qV(t)}{\eta KT} \right]^2 + \frac{1}{3!}\left[\frac{qV(t)}{\eta KT} \right]^3 + \frac{1}{4!}\left[\frac{qV(t)}{\eta KT} \right]^4 + \cdots$$

For power detection, the diode is excited with a sinusoidal waveform. If $qV(t)/\eta kT = A\cos(\omega t)$, what does $I_d(t)$ look like? Recall that:

$$\cos^2(x) = \frac{1}{2}[\cos(2x) + 1] \qquad \cos^3(x) = \frac{1}{4}[\cos(3x) + 3\cos(x)]$$

$$\cos^4(x) = \frac{1}{8}[\cos(4x) + 4\cos(2x) + 3]\cdots$$

Note that only the even exponent terms contain constant values (i.e., the odd exponent terms have only time-dependent terms in them).

When $qV(t)/\eta kT = A\cos(\omega t)$ is substituted into the series, there will be a constant or dc term along with other time-dependent terms. If $qV(t)/\eta kT \ll 1$, then the current will be approximately:

$$\frac{I_d(t)}{I_o} = e^{\frac{qV(t)}{\eta kT}} - 1 = e^{A\cos(\omega t)} - 1$$

$$\approx A\cos(\omega t) + \frac{1}{2!}[A\cos(\omega t)]^2 + \frac{1}{3!}[A\cos(\omega t)]^3 + \cdots$$

$$\approx A\cos(\omega t) + \frac{1}{2}\frac{A^2}{2!}[\cos(2\omega t) + 1] + \frac{1}{4}\frac{A^3}{3!}[\cos(3\omega t) + 3\cos(\omega t)] + \cdots$$

Notice that the series contains a dc part made up of several terms plus other terms that are at the fundamental and harmonic frequencies of the input voltage waveform. The dc part is:

$$\left.\frac{I_d(t)}{I_o}\right|_{\text{dc part}} = \frac{1}{2}\frac{A^2}{2!} + \frac{3}{8}\frac{A^4}{4!} +$$

If $A \ll 1$, then the first term of the series dominates. For the second term to be less than one-tenth of the first term, then $A < 1.265$ or the peak of the voltage waveform needs to be less than 33 mV if $q/\eta kT = 26$ mV. Note that if the second term in the dc part is negligible, then the current is proportional to A^2 or voltage squared and thus proportional to power. This is termed the square law region of a detector. The time-varying portions of the current are filtered out with a low-pass filter.

If A is expressed as:

$$A = A_o + B_o \cos(\omega_m t)$$

where ω_m is an amplitude modulation frequency, the square law term for current from the detector is then:

$$A^2 = [A_o + B_o \cos(\omega_m t)]^2 \cos^2(\omega t)$$

$$\approx \frac{1}{2}[A_o^2 + 2A_o B_o \cos(\omega_m t) + B_o^2 \cos^2(\omega_m t)][1 + \cos(2\omega t)]$$

If $B_o/A_o \ll 1$, then:

$$A^2 = [A_o + B_o \cos(\omega_m t)]^2 \cos^2(\omega t)$$

$$\approx \frac{1}{2}[A_o^2 + 2A_o B_o \cos(\omega_m t)][1 + \cos(2\omega t)]$$

The low frequency or "dc" from the detector is then proportional to:

$$A^2\big|_{\text{dc part}} \approx \frac{1}{2}[A_o^2 + 2A_o B_o \cos(\omega_m t)]$$

This result contains a constant term and a term that varies at the modulation frequency rate. If the modulation percentage is high (i.e., if B_o/A_o is not small), then the "dc" part will contain significant energy terms at twice the modulation frequency. A tuned detector amplifier would be able to discriminate against the higher-order terms. The bandwidth and frequency response requirements of the detector amplifier need to be considered then.

The crystal detector circuit might look like the one in Figure 11.1. The capacitor C_{det} might be a coaxial capacitor to allow it to have a very broad bandwidth. It is important to keep in mind that a properly functioning crystal detector circuit works in its square law range. This is in contrast to the peak detector circuit to be discussed later.

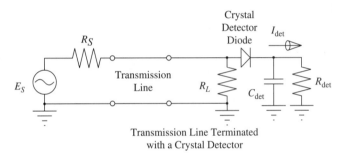

Figure 11.1 Crystal detector circuit.

11.3 THERMAL DETECTORS

It is well known that most metals and some other materials have a relatively linear variation of resistance with temperature. A small piece of this material can be placed across a transmission line and heated with a constant dc current until its resistance matches the line impedance. Assuming that the power absorbed from rf current and that absorbed from dc current heat the material in the same fashion (a good assumption if skin effects are not appreciable), then the power dissipated from rf current absorption and dc current dissipation are additive. A sensing circuit can be used to keep the material at a constant temperature by varying the dc current. The difference between the power generated by dc current with and without rf power incident on the material is then the amount of rf power absorbed by the detector. This type of power detector is called a *thermistor mount or detector* if the material is a thermistor (usually a piece of semiconductor). The material in a thermistor usually has a negative temperature coefficient of resistance. If the material is a metal (like platinum or tungsten, etc.), the detector is called a *barretter detector* and the material usually has a positive temperature coefficient of resistance. The metal may be in the form of a deposited thin film resistor. If the mount is well insulated from external thermal effects, the resistance change is proportional to incident power and the detector is a good *square law* detector. Figure 11.2 shows how a thermal detector might work. Separate compensating circuits for external thermal effects can be designed into the detectors as well.

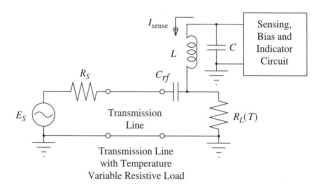

Figure 11.2 One type of thermal detector circuit.

It is possible to use these bolometers without compensating dc currents (a fixed dc current). However, then the *VSWR* or transmission line to device match changes with incident power.

11.4 PEAK DETECTORS

If a low-capacitance diode is placed across a transmission line as shown in Figure 11.3, then C_{det} will charge to approximately the peak value of the voltage waveform if R_{det} is large ($R_{\text{det}} \gg R_L$) and $\tau = R_{\text{det}} C_{\text{det}} \ll T$, where T is the period of the E_s signal. This circuit looks very similar to the crystal detector circuit. The diode is purposely drawn showing the detector diode coming off the main line at an angle. This is to indicate that it does not directly load the line but conducts only on the peaks of the waveform and would not load the line if C_{det} charges up to the peak of the input waveform voltage. This is in contrast to the crystal detector circuit in which the diode conducts throughout the ac waveform cycle. The voltage recorded across R_{det} will be:

$$V_{R_{\text{det}}} = \sqrt{2 P_{R_L} R_L}$$

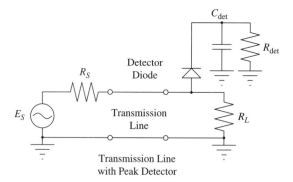

Figure 11.3 Peak detector.

Often the diode has a small amount of dc current imposed upon it to keep it in a conducting state to increase its low level detection ability. When the input ac voltage waveform is much larger than the diode standoff voltage, the bias current is not needed. This type of detector is used in high-power pulse and continuous wave (cw) applications.

11.5 HETERODYNE SCHEMES

Often the signal for which the power is being measured is down-converted to a lower frequency. The signal information may be transferred to a lower frequency carrier by *mixing* or by *sampling*. If a signal turns a device on and off as shown in Figure 11.4, the mathematical description of the output would be the product of the input signal and a square wave. A square

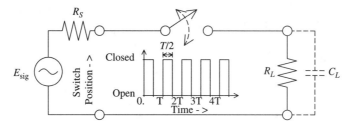

Figure 11.4 A sampler or mixer.

wave has a Fourier series expansion. If the "square wave" is not perfect, it will contain all harmonics of the period switching signal. The switching signal is called a *local oscillator* and its radian frequency is often written as ω_{lo}. If the switching function is described by a Fourier series as:

$$S(t) = \sum_{n=0}^{n=\infty} a_n \cos(n\omega_{lo}t)$$

and the signal is described as:

$$\text{Sig}(t) = b\cos(\omega_{sig}t + \phi_{sig})$$

then the output of the switch is:

$$O(t) = b\cos(\omega_{sig}t + \phi_{sig})S(t) = b\cos(\omega_{sig}t + \phi_{sig})\sum_{n=0}^{n=\infty} a_n \cos(n\omega_{lo}t)$$

$$= \sum_{n=0}^{n=\infty} a_n \cos(n\omega_{lo}t)b\cos(\omega_{sig}t + \phi_{sig})$$

$$= \sum_{n=0}^{n=\infty} \frac{a_n b}{2}[\cos(n\omega_{lo}t + \omega_{sig}t + \phi_{sig}) + \cos(n\omega_{lo}t - \omega_{sig}t - \phi_{sig})]$$

Notice that one of the $n = 1$ outputs is a signal at the difference frequency and contains the amplitude and the phase of the signal. This signal is called a *down-converted signal*. The process is called *mixing*.

This mixing process can be obtained by nonlinear behavior of a circuit element as well. For nonlinear elements, the transfer function of the device is usually written in terms of a power series like that of a detector. The second-order terms then would be similar to:

$$[b\cos(\omega_{sig}t + \phi_{sig}) + a_1 \cos(\omega_{lo}t)]^2$$

Carrying out the squaring operation results in terms at twice the signal and local oscillator frequencies as well as the cross-product term. Expanding the cross-product term yields:

$$2a_1 b\cos(\omega_{sig}t + \phi_{sig})\cos(\omega_{lo}t)$$

$$= a_1 b[\cos(\omega_{sig}t + \phi_{sig} + \omega_{lo}t) + \cos(\omega_{sig}t + \phi_{sig} - \omega_{lo}t)]$$

Notice that the sum and difference frequencies exist here as well. The difference frequency is usually the desired component for power detection purposes. The magnitude of the detected signal is $a_1 b$ where a_1 depends on the nonlinearity coefficient of the device.

If the waveform in Figure 11.4 is a string of very narrow pulses instead of a square wave and $\tau_L = R_L C_L$ is long with respect to the pulse period and $\tau_S = R_S C_L$ is short with respect to a pulse width, then the circuit shown can function as a sampling circuit. The capacitor, C_L, will tend to store the value of the waveform at the time of the narrow pulse width. A similar analysis can be made assuming that the capacitor C_L stores charge proportional to the input signal's value during the switch's on period rather than charging to the voltage of the input signal. The output signal is similar to the output of a mixer circuit except that a_1 is reduced in value because the pulse train is a series of narrow pulses rather than a square wave. The waveform across the load resistance will be the down-converted replica of the E_{sig} waveform. Some network analyzers use this type of down-converting or sampling mixers.

11.6 DETECTOR SATURATION

What has been considered thus far is the detection function. When the detected signal has been recovered, it must be displayed. Noise and saturation effects must be considered. When noise is present (as it always is), small signal levels are masked by the noise present. In addition, when noise is present, the small signal portion of the detection level will not be linear with power since there are two relatively equal powers present. The power is related to the sum of the noise signal plus the desired signal. The square of that total is only approximately equal to the square of the desired signal when the noise signal is small. At the high end of the input power, the detection device will saturate (if it does not burn out first) and the detected signal will not increase with increasing input power.

It is possible in the case of crystal detectors that the detected current would increase faster than the input power, resulting in an indication of more power present than there really is. A crystal detector will often saturate due to the series resistance of the diode before this "gain expansion" occurs. These effects are shown pictorially in Figure 11.5. It is important to consider the square law range of the detector. This is sometimes termed the *dynamic range* of the detector. This is the range over which the output is linear with input power. The low-level signal needs to be displayed in some display device. It will likely have to be amplified by an amplifier. This amplifier will not be noise free and the additive noise from the amplifier and the amplifier's saturation characteristics will need to be considered in addition to detector saturation and noise characteristics in any display instrument setup. Often the display instrument has an attenuator switch on the output of the detector to limit the displayed dynamic range to a small portion of the detector's dynamic range.

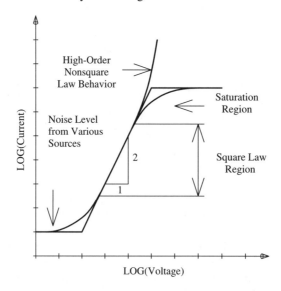

Figure 11.5 A detector response.

11.7 DETECTOR DIODE AND MIXER DIODE SOURCE IMPEDANCE

A Schottky diode is predominately a majority carrier device as opposed to a minority carrier p-n junction. The standoff voltage of the junction is less than that of a p-n junction made out of the same semiconductor material. Since the diode is a predominately majority carrier device, it does not exhibit the large charge storage effects that result from the storage of minority carriers in the junction region of a p-n junction diode. The Schottky diode is often used as a mixer device or as a detector device. Over some range of input power, a diode used as a detector appears to have an rf-dc Thevenin equivalent circuit as shown in Figure 11.6. The value of the dc source shown depends on the value of the input rf voltage. At a constant input rf voltage, the diode will appear to have approximately a fixed value of E_{rf} and R_{th}. Since the values of E_{rf} and R_{th} depend on the diode acting as a detector (i.e., having current flowing through it), the value of E_{rf} cannot be determined by putting an open circuit on the detector. There would be no current in the diode and then the diode would appear to have a very large equivalent circuit impedance. The value of E_{rf} and R_{th} can, however, be determined by perturbation techniques. If the circuit load is varied a small amount around some optimum value R_{ext}, say R_{ext1} and R_{ext2}, then:

$$V_{ext1} = \frac{E_{rf} R_{ext1}}{R_{ext1} + R_{th}}$$

$$V_{ext2} = \frac{E_{rf} R_{ext2}}{R_{ext2} + R_{th}}$$

$$R_{th} = \frac{R_{ext1} R_{ext2}(V_{ext2} - V_{ext1})}{(V_{ext1} R_{ext2} - V_{ext2} R_{ext1})}$$

$$E_{rf} = \frac{V_{ext1} V_{ext2}(R_{ext2} - R_{ext1})}{(V_{ext1} R_{ext2} - V_{ext2} R_{ext1})}$$

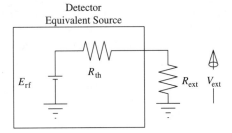

Figure 11.6 Detector equivalent circuit.

This procedure gives the Thevenin-equivalent source voltage and resistance under a given amount of rf voltage input. A curve of E_{rf} and R_{th} can be plotted versus input rf voltage to determine the response of the detector to different input rf voltages. The resultant circuit can then be used to determine the response of the detector to input power and input frequency changes, and so on. All of the diode's parasitic capacitances and inductances would need to be included in a frequency response calculation.

11.8 MIXERS

The mixing phenomenon has already been discussed in the detector section. It is helpful to review that section to refresh the understanding of mixing action. The nonlinearity primarily under consideration for the mixers discussed here is the resistive nonlinearity. A switch is a resistive nonlinearity wherein the resistor value goes from zero ohms to infinite ohms. Other types of switched resistance changes will also result in linear mixing action. Reactance change will also result in mixing action but it can also generate other parametric effects such as parametric gain and subharmonic generation. This section is concerned with variable-resistance mixing action.

11.9 THE SIGNAL AND ITS IMAGE

The large signal that switches the value of the resistance is termed the local oscillator and is abbreviated lo. The small signal that it switches is termed the radio frequency signal (rf). The resultant frequency desired is the intermediate frequency signal (if). Other frequencies exist. In fact, all frequencies:

$$f = \pm(N \cdot f_{lo} \pm M \cdot f_{rf})$$

can exist. Plus and minus signs are necessary since the frequency terms need to be considered for both positive and negative frequencies in the mixing operation.

The desired frequency (the if frequency) is generally:

$$f_{if} = \pm(1 \cdot f_{lo} - 1 \cdot f_{rf})$$

for what is termed a *down converter*. If the plus term is used, the mixer is termed a *high-side injection mixer*, while if the minus term is used, the mixer is termed a *low-side injection mixer*. High-side injection mixers invert FM signal spectrums since as the rf signal frequency increases, the if signal frequency decreases. If it is necessary to preserve the FM spectral components, then low-side injection mixers are necessary for single-conversion processes. In double-conversion processes that use high-side injection, two high-side injection mixing operations are necessary to preserve FM spectral components.

The desired frequency (the if frequency) is:

$$f_{if} = (1 \cdot f_{lo} + 1 \cdot f_{rf})$$

in what is termed an *up converter*. Notice that any mixer will create both a down-converted signal and an up-converted signal.

Another signal, rather than the desired rf signal, called the image signal can also generate the if signal.

$$f_{if} = \pm(1 \cdot f_{lo} - 1 \cdot f_{rf\_desired})$$
$$f_{if} = \pm[1 \cdot f_{lo} - 1 \cdot (f_{rf\_image} \pm 2f_{if})]$$
$$= \pm(1 \cdot f_{lo} - 1 \cdot f_{rf\_image}) + 2f_{if} = \mp(1 \cdot f_{lo} - 1 \cdot f_{rf\_image})$$

Notice that the image rf signal is offset to the opposite side of the lo from the rf signal. In any rf system requiring mixers, it must be determined if the image signal can interfere. If it does, it must be filtered out. Even if no signals exist at the image signal frequency, a filter may be necessary at the input to the mixer to prevent the image frequency from coming into the mixer. Noise signals exist at that frequency and would contribute noise power to the if, degrading the signal-to-noise ratio by 3 dB. Therefore, most systems attempt to get rid of the image frequency with some type of filtering or canceling technique. Mixers that are arranged to add two signal responses and subtract two image responses are called image-canceling mixers. These are discussed in Section 11.11.3.

Higher-order mixer products need to be considered also. One of the most difficult to eliminate is termed the *one-half if signal*. If a system is designed to receive a signal at an rf frequency, then:

$$f_{if} = |f_{lo} \pm f_{rf}|$$

Notice what happens when the rf signal is moved closer to the lo signal. If the difference between the rf and lo signal is equal to:

$$f_{if} = \frac{|f_{lo} \pm f_{rf}|}{N}$$

where N is some integer, then when there is an Nth-order response in the mixer to each of the signals the if signal would be generated.

$$f_{if} = \frac{|N \cdot f_{lo} \pm N \cdot f_{rf}|}{N} = |f_{lo} \pm f_{rf}|$$

Since the second-order response is the strongest, the 2×2 product is the largest. This response can be minimized by balancing the second harmonics of the signals out in a mixer as is done in double-balanced mixers. A single-balanced mixer would balance out either the even harmonic of the lo or the even harmonic of the rf. This is described in further detail in the section on single-ended mixers. It is important to note that the 2×2 response is a second-order response in a mixer. The amplitude of the lo signal stays constant and it is only the amplitude of the signal that is varied. This is different from the 2×2 response for intermodulation where both signal amplitudes are varied and the output signal has a fourth-order response. The equivalent of the 2×2 response can be generated in the if section of the system as well. The response for N equal to 2 is at one-half the if frequency. If the first stage of the if amplifier, which is usually put in front of the if amplifier filter, has a small dynamic range, this one-half if signal will double in the first stage, creating an if signal that will go through the if filter. When trying to determine whether the signal is generated in the mixer or in the if amplifier stage, separate intermodulation curves need to be generated for each stage. Then one can look at the signal levels involved and determine where the interfering signal is generated.

11.10 SINGLE-ENDED MIXERS

The circuits shown in Figure 11.7, either the series or shunt version, are termed single-ended mixer circuits. The upper two parts of Figure 11.7 show approximately what happens to the signal and if circuits when the diode is considered to be a switch and switched at the lo frequency. Notice that there is no isolation between the signal (rf) and the local oscillator (lo) signals and very little isolation between these signals and the if. If the if filters shown in Figure 11.7 did not exist, there would be no isolation between the rf and if and between the rf and lo. Filters would have to be added in the rf and lo port to provide isolation between the rf and lo signals.

Consider the upper-right circuit shown in Figure 11.7. Assuming the diode acts as a switch (in reality the diode acts more like a switched resistor in parallel with a capacitor), the approximate current impressed on the if circuit from the lo and rf sources would contain terms:

$$A \cos(\omega_{rf}t + \psi_{rf}) + B \cos(\omega_{lo}t + \psi_{lo})$$

during the time the switch is on and zero when the switch is off. The switch is assumed to work at the lo frequency. The switch acts as a multiplier with a magnitude of 1 during the on portion and a magnitude of zero during the off portion. This can be characterized as a square wave with an average value of one-half. The square wave can be characterized as a Fourier series. The Fourier series multiplied times the waveform above gives:

$$[A \cos(\omega_{rf}t + \psi_{rf}) + B \cos(\omega_{lo}t + \psi_{lo})] \left[\frac{1}{2} + \sum_{n=1}^{n=\infty} a_n \cos(n\omega_{lo}t + \psi_{lo\_n}) \right]$$

The product of the cosine terms gives the sum and difference frequencies as well as all the harmonic terms with the sums and differences. The actual current would be offset by the

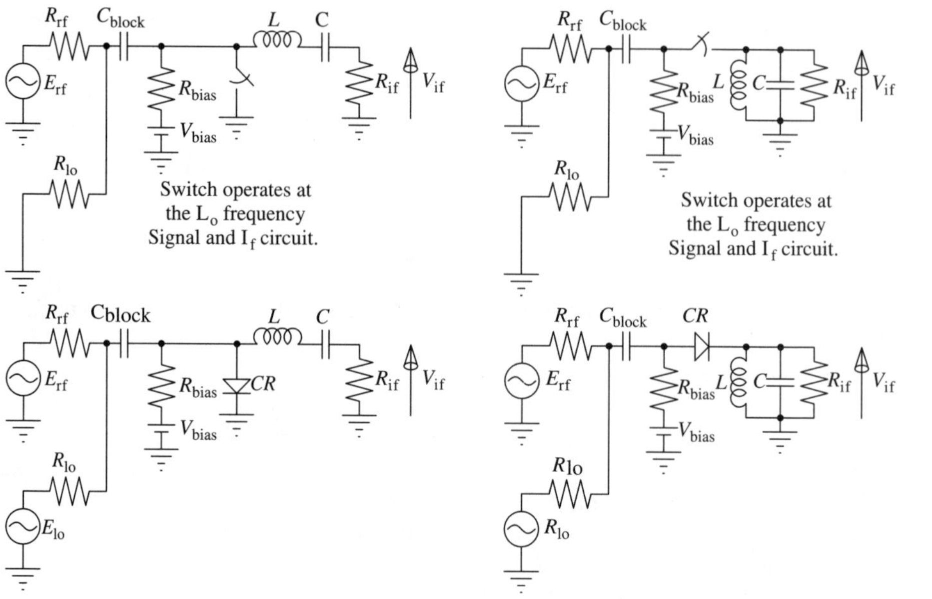

Figure 11.7 Some single-ended mixer circuits.

value of the if voltage on the if port. This somewhat modifies the approximate current given above. A more complete analysis can be done with nonlinear circuit analysis techniques. However, the above function gives a first-order explanation of how the different frequencies are generated.

An if frequency is generated from the difference frequency that comes from the product of the rf cosine term and the fundamental lo cosine term. The same if frequency but with a 180-degree difference in phase comes when the rf frequency is the same frequency offset from the lo frequency but on the other side of the lo frequency. The different mixers to be considered in the next sections have individual diodes in them that are excited by functions similar to that just described.

11.11 BALANCED MIXERS

In order to minimize coupling between the rf, lo, and if ports over that which exists in the single-ended mixer, several circuit topologies have been developed that use balanced circuits to provide isolation between these three signals. Two of these are the single-balanced and the double-balanced mixer circuits.

11.11.1 Single-Balanced Mixer

Hybrid couplers are discussed in Chapter 12. The scattering matrices for hybrids are given here for mixer analysis. Figure 11.8 shows two diodes on the outputs of a hybrid coupler. The hybrid coupler can be either a 90-degree or a 180-degree hybrid. One possible scattering matrix for a 90-degree hybrid circuit is:

$$(S) = \frac{\sqrt{2}}{2} \begin{pmatrix} 0 & 1 & -j & 0 \\ 1 & 0 & 0 & -j \\ -j & 0 & 0 & 1 \\ 0 & -j & 1 & 0 \end{pmatrix}$$

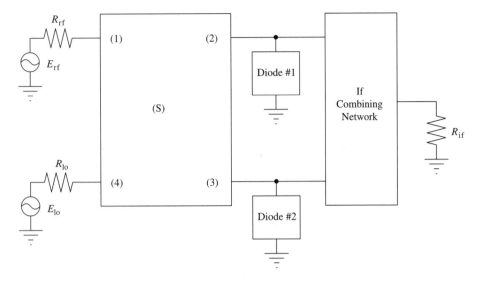

Figure 11.8 A single-balanced mixer.

One possible matrix for a mixer that has 90 degrees phase shift from port one to two and port one to three and has 90 degrees shift from port four to port three but 270 degrees from port four to port two is:

$$(S) = \frac{\sqrt{2}}{2} \begin{pmatrix} 0 & -j & -j & 0 \\ -j & 0 & 0 & +j \\ -j & 0 & 0 & -j \\ 0 & +j & -j & 0 \end{pmatrix}$$

One possible scattering matrix for a 180-degree circuit is:

$$(S) = \frac{\sqrt{2}}{2} \begin{pmatrix} 0 & 1 & -1 & 0 \\ 1 & 0 & 0 & -1 \\ -1 & 0 & 0 & 1 \\ 0 & -1 & 1 & 0 \end{pmatrix}$$

The block shown as DIODE in Figure 11.8 could also be inserted in series with the if combining network in some circuits. Just as in the single-ended mixer, the diode can be in series or in shunt depending on layout, matching networks, and system requirements. Depending on the scattering parameters of a specific input coupler, some topologies may require each diode to have the same polarity or may require one diode with one polarity and the other with the other polarity. The if signal is combined differently depending on whether each diode has the same polarity or each diode has a different polarity. There is also a major difference between the 90 degree and the 180-degree mixer concerning where the image frequency exists and which port it attempts to come out of. This has a major affect on the input *VSWR* at each of the rf or lo ports of a mixer. In addition, the image signal may remix in each diode differently with the local oscillator and cause a miss-balance in the mixer. A miss-balance limits the amount of rejection or cancellation that can be achieved with a specific mixer topology. The user needs to be aware of these effects and take appropriate corrective action or advantage of the effect depending on the system application of the mixer.

Other (S) matrices than those shown in Figure 11.8 will also produce 90- or 180-degree mixers. The relationship between the voltages on the two diodes needs to be plus or minus 90 degrees for a 90-degree mixer and 180 degrees for a 180-degree mixer. Assume the diodes in the circuit shown in Figure 11.8 are in opposite directions (i.e., one diode rectifies when the port voltage is positive and the other diode rectifies when the port voltage is negative).

Assume that the input signals are given by:

$$V_{rf} = |V_{rf}| \cos(\omega_{rf} t)$$

$$V_{lo} = |V_{lo}| \cos(\omega_{lo} t + \psi)$$

where ψ is the phase difference between the local oscillator and signal frequencies at some time $t = 0$. This phase difference is important when considering coherent systems. When the phase difference is not important, it can be dropped out of the analysis but does not harm the analysis by leaving it in.

The signal incident on diode #1 is proportional to:

$$V_{diode\#1} \approx A_{rf} \cos(\omega_{rf} t + \phi_{S_{21}}) + A_{lo} \cos(\omega_{lo} t + \psi + \phi_{S_{24}})$$

where cosine functions are used to describe the incoming signals and the magnitudes are given as A instead of V to account for various loss and VSWR terms. The approximately sign is used in this and the following equations to describe that the outputs or voltages being described depend on the conversion loss or insertion loss of the mixer and its various elements. The signal voltage on diode #2 (if the diodes were linear) would be:

$$V_{diode\#2} \approx A_{rf} \cos(\omega_{rf} t + \phi_{S_{31}}) + A_{lo} \cos(\omega_{lo} t + \psi + \phi_{S_{34}})$$

The diodes are not linear and the impressed local oscillator signal will cause multiplication of the two signals present on the diodes. Assume diode #1 determines the reference phase for the if. Then diode #2 will either rectify in phase with diode #1 or out of phase with diode #1. In the next couple of equations, that will be shown with the plus or minus sign. If the diodes are in the same directions, the diodes will rectify and produce if signals that are out of phase. The signal generated at diode #1 will be related to the following product.

$$\left[A \cos(\omega_{rf} t + \psi_{rf} + \phi_{S_{21}}) + B \cos(\omega_{lo} t + \psi_{lo} + \phi_{S_{24}}) \right] \times$$

$$\left[\frac{1}{2} + \sum_{n=1}^{n=\infty} a_n \cos(n\omega_{lo} t + \psi_{lo\_n} + \phi_{S_{24\_n}}) \right]$$

The if difference frequency generated at diode #1 is proportional to the following sine wave.

$$V_{if\_1} \propto + \cos(\omega_{rf} t + \psi_{rf} + \phi_{S_{21}} - \omega_{lo} t - \psi_{lo} - \phi_{S_{24}})$$

The signal generated at diode #2 will be related to the following product.

$$[A \cos(\omega_{rf} t + \psi_{rf} + \phi_{S_{31}}) + B \cos(\omega_{lo} t + \psi_{lo} + \phi_{S_{34}})] \times$$

$$\left[\frac{1}{2} + \sum_{n=1}^{n=\infty} a_n \cos\left(n\omega_{lo} t + \psi_{lo\_n} + \phi_{S_{34\_n}}\right) \right]$$

The if difference frequency generated at diode #2 is proportional to the following sine wave.

$$V_{if\_2} \propto \pm \cos(\omega_{rf} t + \psi_{rf} + \phi_{S_{31}} - \omega_{lo} t - \psi_{lo} - \phi_{S_{34}})$$
$$= \pm \cos(\omega_{rf} t - \omega_{lo} t + \psi_{rf} - \psi_{lo} + \phi_{S_{31}} - \phi_{S_{34}})$$

The difference in phase between the if signals is:

$$(\omega_{rf} t - \omega_{lo} t + \psi_{rf} - \psi_{lo} + \phi_{S_{21}} - \phi_{S_{24}})$$
$$-(\omega_{rf} t - \omega_{lo} t + \psi_{rf} - \psi_{lo} + \phi_{S_{31}} - \phi_{S_{34}})$$
$$= (\phi_{S_{21}} - \phi_{S_{24}} - \phi_{S_{31}} + \phi_{S_{34}}) = \pm \left(\frac{\pi}{2} \pm \frac{\pi}{2} \right)$$

Notice that the phase difference inside the last parentheses is pi for the first two scattering matrices and zero for the third scattering matrix given above. The first two circuits would need the diodes reversed to give another pi in phase for addition of the signals while the third circuit would use two diodes of the same polarity. However, if the signals between each diode are out of phase, the if network could have an out-of-phase combiner for the first two circuits and use in-phase diodes. The rf and lo signals have isolation. In addition, it can be shown that the amplitude modulation (AM) noise on the local oscillator will also cancel in the if signal. The user should perform this calculation. This is done by assuming the rf signal is zero and the lo signal is:

$$E_{\text{lo}} = A_{\text{lo}}\cos(\omega t) + A_n \cos(\omega t + \omega_{\text{diff}} t) + A_n \cos(\omega t - \omega_{\text{diff}} t)$$

These mixers are single balanced. They will partially cancel out the even-order (N or M even) products of multiplication for one but not both of the rf or lo signals. The amount of cancellation depends on the symmetry of the circuit and the balance of the diodes. The even harmonics of current in the diodes can be made zero by putting a shorted quarter-wavelength of line at the rf and lo frequencies in series with a diode.

Some mixers have phasing and filtering circuits in them to reflect the image frequencies back to the diodes for remixing that energy to decrease the conversion loss. These mixers are called *image-enhanced mixers*. If the image signal is not reflected back into the mixer, the energy associated with that signal is dissipated in a load. The conversion loss of the mixer increases to account for the termination of that energy. However, care has to be exercised when designing an image-enhanced mixer since when the image signal remixes in each diode differently, the balance is disturbed and the rejection of unwanted signals will suffer.

11.11.2 Double-Balanced Mixers

It is desirable that the even harmonics cancel on both the signal and the local oscillator as well as having isolation between all three ports. The transformer circuits shown in Figure 11.9 provide the isolation between the rf and if and lo and if ports. The symmetry of the circuit provides isolation between the rf and lo ports. This mixer configuration is called a double-balanced mixer. It has excellent isolation and balance. Double-balanced mixers balance the even-order harmonics of both the signal and the local oscillator so that responses from the $M \times f_{\text{lo}}$ plus or minus $N \times f_{\text{rf}}$ for M and N both not equal to 1, but either M, N, or both even are minimized. These responses are present in every mixer and either have to be minimized, balanced, or filtered out. It is not always possible to filter out these responses. Therefore, when they are balanced out, filter requirements are minimized. Generally, the conversion loss of a double-balanced mixer is higher than that of a single-balanced mixer.

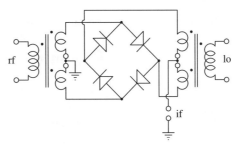

Figure 11.9 A double-balanced mixer.

The mixer shown in Figure 11.9 has a ring diode quad in it. That diode quad could be replaced with a star diode quad. In addition, if a higher power level is desired from a mixer, each of the diodes can be replaced with two or more in series. This increases the amount of lo drive needed for the mixer but decreases the intermodulation product levels for some signals as will be apparent in the section on intermodulation products from a mixer.

11.11.3 Image-Canceling Mixers

If two mixers have the same lo signal but have the rf signals 90 degrees out of phase between them, the if signals will also be 90 degrees out of phase. Suppose that two different rf signals are used. One that is at a lower frequency than the lo frequency will be called rf_a and one that is at a higher frequency than the lo frequency will be called rf_b. These two rf signals will create two if signals out of each mixer. Let the two mixers be called the top mixer and the bottom mixer. Let the top mixer be driven by the rf with zero phase and let the bottom mixer be driven with an additional 90-degree phase lag. The if signals out of the mixers are given by:

$$if_{top\_a} = \cos(\omega_{lo}t - \omega_{rf_a}t)$$

$$if_{bot\_a} = \cos\left(\omega_{lo}t - \omega_{rf_a}t + \frac{\pi}{2}\right)$$

$$if_{top\_b} = \cos\left(\omega_{rf_b}t - \omega_{lo}t\right)$$

$$if_{bot\_b} = \cos\left(\omega_{rf_b}t - \frac{\pi}{2} - \omega_{lo}t\right)$$

Subscript a is for rf_a and subscript b is for rf_b. The signals will be put into a hybrid with the following scattering parameters.

$$(St) = \frac{\sqrt{2}}{2}\begin{pmatrix} 0 & 1 & -j & 0 \\ 1 & 0 & 0 & -j \\ -j & 0 & 0 & 1 \\ 0 & -j & 1 & 0 \end{pmatrix}$$

Port one is put on the top mixer, port four on the bottom mixer, and the outputs are taken off ports two and three. The phase shift from ports one to two and from ports four to three is zero degrees and from ports one to three and ports two to four is minus 90 degrees. The signal out of the ports is given by:

$$if_{top\_a} = \cos(\omega_{lo}t - \omega_{rf_a}t)$$

$$if_{bot\_a} = \cos\left(\omega_{lo}t - \omega_{rf_a}t + \frac{\pi}{2}\right)$$

$$if_{top\_b} = \cos(\omega_{rf_b}t - \omega_{lo}t)$$

$$if_{bot\_b} = \cos\left(\omega_{rf_b}t - \frac{\pi}{2} - \omega_{lo}t\right)$$

$$if_2 \propto \cos(\omega_{lo}t - \omega_{rf_a}t) + \cos\left(\omega_{lo}t - \omega_{rf_a}t + \frac{\pi}{2} - \frac{\pi}{2}\right)$$

$$+ \cos(\omega_{rf_b}t - \omega_{lo}t) + \cos\left(\omega_{rf_b}t - \frac{\pi}{2} - \omega_{lo}t - \frac{\pi}{2}\right)$$

$$= 2\cos(\omega_{lo}t - \omega_{rf_a}t)$$

$$if_3 \propto \cos\left(\omega_{lo}t - \omega_{rf_a}t - \frac{\pi}{2}\right) + \cos\left(\omega_{lo}t - \omega_{rf_a}t + \frac{\pi}{2}\right)$$

$$+ \cos\left(\omega_{rf_b}t - \omega_{lo}t - \frac{\pi}{2}\right) + \cos\left(\omega_{rf F_b}t - \frac{\pi}{2} - \omega_{lo}t\right)$$

$$= 2\cos\left(\omega_{rf_b}t - \omega_{lo}t - \frac{\pi}{2}\right)$$

Notice that the if signal out of port two contains only the difference frequency from rf signal a and the if signal out of port three contains only the difference frequency from rf signal b. This mixer arrangement can be used to separate the upper and lower sidebands from a signal. This type of mixer is called an image-canceling mixer.

11.12 INTERMODULATION PRODUCTS

11.12.1 Mixer Intermodulation Products

When two sinusoidal signals are multiplied together, their amplitudes are multiplied and their frequencies are added and subtracted so that two new signals result. If at the same time that the signals are multiplied together they are also generating harmonics, various higher-order terms will exist. In intermodulation analysis these responses are sometimes called *birdies* as a result of what interfering signals sound like in the audio portion of the system. The intermodulation products will also have a nonlinear response.

The input amplitude versus output amplitude response is not linear (not a 1 : 1 slope on a log-log plot). Figure 11.10 shows a hypothetical log-log response of a mixer. This is the response to a single signal entering the mixer. These intermodulation curves are different from two-tone intermodulation curves. The 1 : 1 slope is the desired response. What are the responses shown with a 2 : 1, 3 : 1, and 4 : 1 slope? These are intermodulation response curves. The intercept point of a given slope with the extended 1 : 1 slope is the intercept point for that slope. On the following figure, all curves are shown with the same intercept point. That is not always the case. The response to an interfering signal will produce an output containing frequencies:

$$f_{\text{out}} = \pm(N f_{\text{lo}} \pm M f_{\text{interfering}})$$

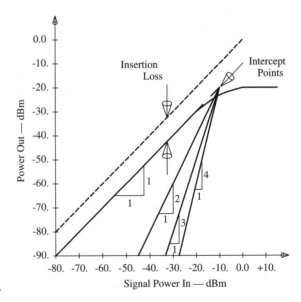

Figure 11.10 Mixer response curves.

The amplitude of each interfering signal will have a slope on a log-log response equal to $M + N$ if each signal increases in magnitude at the same rate. In a mixer, the lo is held at constant level and therefore the slope of interfering signal responses will only be M. The intermodulation or birdie curves shown on the response then are a result of the M = 2nd, 3rd, or 4th signal harmonics mixing with all harmonics of the lo at whatever power level the lo is at. Amplifier intermodulation is usually measured using equal amplitudes of both signals and is often confused with single-tone mixer intermodulation values. Notice that one needs to be careful to specify intermodulation intercept values as a function of input power or as a function of output power. The dB difference between these two reference methods is the conversion loss. However, it will make a considerable difference on the amount of input power that is really being specified. For example, for the fourth-order ($M = 4$) curve, the slope is very large and the conversion loss for that intermodulation is very large at small signal levels.

11.12.2 Linear Circuit Intermodulation

The title of this section is somewhat of a misnomer since a linear circuit by definition cannot produce harmonics. What is meant is that the linear circuit becomes nonlinear when a signal with too large of an amplitude is introduced into it. A circuit with saturation will create harmonics. When intermodulation between two input signals is considered, each of the signals is considered to have the same amplitude. Therefore, the response that exists between the second harmonic of one signal and the fundamental of the other signal has a slope of three to one when plotted as output power versus input power. As noted in the previous section, this is different from the intermodulation that exists between the local oscillator of a mixer and the input signal. The intermodulation in this section is called *two-tone intermodulation*. Two tones of equal amplitude are used. If the frequencies of the signals differ by a small amount df, then the third-order intermodulation response is:

$$f = |2(f + df) - f| = f + 2df$$
$$f = |f + df - 2f| = f - df$$

Both of these signals are in the passband and will pass through the amplifier. Their response can be seen with a spectrum analyzer. The difference frequency df will also be generated by the nonlinear action of the device. This signal will appear across the bias network of the device. The termination of the difference frequency in the bias network can enhance the third-order intermodulation response by mixing action as well. If the bias circuit has a negative resistance in it at the difference frequency, the signal will be amplified in the bias network. That signal can mix again with one of the signals to enhance the third-order response.

The output power level response versus input power level response for a circuit with 30 dB of gain and an output saturation level of 0 dBm is shown in Figure 11.11. A third-order intercept point is also shown on that curve.

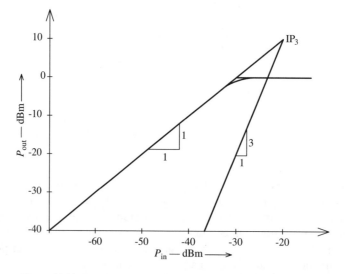

Figure 11.11 P_{out} versus P_{in} resonse for a circuit with a saturation level.

By minimizing the second harmonic response of the active device, the third-order intermodulation point can be made quite high. Balanced circuits can be used to raise the level as well. Figure 11.12 shows the gain of the circuit described in Figure 11.12. The 1-dB compression point for the amplifier is about −29 dBm for the response shown. There is 30 dB of linear gain. The gain has dropped by one dB at the 1-dB compression point. The 1-dB compression point is often specified for amplifiers and other circuits. The relationship between the 1-dB compression point and the third-order intercept point depends on harmonic balance and terminations within the circuit and bias line terminations.

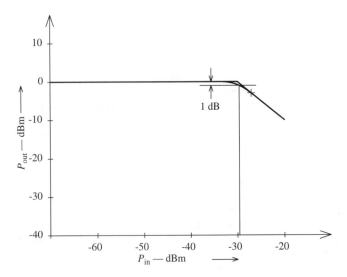

Figure 11.12 Gain versus P_{in} for the previous figure.

A mixer acts as a linear circuit as far as the input and output amplitude response is concerned when it is in its linear region of operation. When two signals of equal level are introduced to the mixer, these two signal will also mix together to create third-order intermodulation signals the same way the signals are created in an amplifier. A manufacturer of a mixer often does not give these third-order intermodulation curves for two equal input signals. The curves that are most often given by the manufacturer of a mixer are the single signal intermodulation curves. The two effects are related and a mixer that has a high degree of balance and thus rejection of the second harmonic would be expected to have a higher third-order intermodulation intercept.

11.12.3 Cross-Modulation Effects

Consider a circuit that is passing two signals. Suppose one of the signals is an amplitude modulated (AM) signal with 100% modulation. Assume the other signal is a small signal greater than 10 dB down from the carrier level of the AM signal. The second signal could be amplitude modulated or frequency modulated. Further assume the amplifier response shown in Figure 11.11 is the amplifier response of the amplifier. Assume the carrier level of the AM signal is at −33 dBm. When the AM signal is at its minimum modulation value, there is no effect on the small signal if the amplifier does not have memory effects in its bias circuits. When the AM signal is at the carrier level, there is very little effect on the gain of the amplifier. However, when the AM level is at its highest level (carrier plus 6 dB), the AM signal level is at −27 dBm and the gain of the amplifier is reduced to 27 dB. That will also be close to the gain experienced by the small signal. The small signal at the output of the amplifier now has amplitude modulation on it caused by the amplitude modulation of the large signal. It has a down modulation of three dB when the large signal is up 6 dB above its carrier. The effect is even worse when the large signal level is higher. The modulation has crossed from one signal to the other and this effect is called *cross modulation*. When both signals are of the same level and they are both amplitude modulated, the peak voltage from each of the signals will add in phase at some point in time. The peak voltage will then be four times the voltage at the carrier level of one signal. This is a peak increase of 12 dB. One can see from the gain compression curve that the output of the amplifier will be greatly compressed at that point.

EXERCISES

11-1 A 50-ohm transmission line is transmitting 50 watts of ac power into a matched load at a fixed frequency. If a low-capacity Schottky diode with a forward voltage of 0.3 volts is placed with one end on the transmission line and the other end on a large capacitor in parallel with a large resistor, what would the dc voltage be across the capacitor? For this exercise, a low-capacity Schottky diode means one that has a negligible effect on loading the transmission line. A large capacitor means one that is at least two orders of magnitude larger than the Schottky diode capacity. A large resistor means one that will not discharge the large capacity appreciably during a period of the frequency of the power on the transmission line.

11-2 A single-balanced mixer is used to down convert a 1 GHz signal to a 100 MHz intermediate frequency. The local oscillator is at 900 MHz. The image and other spurious responses at the output need to be less than −80 dBm. The largest signal level to be considered is −20 dBm. The conversion loss of the mixer is 6 dB. The 2 × 2 (lo × sig) response of the mixer is −106 dBm when the input signal is at −40 dBm. Determine the filter specification for the filter that needs to be put in front of the mixer to prevent the image and 2 × 2 response from being out of specification. Would a double-balanced mixer help minimize the amount of filtering for either the image or the 2 × 2 signal?

11-3 Two signals are within the signal passband of a filter in front of a mixer. The signals are of the same level. The output response of the mixer needs to be below −80 dBm. The largest signal level is −20 dBm. The third-order intercept point for the mixer is +20 dBm. Is the mixer satisfactory for the application?

11-4 A diode is used to detect a small amplitude modulation on a 10 mW carrier signal. When biased with 10 mW of local oscillator power, the diode circuit produces 100 mV dc into a 1-k ohm load. It produces 90 mV dc into an 800-ohm load under the same conditions. What should the detector load impedance be?

11-5 In Exercise 11-2, what option exists besides filtering and a double-balanced mixer?

12

Microwave Components

12.1 MICROWAVE COMPONENTS

This chapter presents a description of some components other than resistors, capacitors, and inductors used in microwave circuits. When semiconductor devices are discussed, the discussion will concentrate on a phenomenological behavior of the components. The physics of semiconductors necessary to understand the characteristics for these devices can be found in semiconductor textbooks. Where it is necessary for an understanding of a device's characteristics, some limited amount of semiconductor physics is referred to. Couplers and hybrid-coupled amplifiers are discussed toward the end of this chapter. A full treatment of the various components given in this chapter would fill several books [1,2,3,4,5]. This summary treatment is included to give the reader an introduction to the components that are available to perform different functions in microwave circuitry. The analysis of PIN diodes used as high-power switches, special applications of transformers, an analysis of hybrid coupled amplifiers, a treatment of ferrite beads, and an analysis of bonding wires should be of special interest to the reader.

12.2 DIODES

Schottky diodes were discussed in the chapter on detectors and mixers. This section discusses several other types of diodes that have important application in microwave circuits.

12.2.1 Circuit Analysis for Diodes

Several types of analyses are made on circuits. If the circuit under consideration is linear, one can perform a linear dc analysis of that circuit. In this type of analysis, all the inductors are replaced by short circuits, all the capacitors are replaced by open circuits and then a voltage and current analysis of the circuit is made for each dc value assumed for the various dc sources. If the circuit under consideration is linear, a linear ac analysis can be made similar to the dc analysis except that all capacitors and inductors are included in the circuit. For linear circuits, using the principle of superposition, the dc and the ac analyses can be added to get a combined analysis of the circuit for both dc and ac excitations. However, when the circuit is nonlinear, either a full nonlinear analysis needs to be done or several other types of approximate analysis can be performed. If the ac signal is small with respect to the dc signal, then a small signal analysis using differential ac analysis is often performed. The circuit is analyzed at a particular dc value and then the ac signals are added to the dc values using differential or small signal analysis of the circuits. These techniques are usually familiar to the microwave engineer. However, when the ac signal is large with respect to the dc signal values, another technique is followed if the ac signal does not appreciably change the dc operating conditions. This is particularly applicable for circuits that have charge storage devices in them. One of the charge storage devices is the PIN diode. The amount of charge stored in the device is determined by small signal dc analysis and the large signal performance is performed about that small signal dc condition under the assumption that the small signal dc value does not change over a cycle of the ac large signal. The validity of this assumption needs to be checked just as the small signal assumption is checked in the ac analysis of small signal nonlinear circuits. This should become evident in the discussion of the PIN diode.

12.2.2 Diode Packages

Two terminal diode devices are often inserted in a microwave package called a *pill*. The pill diode comes in various sizes and sometimes comes with prongs and sometimes without prongs. There are several versions of this package. However, the parasitics are all similar and need to be incorporated into a design. Figure 12.1 shows a typical diode package. This is shown for the purposes of showing the various parasitics that are associated with a diode package. Other types of packages exist. When the leads directly attach to the diode as shown in Figure 12.2, the inductance between the package capacity and diode disappears. The lead inductance still exists but the parasitic package capacity appears almost directly across the device. The diode capacity is also reduced with this construction.

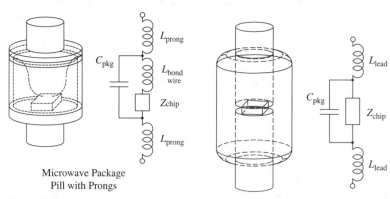

Microwave Package
Pill with Prongs

Figure 12.1 Diode Package and its para-sitics.

Figure 12.2 Diode package formed by leads.

12.2.3 Varactor Diode

A varactor (variable capacitor) diode is a semiconductor junction diode. The parasitic series resistance of a varactor diode is of primary consideration in maximizing the Q of the diode. The depletion region of the semiconductor junction acts as a capacitor. The dc voltage across the diode determines the depletion width and thus the depletion capacitance of the varactor. The rf voltage across this depletion region must be small with respect to the dc voltage across the diode if harmonic generation is to be minimized. However, often, harmonic generation is the desired result in frequency multiplier circuits. A schematic of a varactor chip is shown in Figure 12.3. The chip inductance is not shown. That inductance is often smaller than the inductance of the package that the chip is inserted into. The figure also assumes that the P region is doped much higher than the N region so that the depletion region is primarily in the N region. The depletion region capacity of a varactor diode is given by:

$$C \approx \frac{K}{(V + V_{bi})^s} \qquad s = \frac{1}{m + 2}$$

where V_{bi} is the built-in potential and V is assumed to be a back bias voltage across the diode. In the n side of the region of the junction, the diffusion profile, and thus the charge variation of the depletion region is often approximated by:

$$n(x) = Bx^m$$

where m is the diffusion slope parameter used in the capacity slope equation. A $\log(C)$ versus $\log(V + V_{bi})$ plot will have a slope s given by:

$$\log(C) = \log(K) - s \log(V + V_{bi})$$

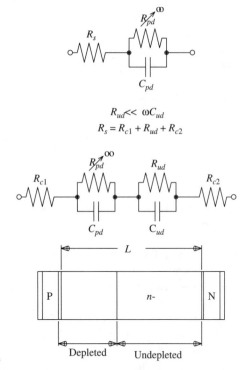

Figure 12.3 Varactor diode equivalent circuit.

An abrupt junction diode has $m = 0$, $s = 1/2$, a linear graded junction has $m = 1$, $s = 1/3$, and a hyperabrupt junction has $m < -1$, $s > 1$. Each of these types of diodes has applications in different types of rf circuits.

A typical varactor diode used in the 100-MHz to 10-GHz frequency range would have package inductances on the order of 0.5 nH, package capacities of several tenths of picofarads, and series resistances of less than one-half ohm. For example, a GaAs abrupt junction varactor might have a Q of 250 at 1 GHz and a junction capacity of 4 pF at 7 V back bias. This would give:

$$R_S = \frac{1}{(250)(2\pi\,10^9)(4.\times 10^{-12})} = 0.16\Omega$$

One limitation on the frequency where the varactor can be used is the resonance caused by L_{pkg} in series with the series combination of the junction capacitance and C_{pkg}. For a $C_{pkg} = 0.5$ pF, $L_{pkg} = 0.5$ nH, the above varactor diode would be used at frequencies much below 11 GHz. However, the mounting structure for the diode package can easily add 1 pF of parasitic capacity if care is not given to the structure. This capacity will be in parallel with C_{pkg} and can limit the upper frequency over which the packaged diode is useful.

Some varactor diodes used for harmonic generation are hybrids between pure varactor and PIN diodes. These hybrid diodes are often called *bimode* diodes. The large capacity variation and large forward voltage swings available with bimode diodes combine to provide high-power multipliers [6]. The function of a PIN diode used as a switch and not for harmonic generation will be discussed next.

12.2.4 PIN Diode Characteristics

The PIN diode comes from considering three different doped regions of a semiconductor put together. These are the p region (P), the intrinsic region (I), and the n region (N). Typically, the N region is the substrate region and the I region and P region are grown on top of the substrate. The diodes can also be fabricated in an *NIP* format where the structure is upside down. The diodes are generally fabricated vertically and often by epitaxial growth. The high breakdown voltage 1N4007 diode has many of the characteristics of a PIN diode. It has a long low doped region with higher doped contacts. At lower frequencies, this diode can often be used to demonstrate many of the features of a PIN diode. A word of caution is in order. The 1N4007 diode is a generic diode and is not manufactured to an rf PIN diode specification. Therefore, a specific diode from a specific vendor may not act the same as a specific diode from another vendor.

12.2.4.1 General Diode Considerations.   This section discusses some of the properties of PIN diodes. The PIN diode is often used as a switch. The switch may be used for low-power or high-power signals. If it is used for high-power signals, the rf current through the diode can be many times larger than the dc bias current through the diode. The diode can also withstand rf voltages across it that cause the net voltage across the diode to be very large in the forward direction. The familiar exponential V-I characteristic of a diode is the equilibrium or steady-state characteristic of a diode. Under large voltage swings associated with rf switching, the transient V-I characteristics of a diode are much different. However, the reverse breakdown of a diode under transient conditions is very close to the dc breakdown of a diode since breakdown is a very fast phenomenon. The objective of this discussion is to determine how much power can be safely switched by the PIN diode. The analysis in this section will be a first-order analysis and is good enough for many calculations. A PIN diode with an undepleted I region is depicted in Figure 12.4. As shown in Figure 12.4, the PIN diode has a P region, an I region, and an N region. The I region may be short or long relative to diffusion profiles of the P and N regions depending on which type of diode is preferred. The I region of the diode may be fully depleted (and thus containing no free carriers) or partially depleted (and thus containing free carriers) depending on the diode structure. In order to understand the equivalent circuit of each region, consider a block of semiconductor as shown in Figure 12.5. If the carriers that exist in the volume of material shown in Figure 12.5 are not moving at a velocity-saturated velocity, the current I across the volume is:

$$I = JA = Aq(v_n n + v_p p) = AqE(\mu_n n + \mu_p p)$$

where J is the current density, A is the cross-sectional area, q is the magnitude of the electronic charge, n is the electron concentration, p is the hole concentration, v_n is the electron velocity, v_p is the hole velocity, E is the electric field, μ_n is the electron mobility, and μ_p is the hole

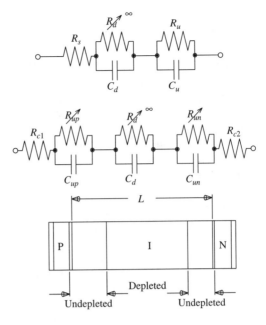

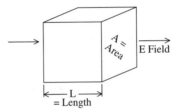

Figure 12.4 PIN diode cross section and equivalent circuit.

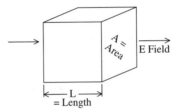

Figure 12.5 A block of semiconductor material.

mobility. The resistivity ρ of the material is given by:

$$\rho = \frac{1}{q(n\mu_n + p\mu_p)}$$

The resistance and capacitance of the block of material are related to the material parameters as:

$$R = \frac{\rho L}{A} \quad C = \frac{\varepsilon_r \varepsilon_o A}{L} \quad RC = \rho \varepsilon_r \varepsilon_o$$

where ε_r and ε_o are the relative dielectric constant and free space permittivity respectively. The frequency (relaxation frequency) occurs when the block of material has a conductance equal to its susceptance or when $\omega = 2\pi f_r = 1/RC$. For silicon this occurs when:

$$f_r = \frac{1}{2\pi \rho \varepsilon_r \varepsilon_o} = \frac{153 \text{ GHz}}{\rho(\Omega - cm)}$$

where $\varepsilon_r = 11.8$ was used. For $100 \ \Omega - cm$ material, $f_r = 1.53$ GHz.

When the applied dc voltage across the PIN diode is changed, the amount of depleted region in the PIN structure changes. Therefore, C_u and C_d change in relation to each other. Meaningful capacitance measurements of the diode I region can be made only when the diode is back biased enough for the I region to be completely depleted. In other words, the undepleted region then effectively disappears, making R_u in the equivalent circuit a short circuit. The value of R_s, which is the parasitic contact resistance, and so on, generally needs to be measured at forward bias in the 100-MHz to 2-GHz range to get meaningful measurements.

The PIN diode is often used as a switch. When the diode is full of carriers, the I region appears as a resistor R. The value of R is determined by the number of carriers in the I region of the diode. The I region of the diode resembles the block of semiconductor discussed above. When the diode is depleted of carriers, the I region appears as a capacitor C. The diode can be changed from a low-impedance state where it appears as a resistor R to a high-impedance state where it appears as a capacitor C just by changing the charge state of the diode. Filling the diode with carriers makes it a low impedance. Removing the charges from the diode makes it a high impedance. What is required to maintain the diode in a state full of charge? What is required to maintain the diode in a state void of charge? How long does it take to fill the diode full of charge or to remove the charge it has in it? How much ac current can be put through the diode when it appears as a resistor? All of these questions can be answered by looking at the switching behavior of a PIN diode.

12.2.4.2 RF Switching Behavior.
In order to determine the value of rf current that the diode can switch, we consider the total amount of charge that exists in the I region. This allows us to determine whether the I region acts as a capacitor or as a resistor. If R, as calculated above, is very large with respect to the reactance of C, then the I region acts as capacitor. If R is very small with respect to the reactance of C, then the I region acts as a resistor. How much rf current can the I region conduct and still be considered a resistor? If the I region has a dc current passing through it and if the rf current peaks are very large, will the I region become depleted during part of the rf cycle? Consider the V-I curves shown in Figure 12.6. The current and voltage waveforms are shown much larger than the dc reverse current or dc forward voltage drop. Remember that the dc characteristics are equilibrium characteristics and not characteristics associated with transient signals or large amplitude sinusoidal signals. The diode V-I characteristics under sinusoidal or transient conditions have to be reconsidered. As indicated by the curves in Figure 12.6, it is possible to have large rf current swings in both in the positive and negative directions under some conditions and also have large voltage swings both in the positive and negative directions under other conditions. These conditions are not applied simultaneously. If they were, the diode would likely burn out from the very large power dissipation.

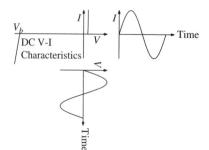

Figure 12.6 V-I curves—large signal current and voltage waveforms impressed on dc curves of a PIN diode.

For step changes in the diode current, the total charge in the diode I region can be approximated by:

$$\frac{dq}{dt} = -\frac{q}{\tau} + I_d \qquad q(t) = Q_o e^{-\frac{t}{\tau}} + I_d \tau$$

where Q_o and I_d are determined from time boundary conditions and τ is the minority carrier lifetime. Carriers will exist for a time in the I region after a forward current is turned off. Their number will gradually decay away. How long will the carriers last? How can τ be measured? If the diode is physically short, then when the forward current is suddenly reversed, most of the carriers would be able to be extracted before very much recombination occurs. For a diode initially forward biased for a long time at current I_f, the initial total charge in the diode is:

$$Q_{initial} = I_f \tau$$

If this charge is to be removed in a short time $\Delta t \ll \tau$, the reverse current necessary is:

$$|I_r| = \frac{Q_{initial}}{\Delta t}$$

These two equations relate the minority carrier lifetime to the amount of current required to remove the charge in a small period of time as shown in Figure 12.7. Notice that the current goes to zero after the short time Δt since one cannot extract charge that does not exist. However, it is not always possible to extract the stored charge from a diode this quick due to inductive and capacitive parasitics. If the bias current through the diode changes from I_1 to I_2, the charge as function of time is:

$$q = (I_1 - I_2)\tau e^{\frac{t}{\tau}} + I_2\tau = Q_o e^{\frac{t}{\tau}} + I_2\tau$$

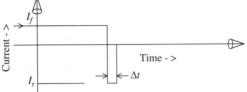

Figure 12.7 Forward current and reverse current waveforms.

For an initial forward current of I_1 that has been flowing in the diode for a long time, setting this charge equal to zero gives the length of time it takes to deplete the I region of charge.

$$q = 0 = (I_1 - I_2)\tau e^{-\frac{t}{\tau}} + I_2\tau \qquad I_2 = \frac{I_1 e^{-\frac{t}{\tau}}}{e^{-\frac{t}{\tau}} - 1}$$

$$e^{-\frac{t}{\tau}} = -\frac{I_2}{(I_1 - I_2)} \qquad t = -\tau \ln\left(\frac{-I_2}{(I_1 - I_2)}\right) = -\tau \ln\left(\frac{I_2}{(I_2 - I_1)}\right)$$

If the time required to deplete the charge is set equal to τ, the value of I_2 can be determined in terms of I_1.

$$I_2 = \frac{I_1 e^{-\frac{\tau}{\tau}}}{e^{-\frac{\tau}{\tau}} - 1} = \frac{I_1 e^{-1}}{e^{-1} - 1} = -\frac{I_1}{e - 1}$$

Typical diode lifetimes are on the order of 0.1 μs to 10 μs. Consider a diode that has a 1-μs lifetime. If I_1 is 100 mA, how much charge is depleted from the diode when the reverse current, I_2, is −100 A for 500 ps? This is one half of a period at 1 GHz. The time 500 ps is much less than the lifetime of 1 μs. From the equation above, the charge at 500 ps is:

$$q(500 \text{ ps}) = (0.1 - (-100))(10^{-6})e^{-\frac{500 \text{ ps}}{1 \text{ us}}} - 100(10^{-6}) = 49.96251 \text{ nC}$$

compared to an initial charge of:

$$q(0) = 0.1 \times 1 \times 10^{-6} = 100 \text{ nC}$$

The final charge represents about 50% of the initial charge. Therefore, 50% of the charge was removed. The diode still acts as a resistance since it still has 50% of the original charge carriers left in it.

When a new current is used to bias the diode, the charge that is remaining at a specified time often needs to be determined. This can be done with the following equation.

$$q_{\text{remain}} = (I_1 - I_2)\tau e^{-\frac{t}{\tau}} + I_2\tau$$

$$e^{-\frac{t}{\tau}} = \frac{\dfrac{q_{\text{remain}}}{\tau} - I_2}{I_1 - I_2} \qquad t = -\tau \ln\left(\frac{\dfrac{q_{\text{remain}}}{\tau} - I_2}{I_1 - I_2}\right)$$

In order to turn on the diode quickly so that it becomes a conductor quickly, a two-step current can be used. A high value of current can be used to initially fill the diode with charge and then a maintenance current can be used to maintain that charge. Suppose a maintenance charge of $I_f\tau$ is needed to allow the diode to reach a given value of resistance. If a forward current were applied to the diode when the diode is initially at zero charge, it would take 2.3 lifetimes for the diode to reach 90% of its charge. For a one-microsecond lifetime diode, this is 2.3 microseconds. That amount of time may be too long after the decision is made to make the diode become a low impedance. The value of current needed to initially fill the diode with a charge of $I_f\tau$ from a unfilled state in time Δt is:

$$I = \frac{I_f}{1 - e^{-\frac{\Delta t}{\tau}}} \approx \left.\frac{\tau}{\Delta t}I_f\right|_{\Delta t \ll \tau}$$

This amount of current can be put into the diode for time equal to Δt and then the current can be switched to a amount equal to I_f for as long as the diode is kept in a low-impedance state. Figure 12.8 shows a time plot of current and charge in a diode as it is turned off.

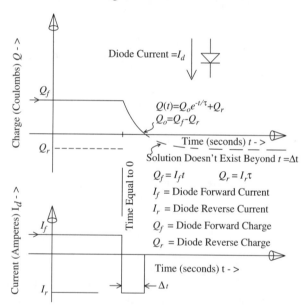

Figure 12.8 General step response of a PIN diode.

While the diode is in a low-impedance state, the amount of charge that is removed and added by a large ac current is:

$$q(t) = \int_0^t i(t)\,dt = \int_0^t [i_{\text{dc}} + i_{\text{ac}}\cos(\omega t)]\,dt$$

$$= i_{\text{dc}}\int_0^t dt + i_{\text{ac}}\int_0^t \cos(\omega t)\,dt$$

$$= i_{\text{dc}}\tau - \frac{i_{\text{ac}}}{\omega}\sin(\omega t) = i_{\text{dc}}\tau - \frac{i_{\text{ac}}}{2\pi f}\sin(\omega t) = i_{\text{dc}}\tau - \frac{i_{\text{ac}}T}{2\pi}\sin(\omega t)$$

The first integral was set equal to the diode lifetime. This is superposition analysis for small signal dc and large signal ac analysis. Notice that the magnitude of the time-varying charge is related to the period T of the ac signal divided by two pi. Even if the period is equal to the lifetime, only about 16% of the charge is removed and added if the magnitude of the ac current is equal to the dc current. However, there is usually many orders of magnitude difference in the currents. For instance, if the bias current is 100 mA, the frequency is 1 GHz, and the lifetime is 1 μs, then an ac current of 100 A only removes 16% of the charge. The charge is added and subtracted from the diode leaving the average charge equal to the dc charge.

When the diode is back biased and the I region is depleted, how much voltage can be put across the diode to prevent the diode from becoming a conductor? Several things take place when the diode is back biased. In order to flood the diode with carriers, the carriers must first be injected into the I region from the contacts and proceed into the diode. This takes time. Assuming the carriers travel at a saturated velocity value (approximately 0.1 micron per picosecond), it takes approximately one nanosecond for a carrier to cross the I region for a 100-micron-thick diode. Before the carrier has a chance to get to the opposite contact and recombine, it is turned around and called back. For those carriers that do cross the I region and combine in forward injection, they create a small forward current that needs to be removed by the bias voltage. The dc bias voltage supply needs to have a low internal resistance to be able to supply the current necessary to keep the average injected charge low. A second source of forward current is impact ionization. When the diode is back biased, it appears as a capacitor. The current into the diode contacts is limited by the amount of capacitive current that flows into the diode. This injected charge is:

$$\int I\, dt = \int C \frac{dV(t)}{dt} dt = CV(t)$$

Only a portion of that charge will be injected since it takes a finite time to build up the charge across the PI and IN junctions. The charge must build up before it can be injected across those junctions. If only a limited number of carriers are injected into the I region, the I region will remain a capacitor or high impedance and not become a resistor.

12.3 SCHOTTKY DIODES

Schottky diodes were discussed in Chapter 11 on detectors and mixers.

12.4 GUNN DIODES

The Gunn diode was named after an investigator named Gunn [7]. Because of the negative differential mobility of III-V compounds, such as GaAs, the V-I characteristic also has a negative resistance. The details of this theory can be found in microwave device textbooks. The negative resistance region results from multivalley energy diagrams that exist in some III-V semiconductors. The transition energy from the conduction band to valence band has two distinct values. However, it is easy to see phenomenologically where the small signal negative resistance comes from. An electric field-velocity plot of a III-V semiconductor such as n-type GaAs demonstrates this. A plot for n-type GaAs with approximate physical values is shown in Figure 12.9. Notice that if a block of n-type material of fixed length and constant doping (n = constant) is biased with a voltage across it, a negative resistance would appear as shown in the upper curve in Figure 12.9, where the region of negative conductance is shown. The Gunn diode placed in parallel with an appropriate resonator and load produces microwave power out for dc power in. The circuit oscillates at the frequency of the resonator. A resonator in parallel with the Gunn diode is shown in Figure 12.10. The dashed components represent the device equivalent circuit elements while the solid line components represent the equivalent

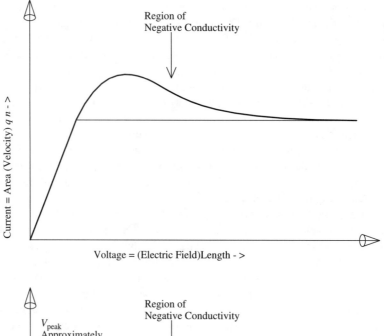

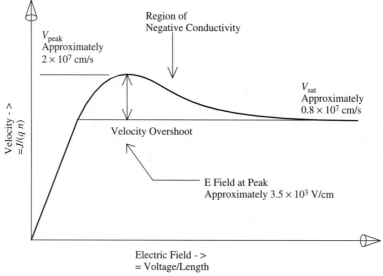

Figure 12.9 *V-I* and *E-v* characteristic of *n*-type GaAs for a Gunn diode.

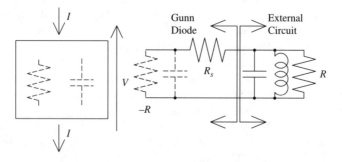

Figure 12.10 Equivalent circuit of a Gunn diode.

circuit of external components put on the device to cause it to oscillate at a given frequency. The parasitic inductance from a bond wire and the package capacity from a package are not shown. The package capacity can be combined with the resonant circuit. Parasitics from the chip to the output circuit need to be minimized. The package for the Gunn chip must be small to minimize inductive parasitics. With some care, the bond wire inductance can be used to help impedance match the resonator to the chip but it will also introduce another resonance in the circuit that might be detrimental.

12.5 IMPATT DIODE

The next device under consideration is the IMPATT diode. The word IMPATT is an acronym for *imp*act ionization *a*valanche *t*ransit *t*ime. This diode will be discussed from a phenomeno-logical basis similar to the discussion for the Gunn diode. Consider what happens when a region of semiconductor goes into breakdown. In the breakdown region, the events happen as follows. First the electric field increases on the region. Next avalanche multiplication carriers are generated. These carriers cannot recombine immediately once they are generated. Most of them drift across a drift region and then recombine at a contact. If the drift time is on the order of a half of period at a given frequency, then the conduction current from these carriers will stop one-half cycle after the carriers are generated. This represents a 180-degree phase shift from when they were generated. This delay will cause the device to act like a negative resistance in parallel with the avalanche region. The avalanche region is a physical region, and therefore has capacity. The carriers that are generated by the avalanche breakdown process will continue to drift due to the impressed dc field even after the rf field turns around. This is similar to an inductive effect since the current is lagging the voltage or electric field that produced it. Therefore, the avalanche region appears as a parallel R-L-C as shown in the schematic of an IMPATT diode shown in Figure 12.11. The drifting carriers getting to the contact a half cycle late generate a negative resistance.

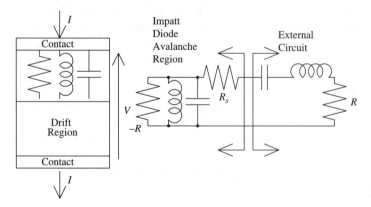

Figure 12.11 Equivalent circuit of an IMPATT diode.

This schematic and that of the Gunn diode do not show the parasitics of the package. These parasitics need to be incorporated into the external circuit. If the package parasitics are too large, anomalous parasitic resonance paths will prevent the IMPATT diode or the Gunn diode from performing as they should. This can be seen in Figure 12.12, which illustrates a diode chip operating into a load in the top part of the figure and a packaged diode working into a load in the bottom part of the figure.

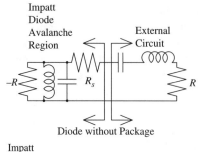

Diode without Package

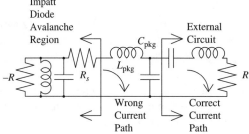

Diode with Package

Figure 12.12 Impatt diode and package equivalent circuit.

12.6 BIPOLAR TRANSISTORS

A cross section of a bipolar device is shown in Figure 12.13 along with an equivalent circuit derived from this cross section. This complex circuit is usually simplified for purposes of roughly determining some figures of merit that will be described. This circuit can be simplified (1) by combining all of the base parasitic R's and C's into a single R and a single C, (2) by ignoring the terminal inductances, (3) by setting the collector resistances to zero, and (4) by ignoring the bond pad capacitances. The circuit can then be redrawn as shown in Figure 12.14. The second portion of Figure 12.14 shows the circuit redrawn in the form of the *hybrid pi* format. The bottom circuit shown in Figure 12.14 shows the central portion of the bipolar transistor without the input series resistance into the base. This is the simplified bipolar circuit often seen in electronics books.

There are several *figures of merit* that are used to compare transistors. One of these is called the f_t (pronounced *f-sub-tee* or simply *f-tee* or sometimes *f-tau*) of the device. This figure of merit is defined as the frequency for which the extrapolated short-circuited common emitter current gain goes to zero. This is determined from the h parameters of the device. Whenever $V_2 = 0$, then $h_{fe} = h_{21} = I_2/I_1$. The lowercase subscript denotes that this is the incremental parameter while h_{FE} indicates the dc parameter for the device. At dc, an incremental analysis shows:

$$i_b = \frac{V_b}{r_{bb'} + r_{b'e}} = \frac{V_{b'e}}{r_{b'e}}$$

and:

$$i_c = g_m V_{b'e} = g_m i_b r_{b'e} = h_{FE} i_b \Rightarrow h_{FE} = g_m r_{b'e}$$

At higher frequencies where the capacitances affect the current gain:

$$i_b = V_{b'e} \left[\frac{1}{r_{b'e}} + j\omega(C_{b'e} + C_{b'c}) \right]\Bigg|_{V_c=0}$$

$$i_c = g_m V_{b'e} = \frac{g_m i_b}{\dfrac{1}{r_{b'e}} + j\omega(C_{b'e} + C_{b'c})} = h_{fe}(\omega) g_m|_{V_c=0}$$

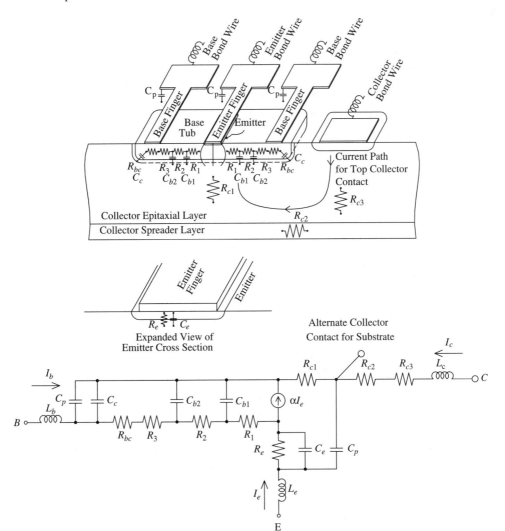

Figure 12.13 Bipolar transistor cross section.

or:

$$h_{fe}(\omega) = \frac{g_m}{\dfrac{1}{r_{b'e}} + j\omega(C_{b'e} + C_{b'c})} = \frac{g_m r_{b'e}}{1 + j\omega r_{b'e}(C_{b'e} + C_{b'c})} = \frac{h_{FE}}{1 + j\dfrac{f}{f_\beta}}$$

$$f_\beta = \frac{1}{2\pi f r_{b'e}(C_{b'e} + C_{b'c})} \approx \frac{1}{2\pi f r_{b'e} C_{b'e}}$$

the frequency for which the magnitude of $h_{fe} = 1$ is called f_t. If:

$$f_t > f_\beta \qquad g_m r_{b'e} > 1$$

$$\Rightarrow f_t = \frac{g_m r_{b'e}}{2\pi C_{b'e} r_{b'e}} = \frac{g_m}{2\pi C_{b'e}} = h_{fe} f_\beta$$

The cutoff frequency f_t is related to the transition time it takes a carrier to cross the device. The total time can be expressed as the time it takes for the carrier to charge the emitter base capacity, cross the emitter into the base, drift across the base, accelerate across the collector base depletion region, and charge up the collector capacity. The total time is τ.

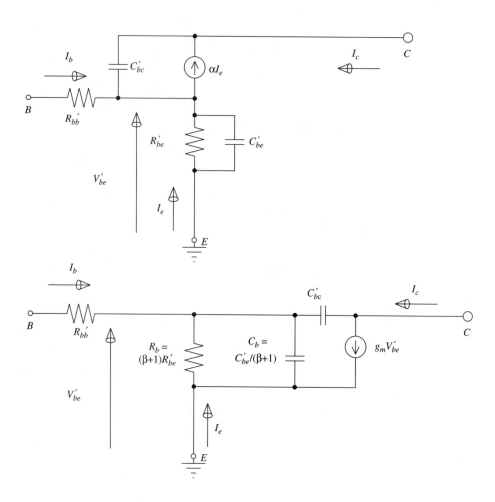

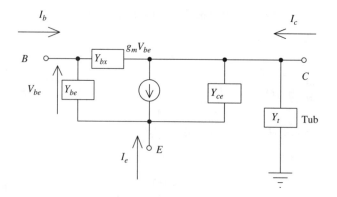

Bipolar Transistor
Intrinsic Circuit Model

Figure 12.14 Simplified bipolar transistor equivalent circuit.

The cutoff frequency related to this time is:

$$f_\tau = \frac{1}{2\pi\tau} > f_t$$

The cutoff frequency f_t is smaller than f_τ since the cutoff frequency also incorporates in it delays associated with charging the base to emitter capacity and base to collector capacity.

The bipolar device can be built from a single type of semiconductor (e.g., silicon with the emitter, base, and collector regions having different doping levels and types but all in a background of silicon). This type of transistor was first built using germanium before silicon processing was perfected. There is another type of bipolar transistor is use. This type of bipolar transistor uses different semiconductor compositions for the emitter and base, or the emitter and base and the base and the collector, or the base and collector regions. The emitter may be fabricated from a wider bandgap material than the base (e.g., Al_xGa_yAs for the emitter and GaAs for the base).

The subscripts x and y represent the amount of Al and Ga respectively. The subscripts x plus y add up to 1. Alternatively, the base may be made out of a lower bandgap material (e.g., Ge_xSi_y) and the emitter and collector from Si. The Ge-Si base has a lower bandgap than silicon alone. The crystal lattice at the junction must be lattice matched to limit carrier recombination. In order to create lattice match, many of the junctions commonly used have the lattices in strain and are called *strained layer lattices*. All of these bipolar devices that are made from different semiconductor types on each side of a junction are called *heterojunction transistors*. The models for these devices are slightly modified with regard to capacitances, and so on, but the model is still generally the same as shown in Figure 12.14 for a uniform material or homojunction transistor.

The circuit in the bottom part of Figure 12.14 shows a tub admittance but ignores the resistance from the outside base terminal to the internal or intrinsic base terminal. The tub admittance contains the capacity of the collector tub (diffusion) to ground and is an important parameter in MMIC circuits. The matrix that describes the circuit is:

$$\begin{pmatrix} i_b \\ i_c \\ i_e \end{pmatrix} = \begin{pmatrix} y_{cb} + y_{be} & -y_{cb} & -y_{be} \\ -y_{cb} + g_m & y_{cb} + y_{ce} + y_t & -g_m - y_{ce} \\ -y_{be} - g_m & -y_{ce} & y_{be} + g_m + y_{ce} \end{pmatrix} \begin{pmatrix} v_b \\ v_c \\ v_e \end{pmatrix}$$

In order to add back the input base resistance, this matrix can be converted to an impedance matrix and the base resistance added to the one-one term of the impedance matrix. Notice that the voltages are with respect to ground and the currents are into the respective terminals. This is not an indefinite matrix since the tub admittance goes to ground. Often the tub admittance is ignored and then the matrix becomes an indefinite admittance matrix. In order to use this matrix for a common emitter amplifier the emitter voltage is set equal to zero. Likewise for a common base amplifier the base voltage is set equal to zero, and for a common collector amplifier the collector voltage is set equal to zero.

After one of the ports is grounded to make a two-port circuit, loads are

$$\begin{pmatrix} i_{in} \\ i_{out} \end{pmatrix} = \begin{pmatrix} y_{11} & y_{12} \\ y_{21} & y_{22} \end{pmatrix} \begin{pmatrix} v_{in} \\ v_{out} \end{pmatrix} + \begin{pmatrix} y_{in} & 0 \\ 0 & y_{out} \end{pmatrix} \begin{pmatrix} v_{in} \\ v_{out} \end{pmatrix}$$

placed on the ports. Then the ratio of the output voltage to the input voltage is the voltage gain of the circuit with those loads. The two-port matrix that results from shorting one of the ports to ground is added to the port admittance matrix. The external currents to the circuit are zero after the port admittances are added to the net circuit. This gives the following equation.

$$\begin{pmatrix} i_{in} \\ i_{out} \end{pmatrix} = \begin{pmatrix} y_{11} & y_{12} \\ y_{21} & y_{22} \end{pmatrix} \begin{pmatrix} v_{in} \\ v_{out} \end{pmatrix} + \begin{pmatrix} y_S & 0 \\ 0 & y_L \end{pmatrix} \begin{pmatrix} v_{in} \\ v_{out} \end{pmatrix}$$

The input-to-output voltage gain as a function of the load admittance is determined by using the bottom row of the matrix equation with the output current set equal to zero.

In the literature associated with bipolar devices, the input and feedback terms in the admittance matrix are sometimes identified as shown in the next equation.

$$y_{be} = y_\pi$$

$$y_{cb} = y_\mu$$

12.7 MESFET TRANSISTORS

There is another type of transistor that is used at high frequencies. Instead of depending on minority carrier transport such as used in a bipolar device, this device uses majority carrier transport. A cross section of a MESFET is shown in Figure 12.15. It is a field-effect transistor (FET). At microwave frequencies, the FET is often made from a gallium arsenide (GaAs) or indium phosphide (InP) semiconductor. It uses a Schottky diode junction depletion region [metal-semiconductor (MES)] to establish the gate depletion region rather than using the depletion region of the capacity of a metal oxide semiconductor (MOS) to establish the gate. The MOS gate is used in the popular MOSFET. The microwave FET using a metal-semiconductor junction for a gate is termed a *MESFET*.

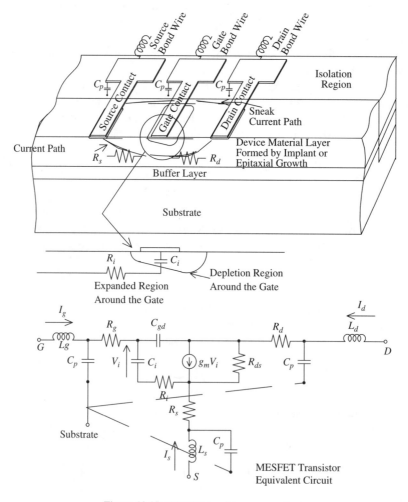

Figure 12.15 MESFET transistor cross section.

An equivalent circuit for a MESFET is shown in Figure 12.15. The cutoff frequency for the MESFET can be defined if some approximations are made. If:

$$C_{gd} < C_i$$

$$R_i < \frac{1}{\omega C_i}$$

$$\Rightarrow V_i = \frac{I_g}{j\omega C_i}$$

When $V_{ds} = 0$:

$$g_m V_i = \frac{I_g}{j\omega C_i} g_m \approx I_d$$

When I_g is equal to I_d, the current gain is equal to 1. The frequency for which the common source short-circuit current gain is equal to 1 is the current gain cutoff frequency and is equal to:

$$f_t = \frac{g_m}{\omega C_i}$$

For C_{gd} approximately zero and a conjugate match on the output ($R_{load} = R_{ds}$):

$$P_{out} = \left| \frac{g_m V_i}{2} \right|^2 R_{ds}$$

The input power under these conditions is:

$$P_{in} = \left| I_g \right|^2 R_i = V_i^2 \omega^2 C_i^2 R_i$$

Setting input power equal to output power gives:

$$P_{out} = P_{in} = V_i^2 \omega^2 C_i^2 R_i = \frac{|g_m V_i|^2}{4} R_{ds}$$

$$\omega^2 C_i^2 R_i = \frac{g_m^2}{4} R_{ds}$$

$$f_{max} = \frac{g_m}{4\pi C_i} \sqrt{\frac{R_{ds}}{R_i}} = \frac{f_t}{2} \sqrt{\frac{R_{ds}}{R_i}}$$

There is a finite length that the carrier has to travel to get from the source to the drain. The fastest it can get there is at the saturated velocity unless there is a velocity overshoot over the steady-state saturated velocity. The shortest distance it can go is the length of the gate finger. The time τ_c to go that distance and the cutoff frequency is:

$$\tau_c = \frac{length}{v_{sat}}$$

$$f_t = \frac{1}{2\pi \tau_c} = \frac{v_{sat}}{2\pi \, length} = \frac{g_m}{2\pi C_i}$$

The equivalent circuit of the MESFET shown in Figure 12.15 does not show the back gating effect that exists in MESFET technology. The effect can be added to the circuit similar to the back gating current source that is in the equivalent circuit for the MOSFET (shown later in Figure 12.17).

12.8 DUAL-GATE MESFET TRANSISTORS

The MESFET is often made with two gates as shown in Figure 12.16. It is called a *dual-gate* MESFET. The dual-gate MESFET has several applications such as devices used for mixers and gain control blocks in much the same way separate discrete device cascode configurations

Dual-Gate FET Configuration

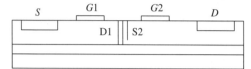

Dual-Gate FET Configuration
with Pseudo *D*1 and *S*2

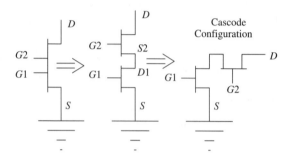

Figure 12.16 Dual-gate MESFET transistor.

are used. The dual-gate MESFET is slightly harder to bias than two separate devices since the D1-S2 terminal is not accessible outside the device—it is only a virtual terminal. The bias conditions for a dual-gate FET can be determined by superposing the *V-I* curves for the two FETs on the same graph. When doing that, the first FET's drain to source voltage is subtracted from the power supply voltage to give the second FET's drain to source voltage.

12.9 MOSFET TRANSISTORS

There is current interest in doing microwave circuits with silicon MOSFET integrated circuits. Figure 12.17 shows a cross section of a MOSFET transistor. The gate is shown as overlapping the source and drain contacts. This increases the capacity between the source and gate and source and drain. In addition, the gate is made from a polysilicon layer rather than a metal layer. Careful consideration of these effects allows the circuit designer to fabricate microwave analog and digital circuits using MOSFET technology in large-scale integrated circuits. The equivalent circuits shown in Figure 12.17 include the back gating current source to account for the possibility that the body of the FET may not be kept at ground. This effect exists in MESFET technology as well. Whenever an electric field modifies the charge that exists on the substrate/body to source capacitor, the channel depletion region is slightly affected. This effect takes place on the opposite side of the depletion region from that of the gate but is in the gate body region.

The lower row of circuits shown in Figure 12.17 show an equivalent circuit for the intrinsic portion of the MOSFET. The equivalent circuit in the lower right is given in terms of the admittance parameters. The admittance matrix for that circuit is:

$$\begin{pmatrix} i_g \\ i_d \\ i_s \end{pmatrix} = \begin{pmatrix} y_{dg} + y_{gs} & -y_{dg} & -y_{gs} \\ -y_{dg} + g_m & y_{dg} + y_{ds} + y_{db} & -g_m - g_{mb} - y_{ds} \\ -y_{gs} - g_m & -y_{ds} & y_{gs} + g_m + g_{mb} + y_{ds} + y_{sb} \end{pmatrix} \begin{pmatrix} v_g \\ v_d \\ v_s \end{pmatrix}$$

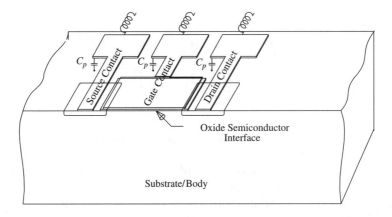

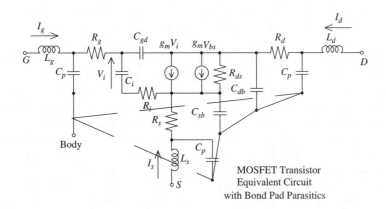

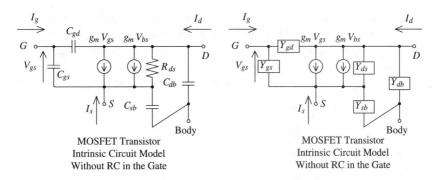

Figure 12.17 MOSFET transistor cross section.

The body is assumed to be at the ground potential. Common gate, drain, or source amplifier matrices are easily determined from this matrix by setting the gate, drain, or source voltages to zero respectively.

12.10 TRANSFORMERS AND BALUNS

12.10.1 Transformers with Cores

An equivalent circuit of a transformer used at rf and microwave frequencies is shown in Figure 12.18. This circuit assumes the wire lengths are short enough so that distributed effects

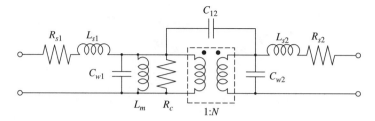

Figure 12.18 Equivalent circuit of a transformer.

can be ignored. When distributed effects become important, then this equivalent circuit becomes of less value. The series inductances account for the amount of magnetic field energy that exists in the transformer assembly but not in the core. Since it appears to leak out of the core, it is called *leakage inductance*. One source of this leakage inductance is shown in Figure 12.19. The x's represent flux that is in the core. The $+$'s represent flux that does not enter the core. Flux from one winding that does not enter the core is not coupled to the opposite winding. The magnetic energy for the flux that does not enter the core is represented by the leakage inductance or series inductances shown in Figure 12.18. The series resistances are the resistances of the winding. These are a function of frequency due to the skin effect. This equivalent circuit assumes that the magnetic field totally penetrates the core (i.e., there is no skin depth associated with the magnetic field in the core). If there is skin effect in the core, the magnetizing inductance needs to be changed to several transformers in parallel. The parallel inductance is the inductance associated with a winding on the magnetic core of the transformer. This inductance could be associated with either the input (primary) or the output (secondary) side of the transformer but is shown only on the input in this figure to account for the magnetic energy in the core. The ideal transformer shown with the conventional dot symbols and the turns ratio N accounts for the energy transfer between port one and port two. The resistance R_c primarily accounts for energy lost in the core of the transformer. The capacitors on either side of the transformer account for the capacitive energy associated with the fact that individual turns of a winding lay close to each other. The C_{12} term accounts for the capacitive energy between the two sets of windings. This equivalent circuit of the transformer is a nondistributed circuit. When the frequency gets high enough, distributed effects would have to be considered. At rf and submicrowave frequencies, the transformer is often toroidal in shape as depicted in Figure 12.20.

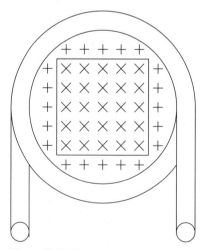

Figure 12.19 One source of leakage inductance.

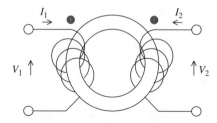

Figure 12.20 Toroidal transformer.

12.10.2 Transmission Line Transformers

Other types of transformers can be used at microwave frequencies. Several of them are shown in Figure 12.21. In analyzing the circuits in Figure 12.21, the coaxial lines are considered to be very short at the frequency of interest. The current that goes into the center of a coaxial line has a mirroring current that leaves the inside of the outer conductor. Current could also travel on the outside of the coax unless it is somehow prevented from flowing. If the outside of the coax represents a transmission line that is a quarter-wavelength long, then at that frequency, the current entering the outside of the coax would be zero. For circuits that are broader band another solution needs to be found. A ferrite bead placed on the outside of the coaxial line helps prevent any current from flowing on the outside of the outer conductor of the coaxial line by presenting a large inductive reactance to any current attempting to flow on the outside of the conductor. Then the current leaving the inside of the coaxial line must go through the external connection rather than down the outside of the coax line. The voltage between the center conductor and outer conductor on one end of a short coaxial line is the same as the voltage between the center conductor and outer conductor on the other end. The 9 : 1 transformer has three of the same voltages in series at the output and the same current through them. At the input, the transformer has only one of the same voltages but there are three currents in parallel across that voltage. The input voltage to current ratio is $3V : I$ whereas the output voltage to current ratio is $V : 3I$, resulting in an impedance ratio of $[3V : I]/[V : 3I]$ or 9 : 1.

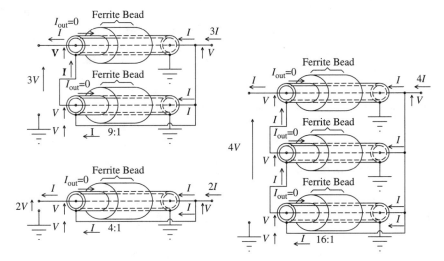

Figure 12.21 Transmission line transformers.

It is often necessary to change unbalanced transmission circuits to balanced transmission circuits and vice-versa. The circuit that performs this function is a *balun* (short for *bal*anced to *un*balanced). Unbalanced circuits are those similar to standard coaxial cable, microstrip line, and so forth. Balanced circuits are those that have both positive and negative current carrying conductors in a balanced configuration with respect to the ground plane or conductor. Twin-lead transmission line once commonly used for FM radio or VHF television antenna connections is an example of a balanced transmission line. There are many circuits that function as a balun including a center tapped transformer. However, at microwave frequencies there are several types that are commonly used as shown in Figure 12.22.

The baluns shown in Figure 12.22 are constructed from transmission lines and are often referred to as 1 : 1 baluns. This is a result of the input impedance being equal to the sum of

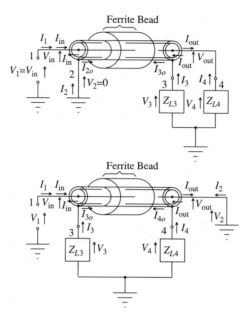

Figure 12.22 Transmission line baluns.

the two output impedances. The two output impedances are considered as a single impedance in a balanced circuit. These baluns are similar to the transformer baluns described below. The baluns can be used as push-pull transformers with the center tap grounded. They then function as 2 : 1 transformers. The current through the loads is the same as the input current and the voltage across each load (when grounded) is equal to one-half of the input voltage. The balun on the bottom of Figure 12.22 depends on its line length being short but gives dc isolation from the source to the loads. The balun on the top part of Figure 12.22 does not depend on its length being short but does require that the coaxial line impedance be equal to the sum of the load impedances if the line is not short. The balun on the top has a different dc and low frequency termination for each of the loads.

12.10.3 Transformer Baluns

A balun can also be formed from a single transformer. Consider the structure shown in Figure 12.23. The current and voltage relationships for this circuit are:

$$I1 = -I2 = -I3 = I4$$
$$V3 - V1 = V4 - V2$$

This results in the input impedance into the circuit as:

$$Z_{in} = \frac{V1}{I1} = Z2 + Z3 + Z4$$

This structure can be used as a two- or three-way power splitter. Notice that if $Z3$ and $Z4$ are two separate resistive loads, $Z2$ could be used to tune out the reactance of the source. When $Z3 = Z4$, $V3 = -V4$. Under this condition, an impedance equal to $Z3 + Z4$ could be put

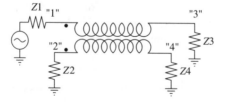

Figure 12.23 Single transformer balun.

between $V3$ and $V4$ to form a balanced load and the structure functions as a balun. $Z2$ can still be used to reactively match the source to the circuit. If $Z2$ is set equal to zero, the circuit performs as a simple balun.

When $Z2 = Z3$, $V2 = V3$. Under this condition, on can tie ports two and three together. Whenever $V1$ is equal to $-V4$, no current would flow in the parallel loads $Z2$ and $Z3$ since then $V2 = V3 = 0$. However, if $V1$ is not equal to the negative of $V4$, then a net voltage will appear at ports two and three and current will flow through $Z2$ and $Z3$. The structure connected in this manner can be used to terminate unbalanced power in "balanced" structures.

12.11 COUPLED LINES

A pair of coupled transmission lines is shown schematically in Figure 12.24. The Y matrix of a pair of coupled lines is [8, 9]:

$$\begin{pmatrix} I_1 \\ I_2 \\ I_3 \\ I_4 \end{pmatrix} = \begin{pmatrix} Y_{11} & Y_{12} & Y_{13} & Y_{14} \\ Y_{12} & Y_{11} & Y_{14} & Y_{13} \\ Y_{13} & Y_{14} & Y_{11} & Y_{12} \\ Y_{14} & Y_{13} & Y_{12} & Y_{11} \end{pmatrix} \begin{pmatrix} V_1 \\ V_2 \\ V_3 \\ V_4 \end{pmatrix}$$

where:

$$Y_{11} = -j\frac{Y_{0o}}{2}\cot(\theta_o) - j\frac{Y_{0e}}{2}\cot(\theta_e)$$

$$Y_{12} = +j\frac{Y_{0o}}{2}\cot(\theta_o) - j\frac{Y_{0e}}{2}\cot(\theta_e)$$

$$Y_{13} = +j\frac{Y_{0o}}{2}\csc(\theta_o) + j\frac{Y_{0e}}{2}\csc(\theta_e)$$

$$Y_{14} = -j\frac{Y_{0o}}{2}\csc(\theta_o) + j\frac{Y_{0e}}{2}\csc(\theta_e)$$

θ_o and θ_e are the odd-mode and even-mode electrical length respectively of the coupled-line segment. Note that ports three and four are interchanged from what is often given in the literature. When the even- and odd-mode velocities are the same, and the lines are a quarter-wavelength long, the characteristic admittances are related to the coupling coefficient, c, by the following formulas.

$$Y_{0o} = Y_0\sqrt{\frac{1+c}{1-c}} \quad Y_{0e} = Y_0\sqrt{\frac{1-c}{1+c}} \quad Z_{0o} = Z_0\sqrt{\frac{1-c}{1+c}} \quad Z_{0e} = Z_0\sqrt{\frac{1+c}{1-c}}$$

$$Y_0 = \sqrt{Y_{0o}Y_{0e}} \quad c = \frac{Y_{0o}-Y_{0e}}{Y_{0o}+Y_{0e}} = \frac{Z_{0e}-Z_{0o}}{Z_{0e}+Z_{0o}} \quad Z_0 = \sqrt{Z_{0o}Z_{0e}}$$

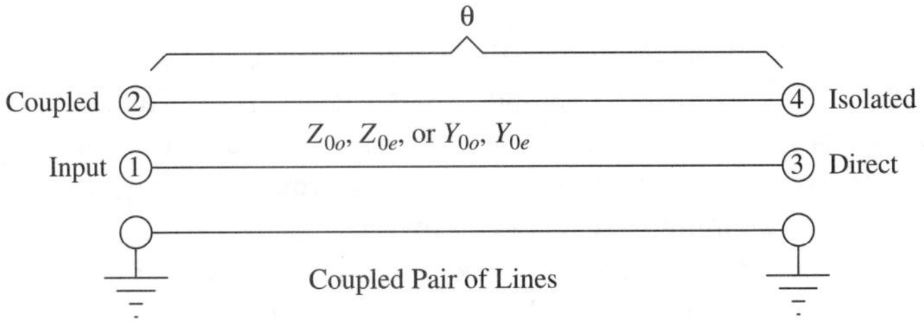

Figure 12.24 A pair of coupled lines.

The relationship between coupling c and dB coupling is:

$$c = 10^{\left(\frac{dB}{-20}\right)}$$

where the negative sign is used in the exponent if dB is expressed in dB down. The coupling factor c must be less than 1 in magnitude.

The Z matrix for the same pair of coupled lines is:

$$\begin{pmatrix} V_1 \\ V_2 \\ V_3 \\ V_4 \end{pmatrix} = \begin{pmatrix} Z_{11} & Z_{12} & Z_{13} & Z_{14} \\ Z_{12} & Z_{11} & Z_{14} & Z_{13} \\ Z_{13} & Z_{14} & Z_{11} & Z_{12} \\ Z_{14} & Z_{13} & Z_{12} & Z_{11} \end{pmatrix} \begin{pmatrix} I_1 \\ I_2 \\ I_3 \\ I_4 \end{pmatrix}$$

where:

$$Z_{11} = -j\frac{Z_{0e}}{2}\cot(\theta_e) - j\frac{Z_{0o}}{2}\cot(\theta_o)$$

$$Z_{12} = -j\frac{Z_{0e}}{2}\cot(\theta_e) + j\frac{Z_{0o}}{2}\cot(\theta_o)$$

$$Z_{13} = -j\frac{Z_{0e}}{2}\csc(\theta_e) - j\frac{Z_{0o}}{2}\csc(\theta_o)$$

$$Z_{14} = -j\frac{Z_{0e}}{2}\csc(\theta_e) + j\frac{Z_{0o}}{2}\csc(\theta_o)$$

The input immittances into a two-port described in terms of Y and Z parameters are:

$$Y_{\text{in}} = \frac{\Delta_Y + Y_{11}Y_{\text{load}}}{Y_{22} + Y_{\text{load}}} \qquad Z_{\text{in}} = \frac{\Delta_Z + Z_{11}Z_{\text{load}}}{Z_{22} + Z_{\text{load}}}$$

If Y_{11} and Y_{22} are equal to zero, then the two-port acts as an admittance inverter. If Z_{11} and Z_{22} are equal to zero, the two-port acts as an impedance inverter. Various combinations of terminations on the four-port can make the coupled-line pair into impedance or admittance inverters.

If ports three and four are shorted on a coupled-line pair, then the resulting two-port Y matrix is as follows.

$$\begin{pmatrix} I_1 \\ I_2 \end{pmatrix} = \begin{pmatrix} -j\frac{Y_{0o}}{2}\cot(\theta_o) - j\frac{Y_{0e}}{2}\cot(\theta_e) & +j\frac{Y_{0o}}{2}\cot(\theta_o) - j\frac{Y_{0e}}{2}\cot(\theta_e) \\ +j\frac{Y_{0o}}{2}\cot(\theta_o) - j\frac{Y_{0e}}{2}\cot(\theta_e) & -j\frac{Y_{0o}}{2}\cot(\theta_o) - j\frac{Y_{0e}}{2}\cot(\theta_e) \end{pmatrix} \begin{pmatrix} V_1 \\ V_2 \end{pmatrix}$$

When the even- and odd-mode velocities are the same, Y_{11} and Y_{22} are equal to zero whenever $\theta = \pi/2$. However, then the y matrix is also zero. This implies that a shorted comb line pair of lines does not couple when the lines are a quarter of a wavelength long. If the even- and odd-mode velocities are different, however, then the quarter-wavelength comb line pair has some net coupling and acts as an inverter.

In the following analysis, the even- and odd-mode velocities are assumed to be equal. Putting susceptances in parallel with ports one and two that cancel Y_{11} and Y_{22} of the network when θ is not equal to $\pi/2$ allows the coupled-line pair to become an inverter. Note that then the pair does not have the same characteristic impedance over its whole length. If the added susceptances are made from transmission lines with negative characteristic admittances, the total susceptance tracks with frequency. The combined circuit has the following two-port y

matrix.

$$\begin{pmatrix} I_1 \\ I_2 \end{pmatrix} = \begin{pmatrix} j\dfrac{(Y_{0o} + Y_{0e})}{2} \cot(\theta) & 0 \\ 0 & j\dfrac{(Y_{0o} + Y_{0e})}{2} \cot(\theta) \end{pmatrix} \begin{pmatrix} V_1 \\ V_2 \end{pmatrix}$$

$$+ \begin{pmatrix} -j\dfrac{(Y_{0o} + Y_{0e})}{2} \cot(\theta) & +j\dfrac{(Y_{0o} - Y_{0e})}{2} \cot(\theta) \\ +j\dfrac{(Y_{0o} - Y_{0e})}{2} \cot(\theta) & -j\dfrac{(Y_{0o} + Y_{0e})}{2} \cot(\theta) \end{pmatrix} \begin{pmatrix} V_1 \\ V_2 \end{pmatrix}$$

$$= \begin{pmatrix} 0 & +j\dfrac{(Y_{0o} - Y_{0e})}{2} \cot(\theta) \\ +j\dfrac{(Y_{0o} - Y_{0e})}{2} \cot(\theta) & 0 \end{pmatrix} \begin{pmatrix} V_1 \\ V_2 \end{pmatrix}$$

This network has the following input-to-output relationship.

$$Y_{\text{in}} = \frac{-\left(+j\dfrac{(Y_{0o} - Y_{0e})}{2} \cot(\theta)\right)^2}{Y_{\text{load}}} = \frac{J^2}{Y_{\text{load}}}$$

$$\Rightarrow J = \frac{(Y_{0o} - Y_{0e})}{2} \cot(\theta)$$

This circuit is implemented as shown in the bottom part of the drawing in Figure 12.25. A ground is inserted between the top part of the comb line pair, preventing capacitive coupling over the top part of the pair. The negative characteristic admittance sections are in parallel with the positive characteristic admittance sections of the resonator, giving a net positive characteristic impedance to the transmission line in that region. The inverter is made from the coupled-line section and the virtual negative characteristic admittance sections.

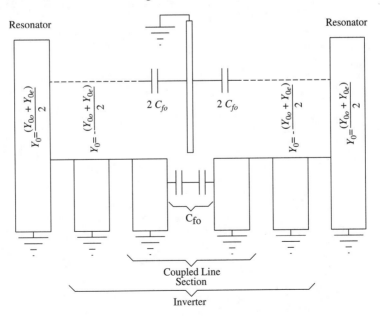

Figure 12.25 Inverter on the bottom side of a comb line pair.

If the coupled-line section is open circuited (i.e., ports three and four are left open), then the resulting two-port Z matrix is:

$$\begin{pmatrix} V_1 \\ V_2 \end{pmatrix} = \begin{pmatrix} -j\dfrac{Z_{0e}}{2}\cot(\theta_e) - j\dfrac{Z_{0o}}{2}\cot(\theta_o) & -j\dfrac{Z_{0e}}{2}\cot(\theta_e) + j\dfrac{Z_{0o}}{2}\cot(\theta_o) \\ -j\dfrac{Z_{0e}}{2}\cot(\theta_e) + j\dfrac{Z_{0o}}{2}\cot(\theta_o) & -j\dfrac{Z_{0e}}{2}\cot(\theta_e) - j\dfrac{Z_{0o}}{2}\cot(\theta_o) \end{pmatrix} \begin{pmatrix} V_1 \\ V_2 \end{pmatrix}$$

When the even- and odd-mode velocities are the same, Z_{11} and Z_{22} are equal to zero whenever $\theta = \pi/2$. However, then the Z matrix is zero. This also implies that an opened comb line pair of lines does not couple when they are a quarter of a wavelength long. When θ is not equal to $\pi/2$, Z_{11} and Z_{22} of a new network can be made equal to zero by using a coupled-line pair in parallel with admittances that cancel Z_{11} and Z_{22}, similar to the case for shorted lines above. It is also possible to make Z_{11} and Z_{22} equal to zero by putting series stubs in series with ports one and two.

Using equal even- and odd-mode velocities, inverting the above two-port z matrix to form a two-port y matrix gives:

$$\begin{pmatrix} I_1 \\ I_2 \end{pmatrix} = \frac{j\tan(\theta)}{2} \begin{pmatrix} (Y_{0e}+Y_{0o}) & (Y_{0e}-Y_{0o}) \\ (Y_{0e}-Y_{0o}) & (Y_{0e}+Y_{0o}) \end{pmatrix} \begin{pmatrix} V_1 \\ V_2 \end{pmatrix}$$

This network can be represented as the top part of the circuit shown in Figure 12.26. The inverter in the top part of Figure 12.26 has the following inversion properties.

$$Y_{\text{in}} = \frac{-\left(\dfrac{j(Y_{0o}-Y_{0e})\tan(\theta)}{2}\right)^2}{Y_{\text{load}}} = \frac{J^2}{Y_{\text{load}}}$$

$$\Rightarrow J = \frac{(Y_{0o}-Y_{0e})\tan(\theta)}{2}$$

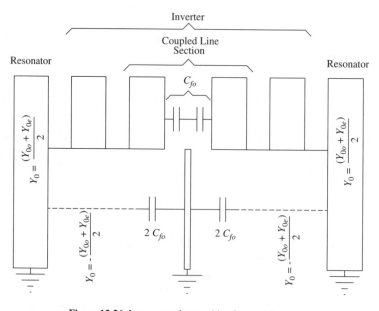

Figure 12.26 Inverter on the top side of a comb line pair.

Whenever ports two and three of the four-port coupled-line circuit are shorted, forming an interdigital coupling section, then:

$$\begin{pmatrix} I_1 \\ I_4 \end{pmatrix} = \begin{pmatrix} -j\dfrac{Y_{0o}}{2}\cot(\theta_o) - j\dfrac{Y_{0e}}{2}\cot(\theta_e) & -j\dfrac{Y_{0o}}{2}\csc(\theta_o) + j\dfrac{Y_{0e}}{2}\csc(\theta_e) \\ -j\dfrac{Y_{0o}}{2}\csc(\theta_o) + j\dfrac{Y_{0e}}{2}\csc(\theta_e) & -j\dfrac{Y_{0o}}{2}\cot(\theta_o) - j\dfrac{Y_{0e}}{2}\cot(\theta_e) \end{pmatrix}\begin{pmatrix} V_1 \\ V_4 \end{pmatrix}$$

Consider ports one and four as forming a two-port with port one as the input and port four as the load. The input admittance at port one for equal even- and odd-mode velocities and for $\theta = \pi/2$ is:

$$Y_{\text{in}} = \frac{-\left(\dfrac{j(Y_{0o} - Y_{0e})}{2}\right)^2}{Y_{\text{load}}} = \frac{J^2}{Y_{\text{load}}}$$

$$\Rightarrow J = \frac{(Y_{0o} - Y_{0e})}{2}$$

Now let ports one and four form a two-port with ports two and three terminated in an open circuit. Then:

$$\begin{pmatrix} V_1 \\ V_4 \end{pmatrix} = \begin{pmatrix} -j\dfrac{Z_{0e}}{2}\cot(\theta_e) - j\dfrac{Z_{0o}}{2}\cot(\theta_o) & -j\dfrac{Z_{0e}}{2}\csc(\theta_e) + j\dfrac{Z_{0o}}{2}\csc(\theta_o) \\ -j\dfrac{Z_{0e}}{2}\csc(\theta_e) + j\dfrac{Z_{0o}}{2}\csc(\theta_o) & -j\dfrac{Z_{0e}}{2}\cot(\theta_e) - j\dfrac{Z_{0o}}{2}\cot(\theta_o) \end{pmatrix}\begin{pmatrix} I_1 \\ I_4 \end{pmatrix}$$

The input impedance at port one, for equal even- and odd-mode velocities and for $\theta = \pi/2$, is:

$$Z_{\text{in}} = \frac{-\left(\dfrac{-j(Z_{0e} - Z_{0o})}{2}\right)^2}{Z_{\text{load}}} = \frac{K^2}{Z_{\text{load}}}$$

$$\Rightarrow K = \frac{(Z_{0e} - Z_{0o})}{2}$$

12.12 TRIAX

A triax line has properties similar to a coupled pair of lines. The admittance matrix is somewhat different than for coupled lines but it can be used to define several types of circuit functions. The triax line is shown schematically in Figure 12.27. The admittance matrix of the triax line is:

$$\begin{pmatrix} I_1 \\ I_2 \\ I_3 \\ I_4 \end{pmatrix} = \begin{pmatrix} Y_{11} & Y_{12} & Y_{13} & Y_{14} \\ Y_{12} & -Y_{12} & Y_{14} & -Y_{14} \\ Y_{13} & Y_{14} & Y_{11} & Y_{12} \\ Y_{14} & -Y_{14} & Y_{12} & -Y_{12} \end{pmatrix}\begin{pmatrix} V_1 \\ V_2 \\ V_3 \\ V_4 \end{pmatrix}$$

$$Y_{11} = -jY_{0o}\cot(\theta_o) - jY_{0i}\cot(\theta_i)$$

$$Y_{12} = +jY_{0i}\cot(\theta_i)$$

$$Y_{13} = +jY_{0o}\csc(\theta_o) + jY_{0i}\csc(\theta_i)$$

$$Y_{14} = -jY_{0i}\csc(\theta_i)$$

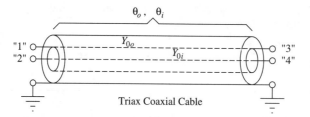

Figure 12.27 Triax line.

Y_{0o} is the outer coaxial line characteristic admittance, Y_{0i} is the inner coaxial line characteristic admittance, θ_o is the outer coaxial line electrical length, and θ_i is the inner coaxial line electrical length. Note that if both thetas are a quarter wavelength, $Y_{11} = Y_{12} = 0$. When the inner and outer thetas are different, there will be some frequency for which Y_{11} or Y_{12} is equal to zero. At those frequencies, the line will act as a resonant circuit. The resonance can be used for some application or could result in some parasitic circuit coupling.

12.13 DIRECTIONAL COUPLERS

Directional couplers are three- or four-port networks as shown in Figure 12.28. Directional couplers are used to sample waves traveling on a transmission line. The ratio of a_2 to b_2 in Figure 12.28 is the reflection coefficient of a load or termination put on port two.

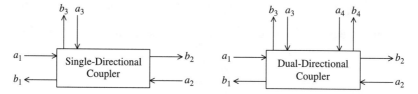

Figure 12.28 Directional couplers.

12.13.1 Ideal Directional Couplers

A matched, lossless, ideal directional coupler has the following (S) matrix:

$$\begin{pmatrix} b_1 \\ b_2 \\ b_3 \\ b_4 \end{pmatrix} = \begin{pmatrix} 0 & L_1 & c_1 & 0 \\ L_1 & 0 & 0 & c_2 \\ c_1 & 0 & 0 & L_2 \\ 0 & c_2 & L_2 & 0 \end{pmatrix} \begin{pmatrix} a_1 \\ a_2 \\ a_3 \\ a_4 \end{pmatrix}$$

$$|L_1|^2 + |c_1|^2 = 1 \qquad |L_2|^2 + |c_2|^2 = 1$$

When the directional coupler is symmetric, L_1 is equal to L_2. A directional coupler has c_1 equal to c_2 when it is symmetric and c_1 equal to minus c_2 when it is antisymmetric. Under some conditions, the ratio of b_4 to b_3 can be used to determine the ratio of a_2 to b_2, the output load reflection coefficient. Let:

$$a_3 = a_4 = 0$$
$$a_2 = \Gamma_2 b_2$$

where Γ_2 represents a load reflection coefficient on port two. Then:

$$b_1 = L_1 a_2 = L_1 \Gamma_2 b_2$$
$$b_2 = L_1 a_1$$
$$b_3 = c_1 a_1$$
$$b_2 = c_2 a_2 = c_2 \Gamma_2 b_2$$

Let the ratio of b_4 to b_3 be the measured value of Γ. Substituting for b_2 in the equation gives:

$$b_1 = L_1 a_2 = L_1^2 \Gamma_2 a_1$$
$$b_2 = L_1 a_1$$
$$b_3 = c_1 a_1$$
$$b_4 = c_2 a_2 = c_2 L_1 \Gamma_2 a_1$$
$$\Gamma_{\text{meas}} = \frac{b_4}{b_3} = \frac{c_2 L_1 \Gamma_2}{c_1}$$

12.13.2 Nonideal Directional Couplers

Now assume the coupler has some parasitic or residual reflection coefficient Γ_p in addition to the load reflection coefficient Γ_L. Then:

$$\Gamma_{\text{meas}} = \frac{b_4}{b_3} = \frac{c_2 L_1 (\Gamma_{\text{load}} + \Gamma_p)}{c_1}$$

If $\Gamma_L = -\Gamma_p$, the coupler would wrongly indicate that the load is a perfect load. Also, if $\Gamma_L = 0$, the coupler would wrongly indicate that the measured load reflection coefficient is not perfect. The quantity:

$$D = -20 \log_{10} \left(\left| \frac{b_4}{b_3} \right| \right) \Bigg|_{\Gamma_{\text{load}}=0} \text{dB} = -20 \log_{10} \left(\left| \frac{c_2 L_1}{c_1} \Gamma_p \right| \right) \Bigg|_{\Gamma_{\text{load}}=0} \text{dB}$$

is called the *directivity* of the coupler. It is readily seen that D is extremely hard to measure. The load reflection coefficient Γ_L needs to be known to a higher degree of accuracy than Γ_p. Poor couplers have directivities in the -10 to -20 dB range while good couplers would have directivity magnitudes more negative than -30 dB. Put simply, there is an output at the "uncoupled" port when there should not be. It is very important that this extraneously coupled power be considered when using directional couplers. For example, the extraneous signal adds vectorially to the actual reflected signal and gives "wrong" readings of the reflected power. Mismatches can be considered on all of the ports. The analysis is similar but the mathematics rapidly gets more involved when considering mismatches on more than one port at a time. A quantity called the isolation I of the coupler is given as:

$$I = -20 \log_{10} \left(\left| \frac{b_4}{b_1} \right| \right) \Bigg|_{\Gamma_{\text{load}}=0} \text{dB}$$

The coupling factor of the coupler C is given as:

$$C = -20 \log_{10} \left(\left| \frac{b_3}{b_1} \right| \right) \Bigg|_{\Gamma_{\text{load}}=0} \text{dB}$$

Using quantities expressed in dB, $I = C + D$.

12.14 DUAL-DIRECTIONAL COUPLERS WITH INTERNAL TERMINATIONS

Normally directional couplers used to make reflection coefficient measurements are constructed with internal termination resistors that terminate a_3 or a_4 internal to the coupler. If one terminates one of the ports three or four, then the resultant three-port is termed a single-directional coupler or simply a directional coupler. A coupler made from two such single-directional couplers is sometimes called a *reflectometer*. Assume that port four is terminated and that the coupler is used to detect a forward wave. If the terminated port's load is perfect, the following scattering matrix results from the coupler.

$$\begin{pmatrix} b_1 \\ b_2 \\ b_3 \end{pmatrix} = \begin{pmatrix} 0 & L & c \\ L & 0 & 0 \\ c & 0 & 0 \end{pmatrix} \begin{pmatrix} a_1 \\ a_2 \\ a_3 \end{pmatrix}$$

Notice that this coupler no longer satisfies the lossless conditions. There is some power dissipated inside the coupler. If one cascades two such couplers, one as just discussed and the other with port three terminated, then a different dual-directional coupler is formed. There are two internal load terminations as shown in Figure 12.29 (indicated by the first $a_4 = 0$ and the second $a_3 = 0$). The caret quantities are for the second coupler. The scattering matrix for the new network shown inside the dashed line is:

$$\begin{pmatrix} b_1 \\ \hat{b}_2 \\ b_3 \\ \hat{b}_4 \end{pmatrix} = \begin{pmatrix} 0 & L^2 & c_1 & 0 \\ L^2 & 0 & 0 & c_2 \\ c_1 & 0 & 0 & 0 \\ 0 & c_2 & 0 & 0 \end{pmatrix} \begin{pmatrix} a_1 \\ \hat{a}_2 \\ a_3 \\ \hat{a}_4 \end{pmatrix}$$

Notice that the lower right-hand corner of this matrix is filled with zeroes rather than with two L's. The net result of this is that it is easier to achieve a higher degree of directivity when the load reflection coefficients on the internal and external coupled ports are not exactly zero.

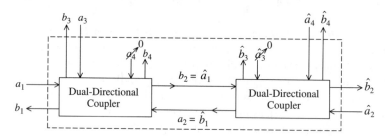

Figure 12.29 Dual-directional coupler formed from two lossless couplers.

12.14.1 Transformer Couplers

A coupled winding transformer without a ground reference terminal (an indefinite matrix circuit) is shown in Figure 12.30. This circuit has a Y matrix:

$$\begin{pmatrix} I_1 \\ I_2 \\ I_3 \\ I_4 \end{pmatrix} = \begin{pmatrix} Y_{11} & Y_{12} & -Y_{11} & -Y_{12} \\ Y_{12} & Y_{22} & -Y_{12} & -Y_{22} \\ -Y_{11} & -Y_{12} & Y_{11} & Y_{12} \\ -Y_{12} & -Y_{22} & Y_{12} & Y_{22} \end{pmatrix} \begin{pmatrix} V_1 \\ V_2 \\ V_3 \\ V_4 \end{pmatrix}$$

$$\text{DEN} = 2\pi f (L_{11}L_{22} - L_{12}L_{12}) \quad Y_{11} = \frac{-jL_{22}}{\text{DEN}} \quad Y_{12} = \frac{+jL_{12}}{\text{DEN}} \quad Y_{22} = \frac{-jL_{11}}{\text{DEN}}$$

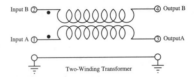

Figure 12.30 Transformer used as a coupler.

L_{11} is the inductance of the coil on port one, L_{22} is the inductance of the coil on port two, and L_{12} is the mutual inductance between the two windings with the dot convention shown on ports one and three. Notice that the rows and columns of the Y matrix sum to zero since this is an indefinite circuit matrix. There is no return of any of the currents to ground except through external terminations. Since $I_1 = -I_3$ and $I_2 = -I_4$, the first and third rows are negative of each other and the second and fourth rows are negative of each other. These facts and the use of reciprocity allows the full four-port admittance matrix to be determined from only the upper-left 2×2 transformer matrix.

If coupling and termination capacitors are added to the two winding transformers as shown in Figure 12.31, they can become directional couplers. The admittance matrix becomes:

$$
\begin{pmatrix} I_1 \\ I_2 \\ I_3 \\ I_4 \end{pmatrix} = \frac{1}{\omega(L_{11}L_{22} - L_{12}L_{12})} \begin{pmatrix} -jL_{22} & +jL_{12} & +jL_{22} & -jL_{12} \\ +jL_{12} & -jL_{11} & -jL_{12} & +jL_{11} \\ +jL_{22} & -jL_{12} & -jL_{22} & +jL_{12} \\ -jL_{12} & +jL_{11} & +jL_{12} & -jL_{11} \end{pmatrix} \begin{pmatrix} V_1 \\ V_2 \\ V_3 \\ V_4 \end{pmatrix} +
$$

$$
\begin{pmatrix} +j\omega(C_{11}+C_{12}+C_{14}) & -j\omega C_{12} & 0 & -j\omega C_{14} \\ -j\omega C_{12} & +j\omega(C_{22}+C_{12}+C_{23}) & -j\omega C_{23} & 0 \\ 0 & -j\omega C_{23} & +j\omega(C_{33}+C_{23}+C_{34}) & -j\omega C_{34} \\ -j\omega C_{14} & 0 & -j\omega C_{34} & +j\omega(C_{44}+C_{14}+C_{34}) \end{pmatrix} \begin{pmatrix} V_1 \\ V_2 \\ V_3 \\ V_4 \end{pmatrix}
$$

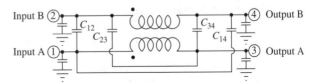

Figure 12.31 Transformer and capacitors as a directional coupler.

For this matrix to represent a directional coupler similar to the quarter-wavelength coupled transmission line coupler, $Y_{11} = Y_{12} = 0$. If the coupler is made from discrete inductors, $L_{12} = 0$ and if $L_{12} = 0$, then $C_{12} = 0$. With the correct choice of the other capacitive reactances, the new network can act as a directional coupler at a given frequency. The sum of the original Y matrix and the admittance matrix with the capacitors is the total admittance matrix. If L_{12} is not equal to 0, then C_{12} is required and the coupling is adjusted by using the other capacitors.

12.14.2 Transformer Couplers—$L_{12} = 0$

A special case of the directional coupler can be built using discrete L's and C's using the coupler just analyzed. This is the special case for which $L_{12} = 0$, $L_{11} = L_{22}$, $C_{14} = C_{23}$, $C_{12} = C_{34} = 0$, and $C_{11} = C_{22} = C_{33} = C_{44}$. This case is compared with the directional coupler made from a pair of transmission lines with odd- and even-mode admittances Y_{0o} and

Y_{0e}. The admittance matrix for this case is as follows.

$$Y_{11} = \frac{-j}{\omega L_{11}} + j\omega(C_{11} + C_{14})$$

$$Y_{13} = \frac{+j(Y_{0o} + Y_{0e})}{2} = \frac{+j}{\omega L_{11}}$$

$$Y_{14} = \frac{+j(Y_{0o} - Y_{0e})}{2} = -j\omega C_{14}$$

$$L_{11} = \frac{2}{\omega(Y_{0o} + Y_{0e})}$$

$$C_{14} = \frac{Y_{0o} - Y_{0e}}{2\omega}$$

$$C_{12} = \frac{1}{\omega^2 L_{11}} - C_{14} = \frac{Y_{0e}}{\omega}$$

In some applications, some of the ports might need to be terminated in an impedance other than Z_0. In order to change to a different reference impedance on a port, consider that a Y matrix with all ports normalized in the same Z_0 is:

$$(Y) = (\sqrt{Y_0})[(S) + (U)]^{-1}[(U) - (S)](\sqrt{Y_0})$$

This means that:

$$[(S) + (U)]^{-1}[(U) - (S)] = (\sqrt{Z_0})(Y)(\sqrt{Z_0})$$

The left term above is a normalized Y matrix with all ports normalized to a characteristic impedance of Z_0. A new Y matrix derived from a scattering matrix normalized to other port impedances will give a new directional coupler that has different termination impedances on all the ports. This matrix is:

$$(Y_{\text{new}}) = (\sqrt{Y_i})(\sqrt{Z_0})(Y)(\sqrt{Z_0})(\sqrt{Y_i}) = (\sqrt{Y_i})(Y_{\text{norm}})(\sqrt{Y_i})$$

This allows a lumped-constant directional coupler to be fabricated where some of the ports are not normalized to the same characteristic impedance as other ports. The procedure is to normalize the Y matrix and then premultiply and postmultiply the normalized Y matrix by the diagonal matrices containing the square root of the new normalization admittances. The diagonal normalization matrices look like:

$$(\sqrt{Y_i}) = \begin{pmatrix} \sqrt{Y_1} & 0 & 0 & \dots & 0 \\ 0 & \sqrt{Y_2} & 0 & \dots & 0 \\ & & \dots & & \\ 0 & \dots & 0 & \sqrt{Y_{n-1}} & 0 \\ 0 & \dots & 0 & 0 & \sqrt{Y_n} \end{pmatrix}$$

$$(\sqrt{Z_0}) = \begin{pmatrix} \sqrt{Z_0} & 0 & 0 & \dots & 0 \\ 0 & \sqrt{Z_0} & 0 & \dots & 0 \\ & & \dots & & \\ 0 & \dots & 0 & \sqrt{Z_0} & 0 \\ 0 & \dots & 0 & 0 & \sqrt{Z_0} \end{pmatrix}$$

12.15 TWO CORE TRANSFORMER-BASED DIRECTIONAL COUPLERS

Two cross-connected transformers as shown in Figure 12.32 are often used as a directional coupler. Theoretically the directivity of this structure is not infinite if $R_2 = R_3$, but practically the directivity is quite good if the coupling value is small (large negative dB). Real transformers have parasitics not included in the diagram and those parasitics are not included in the analyses to follow. Since the structure is not symmetrical, a forward and a reverse analysis will be done. In the analysis to follow, currents are referenced into the ports and voltages are from the ports to ground. Due to transformer action, the following are the circuit constraints:

$$I_1 = n(I_2 + I_4)$$

$$I_2 = n(I_1 + I_3)$$

$$V_3 = n(V_4 - V_2)$$

$$V_4 = n(V_3 - V_1)$$

where n is the turns ratio of each transformer.

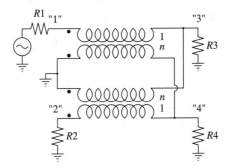

Figure 12.32 Transformer-based directional coupler.

12.15.1 Forward Analysis—Coupler

For the forward analysis, R_3 is the load resistor, R_2 is the load on the coupled port, and R_4 is the load on the isolated port. The following port currents and voltages are expressed in terms of the input current I_1.

$$I_2 = +\frac{n(R_3 + R_4)}{n^2(R_2 + R_4) + R_3} I_1 \qquad V_2 = -R_2 \frac{n(R_3 + R_4)}{n^2(R_2 + R_4) + R_3} I_1$$

$$I_3 = -\frac{n^2(R_2 + R_4) - R_4}{n^2(R_2 + R_4) + R_3} I_1 \qquad V_3 = +R_3 \frac{n^2(R_2 + R_4) - R_4}{n^2(R_2 + R_4) + R_3} I_1$$

$$I_4 = +\frac{n^2(R_2 - R_3) + R_3}{n[n^2(R_2 + R_4) + R_3]} I_1 \qquad V_2 = -R_4 \frac{n^2(R_2 - R_3) + R_3}{n[n^2(R_2 + R_4) + R_3]} I_1$$

Since $V_1 = V_3 - V_4/n$, V_1 and Z_{in} are equal to:

$$V_1 = \frac{n^4 R_3(R_2 + R_4) - n^2(2R_3 R_4 - R_2 R_4) + R_3 R_4}{n^2[n^2(R_2 + R_4) + R_3]} I_1$$

$$Z_{in} = \frac{n^4 R_3(R_2 + R_4) - n^2(2R_3 R_4 - R_2 R_4) + R_3 R_4}{n^2[n^2(R_2 + R_4) + R_3]}$$

There are some interesting substitutions one can make in the above equations. For the directivity of the coupler to be infinite, V_4 should be zero whenever R_3 is equal to the characteristic load impedance. If all the termination resistors are equal in value, then there is no solution for $V_4 = 0$. If all the resistors are set to the same value R, we have:

$$V_1 = \frac{(2n^4 - n^2 + 1)R}{n^2(2n^2 + 1)} I_1$$

$$V_2 = \frac{-2nR}{(2n^2 + 1)} I_1$$

$$V_3 = \frac{(2n^2 - 1)R}{(2n^2 + 1)} I_1$$

$$V_4 = \frac{-R}{n(2n^2 + 1)} I_1$$

These equations can be used to determine the coupling coefficient of the structure as a coupler. The ratio of V_4 to V_2 is a measure of the directivity of the coupler. Notice that it is related to 1 over n squared and will be quite small for large n.

12.15.2 Forward Analysis—Power Divider

If one wishes to use this structure as a power divider with $R_2 = R_3 = R$, then R_4 depends on the choice of R. There are several choices of R and n which will give power divider performance. Equating the power in loads R_2 and R_3, the turns ratio n and R_4 are related by the following equations.

$$n = 0.5 + 0.5\sqrt{1 + \frac{4R_4}{R + R_4}}$$

$$R_4 = R\frac{n - n^2}{n^2 - n - 1}$$

This equation for R_4 is valid only for $1 < n < 0.5 + 0.5\sqrt{5}$. R_4 will range from zero to infinity as n is varied over this range. It is interesting to note that equal power split takes place for various values of n. The input impedance is not always equal to R but is also a function of n. These values are summarized in the table below. Substituting the above relationships back into the voltage equations gives the following results when $R_2 = R_3 = R$.

$$\frac{V_1}{I_1} = R\frac{n^2 + n - 1}{n(n + 1)}$$

$$\frac{V_2}{I_2} = R\frac{-n}{n + 1}$$

$$\frac{V_3}{I_3} = R\frac{n}{n + 1}$$

$$\frac{V_4}{I_4} = R\frac{n - 1}{n + 1}$$

The power into port one and out of ports two, three, and four are then given as:

$$P_{\text{in}} = I_1^2 R \frac{n^2 + n - 1}{n(n+1)}$$

$$P_2 = I_1^2 R \left(\frac{n}{n+1}\right)^2$$

$$P_3 = I_1^2 R \left(\frac{n}{n+1}\right)^2$$

$$P_4 = I_1^2 R \frac{-n^3 + 2n^2 - 1}{n(n+1)^2}$$

The efficiency of the power divider is computed by determining the total power out of ports two and three and dividing this result by the input power. The remaining power is dissipated in port four. Although the efficiency of the power divider does not vary much over the range of n, there is a minimum value of efficiency η as n varies. The efficiency of the circuit with $(R_2 = R_3 = R)$ used as a power divider is:

$$\eta = \frac{2n^3}{(n+1)(n^2 + n - 1)}$$

The following table shows these various parameters.

Parameter	Minimum n	$n = \sqrt{2}$	Equal R's	Minimum η	Maximum n
n	1	$\sqrt{2}$	$0.5 + 0.5\sqrt{3}$	$\sqrt{1.5}$	$0.5 + 0.5\sqrt{5}$
efficiency	100%	97.056%	96.534%	95.755%	100%
Z_{in}	0.5 R	$\dfrac{\sqrt{2}}{2} R$	$\left(3 - \sqrt{\dfrac{16}{3}}\right) R$	$\left(-1 + \sqrt{\dfrac{8}{3}}\right) R$	$(3 - \sqrt{5})R$
R_4	0	$\sqrt{2} R$	R	$\dfrac{-3 + 2\sqrt{6}}{5} R$	infinite

12.15.3 Scattering Matrix Analysis

The scattering matrix for the coupled transformer is given below. The scattering matrix is normalized to equal value resistances on all ports. One can use the S parameter impedance transformation equations if one desires to change the normalization impedance on the ports to other values, for instance, to various values of R_4 as shown in the above table. Notice from the matrix that there is no natural directivity in the structure (i.e., none of the values in the matrix is equal to zero). The apparent directivity of the structure depends on the difference of powers of n.

$$(S) = \frac{1}{4n^4 + 1} \begin{pmatrix} -2n^2 + 1 & -4n^3 & 2n^2(2n^2 - 1) & -2n \\ -4n^3 & -2n^2 + 1 & -2n & 2n^2(2n^2 - 1) \\ 2n^2(2n^2 - 1) & -2n & 2n^2 - 1 & 4n^3 \\ -2n & 2n^2(2n^2 - 1) & 4n^3 & 2n^2 - 1 \end{pmatrix}$$

Notice that the upper-left 2×2 portion of the matrix is different from the lower-right 2×2 portion. They are different by a minus sign.

12.15.4 Reverse Analysis—Coupler

If the structure is excited at the third port, the following relationships of the port currents and voltages apply.

$$I_1 = \frac{-n^4(R_2 + R_4) + n^2 R_4}{n^4(R_2 + R_4) + n^2(R_1 - 2R_4) + R_4} I_3$$

$$I_2 = \frac{n^3(R_1 - R_4) + n R_4}{n^4(R_2 + R_4) + n^2(R_1 - 2R_4) + R_4} I_3$$

$$I_4 = \frac{-n^3(R_1 + R_2)}{n^4(R_2 + R_4) + n^2(R_1 - 2R_4) + R_4} I_3$$

$$V_1 = \frac{n^4(R_2 + R_4) - n^2 R_4}{n^4(R_2 + R_4) + n^2(R_1 - 2R_4) + R_4} R_3 I_3$$

$$V_2 = \frac{-n^3(R_1 - R_4) - n R_4}{n^4(R_2 + R_4) + n^2(R_1 - 2R_4) + R_4} R_2 I_3$$

$$V_3 = \frac{n^4(R_2 + R_4)R_1 + n^2 R_2 R_4}{n^4(R_2 + R_4) + n^2(R_1 - 2R_4) + R_4} I_3$$

$$V_4 = \frac{n^3(R_1 + R_2)}{n^4(R_2 + R_4) + n^2(R_1 - 2R_4) + R_4} R_4 I_3$$

The output impedance of the structure is V_3/I_3.

12.15.5 Coupled Transformer Pair—Balun

If R_4 is set equal to zero then:

$$I_2 = \frac{n R_3}{n^2 R_2 + R_3} I_1$$

$$I_3 = \frac{n^2 R_2}{n^2 R_2 + R_3} I_1$$

$$I_4 = \frac{n^2(R_2 - R_3) + R_3}{n(n^2 R_2 + R_3)} I_1$$

If $R_2 = R_3 = R$ (the left column in the table above), $I_2 = -I_3 = 0.5I_1$ and $Z_{in} = 0.5R$. Since the voltages on ports two and three are out of phase, a resistor equal to $2R$ could be placed between ports two and three. A virtual ground would exist halfway through the resistor. This new load, $2R$, is four times the input impedance of $0.5R$. The structure therefore functions as a balun with a $4 : 1$ impedance transformation.

Various other substitutions can be made into the formulas to demonstrate other circuit functions.

12.16 COUPLED-LINE COUPLERS AND HYBRIDS

12.16.1 Three-dB Hybrid

The 3-dB hybrid is a subset of the directional couplers discussed previously. The directional coupler that has 3 dB of coupling has some special properties. The scattering matrix of a quarter-wavelength-long directional coupler with 3 dB of coupling is:

$$\begin{pmatrix} b_1 \\ b_2 \\ b_3 \\ b_4 \end{pmatrix} = \frac{\sqrt{2}}{2} \begin{pmatrix} 0 & 1 & -j & 0 \\ 1 & 0 & 0 & -j \\ -j & 0 & 0 & 1 \\ 0 & -j & 1 & 0 \end{pmatrix} \begin{pmatrix} a_1 \\ a_2 \\ a_3 \\ a_4 \end{pmatrix}$$

The top view and cross-sectional view of a coupler constructed out of stripline are shown in Figure 12.33. If equal terminations are put on ports two and three, then the input reflection coefficient looking into port one is zero. Consider:

$$b_1 = +\frac{\sqrt{2}}{2}a_2 - j\frac{\sqrt{2}}{2}a_3$$

$$b_2 = +\frac{\sqrt{2}}{2}a_1 - j\frac{\sqrt{2}}{2}a_4$$

$$b_3 = -j\frac{\sqrt{2}}{2}a_1 + \frac{\sqrt{2}}{2}a_4$$

$$b_4 = -j\frac{\sqrt{2}}{2}a_2 + \frac{\sqrt{2}}{2}a_3$$

Let the terminations on ports two, three, and four be:

$$\Gamma_2 = \frac{a_2}{b_2}$$

$$\Gamma_3 = \frac{a_3}{b_3} = \Gamma_2$$

$$\Gamma_4 = 0$$

Substituting these into the above equations gives:

$$b_1 = \frac{\sqrt{2}}{2}\Gamma_2 b_2 - j\frac{\sqrt{2}}{2}\Gamma_2 b_3$$

$$b_2 = \frac{\sqrt{2}}{2}a_1$$

$$b_2 = -j\frac{\sqrt{2}}{2}a_1$$

$$b_4 = -j\frac{\sqrt{2}}{2}\Gamma_2 b_2 + \frac{\sqrt{2}}{2}\Gamma_2 b_3$$

or:

$$b_1 = \frac{1}{2}\Gamma_2 a_1 - \frac{1}{2}\Gamma_2 a_1 = 0$$

$$b_2 = \frac{\sqrt{2}}{2}a_1$$

$$b_2 = -j\frac{\sqrt{2}}{2}a_1$$

$$b_4 = -j\frac{1}{2}\Gamma_2 a_1 - j\frac{1}{2}\Gamma_2 a_1 = -j\Gamma_2 a_1$$

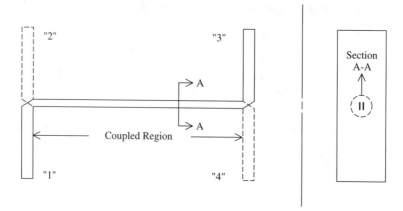

Figure 12.33 Three-dB 90-degree hybrid in stripline.

This characteristic of a 3-dB 90-degree hybrid is used in many "hybrid coupled" amplifier designs to control the input impedance of the amplifiers. If matched amplifiers are used, the input impedance into the matched set is very low when using a 90-degree coupler as an input power splitting network. Hybrid coupled amplifiers will be discussed later in this chapter. The scattering parameters of the coupler vary with frequency. The analysis of coupled lines in Section 12.11 can be used to determine the frequency variation of the different terms of the scattering matrix. A 3-dB coupler formed from coupled lines has an even-mode impedance of 120.71 ohms and an odd-mode impedance of 20.71 ohms.

12.16.2 Transmission Line Hybrids

12.16.2.1 Branch Line Ninety-Degree Hybrid. A 90-degree hybrid can be built using a set of branch lines as shown in Figure 12.34 to construct a *branch line hybrid*. This circuit has a scattering matrix similar to the coupled line hybrid but with added 90-degree phase shifts. The scattering matrix of the branch line hybrid at the frequency for which the lines are a quarter-wavelength long is:

$$\begin{pmatrix} b_1 \\ b_2 \\ b_3 \\ b_4 \end{pmatrix} = \frac{\sqrt{2}}{2} \begin{pmatrix} 0 & -j & -1 & 0 \\ -j & 0 & 0 & -1 \\ -1 & 0 & 0 & -j \\ 0 & -1 & -j & 0 \end{pmatrix} \begin{pmatrix} a_1 \\ a_2 \\ a_3 \\ a_4 \end{pmatrix}$$

This hybrid also has the *isolating* properties when used with equal reflection coefficients on ports two and three. The frequency dependence of the matrix can be derived by forming the four-port y matrix of the network from four two-port y matrices on the transmission lines. The two-port y matrix of a transmission line is given in Section 2.6.2. The four-port y matrix can then be transformed to a four-port scattering matrix using the equation in Section 2.3.3.

12.16.2.2 Transmission Line Rat Race. There is another type of hybrid termed a 180-degree hybrid. An example is shown in Figure 12.35. This example is termed the *rat race* hybrid. This hybrid has a scattering matrix at the quarter-wavelength frequency given by:

$$\begin{pmatrix} b_1 \\ b_2 \\ b_3 \\ b_4 \end{pmatrix} = \frac{\sqrt{2}}{2} \begin{pmatrix} 0 & -j & 0 & +j \\ -j & 0 & -j & 0 \\ 0 & -j & 0 & -j \\ +j & 0 & -j & 0 \end{pmatrix} \begin{pmatrix} a_1 \\ a_2 \\ a_3 \\ a_4 \end{pmatrix}$$

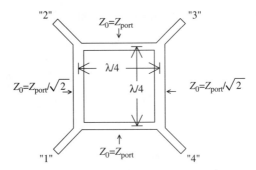

Figure 12.34 Branch line 90-degree hybrid.

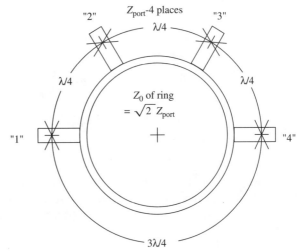

Figure 12.35 Rat race 180-degree hybrid.

The frequency dependence of the matrix elements can also be found by using the y parameters for four transmission lines and converting the four-port y matrix to a four-port scattering parameter matrix.

If equal terminations are put on ports one and three, then the input reflection coefficient looking into port two is equal to the reflection coefficient of the terminations on port one and three. Consider:

$$b_1 = -j\frac{\sqrt{2}}{2}a_2 + j\frac{\sqrt{2}}{2}a_4$$

$$b_2 = -j\frac{\sqrt{2}}{2}a_1 - j\frac{\sqrt{2}}{2}a_3$$

$$b_3 = -j\frac{\sqrt{2}}{2}a_2 - j\frac{\sqrt{2}}{2}a_4$$

$$b_4 = +j\frac{\sqrt{2}}{2}a_1 - j\frac{\sqrt{2}}{2}a_3$$

Let the terminations on ports one, three, and four be:

$$\Gamma_1 = \frac{a_1}{b_1}$$

$$\Gamma_3 = \frac{a_3}{b_3} = \Gamma_1$$

$$\Gamma_4 = 0$$

Substituting these into the above equations gives:

$$b_1 = -j\frac{\sqrt{2}}{2}a_2$$

$$b_2 = -j\frac{\sqrt{2}}{2}\Gamma_1 b_1 - j\frac{\sqrt{2}}{2}\Gamma_1 b_3$$

$$b_3 = -j\frac{\sqrt{2}}{2}a_2$$

$$b_4 = +j\frac{\sqrt{2}}{2}\Gamma_1 b_1 - j\frac{\sqrt{2}}{2}\Gamma_1 b_3$$

or:

$$b_1 = -j\frac{\sqrt{2}}{2}a_2$$

$$b_2 = -\frac{1}{2}\Gamma_1 a_2 - \frac{1}{2}\Gamma_1 a_2 = -\Gamma_1 a_2$$

$$b_3 = -j\frac{\sqrt{2}}{2}a_2$$

$$b_4 = +\frac{1}{2}\Gamma_1 a_2 - \frac{1}{2}\Gamma_1 a_2 = 0$$

Now, instead of a matching input reflection coefficient, the 180-degree hybrid has an input reflection coefficient equal to the equal reflection coefficient on the symmetrical ports. This hybrid does not give a good input reflection coefficient when the other ports are not matched.

12.16.3 Hybrid with Transmission Lines and a Coupled Line [10]

The rat race has a three-quarter-wavelength line between ports one and four. The y matrix for that line is:

$$(Y) = \frac{\sqrt{2}}{2}\begin{pmatrix} 0 & -jY_0 \\ -jY_0 & 0 \end{pmatrix}$$

at the center frequency and Y_0 is the port impedance. The quarter-wavelength-long coupled-line section that has shorts on ports two and three has an admittance matrix at the center frequency of:

$$\begin{pmatrix} I_1 \\ I_4 \end{pmatrix} = \begin{pmatrix} 0 & -j\frac{Y_{0o}}{2} + j\frac{Y_{0e}}{2} \\ -j\frac{Y_{0o}}{2} + j\frac{Y_{0e}}{2} & 0 \end{pmatrix}\begin{pmatrix} V_1 \\ V_4 \end{pmatrix}$$

This matrix can be made to equal the matrix of the three-quarter-wavelength long line if:

$$Y_{0o} - Y_{0e} = \sqrt{2}Y_0 \qquad Y_{0o}Y_{0e} = \frac{1}{2}Y_0^2$$

$$Y_{0o}Y_{0e} - Y_{0e}^2 - \sqrt{2}Y_0 Y_{0e} = \frac{1}{2}Y_0^2 - Y_{0e}^2 - \sqrt{2}Y_0 Y_{0e} = 0$$

$$\Rightarrow Y_{0e} = \frac{-\sqrt{2}+2}{2}Y_0 \qquad Y_{0o} = \frac{\sqrt{2}+2}{2}Y_0$$

$$\Rightarrow Z_{0e} = (2+\sqrt{2})Z_0 \qquad Z_{0o} = (2-\sqrt{2})Z_0$$

These values of even- and odd-mode impedance are difficult to implement but can be accomplished. This circuit is shown in Figure 12.36. Since the coupled-line section is the same physical length as the other lines in the coupler, the time delay and thus the phase will track over a much wider frequency range. This is responsible for this coupler being broader in bandwidth than the rat race coupler.

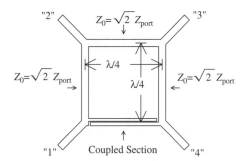

Figure 12.36 A 180-degree hybrid using coupled section.

12.17 RESISTIVE POWER SPLITTERS

Sometimes it is desirable to split signal power from one source into two loads when loss is not a real issue. The left side of Figure 12.37 shows a circuit that will split a signal from a Z_0 source to two Z_0 loads. The circuit puts half of the input port voltage on each load port when the load ports are terminated in Z_0. The circuit is then matched at the input but each load port is not matched. Looking back from each load one sees Z_0 plus some impedance on port one. If one adds a resistor from port two to port three with a value of Z_0 as shown in the right side of Figure 12.37, then each output port is also matched when all ports have matched loads. The device is also then symmetrical. If each of the output load impedances is the same, then there is no current in the resistor that is inserted between ports two and three since the voltage on each load is then the same. It can be shown that the scattering matrix of the power splitter shown in Figure 12.37 is:

$$\begin{pmatrix} b_1 \\ b_2 \\ b_3 \end{pmatrix} = \frac{1}{2} \begin{pmatrix} 0 & 1 & 1 \\ 1 & 0 & 1 \\ 1 & 1 & 0 \end{pmatrix} \begin{pmatrix} a_1 \\ a_2 \\ a_3 \end{pmatrix}$$

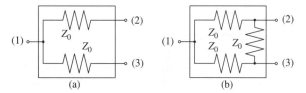

Figure 12.37 Resistive power splitter (a) matched at input only; (b) matched at all three ports.

The power output into each load is 6 dB down. Notice that this power splitter does not have isolation; even when the device is terminated in matched impedances, putting power in either port two or three results in power coming out of port three or two respectively as well as out of port one. One would expect this since the part is three-way symmetric. What is the effect of load mismatch on the input impedance? It can be shown that the input reflection coefficient is:

$$\Gamma_{in} = \frac{\Gamma_2 + \Gamma_3 + \Gamma_2 \Gamma_3}{4 - \Gamma_2 \Gamma_3}$$

where Γ_2 and Γ_3 are the load reflection coefficients on ports two and three respectively. If

$\Gamma_2 = \Gamma_3 = \Gamma$, then:

$$\Gamma_{\text{in}} = \frac{\Gamma}{2 - \Gamma}$$

Notice that when $\Gamma = 1$, an open circuit, then Γ_{in} is also equal to 1. However, when $\Gamma = -1$, a short circuit, Γ_{in} is equal to $-1/3$, representing an input impedance of $Z_0/2$ ohms. Now when $\Gamma_2 = \Gamma_o + \Delta\Gamma$ and $\Gamma_3 = \Gamma_o - \Delta\Gamma$ and $\Gamma_o = 0$:

$$\Gamma_{\text{in}} = \left.\frac{2\Gamma_o + \Gamma_0^2 - \Delta\Gamma^2}{4 - \Gamma_0^2 + 2\Gamma_o\Delta\Gamma + \Delta\Gamma^2}\right|_{\Gamma_o=0} = \frac{-\Delta\Gamma^2}{4 + \Delta\Gamma^2}$$

If $\Delta\Gamma$ is small, then:

$$\Gamma_{\text{in}} \approx \frac{-\Delta\Gamma^2}{4}$$

Note that this represents a very small input reflection coefficient. This type of power divider works very well when the load reflection coefficients are small but different by 180 degrees. However, when the load reflection coefficients are small but the same, the input reflection coefficient magnitude is reduced only by an approximate factor of two or approximately an additional 6 dB of return loss.

For this power splitter, the Y matrix and its determinant are:

$$(Y) = \begin{pmatrix} 2Y_0 & -Y_0 & -Y_0 \\ -Y_0 & 2y_0 & -Y_0 \\ -Y_0 & -Y_0 & 2Y_0 \end{pmatrix}$$

$$\Delta_Y = 8Y_0^3 - Y_0^3 - Y_0^3 - 2Y_0^3 - 2Y_0^3 - 2Y_0^3 = 0$$

This matrix is indefinite as expected. There is no ground in the circuit. In addition, the determinant of the Y matrix is zero and therefore the Z matrix of the network does not exist.

Applying the delta-to-wye transformation to the resistive power divider shown above, the power divider shown in Figure 12.38 is derived. It has the same scattering and Y matrix.

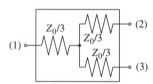

Figure 12.38 Tee (wye) version of the resistive power splitter.

12.18 WILKINSON POWER SPLITTERS

If the lower-right four terms of the scattering matrix for a resistive power splitter could be made equal to zero, then the power splitter would have isolation between the two output ports. The *Wilkinson power splitter* does that in addition to reducing the insertion loss to zero in the lossless case. Figure 12.39 shows a general type of Wilkinson power splitter/combiner. When the J and K inverter boxes consist of quarter-wave transmission lines, then the part is a symmetrical Wilkinson power splitter. The J and K inverters were discussed in Chapter 9. The J and K inverter boxes can consist of lumped-constant inverters resulting in a lumped-constant Wilkinson. They do not have to be identical. The only requirement is that their phase shifts and J or K values are the same at the design frequency. If one of the inverters is a J inverter and the other inverter is a K inverter (i.e., a pi circuit and a tee circuit), the input *VSWR* has a broader bandwidth than when both inverters are of the same type. However, then the output balance at the loads is different when the frequency is varied from the center frequency.

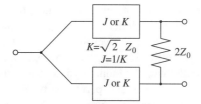

Figure 12.39 Wilkinson-type in-phase power splitter.

These choices represent a different engineering trade-off. The J and K inverters do not have to be low-pass structures. They can be high-pass structures as well (e.g., both inverters could be high-pass inverters. Likewise, then one could be a J inverter or one a K inverter. At the center frequency (the frequency for which the transmission lines are a quarter of a wavelength long), the scattering matrix of the Wilkinson is:

$$(S) = \frac{\sqrt{2}}{2}\begin{pmatrix} 0 & \mp j & \mp j \\ \mp j & 0 & 0 \\ \mp j & 0 & 0 \end{pmatrix}$$

where the minus signs are used when the inverters are low-pass inverters and the plus signs are used when the inverters are high-pass inverters. The transmission line is a low-pass structure and the minus signs are used. The structure can be extended to unequal power split between the ports and to more than two ports on the output.

Notice that if a 90-degree phase shift is added to the output of one of the ports of the Wilkinson, the resultant scattering matrix is the same as the scattering matrix of the terminated four-port 90-degree hybrid. In the transmission line version, adding a Z_0 impedance 90-degree line on one of the ports accomplishes this. In the lumped-constant versions, the added phase shift could be made from a J or K inverter of Y_0 or Z_0 in value respectively. When the added phase shift is put on port three, the resultant scattering matrix is:

$$(S) = \frac{\sqrt{2}}{2}\begin{pmatrix} 0 & \mp j & \pm 1 \\ \mp j & 0 & 0 \\ \pm 1 & 0 & 0 \end{pmatrix}$$

where the sign in front of the "1" depends on the type of inverter originally used and the type added. The resistor that exists between the original ports two and three remains at that position. The added phase shift is added to the original port. The reflected power from equal loads on the output ports is then dissipated in the internal resistor, whereas the reflected power from unequal load is dissipated in the internal resistor for the Wilkinson without the added phase shift.

12.19 FERRITE BEAD

The derivation of the approximate impedance of a ferrite bead given here relies on the material parameters of ferrite material. In field theory, complex magnetic permeability is given as:

$$\mu = \mu' - j\mu''$$

From solid-state theory, the complex permeability is related to magnetic dipole resonances and relaxation effects in material given by the following formula.

$$\frac{\mu}{\mu_o} = 1 + \frac{K_1}{1 + j\omega\tau} + \frac{K_2}{\omega_o^2 - \omega^2 + j\omega K_3} + \cdots$$

The term containing K_1 is related to relaxation effects of a magnetic dipole, and the term containing K_2 is related to magnetic dipole resonance in the material. There may be more terms like the term containing K_2, although for this ferrite bead analysis often only one of the K_i terms will be included. The magnetic equivalent circuit for material satisfying the above equation around ω_o is given in Figure 12.40. This model is not necessarily valid at some distance from the resonance. It attributes all loss mechanisms to the magnetic effect (the resistance in parallel with the inductor) and none to the capacitors. The circuit of Figure 12.41 matches the measured loss and permeability of ferrite material over very large frequency ranges. It attributes losses to the magnetic effect as well as other material loss effects. The two capacitors, each with their separate resistances, model the separate effects of domain rotation and domain wall strain. Some magnetic materials have a relaxation effect [11], and for those materials the parallel L-R section is needed. Magnetic materials may be somewhat conductive and eddy currents may result. The losses from that effect would be included in the permittivity equivalent circuit for the volume. If these losses are not substantial, those losses would appear to modify the resistances of the magnetic circuit but that would be mixing two phenomena into one equivalent circuit. For ease of analysis, that might be appropriate. The material parameters are affected by the magnitude of the total magnetic field in the material and specifically by the magnitude of a dc field in the material. An axial dc current produces a magnetic field in the circumferential direction. The magnitude of the field at a radius r due to a dc axial current through the bead is:

$$H_\phi = \frac{I_{dc}}{2\pi r}$$

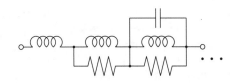

Figure 12.40 Magnetic equivalent circuit of magnetic material parameters.

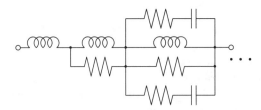

Figure 12.41 Wide-bandwidth equivalent circuit of the magnetic material parameters.

When the bead is used in an application, this field needs to be kept well below the magnetic saturation point in order to maintain the rf parameters of the material. Notice that the magnetic equivalent circuit produces a parallel R-L-C impedance function apart from winding capacitance, and so on. As discussed in the inductor model in Chapter 2, an inductor also has a parallel R-L-C equivalent circuit due to the capacity between windings. The material effects that give a parallel R-L-C are quite often at a higher frequency than the distributed capacity effects of the winding but cannot be ignored in a complete model of a ferrite bead. The magnetizing inductance of an inductor core has both inductance and capacitance responses when the permeability is inserted in the inductance equation. The winding capacity is in parallel with the magnetizing core impedance function.

If the material used also has dielectric loss, the dual of the above general formula is used to characterize the dielectric loss. The complex permittivity is given as:

$$\varepsilon = \varepsilon' - j\varepsilon''$$

From solid-state theory, the complex permittivity is related to electric dipole resonances and relaxation effects in the material given by the following formula.

$$\frac{\varepsilon}{\varepsilon_o} = 1 + \frac{K_4}{1 + j\omega\tau} + \frac{K_5}{\omega_0^2 - \omega^2 + j\omega K_6} + \cdots$$

The term containing K_4 is related to relaxation effects of an electric dipole, and the term containing K_5 is related to electric dipole resonance in the material. There may be more terms like the term containing K_5. The electric equivalent circuit for material satisfying the above equation is given in Figure 12.42. Conductance of the material appears in parallel with that equivalent circuit.

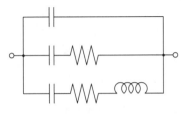

Figure 12.42 Electric equivalent circuit of electric material parameters.

In order to understand the behavior of very large ferrite beads for which the permeability has a relaxation effect and perhaps is made from material with appreciable conductivity, the skin depth of penetration of the electromagnetic field into the ferrite material needs to be calculated. The assumption is made that the skin depth is much smaller than the thickness of the bead. This is usually true only for very large beads. The ferrite bead then acts approximately like an inductor in series with a resistor. Consider the circuit shown in Figure 12.43. The magnetic field around path A encloses current I. The magnetic field around path B encloses no current since equal and opposite current flows in the line and in the skin depth inside the diameter of the bead. The path around C encloses current I. That current flows on the outside of the bead. The current is passing through a region equal to the skin depth δ thick on the outside, the inside, and the ends. Assuming that the inner and outer diameters of the bead are much larger than the skin depth, the sum of the resistances, including the outer skin, the inner skin, and the two end skins is:

$$R = \frac{\rho W}{\delta\pi D_o} + \frac{\rho W}{\delta\pi D_i} + 2\frac{\rho}{\delta\pi}\ln\left(\frac{D_o}{D_i}\right)$$

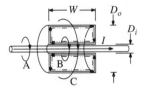

Figure 12.43 Cross section of a ferrite bead on a wire.

The dimensions can be measured. The quantities δ and ρ need to be determined. Starting from the formula for the magnetic effect and using just the free space term (the unity term) and the relaxation term, the skin depth can be determined from the propagation constant. The propagation constant gamma is:

$$\gamma = \sqrt{-(\sigma + \omega\varepsilon'' + j\omega\varepsilon')(\omega\mu'' + j\omega\mu')}$$

Assuming that the material is nonconducting, that the dielectric loss is not significant, and that the magnetic loss is small, the propagation constant is approximately:

$$\gamma = \sqrt{-(j\omega\varepsilon)(\omega\mu'' + j\omega\mu')} \approx \omega\sqrt{\varepsilon\mu'}\sqrt{1 - \frac{j\mu''}{\mu'}} \approx \omega\sqrt{\varepsilon\mu'}\left(1 - \frac{j\mu''}{2\mu'}\right)$$

The skin depth is the inverse of the imaginary part of the propagation constant. Therefore:

$$\gamma = \sqrt{-(j\omega\varepsilon)(\omega\mu'' + j\omega\mu')} \approx \omega\sqrt{\varepsilon\mu'}\sqrt{1 - \frac{j\mu''}{\mu'}} \approx \omega\sqrt{\varepsilon\mu'}\left(1 - \frac{j\mu''}{2\mu'}\right)$$

$$\delta = \frac{2\mu'}{\omega\sqrt{\varepsilon\mu'}\mu''} = \frac{2\sqrt{\mu'}}{\omega\sqrt{\varepsilon}\mu''}$$

From the relaxation portion of the magnetic permeability, this is:

$$\frac{\mu}{\mu_o} = 1 + \frac{K_1}{1 + j\omega\tau} = 1 + K_1\frac{1 - j\omega\tau}{1 + (\omega\tau)^2} = \frac{1 + (\omega\tau)^2 + K_1}{1 + (\omega\tau)^2} - j\frac{K_1\omega\tau}{1 + (\omega\tau)^2}$$

$$\delta = \frac{2\sqrt{\mu'}}{\omega\sqrt{\varepsilon}\mu''} = \frac{2\sqrt{\dfrac{1 + (\omega\tau)^2 + K_1}{1 + (\omega\tau)^2}}}{\omega\sqrt{\varepsilon\mu_0}\dfrac{K_1\omega\tau}{1 + (\omega\tau)^2}} \approx \frac{2\tau}{K_1\sqrt{\varepsilon\mu_0}}\Bigg|_{\omega\tau \gg 1 + K_1}$$

When the frequency is high enough, then the skin depth is a constant depth. Then the resistance presented in series with the bead is a constant and determined from the resistivity of the material. The resistivity is not due to the conductance of the ferrite but comes from the loss of the ferrite material. The value of the resistance in the circuit shown in Figure 12.40 depends on the value of the inductance. The ratio between the inductance and resistance is given from the material parameters.

The inductor shown in the left part of Figure 12.40 without any resistance is related to the free space inductance. The inductor with a relaxation resistor is $\mu_r - 1$ times this value if only relaxation effects are important. The input impedance into the circuit with only the left two inductances is:

$$\frac{Z}{j\omega L_o} = 1 + \frac{(\mu_r - 1)R}{R + j\omega(\mu_r - 1)L_o} = 1 + \frac{(\mu_r - 1)}{1 + j\omega\tau}$$

$$\tau = \frac{(\mu_r - 1)L_o}{R}$$

When the frequency is very low, the impedance starts as if the inductance is μ_r times the inductance of the part without any ferrite. As the frequency goes up, the impedance changes to:

$$Z = j\omega L_o\left[1 + \frac{(\mu_r - 1)}{1 + j\omega\tau}\right] \approx j\omega L_o\left[1 + \frac{(\mu_r - 1)}{j\omega\tau}\right] = \frac{(\mu_r - 1)L_o}{\tau} + j\omega L_o$$

This is a constant resistance in series with a small air inductance. The constant skin depth and constant resistance value for the impedance of the medium leads to a small inductive reactance in series with a constant resistance when the frequency gets high.

At low frequencies, the impedance locus starts as if it were an inductance of value $\mu_r L_o$. Then the value of the inductance drops to L_o and the resistance ends up at a constant value. This is typical of bead performance from dc to frequencies for which distributed effects take over. Notice from Figure 12.43 that a bead with a skin depth that is less than the bead thickness can be analyzed at high frequencies as if it were a triax line. The line would have an inner diameter equal to the inner diameter of the bead plus approximately the skin depth and an outer diameter roughly the diameter of the bead.

When the material parameters do not have a relaxation effect, the skin depth is usually quite large and the fields permeate the whole bead. The material parameters often still result in a constant resistance for the bead over wide frequency ranges. The model of a bead shown in Figure 12.44 is for a bead made from a material permeability of 850. The bead has an

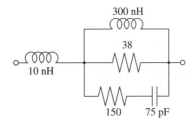

Figure 12.44 Model of a ferrite bead with a permeability of 850.

outer diameter of 3.5 mm, and inner diameter of 1.3 mm, and a length of 3.25 mm. This model works well up into the hundreds of MHz almost to 1 GHz. The model shown in Figure 12.45 shows a normalized impedance for a nickel ferrite material [12]. That model works up into the several-GHz region. The low-capacity series *R-C* is needed to set the shape of the loss curve at frequencies past magnetic resonance. The resistance in series with the larger capacitor and the parallel resistor shape the loss curve below resonance. Specific materials require different choices of the model parameters but the model is physically based and usually tracks the bead response well. Some material parameters show a resonance with a negative effective permeability at high frequencies. Some materials show a resonance with the resistance dropping after resonance. The choice of the loss resistor in series with the capacitors affects this behavior. The resistors in series with each of the capacitors model the loss of the domain rotation and domain wall bowing.

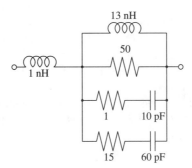

Figure 12.45 Model of nickel ferrite material.

12.20 BOND WIRE

A bond wire might not appear to be a microwave component. However, it is a very important microwave component. The discussion in Chapter 5 regarding how much gain one can have in a module when the current is returned through a bond wire inductance shows that importance. In order to minimize common mode inductance, one is tempted to put two or more bond wires in parallel. What effect does putting bond wires in parallel have? If the bond wires are installed at right angles to each other, their magnetic fields along their length do not interact and the inductance is roughly reduced to one-half of the inductance of a single bond wire. However, when they are installed in parallel, the inductance is not really reduced as much as anticipated. A plot of the inductance of two bond wires in parallel is shown in Figure 12.46. This plot is for two one-mil (25.4-micron) diameter wires placed 10 mils (254-micron) above a ground plane. The curve was generated from a modified method of moments calculation. In the calculation, fields from 360 times 4 lines of current were summed. These total fields produced an equal potential surface equivalent to the surface of the wires. It can be seen that the inductance for two wires widely separated is approximately one-half that for a single wire. However, when they are spaced closely together, the inductance is roughly the same as for a single wire. The

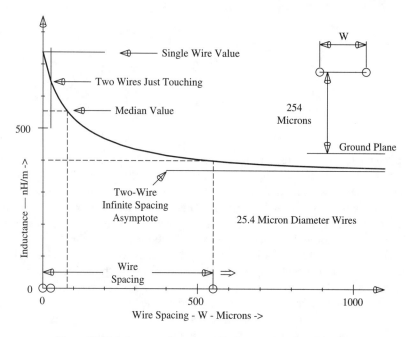

Figure 12.46 Inductance of two one-mil wires as a function of spacing.

perceived benefit may not be achieved. It is hard to put two one-mil wires very far apart when the bond pad they are bonded to is only three or four mils square.

An approximate analysis of two bond wires in parallel can be accomplished by considering the inductance of two thin lines of current above a ground plane. The analysis is done using the method of images. The potential at a point (x, y) from four symmetrically situated lines of current is as follows. The four lines of current are $+\rho$ at (a, b), $+\rho$ at $(-a, b)$, $-\rho$ at $(a, -b)$, and $-\rho$ at $(-a, -b)$, where ρ is a line charge of a given coulombs per meter. The potential ϕ is given as:

$$\phi = -\frac{\rho}{2\pi\varepsilon} \ln \sqrt{\frac{[(x+a)^2 + (y-b)^2][(x-a)^2 + (y-b)^2]}{[(x+a)^2 + (y+b)^2][(x-a)^2 + (y+b)^2]}}$$

$$= -\frac{\rho}{4\pi\varepsilon} \ln \left\{ \frac{[(x+a)^2 + (y-b)^2][(x-a)^2 + (y-b)^2]}{[(x+a)^2 + (y+b)^2][(x-a)^2 + (y+b)^2]} \right\}$$

The upper-right part of the equation is for the space around (a, b). In image theory, a plane can be inserted at $y = 0$ and the fields for the two conductors above the ground plane (for $y > 0$) are the same as the original fields for the four lines without a ground plane. If a conductor of radius r is placed at (a, b), and the distance from the ground plane and the other conductor is large with respect to r, then the potential on the conductor is given approximately by the next formula.

$$\phi \approx -\frac{\rho}{4\pi\varepsilon} \ln \left\{ \frac{a^2 r^2}{4(a^2 + b^2)b^2} \right\}$$

The capacity per unit length D for each line is the total charge on the line per unit length divided by the voltage.

$$C = \frac{\rho}{\phi} \approx \frac{4\pi\varepsilon}{\ln \left\{ \dfrac{4(a^2 + b^2)b^2}{a^2 r^2} \right\}}$$

The characteristic impedance of a transmission line is related to the capacity per unit length C, the inductance per unit length L, and the velocity on the transmission line. Using these relationships, the inductance per unit length of each wire is given as follows.

$$v = \frac{1}{\sqrt{LC}} \Rightarrow v^2 = \frac{1}{LC} = \frac{1}{\mu_r \mu_o \varepsilon_r \varepsilon_o}$$

$$Z_0 = \frac{1}{vC} = vL \Rightarrow L = \frac{1}{v^2 C} = \frac{\mu_r \mu_o \varepsilon_r \varepsilon_o}{C}$$

$$L \approx \frac{\mu_r \mu_o}{4\pi} \ln \left\{ \frac{4(a^2 + b^2)b^2}{a^2 r^2} \right\} = 100 \, \mu_r \ln \left\{ \frac{4(a^2 + b^2)b^2}{a^2 r^2} \right\} \text{nH}/m$$

When a is much greater than b, as it would be for an isolated line,

$$L \approx 100 \, \mu_r \ln \left\{ \frac{4(a^2 + b^2)b^2}{a^2 r^2} \right\} \approx 100 \, \mu_r \ln \left\{ \frac{4b^2}{r^2} \right\}$$

This inductance is just the inductance of a single wire above a ground plane. However, when b is much greater than a, and r is equal to some constant k times a,

$$L \approx 100 \, \mu_r \ln \left\{ \frac{4(a^2 + b^2)b^2}{a^2 r^2} \right\} \approx 100 \, \mu_r \ln \left\{ \frac{4b^4}{k^2 a^4} \right\}$$

$$= 100 \, \mu_r \left(-\ln(4k^2) + \ln \left\{ \frac{4^2 b^4}{a^4} \right\} \right) = 100 \, \mu_r \left(-2\ln(2k) + 2\ln \left\{ \frac{4b^2}{a^2} \right\} \right)$$

The logarithm of $2k$ is a slow function. The inductance L for a single wire in close proximity to another identical wire is now approximately twice the inductance for lines spaced far from each another.

Another interesting comparison is between a five-mil ribbon wire, one-mil thick, and two one-mil wires spaced four mils apart on their centers as shown in Figure 12.47. The ribbon is assumed to have rounded edges equivalent to the cross section of a semicircle. The centers of the semicircles are four mils apart. The wires are placed 10 mils above the ground plane. The current density in the ground plane between the two different configurations is virtually identical. The same voltage is used on the conductors. The inductance of the five-mil ribbon is 504 nH/m and the combined inductance of the two wires is 530 nH/m. A single one-mil wire according to the formula above is 740 nH/m. Two wires in parallel in this situation give 70% of the inductance rather than the expected 50%. The curved lines around the conductors are plots of the magnitude of the surface current density. At each point on the conductor, an axis that is normal to the surface is drawn and the normalized current density is plotted. If there were no surface current, the plot would touch the surface.

When two wires are very close as shown in Figure 12.48, the current on the adjacent surfaces of the two wires goes to zero. This is shown by the normalized current density plots in the blown up section of Figure 12.48. When the wires approach the ground plane, the current is concentrated on the side toward the ground plane as anticipated.

How current density exists in ribbons and bond wires needs to be considered carefully. Power is related to current density squared integrated over the surface. Figure 12.49 shows the same data presented in two different ways. The top part of the figure shows the relative current density at any x position. This assumes all the current is either on top or on the bottom of the conductor and the two are summed and plotted as one value. In reality, the conductor has some thickness and the current on the ends cannot go to infinity or to a very high density since there is a skin depth limit from the outer ends. If one assumes all the current that flows on the top and bottom at the end is distributed in the skin depth, then the lower part of the upper curve applies. The bottom part of the curve presents the same data but looks at the surface

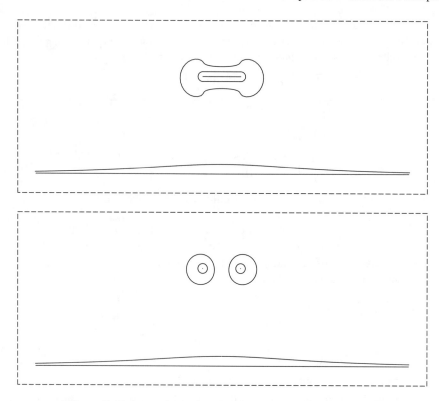

Figure 12.47 Current density for a five-mil ribbon vs. two one-mil wires.

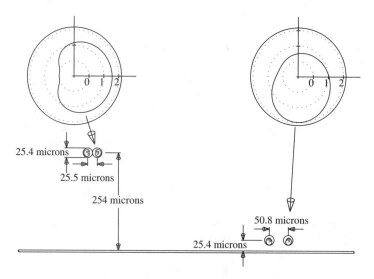

Figure 12.48 Current density in closely spaced wires.

current density and plots the magnitude orthogonal to the surface. The current density does not rise nearly as high plotted that way and the skin depth value in the top curve is similar to the current density shown in the bottom presentation. Granted some conductors may have a sharper corner than that shown in the figure but it demonstrates that some reason must be applied to current density curves for microstrip lines.

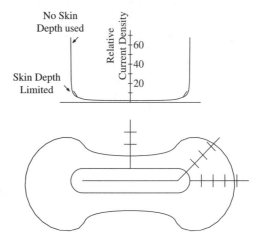

Figure 12.49 Different methods of looking at current density.

12.21 GROUND CONTACT

It is not often appreciated that a ground contact is a microwave circuit element. It is, however, quite important in MMIC and PCB design, especially for circuits that have both low- and high-frequency signals on them. MMIC and PCB circuits that contain both analog and digital signals are especially susceptible to the wrong application of this circuit element.

The resistance R between two perfectly conducting one-meter-long cylindrical electrodes of radii R_1 and R_2 separated by a distance D between their centers and embedded in a medium of resistivity equal to ρ is [13]:

$$R = \frac{\rho}{2\pi} \cosh^{-1}\left(\pm \frac{D^2 - R_1^2 - R_2^2}{2R_1 R_2} \right)$$

Consider the two cylinders as shown in Figure 12.50. The above equation is a result of an assumption that the length of the cylinders is infinite, the region is infinite in extent, and the frequency is dc. Since the total current flow under this assumption in the resistive volume has no z component, a ground plane of thickness T can be cut out of the infinite space. This assumes that the skin depth is much larger than the thickness T and that the region outside the ground plane is nonconducting. The equations to be derived will be used when the frequency is low and the skin depth is large. Then the resistance between the two contacts on the ground plane is:

$$R_g = \frac{\rho}{2\pi T} \cosh^{-1}\left(\pm \frac{D^2 - R_1^2 - R_2^2}{2R_1 R_2} \right) = \frac{R_\square}{2\pi} \cosh^{-1}\left(\pm \frac{D^2 - R_1^2 - R_2^2}{2R_1 R_2} \right)$$

where $R_\square$ is ρ/T and is called *resistance in ohms per square*. A square has an equal length and width. Let both contacts have the same radius R_c and assume that the contacts do not overlap (i.e., $D > R_1 + R_2 = 2R_c$). Under these conditions, the plus sign is used in the parentheses term.

$$R_g = \frac{R_\square}{2\pi} \cosh^{-1}\left(+ \frac{D^2 - 2R_c^2}{2R_c^2} \right) = \frac{R_\square}{2\pi} \ln\left(\frac{D^2 - 2R_c^2}{2R_c^2} + \sqrt{\left(\frac{D^2 - 2R_c^2}{2R_c^2} \right)^2 - 1} \right)$$

When $D \gg R_c$, then:

$$R_g \approx \frac{R_\square}{2\pi} \ln\left(\frac{D^2}{R_c^2} \right) = \frac{R_\square}{\pi} \ln\left(\frac{D}{R_c} \right)$$

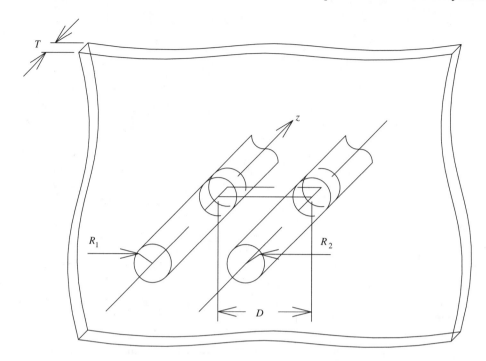

Figure 12.50 Model for resistance in grounds.

The resistance increases only slowly as the distance between the contacts increases. The contacts on the ground plane may represent ground returns on a transmission line circuit. How the ground goes from one point in the ground to another point in the ground is important in MMIC and PCB design. For instance, if an amplitude-modulated signal exists in the circuitry, and the circuitry is nonlinear, low-frequency currents will flow in the ground plane as well as in the circuitry. If the signals include digital signals with different bit rates, there will be low-frequency signals in the circuits related to the bit pattern. These signals will also exist in the ground plane. Do the low-frequency signals travel the same path that the high-frequency signals travel? The path in the ground plane may be entirely different and the resulting crosstalk effect will be quite different depending on the data pattern or carrier modulation pattern.

Current returning from one end of a transmission line tends to go to a return at the other end of a transmission line over the path of least impedance. In the ground plane this might be directly from one contact to the other rather than flowing underneath the transmission line. Consider a circuit like that shown in Figure 12.51. There are two configurations shown in the figure. The configuration on the right side might result when a top circuit trace is routed around some component. The current in the ground does not necessarily follow the same route. The total per-unit-length circuit impedance is the wire resistance per unit length, the inductance per unit length related to the area between the wire and the ground, and the ground resistance per unit length (ignoring internal inductance of the conductors). The inductance for ground current in the path shown as "b" is nominally related to the area of the loop divided by the width w (μhl/w) over which one considers the current to flow on the ground plane. Notice that as the path "b" is moved closer to directly under the wire (path "a"), the resulting area decreases and thus the inductance is smaller and the amount of current is higher than it was farther away from the line. The total circuit inductance results from paralleling many such small differential inductances and resistances. Since the same potential exists across each inductance, the current decreases (as the area gets larger) the farther the path on the ground plane is from directly under the wire. This argument assumes a TEM or quasi-TEM field

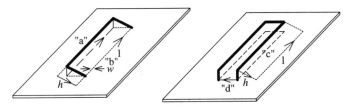

Figure 12.51 Ground current under microwave circuitry.

structure and thus the fields are normal to the direction of the wire. The ground resistance along path "b" can be approximated from equation given above. At high frequency when the inductive reactance for path "b" is high, the current will tend to follow path "a" under the wire. However, as the inductive reactance decreases, more current tends to travel along path "b" since the resistance going along path "a" is higher. At dc, there is no inductive reactance and then most of the current will tend to flow along path "b". This is a very important consideration in MMIC design. Any other component that might be placed in or close to path "b" will couple to the current from path "b" and crosstalk results. This is especially important for the configuration of the circuit on the right where at dc the current takes path "d".

12.22 CIRCULATOR

A nonreciprocal three-port device used for isolation and for directing energy around microwave systems is a *circulator*. It is shown schematically in Figure 12.52. It can be shown that due to energy conservation considerations, any matched, lossless, nonreciprocal three-port has an S matrix equal to:

$$(S) = \begin{pmatrix} 0 & 0 & S_{13} \\ S_{21} & 0 & 0 \\ 0 & S_{32} & 0 \end{pmatrix} \qquad |S_{ij}| = 1$$

This S matrix is for the direction of rotation as shown in Figure 12.52. In reality, the device has some loss. If the device is assumed to have threefold symmetry, but has some loss, the S

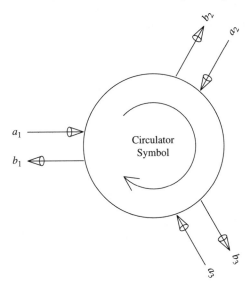

Figure 12.52 Circulator schematic.

matrix appears as:

$$(S) = \begin{pmatrix} 0 & S_{12} & S_{13} \\ S_{13} & 0 & S_{12} \\ S_{12} & S_{13} & 0 \end{pmatrix} \qquad |S_{12}|^2 + |S_{13}|^2 < 1$$

where S_{12} is used to designate the isolation term in the direction opposed to the arrow and S_{13} to designate the forward loss term in the direction of the arrow. Typical magnitudes are 0.1 for S_{12} and 0.95 for S_{13}.

12.23 CRYSTAL EQUIVALENT CIRCUIT

The *piezoelectric crystal* (or *crystal* for short) is a device often used to control frequency or used for a narrowband filter. The equivalent circuit for the crystal is shown in Figure 12.53.

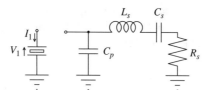

Figure 12.53 Crystal equivalent circuit.

This circuit has a series type of resonance and a parallel type of resonance. The series resonance is approximately where the series elements L_s and C_s series resonate. The parallel resonance is where L_s resonates with the series combination of C_s and C_p. Depending on the material used to make the crystal, the Q's of the device can vary from the low hundreds to the high hundred thousands. The ratio of the parasitic capacity C_p to C_s determines the difference between the parallel resonance and series resonance of the circuit. These capacities are related to physical constants of the material.

12.24 YIG EQUIVALENT CIRCUIT

The YIG resonator is a circuit that looks like the dual of the crystal from a circuit topology standpoint. The YIG resonator is a sphere of magnetic material made from *yttrium iron garnet*. The magnetic resonance of this device is tunable with an external dc magnetic field. The equivalent circuit of the device is shown in Figure 12.54. The YIG resonator circuit also has a series and parallel resonance. The value of the resonant frequency of the sphere is $\omega_o = 2\pi \gamma H_o$, where ω_o is the resonant radian frequency of the sphere, H_o is the value of the dc magnetic field impressed on the sphere, and γ is 2.8MHz/G. The Q of the sphere has a high value only above a frequency given by $2\gamma 4\pi Ms/3$. L_p is directly proportional to the volume of the sphere and the saturation magnetism of the sphere and inversely proportional to the resonant frequency of the sphere. The YIG sphere is used in tuned filters and oscillators in the microwave frequency range. The resonant frequency is tuned by varying a dc magnetic field on the YIG material.

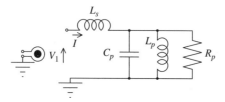

Figure 12.54 YIG equivalent circuit.

12.25 GYRATOR

A circuit that is often encountered in filter theory or device theory is the gyrator circuit. It is shown schematically in Figure 12.55. The gyrator has the following immittance and scattering matrices:

$$(Y) = \begin{pmatrix} 0 & G \\ -G & 0 \end{pmatrix} \quad (Z) = \begin{pmatrix} 0 & -R \\ R & 0 \end{pmatrix} \quad (S) = \begin{pmatrix} 0 & -1 \\ 1 & 0 \end{pmatrix}$$

where the S matrix is normalized to R, the value used in the impedance matrix definition. The direction of the arrow is important since it determines the sign of the off diagonal components of the above matrices. Using the S matrix, the input reflection coefficient is:

$$\Gamma_{in} = \frac{S_{11} - \Delta_S \Gamma_{load}}{1 - S_{22}\Gamma_{load}} = \frac{0 - (+1)\Gamma_{load}}{1 - 0} = -\Gamma_{load}$$

since $\Delta_S = 1$. Therefore, a gyrator circuit exhibits impedance inversion properties. If the load on the gyrator circuit is a capacitor, the input terminals look like an inductor and vice-versa. This circuit is often used to synthesize inductors from capacitors in active filter design.

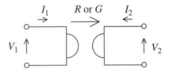

Figure 12.55 Gyrator equivalent circuit.

12.26 HYBRID COUPLED AMPLIFIERS

Figure 12.56 contains a drawing of a hybrid coupled amplifier. The input power divider and the output power combiner are assumed to be identical. This analysis assumes that the couplers are matched and split the power equally to each output port, and that there is a 90-degree difference in the phase between each output port. The a and b vectors will be described with respect to the divider and combiner rather than with respect to the amplifiers A1 and A2. The input power divider has the following S matrix.

$$(S) = \begin{pmatrix} 0 & \dfrac{\sqrt{2}}{2} & -j\dfrac{\sqrt{2}}{2} \\ \dfrac{\sqrt{2}}{2} & 0 & 0 \\ -j\dfrac{\sqrt{2}}{2} & 0 & 0 \end{pmatrix}$$

Ports two and three are loaded with the input impedances of the amplifiers A1 and A2. The S matrices for the top and bottom amplifiers are:

$$(S_t) = \begin{pmatrix} S_{11t} & S_{12t} \\ S_{21t} & S_{22t} \end{pmatrix} \quad (S_b) = \begin{pmatrix} S_{11b} & S_{12b} \\ S_{21b} & S_{22b} \end{pmatrix}$$

Then for the top amplifier A1:

$$a_2 = S_{11t} \frac{\sqrt{2}}{2} a_1 + S_{12t} b_4$$

$$a_4 = S_{21t} \frac{\sqrt{2}}{2} a_1 + S_{22t} b_4$$

For the bottom amplifier A2:

$$a_3 = -j S_{11b} \frac{\sqrt{2}}{2} a_1 + S_{12b} b_5$$

$$a_5 = -j S_{21b} \frac{\sqrt{2}}{2} a_1 + S_{22b} b_5$$

The output of the coupled amplifier is:

$$b_6 = -j \frac{\sqrt{2}}{2} a_4 + \frac{\sqrt{2}}{2} a_5$$

$$a_6 = \Gamma_L b_6$$

Note that:

$$b_4 = -j \frac{\sqrt{2}}{2} \Gamma_L b_6$$

$$b_5 = \frac{\sqrt{2}}{2} \Gamma_L b_6$$

$$b_5 = j b_4$$

and b_4 and b_5 are:

$$b_4 = -j \frac{\sqrt{2}}{2} \Gamma_L b_6 = -j \frac{\sqrt{2}}{2} \Gamma_L \left(-j \frac{\sqrt{2}}{2} a_4 + \frac{\sqrt{2}}{2} a_5 \right)$$

$$= -j \frac{\sqrt{2}}{2} \Gamma_L \left[-j \frac{\sqrt{2}}{2} \left(S_{21t} \frac{\sqrt{2}}{2} a_1 + S_{22t} b_4 \right) + \frac{\sqrt{2}}{2} \left(-j S_{21b} \frac{\sqrt{2}}{2} a_1 + S_{22b} b_5 \right) \right]$$

$$b_5 = \frac{\sqrt{2}}{2} \Gamma_L b_6 = \frac{\sqrt{2}}{2} \Gamma_L \left(-j \frac{\sqrt{2}}{2} a_4 + \frac{\sqrt{2}}{2} a_5 \right)$$

$$= \frac{\sqrt{2}}{2} \Gamma_L \left[-j \frac{\sqrt{2}}{2} \left(S_{21t} \frac{\sqrt{2}}{2} a_1 + S_{22t} b_4 \right) + \frac{\sqrt{2}}{2} \left(-j S_{21b} \frac{\sqrt{2}}{2} a_1 + S_{22b} b_5 \right) \right]$$

$$b_4 = \frac{-\frac{\sqrt{2}}{2} \Gamma_L (S_{21t} + S_{21b})}{2 + S_{22t} - S_{22b}} a_1$$

$$b_5 = \frac{-j \frac{\sqrt{2}}{2} \Gamma_L (S_{21t} + S_{21b})}{2 + S_{22t} - S_{22b}} a_1$$

These can be substituted back into the a_4 and a_5 formulas above, allowing one to calculate the effect of amplifier imbalance on the network. However, the formulas are cleaner if the two

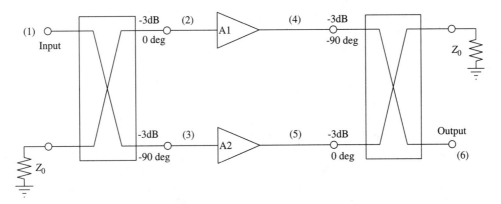

Figure 12.56 Hybrid coupled amplifier.

amplifiers are considered equal. Then,

$$b_4 = -\frac{\sqrt{2}}{2}\Gamma_L S_{21} a_1$$

$$b_5 = -j\frac{\sqrt{2}}{2}\Gamma_L S_{21} a_1$$

$$a_4 = \frac{\sqrt{2}}{2} S_{21} a_1 - S_{22}\frac{\sqrt{2}}{2}\Gamma_L S_{21} a_1 = \frac{\sqrt{2}}{2} S_{21} a_1 (1 - \Gamma_L S_{22})$$

$$a_5 = -j\frac{\sqrt{2}}{2} S_{21} a_1 - j S_{22}\frac{\sqrt{2}}{2}\Gamma_L S_{21} a_1 = -j\frac{\sqrt{2}}{2} S_{21} a_1 (1 + \Gamma_L S_{22})$$

This gives the load reflection coefficients for the amplifiers as:

$$\Gamma_4 = \frac{b_4}{a_4} = \frac{-\Gamma_L}{(1 - S_{22}\Gamma_L)}$$

$$\Gamma_5 = \frac{b_5}{a_5} = \frac{+\Gamma_L}{(1 + S_{22}\Gamma_L)}$$

The input reflection coefficient into the input power divider is:

$$\Gamma_{in} = \frac{b_1}{a_1} = \frac{\sqrt{2}}{2}\frac{a_2 - j a_3}{a_1}$$

$$= \frac{\sqrt{2}}{2}\frac{S_{11t}\frac{\sqrt{2}}{2} a_1 + S_{12t}\dfrac{-\dfrac{\sqrt{2}}{2}\Gamma_L(S_{21t} + S_{21b})}{2 + S_{22t} - S_{22b}} a_1}{a_1}$$

$$-j\frac{\sqrt{2}}{2}\frac{-j S_{11b}\frac{\sqrt{2}}{2} a_1 + S_{12b}\dfrac{-j\dfrac{\sqrt{2}}{2}\Gamma_L(S_{21t} + S_{21b})}{2 + S_{22t} - S_{22b}} a_1}{a_1}$$

$$= \frac{1}{2}\left[S_{11t} + S_{12t}\frac{-\Gamma_L(S_{21t} + S_{21b})}{2 + S_{22t} - S_{22b}} \right] + \frac{1}{2}\left[S_{11b} + S_{12b}\frac{\Gamma_L(S_{21t} + S_{21b})}{2 + S_{22t} - S_{22b}} \right]$$

If the top and bottom amplifiers are identical, then:

$$\Gamma_{\text{in}} = -S_{12}S_{21}\Gamma_L$$

By interchanging the subscripts, the output reflection coefficient of the amplifier is:

$$\Gamma_{\text{out}} = -S_{12}S_{21}\Gamma_S$$

If the part has good isolation, then the input and output reflection coefficients are very good even when the input match into each individual amplifier inside the hybrid power dividers/combiners is quite poor. Note that Γ_{in} is not zero, though, as one anticipates when using hybrid coupled input circuits. However, when one considers the load seen by an individual amplifier, the match conditions are not as good. The magnitude of the reflection coefficient, seen by the top amplifier, needs to be less than 1 in order to give that amplifier a positive real load. Imposing this condition gives the following result.

$$|\Gamma_4| = \left|\frac{b_4}{a_4}\right| = \frac{|\Gamma_L|}{|1 - S_{22}\Gamma_L|} < 1$$

Assuming that the magnitude of S_{22} is less than 1, the value inside the denominator absolute value has a magnitude less than 1. The denominator will be the smallest when the angle of $S_{22}\Gamma_L$ is zero. Under that condition:

$$|\Gamma_L| < |1 - S_{22}\Gamma_L|$$
$$|\Gamma_L|(1 + |S_{22}|) < 1$$
$$|\Gamma_L| < \frac{1}{(1 + |S_{22}|)}$$

When considering the lower amplifier, the result is the same except that the angle of $S_{22}\Gamma_L$ is 180 degrees for the maximum magnitude of the load reflection coefficient. This means that for an open-circuit (current source) amplifier, a high-impedance load on the output of the network can cause possible instability of the structure for the top part. Likewise, low impedances will do the same for the lower part. If the amplifier represents a voltage source output, then low impedances can cause possible instabilities on the upper part while high impedances would cause possible instabilities for the lower part. If the magnitude of S_{22} approaches unity, the magnitude of the load reflection coefficient must be less than 0.5 or less than a $3 : 1$ *VSWR* to prevent an amplifier from seeing a negative real immittance. Small variations of the load impedance also cause large efficiency changes for amplifiers, especially for those designed for power generation rather than for gain, since for each amplifier small output load variations cause large individual amplifier load variations.

12.27 LIMITER

When it is necessary to protect a power-sensitive circuit from the damaging effects of high power, a limiter can be used. Several types of limiters have been used. For high-power pulses, gaseous discharges have been used to provide a short across waveguides. The gaseous discharge is triggered by the high electric field from the incident high-power signal. There is a small amount of time necessary to initiate the discharge. The discharge is essentially an avalanche breakdown of the gases in the limiter tube. During the small amount of time necessary to initiate the breakdown, a spike of leakage comes through the limiter. This spike of leakage can be enough to damage sensitive circuits. Often a gaseous discharge limiter tube has a beta radiation source inside it to provide a source of electrons to keep a small amount of ions available in the tube to shorten the time required for the gas to reach avalanche breakdown. Once the discharge is initiated, it takes a small amount of time for the plasma to quench and allow low-level signals to again pass through the device. There is a then a latent period after interference from a high-power pulse that initiates the breakdown before low-level signals can again pass through the system.

If the damaging power is from a transmitter in the same module as the sensitive circuits, a PIN switch or electrical metal contact switch can be used to protect the circuit since the time frame necessary to protect the circuits is known. However, if the damaging power comes from an unknown source such as a radar set or high-power communications transmitter, then one does not know when to turn on the switch to protect a circuit. That is when a limiter is necessary.

Diode limiters can be used to protect sensitive circuits. Microwave spectrum analyzers often have a diode limiter across the input to protect the sensitive mixer diode that is right in the front end of the instrument. If the amount of power required by the circuit during normal use results in a voltage on a transmission line much smaller than the forward voltage of a Schottky diode, a parallel pair of Schottky diodes can be used to protect a sensitive circuit. Such a parallel pair of diodes is shown in Figure 12.57. The diodes have a path for their forward bias current through each other and there is very little parasitic capacity to limit their turn-on time. If instead of using two Schottky diodes in parallel, one of the diodes is a Schottky diode and the other diode is a PIN diode, the Schottky diode can act as a rectifier and the bias current can turn on the PIN diode. This type of limiter has a small latency time as well—the time necessary to flood the I region of the PIN diode with carriers. The PIN diode also requires a small amount of time after the disappearance of the interfering signal to return to an unbiased condition. The PIN diode is not back biased so that the carriers have to disappear by diffusion or recombination. The PIN diode needs to be chosen so that its I region is free of carriers at zero bias. The Schottky diode is sometime placed a small distance down a transmission line from the PIN diode toward the incoming power so that the voltage across it can be maintained when the PIN diode turns on and places a short across the line. In diode limiter design, power dissipation and maximum current ratings for the diodes need to be observed.

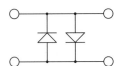

Figure 12.57 Parallel diode limiter.

EXERCISES

12-1 Design a duplexer (a circuit that routes or switches signals from one port to another port based on switching or nonlinear effects) to couple an antenna to both a transmitter and a receiver. The transmitter and the receiver transmit on the same frequencies. The transmitter is pulsed on and off and transmits 1 kW when it is on. The impedance of the transmitter is unknown when it is off. The transmitter should be protected from accidental disconnection of the antenna and bad antenna VSWR. The maximum VSWR the transmitter can have on it is 1.5 : 1. The receiver needs to be protected when the transmitter is on. A signal is available to tell the receiver when the transmitter is about to transmit. The maximum power the receiver can take without burn out is 100 mW. *Hint:* Consider using a three-port circulator, some quarter-wavelength stubs, and some PIN diodes either in series or in shunt.

12-2 Design a −20-dB directional coupler (1) using a transformer, (2) using coupled transmission lines, (3) using a transformer and capacitors, and (4) using just uncoupled inductors and capacitors.

12-3 A PIN diode is used as a single-pole single-throw (SPST) switch in shunt with a transmission line. It is bonded directly across a 50-ohm microstrip line. Calculate the switch insertion loss in dB at 5 GHz for each state of the switch—when the diode is forward biased with 0.020 A dc current and when it is reverse biased with enough bias to fully deplete the I layer. The diode specifications are a series resistance of 0.5 ohm at 0.020 A dc bias, a capacitance of the chip of 0.5 pF, and an inductance of the contacts of 0.2 nH for both.

12-4 A PIN diode is used as a switch for a 10-GHz signal. It is biased at 20 mA dc. The rf current is a sine wave. The peak value of the sine wave signal is 5 A. The PIN diode has a 0.5-microsecond lifetime. What is the percent of charge variation in the diode due to the rf signal?

12-5 A hybrid coupled amplifier is used to minimize the input VSWR for a module. The scattering parameters of each amplifier are: $S_{11} = 0.8$ at 0 deg., $S_{12} = 0.05$ −90 deg., $S_{21} = 4$ at −45 deg., and $S_{22} = 0.95$ at 180 deg.

Determine the input impedance into the hybrid coupled amplifier when the output of the total amplifier is loaded in a matched load, and when the output is loaded in unity reflection coefficient but the phase is varied from zero to 360 degrees. Determine whether either of the internal amplifiers will see a negative resistance load when the unity reflection coefficient is placed on the output of the amplifier.

13

Pulsed Microwave Circuit Analysis

13.1 COMBINED DIGITAL AND ANALOG CIRCUITS

The boundary line between digital and analog circuits is often blurred. When devices are used for digital circuit functions and when the clock rate approaches the current gain limit of the device, the waveform on the device output is not a good square wave. In fact, the waveform may never reach either the zero level or the 1 level within the clock period. When this happens, the device in a gate is performing more like an analog device. When a digital device has a slowly rising clock edge at its input, the device may be in the active region long enough to allow a high-frequency oscillation to build up. This might also happen with a slow clock edge and a heavily reactively loaded output on a gate. During the input period of the slow clock edge, the part may oscillate at a very high frequency. Analog circuit stability analysis can be performed on the device using circuit parameters valid in the active region to determine the load and source impedances that will allow the oscillation to build up. There are other applications of digital and analog devices for which linear analysis is meaningful and other applications where a full nonlinear analysis needs to be performed. Analysis of digital circuits that involves only logic analysis, time delays, and races does not address the analog type of oscillation behavior. The circuit designer should be aware of the possibility of analog effects in digital circuits. The effect of mismatch on physically large circuits should be investigated as well. Physically large circuits are those circuits that have the time of propagation from a driver circuit to a load circuit an appreciable portion of the clock width. For clocks in the gigahertz region, this means delays in the tens of picoseconds or 2500 microns in free space. On a large digital chip, this distance might be as low as 1000 microns due to the slowing effects of the dielectrics involved. For gigahertz clock rates, distances across a moderately sized chip can present delays long enough that the delay becomes an appreciable part of a clock cycle. Clock pulses by their very nature have a high amount of harmonic energy. This harmonic energy needs to travel at the same velocity as the fundamental energy to preserve clock pulse shapes. If gates and unit cells are not resistively terminated, there will be a difference in the amount of reflection at each harmonic. This contributes to signal distortion.

The response calculated by microwave frequency analysis programs is often not a transient response but a response due to continuous wave (cw) inputs. The network transfer function is in general irrational and very few circuits would have closed-form Laplace transforms to use for linear transient analysis. An approximate analysis of some of these circuits is possible when the distributed effect of the interconnection lines is not appreciable.

Some nonlinear microwave circuit analysis programs use a harmonic-balance approach [1, 2]. The active device response portions of the circuit are determined using a Spice type of analysis and the linear portions of the circuit are analyzed using frequency analysis. The total response of the circuit is then determined by finding undetermined coefficients for all the harmonic responses and forcing the voltages and currents at the active devices to match the response resulting from summing all the harmonic responses from the linear circuit analysis. Error minimization algorithms vary the harmonic coefficients to find a solution.

The analysis that is discussed in this chapter uses only linear state analysis for the devices (i.e., each gate is assumed to have only one impedance state over the time period of the analysis but this state can be a function of frequency). Harmonics are summed using the correct phase shift and amplitude response. Linear network analysis programs do not generate the correct phase shift output. A correction needs to be applied to the analysis. The number of harmonics needed for the analysis must be determined.

13.2 PULSE ANALYSIS USING FOURIER SERIES

In digital circuits, the harmonics of clock pulses for clocks running in or above the hundreds of MHz range are well into the microwave range. Clock rates currently used on integrated circuits are in the microwave range.

When distributed circuits are used to transport these high-speed pulse signals from one point of a circuit to another point, it is necessary to determine whether mismatches generate multiple pulses and whether the received and interfering pulses are above digital switching level thresholds. The presence of multiple reflected pulses can in effect raise the noise level present at the input to a gate or other logic element. The analysis in this section assumes that the device driver and load can be considered to be linear over the time associated with the analysis. Although the gate impedance does change versus its logical state, a linear analysis can give considerable insight into the network behavior without using a computationally intensive full nonlinear analysis. Consider the example shown in Figure 13.1. The source is at point "A," and the signal travels to point "B" and some of the signal is reflected from that point. If there are two similar lines at that point, the load impedance seen by the first line is equal to one-half of the characteristic impedance of the lines. Therefore, the reflection coefficient at that point in time is:

$$\Gamma = \frac{\dfrac{Z_0}{2} - Z_0}{\dfrac{Z_0}{2} + Z_0} = -\frac{1}{3}$$

$$V_{\text{refl}} = -\frac{1}{3} V_{\text{for}}$$

$$V_{\text{refl}} = V_{\text{for}} + V_{\text{refl}} = \frac{2}{3} V_{\text{for}}$$

This indicates that the pulse going to each of the gate inputs "C" and "D" is $\frac{2}{3}$ of the initial voltage. If the pulse width is less than the propagation time between the source and the load, the voltage will not rise higher than this value. Unless the circuit is symmetrical about the point of intersection of the lines, each of the pulses will generally reach the gate inputs at different times. There will be a reflection of each of the pulses if the gate inputs are not impedance matched. These reflected waves will come back to point "B" and re-reflect and some wave energy will be transmitted past point "B" in both directions. If the reflection coefficient at the gate inputs and source outputs is large, the possibility of multiple triggering

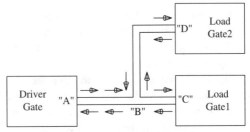

Simple Two-Gate Load Example

Figure 13.1 Digital signal on transmission lines.

of inputs "C" and "D" exists. What do the signals look like in time? A multiple *bang-bang* diagram could be drawn. However, the gate inputs are usually capacitive and the reflected pulses do not have square edges, making it difficult to draw meaningful bang-bang diagrams. There is another way to do this problem using microwave circuit analysis techniques.

The output pulse at point "A" will be considered to be a trapezoidal pulse of amplitude A as shown in Figure 13.2. This pulse repeats every T seconds. If one desires to see the effect of a single pulse on the system, the period T needs to be chosen long enough to allow any multiple reflections to die out. Occasionally it is necessary to perform the calculation once as a test to determine how long it takes for the pulse energies to die out and then choose a final value for T to perform another calculation. The trapezoidal waveform has a cosine Fourier series representation as follows.

$$f(t) = a_o + \sum_{n=1}^{n=\infty} a_n \cos(\omega_n t)$$

$$\frac{a_o}{A} = \frac{t_w + t_r}{T} \qquad a_n = 2a_o \frac{\sin\left(n\pi \dfrac{a_o}{A}\right)}{n\pi \dfrac{a_o}{A}} \frac{\sin\left(n\pi \dfrac{t_r}{T}\right)}{n\pi \dfrac{t_r}{T}}$$

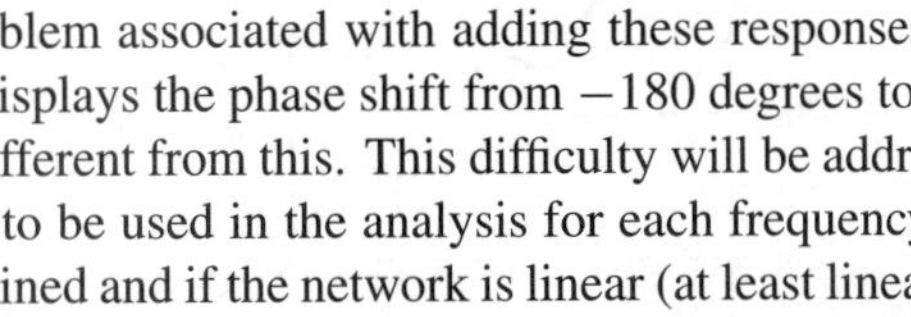

Figure 13.2 Trapezoidal pulse used for circuit analysis.

For each frequency component of the Fourier series, the transfer function of the network is calculated. The responses from each of the terms of the Fourier series are then calculated and added together. There is a problem associated with adding these responses together. A typical network analysis program displays the phase shift from -180 degrees to 180 degrees. In reality the actual phase shift is different from this. This difficulty will be addressed shortly. The correct total phase shift needs to be used in the analysis for each frequency component. When the correct phase shift is obtained and if the network is linear (at least linear in the sense of some synchronized multistate network; i.e., the input and output impedance of each gate is relatively constant over the total time of the analysis even though the voltage state of the gate may be changing), then the response $r(t)$ of the network to the pulse train is:

$$r(t) = b_o + \sum_{n=1}^{\infty} b_n \cos(\omega_n t + \phi_n)$$

where b_n and ϕ_n are the voltage and phase shift for the nth component of the result due to the nth component of the input $a_n \cos(\omega_n t)$ respectively. The quantity b_o is the response of the network to the a_o input.

The phase shift is determined by considering the response of a network to a cw excitation. A single sinusoidal signal into a network may look like that shown in Figure 13.3. The same amplitude point on this waveform is shown at two different times. The bottom waveform needs to be properly referenced to the top waveform. Because of the periodicity of trigonometric functions, the formulas used to determine the phase shift would only give θ as the answer. That is not the correct phase shift for time delay and harmonic analysis. Let the x on the input curve represent a certain "event." The output "event" will be delayed in time by an amount t_d. As shown in Figure 13.3, this amount of delay is:

$$t_d = \frac{2\pi K + \theta}{2\pi} T = \frac{\theta_T}{2\pi} T = \frac{\theta_T}{2\pi f}$$

where f is the frequency of the sinusoidal wave and θ_T is the negative of the calculated phase delay. A frequency-based circuit analysis program gives phase shift only within plus or minus 180 degrees or it might give differential time delay.

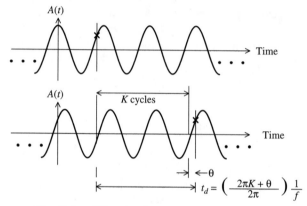

Time Delay of More than Several Cycles

Figure 13.3 Phase shift between two sine waves.

How can the true phase delay through the network be calculated? If the network has some response from dc through to the frequency under consideration, a plot of θ might look like that shown in Figure 13.4. Note that the top curve cannot go positive. A network must

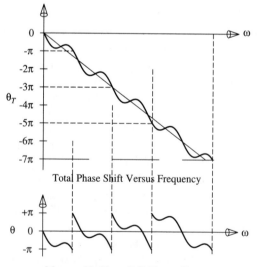

Measureable Phase Shift Versus Frequency

Figure 13.4 Total phase shift of a circuit.

have some time delay. If the total phase shift were positive, the absolute time delay through the network would be less than zero. The bottom curve shown in Figure 13.4 is what a network analyzer would display for phase shift. The phase angle would be between $+180$ degrees and -180 degrees. If all the phase cuts are accounted for from dc to the frequency in question, then the total phase shift of the network can be determined. The time delay is then:

$$t(\omega) = -\frac{\theta_T(\omega)}{\omega}$$

If the network has a high-pass response, it is not possible to measure the phase shift at dc since the response would be zero. In addition, if the phase shift is not known below some frequency, the total phase shift might not be measurable. For this reason, a quantity called the differential time delay has been defined. Differential time delay is:

$$\tau(\omega) = -\frac{d\theta_T(\omega)}{d\omega}$$

If the network's phase shift can be measured dc to the frequency under consideration, then differential time delay can be integrated to get the total time delay.

$$\int_0^\omega \tau(\omega)d\omega = -\int_0^\omega \frac{d\theta_T(\omega)}{d\omega}\,d\omega = -\int_0^\omega d\theta = -\theta_T(\omega) = \omega t_d$$

$$t_d = \frac{1}{\omega}\int_0^\omega \tau(\omega)\,d\omega$$

Notice that differential time delay τ can be either positive or negative. Differential time delay gives relative time delay through a network for one frequency compared to that of another frequency. Even if the absolute time delay cannot be easily calculated, differential time delay gives the relative time delay of one frequency compared to the delay at another frequency close to the original frequency.

In order to use pulse Fourier analysis to calculate the time response through microwave networks, the period of the test pulse train must be longer than the time delay through the network. Consider Figure 13.5. If the time delay through the network is B, the use of a period shorter than B would make the network appear to have a time delay equal to A using pulse Fourier analysis.

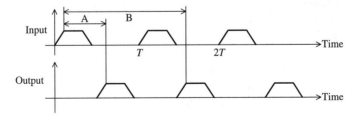

Figure 13.5 Time delay considerations.

Assuming T is chosen large enough, how many harmonics of the period waveform need to be chosen in order to have a faithful reproduction of the signal? If the trapezoidal waveform had an approximately 50% duty factor, one would generally use ten harmonics to approximate the waveform. In order to preserve the same pulse fidelity that is obtained at 50% duty factor, the Fourier series would have to contain the same high-frequency components that are in the waveform for a 50% duty factor. The period of a 50% duty factor waveform would be two times $(t_r + t_w)$. Therefore the number of harmonics needed whenever the duty factor is less than 50% is approximately:

$$N = 10\left[\frac{T}{2(t_r + t_w)}\right]$$

in order to get a faithful reproduction of a waveform that has a duty factor less than 50%. The factor 10 in the equation is from ten harmonics of a square wave. If it is determined that a different number than ten harmonics would meet the required fidelity, that number could be used in the formula. Rise time and pulse overshoot will be different for a smaller number of harmonics and consideration should be given to the number of harmonics necessary to get a given rise time.

The time delay through a circuit is affected by the longest time constant in the circuit. Roughly a period T needs to be chosen five to ten times the longest time constant existing in the circuit. If such a long time constant gives too large a number of harmonics, then if it can be determined that the currents in inductors and voltages across capacitors do not change substantially during the time T used in a calculation, a shorter time T can be used. Short time, but large signal analysis was used previously when PIN diodes were discussed.

When the desired number of harmonics has been chosen, the response is calculated at each of the frequencies in the Fourier series. Outputs at all ports of interest resulting from each of the separate input frequencies are calculated. For each output, all of the frequency components are summed and the result is then the time response of the circuit. The input Fourier series is:

$$\text{Input}(t) = a_o + \sum_{n=1}^{n=N} a_n \cos(n\omega_o t) = a_o + \sum_{n=1}^{n=N} a_n \cos\left(n\frac{2\pi}{T}t\right)$$

The output series is:

$$\text{Output}(t) = b_o + \sum_{n=1}^{n=N} b_n \cos(n\omega_o t + \phi_n)$$

where b_o and b_n are the dc amplitude and harmonic amplitude responses respectively and ϕ_n is the total phase delay at each harmonic.

If instead of a pulse signal, the input is a pulse-amplitude-modulated signal (pulsed rf), the output can also be calculated but the series has to include both positive and negative harmonic terms. When the input is a pulse-modulated carrier, the desired output will also be a pulse-modulated carrier. However, what is often desired is the shape of the output modulation as compared to the shape of the input modulation. If the carrier frequency is ω_c, then the input waveform is:

$$\text{Input}(t) = \left[a_o + \sum_{n=1}^{n=N} a_n \cos(n\omega_o t)\right]\cos(\omega_c t)$$

$$= a_o \cos(\omega_c t) + \sum_{n=1}^{n=N} \frac{a_n}{2}[\cos(\omega_c t + n\omega_o t) + \cos(\omega_c t - n\omega_o t)]$$

where the term in the square brackets is the trapezoidal term and ω_c is the carrier frequency. Notice that this input waveform has frequencies at:

$$\omega = \omega_c t \pm n\omega_o t; \quad n = 0, 1, 2, \ldots, N$$

The output can be shown to be:

$$\text{Output}(t) = \sum_{n=-N}^{n=N} b_n \cos(\omega_c t + n\omega_o t + \phi_n)$$

where b_o is the response to an input $a_o \cos(\omega_c t)$, ϕ_o is total the phase response at ω_c, b_n is the amplitude response to $a_o \cos(\omega_c t + n\omega_o t)$, and ϕ_n is the total phase response at $\omega_c t + n\omega_o t$. The output waveform can be rewritten as:

$$\text{Output}(t)$$

$$= \sum_{n=-N}^{n=N} b_n \cos(\omega_c t + n\omega_o t + \phi_n)$$

$$= \sum_{n=-N}^{n=N} b_n \cos(n\omega_o t + \phi_n) \cos(\omega_c t)$$

$$- \sum_{n=-N}^{n=N} b_n \sin(n\omega_o t + \phi_n) \sin(\omega_c t)$$

$$= \left\{ \sum_{n=-N}^{n=N} b_n \cos(n\omega_o t + \phi_n)] \right\} \cos(\omega_c t)$$

$$- \left\{ \sum_{n=-N}^{n=N} b_n[\sin(n\omega_o t + \phi_n)] \right\} \sin(\omega_c t)$$

The magnitude of the envelope $\text{Env}(t)$ of this waveform is:

$$\text{Env}(t) = \sqrt{ \left\{ \sum_{n=-N}^{n=N} b_n[\cos(n\omega_o t + \phi_n)] \right\}^2 - \left\{ \sum_{n=-N}^{n=N} b_n[\sin(n\omega_o t + \phi_n)] \right\}^2 }$$

The delay of the envelope relative to the envelope of the input signal is often called the *envelope delay* of the signal.

EXERCISES

13-1 A one-half-nanosecond-wide pulsed voltage source has a 100-ohm impedance. The two-volt source is used to supply voltage to three gates. Each gate has an input impedance of 100 fF. The transmission line from the source has a 150-ohm characteristic impedance. The effective dielectric constant of the transmission line is 2.5. At a distance of 1000 microns from the source, the line splits into three 150-ohm transmission lines. The first of these lines is 2500 microns long and is terminated in gate number one. The second of these lines is progressed around the circuit and is 7500 microns long when it is terminated in gate number two. The third line is also 2500 microns long but it is terminated in a 100-micron-long, 1-ohm transmission line (a bond pad). A gate with 100 fF in parallel with a 100-ohm termination is placed at the end of the 1-ohm transmission line. Determine the pulse voltages at each of the three gates as a function of time. If 1.7 volts is needed for a "1" and 0.2 volts is needed for a "0", is there a possibility of faulty logic states?

13-2 Design a three-pole 50-ohm filter with $g_0 = 1$, $g_1 = 1$, $g_2 = 2$, $g_3 = 1$, and $g_4 = 1$ with a 10% bandwidth at 1 GHz. Terminate the filter with 100 ohms (2:1 VSWR) instead of 50 ohms. Use a 25-ohm source also. Use a trapezoidal pulsed carrier at 1 GHz. Let the pulse be 10 ns wide with a period of 50 ns. Calculate the voltage at the input and the output of the filter using envelope analysis. Determine whether you see some of the envelope reflected coming back to the input being reflected from the input and reappearing later at the output. That is termed a triple travel response. One of its causes is mismatched sources and loads.

14

Nonlinear Effects
in Microwave Circuits

14.1 NONLINEAR EFFECTS

This chapter contains some brief discussions of nonlinear effects and the oscillations that they can cause. A full-scale analysis is beyond the scope of this book but the reader is referred to other works that treat the nonlinear analysis in greater detail [1,2,3].

Oscillations due to linear effects in the even mode and the odd mode were discussed in Chapter 5. The reader should review the conditions for stability under those modes of operation. There are three additional types of oscillation that can be encountered in rf and microwave circuits. These are (1) subharmonic and other parametric oscillations, (2) motorboating types of oscillations, and (3) oscillations that require a mixing operation to occur.

14.2 SUBHARMONIC AND PARAMETRIC OSCILLATIONS

Most nonlinear reactive circuits can put out frequencies that are below the frequency of the signal causing the nonlinearity. This is easy to see from an everyday example. Consider a person on a swing as shown in Figure 14.1. This is an example of behavior that is undesirable and should not be attempted. The person standing on the left is pumping the swing. It takes several individual pushes to impart enough energy to the swing to make the system very nonlinear. When enough energy is imparted to the swing as shown in the figure on the right, the swing system becomes quite nonlinear. There is no force left in the rope holding the swing to the stationery structure. The swing and the occupant on the swing will free fall until the rope holding the swing is again pulled to its full length. At that time, the swing and occupant will be jerked into a motion consistent with a constant radius from the fulcrum. The energy of free fall will be largely lost. Energy is being taken out of the system at a subharmonic of the pump frequency. The frequency at which the energy is being taken out does not have to be a direct subharmonic since the free fall energy given up might not allow the swing to come back into position at precisely the same position each time the system becomes nonlinear. The rate at which the energy is taken from the system is not necessarily commensurate with the pump frequency. Similar situations occur in circuits. A resonate circuit may be gaining energy from a pump source. When enough energy is absorbed by the resonate circuit and voltage and current levels build up, some nonlinearity may be reached. A junction may break down

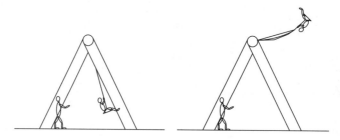

Figure 14.1 Nonlinear oscillation system.

or, as in the case of a transistor circuit, the piecewise linear region may come to its limit and transistor action may cease.

Devices that contain nonlinear reactive components are used in some multiplier and divider circuits. The same devices can generate spurious signals—signals that appear for the same reasons but are not wanted. Multiplier circuits using nonlinear reactive components that have a small amount of nonlinearity often use idler circuits to enhance the power conversion of the multiplier circuit. The circuit designer who is building a power circuit needs to be careful that idler circuits are not inadvertently introduced in the circuit. It is easier than one might think to introduce these unwanted idler circuits. The circuit does not know what frequency it is to amplify. If a current exists at some frequency and gain exists for that current, the current will be amplified. Noise voltage exists at all frequencies. That voltage will drive current if an idler circuit exists. If gain also exists, that current will be amplified.

The circuits that enhance low-frequency nonlinear oscillations are often the bias circuits or impedance-matching circuits. Consider the transistor circuit shown in Figure 14.2. If the bias inductor is much larger in value than the value of L_2, then it will not have a great effect on the resonance of L_2 and C_2. The resonance of L_2 and C_2 is not as important as the parallel resonance of L_2, C_2, and the output circuit reactance. Remember that a parallel circuit is internally series resonant. Idlers are circuits that idle current. Parallel circuits can idle current. A parallel resonance is detected by looking at the node containing the collector terminal and determining where the impedance peaks. It will be in the vicinity of the series resonance of L_2 and C_2. It is usually just higher than that frequency where the bias circuit has a net positive reactance to resonate with the net capacitive reactance of the transistor. Notice that when the coupling capacitor is small, the load resistor does not damp low-frequency resonances very much. When parallel resonance is reached, the collector circuit is loaded with an idler. The idler frequency will almost always be generated when the device goes nonlinear. The way to

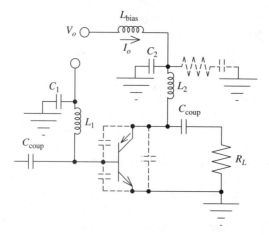

Figure 14.2 Amplifier bias idlers.

dampen that frequency is to place a series R-C at the junction of L_2 and C_2 as shown by the components shown dashed at that junction. When the collector node is parallel resonant, that node is also parallel resonance and thus at a high voltage. A large capacitor and fairly small resistor will dampen that resonance by absorbing energy at the idling frequency. The same test needs to be performed on the input side of the circuit. If there are no parallel resonances at frequencies below the drive frequencies, then the likelihood that a parametric subharmonic will be generated is diminished. If one puts a network analyzer on the input and output of the amplifier device, it is readily apparent when there is a parallel resonance and at what frequency. When there is a parallel resonance, the device is seeing a series resonance and thus an idler.

The same considerations need to be made on the bias elements that are put on a diode multiplier circuit to establish the bias to the diode. The multiplier circuit will likely have idlers introduced in it on purpose at the second or third harmonic of the drive or pump frequency. However, other idler circuits, especially those at a lower frequency than the pump frequency, will almost always produce an oscillation.

A particularly bad condition exists when an idler exists between the drive frequency and one-half the drive frequency and one also exists at a very low frequencies. Mixing then takes place between the very low frequency and the other idler frequency and a whole ensemble of frequencies is put out related to all possible mixing products of the idler frequency, the low-frequency idler frequency, and the pump frequency. This is some times termed a *picket fence spectrum*. The first thing to look for is the very-low-frequency idler circuit. This is often the result of a large inductor running back to some very large bypass or energy storage capacitor and then connected to the device matching circuits. That idler circuit needs to be dampened or large currents and voltage will build up destroying either devices or components. A ferrite bead is often effective in introducing a series resistance in series with the bias line and dampening the resonance.

14.3 MOTORBOATING OSCILLATIONS

Motorboating oscillation sounds like a motorboat when it occurs in audio amplifiers. The type of oscillation referred to is a result not of feedback from the output to the input via the signal path but from the output stage back to the input stage via the bias lines and other paths. This type of oscillation occurs when the devices are driven close to or into saturation. Power amplifiers are purposely biased for operation close to or into saturation. This type of oscillation occurs frequently in power amplifiers. Consider the circuit shown in Figure 14.3. The amplifiers have their own biasing components inside the triangle labeled AMP. The outer components are the components used to connect these amplifiers to the power supply. The circuits will be considered to be Class C. The current for these amplifiers then depends on the input rf drive. Current I_1 to the first-order approximation is a linear function of the input drive. The output power in the interstage is a function of the product of the amplifier voltage on C_1

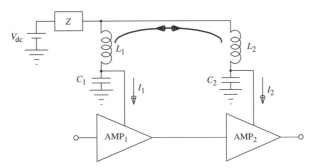

Figure 14.3 Circuit used to consider motorboating oscillation.

and the current I_1. The current I_2 drawn in the second stage is a function of the interstage power. If the impedance shown as Z has a value of zero ohms, there is no feedback. However, that impedance is usually fairly large since it consists of the total impedance clear back to the power supply unless a bypass capacitor is placed close to Z. When Z is not zero, the voltage at the junction of L_1 and L_2 is pulled down when the current is drawn from the supply. Consider the sequence of events leading to a motorboating oscillation. When the input current I_1 is excited when power is applied to the amplifier input, the voltage at the junction between the bias inductors drops a little. However, a very short time later, the current in the second amplifier stage increases and the voltage at the junction of the two inductors drops further. A short time later, determined by the bias components L_1 and C_1, the voltage on the first stage drops and the power out of the first stage drops, resulting in a decrease in current in the second stage, and so forth. The voltage then rises to the first stage and the motorboating cycle is repeated. This results in amplitude modulation on the output of the amplifier, but when the oscillation builds up to a high level the variations can also burn out the amplifiers. A similar fix to that discussed for the idler circuit in the previous section is necessary. Place a series R-C across C_1 or C_2 or put a bead in place of L_2.

Notice that when L_1 and L_2 are identical and also C_1 and C_2 are the same, an odd-mode resonance can exist. The components are not loaded when there is no rf drive into the amplifiers. If the amplifier is excited with a pulse of rf, the current comes on and some while later turns off. The current being drawn by the inductors continues to flow even when the amplifiers are not drawing any current. The energy is given to the capacitors. When the inductors and capacitors are identical, the bias node between these and the impedance Z is a virtual ground and the circuit consisting of Z is not coupled to the energy existing in L_1, L_2, C_1, and C_2. The circuit will ring with a damped oscillation due to losses in these components. However, if the bias supply voltage is ringing high on the input stage when the rf pulse comes back on, the input stage could easily be burned out. In order to prevent this type of oscillation, the values of the components should not be identical. When they are different, then Z loads the circuit. The same solution as already suggested will limit this type of oscillation as well—a bead or a series R-C in parallel with one of the bypass capacitors.

Even if the circuit does not go into an oscillation, an underdamped circuit ringing will cause amplitude modulation on an rf pulse string. The amplifier string does not have to be Class C for these oscillations to exist. When a Class A amplifier is overdriven, the output stage will go nonlinear. If the power supply node is pulled down in voltage, the input stage will put out a smaller amount of power if it too is overdriven. The situation then exists for a motorboating oscillation. That oscillation can burn out one of the stages if the bypassing is not correctly done.

14.4 MIXING OSCILLATIONS

Consider the circuit shown in Figure 14.4. Assume that this is a typical down-converter circuit. The output frequency is a lower frequency than the input frequency. This circuit is similar to the circuit discussed in the previous section except that a mixer exists in the interstage. The circuit may behave normally except when overdriven. When the circuit is overdriven, the output amplifier, although it is operating at a lower frequency, will produce harmonics and then the input current to the output stage will depend on the level of overdrive. If the isolation to the power supply leads is not very high, then some of the harmonic energy will get onto the power supply leads. The overdriven amplifier will likely put out a lot of odd harmonic energy. If the passband of the filter shown between the mixer and the output amplifier is wide enough or if the output frequency of the mixer is a subharmonic of the input frequency, then a loop oscillation may exist. One may not have purposely made the output of the mixer at a

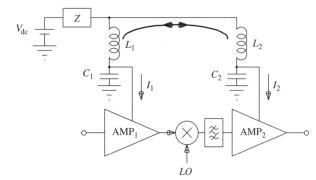

Figure 14.4 Circuit to discuss mixing oscillations.

subharmonic of the input signal. If the filter bandwidth is large enough, there may be some signal output that is a subharmonic of the broadband input passband. Then oscillation may occur. It consists of an oscillation through the output amplifier, the harmonic generation in the output amplifier, leakage in the power supply leads, back into the input amplifier via leakage in the power supply bias leads, amplification in the input stages, back into the mixer, and down converted into the output amplifier, and so on. The isolation necessary to prevent this oscillation can be quite demanding. The total loop gain of this type of circuit can easily be in excess of 60 dB, for instance, when the first stage has 30 dB of gain, and the mixer has 10 dB of loss, and the final amplifier has 50 or 60 dB of gain. The conversion efficiency to harmonic energy may be on the order of 15 dB loss. This gives a total loop isolation necessary through the power supply leads on the order of 60 dB. Often R-C filtering is necessary in the bias supply to prevent parallel and series resonances in the wiring and to reduce the feedback to the very low levels necessary.

Other types of mixing oscillations can also occur. When a circuit is sampled or switched at some given rate, the resulting down-converted signal may be at a subharmonic of the sampled signal. If the down-converted signal subsequently causes some stage to go nonlinear, the harmonics generated may be fed back to the stage that has the high-frequency signal on it. Bias supply isolation and total input-to-output isolation needs to be considered when using digital signals with analog circuitry. This is especially of concern with the present increasing interest in mixed-mode CMOS MMIC design. Digital circuits may inadvertently cause this type of oscillation. The dc power supply on a VLSI chip may shift slightly. This shift might be fed to the supply that sets the level for an analog to digital (A-D) converter. The bit pattern out of the A-D converter might then cause the power supply to shift at a rate that causes the whole system to go into a sampled data feedback oscillation.

14.5 COUPLING CAPACITORS AND BYPASS CAPACITORS

The bypass capacitor and coupling capacitor circuitry is an important part of microwave engineering. Capacitors for bypassing are often series resonant at the frequency of operation. In broadband circuits, the bypass capacitor and bias inductors for an amplifier stage need to be very broadband. Resonances and loading effects of the varying reactances need to be considered. In regard to coupling capacitors, more is not necessarily better. For most transistors, gain drops off considerably with frequency. When very large coupling capacitors are used, the frequency response of the network goes down to very low frequencies so that very large gains may be appear at frequencies where there is a resonance between the large coupling capacitor and some matching or bias inductor. Large gains require high isolation in the bias networks. An engineering tradeoff analysis needs to be done in selecting the value of the coupling capacitor. Consideration for the type of capacitor is also necessary. Parallel resonances at a harmonic of the desired frequency can lead to component failure if the amplifier is overdriven and creates harmonics. The harmonic energy may be absorbed in a coupling capacitor resonance and it is possible for the component to be destroyed in minutes or seconds if it absorbs several watts of power from harmonic generation in a power amplifier.

It is often not possible to include a sufficient amount of bypass capacity on a circuit, especially a MMIC circuit. The amount of energy stored in a bypass capacitor should be at least the amount necessary to support at least one cycle of the lowest frequency under consideration. The impedance back to the next source of energy needs to be low to prevent the circuit from being starved of energy. This limits the amount of isolation that can be put between an amplifier or MMIC circuit and a power supply. Other circuits connecting to the power supply may couple to the circuit under consideration. It is quite difficult to achieve a high degree of isolation on a MMIC chip. For that reason, the amount of gain that can be achieved in any circuit function on a MMIC is also limited. Power supply isolation needs to be as high as or higher than the gain of the circuit.

EXERCISES

14-1 In Figure 14.2, let $L_2 = 30$ nH, $C_2 = 30$ pF, $C_{\text{coup}} = 1$ pF, $R_L = 50$, $L_1 = 30$ nH, $C_1 = 10$ uF. Let the output capacity of the transistor be 1 pF. Calculate the frequencies for which one might expect to see nonlinear effects in the collector circuit when the circuit is overdriven. Describe how to fix this problem. Describe what could be changed in the base circuit. Consider that in the vicinity of where L1 and C1 series resonate, the part becomes a common base part with some base inductance. This might indicate oscillation difficulties.

14-2 How would one test a circuit similar to Figure 14.2 on a network analyzer to determine whether an idler circuit exists?

14-3 The circuit shown in Figure 14.3 uses bias bypass capacitors of 1 uF and bias line inductors of 100 nH. What is the natural resonant frequency of that circuit when the currents in the amplifiers, I_1 and I_2, suddenly go to zero? Notice that I_2 should be considerably bigger than I_1 when the circuit is operating.

14-4 In Figure 14.4, the first amplifier is a 60-MHz amplifier and the second amplifier is a 12-MHz amplifier. What possible condition exists in the bias lines when the input to the first amplifier is suddenly increased 10 dB beyond the level required to saturate the amplifier string? How would this condition be minimized?

15

Amplifier, Oscillator, and Filter Circuit Design Examples

15.1 CIRCUIT DESIGNS

In this chapter, designs of four different types of microwave circuits are given. The first circuit analyzed is an amplifier that is first stabilized and then presented with simultaneous conjugate matches. The second circuit is a shunt oscillator matched for maximum dc-to-rf power conversion. The third circuit is a loop oscillator also matched for maximum dc-to-rf power conversion. The fourth circuit is a microstrip two-pole comb-line filter. The circuits are designed to be fabricated on 0.76-mm (0.030-in.) thick material with a dielectric constant of 3.38. A design frequency of 1.25 GHz was chosen.

The active device's scattering parameters are based on measurements on a small signal bipolar SOT23 device. The device was biased at 10 volts and 20 milliamps. The equations match the scattering parameters very well from about 1 MHz to 6 GHz. The equations are given here so that the reader can use them to practice small signal design on scattering parameters that are close to real values. The SOT23 package has only a single ground. Therefore, the device has a lower K factor than a part that has two grounds. The low K factor also gives the reader practice in stabilizing devices. The components used in the design are 1608 (0603) - 1.6-mm × 0.8-mm (.06" × .03") components.

The parasitics for the resistor and capacitor components are considered to be the inductance and capacitance for a 0.8-mm line on the 0.76-mm-thick substrate described above. This gives an 80-ohm characteristic impedance and a 2.52 effective dielectric for the equivalent TL (transmission line). Extra inductance for the components and capacitance across the components is ignored in this analysis. These extra parasitics would be included when the final design is put in a CAD package for final tuning. However, in the low-GHz range when 1608-size components are used, this is a good first-order approximation. The inductors are considered to have a 0.25-pF parallel capacity across them. This value is on the order of the capacity one calculates for the parasitic capacitance from the self-resonant frequency (SRF) for commercial chip inductors.

The devices are assumed to be mounted to 1.27-mm-wide lines (0.05") that are 2.54 mm long. The effect of this parasitic mounting pad is included in the designs. This microstrip line has a 61.2-ohm characteristic impedance and a 2.65 effective dielectric constant.

15.2 ACTIVE DEVICE SCATTERING PARAMETERS— DEVICE MODEL

The active device is measured with the same ground configuration with which the device will be used in the circuit. A fixture very similar to that shown in Figure 2.38 is used to measure the scattering parameters. The gap for the device is 1.27 mm and the ground for the common terminal is made with a 1.27-mm-wide piece of metal directly to ground. The position of the ground metal and the angle at which it is placed with respect to the transmission lines has some effect on the scattering parameters. The fixture should have a ground configuration very similar to that used on the printed circuit board.

The following equations were used to produce the scattering parameters for the active device. The frequency f is in GHz and angles are in degrees.

$$|S_{11}| = \frac{0.7f^6 + 875}{f^6 + 3500} + 0.5e^{-9f}$$

$$\angle S_{11} = -150 - 34f + 150e^{-6f} + 10(f - 3.6)(1 - \tanh(3.6 - x)) + 10 - 10e^{-f}$$

$$|S_{12}| = 0.13f + 0.14(f - 3.3)(1 - \tanh(3.3 - x)) + 0.001$$

$$\angle S_{12} = 60e^{-25f} - 20e^{-5f} + 75 - 1.75f^2$$

$$|S_{21}| = \frac{42(1 + 0.52f)}{1 + (8.333f)^{1.35}}$$

$$\angle S_{21} = 92 - 23f + 78e^{-7.2f} + 7(f - 3.8)(1 - \tanh(3.8 - f))$$

$$|S_{22}| = \frac{700 + 20f}{2500 + f^6} + 0.25e^{-1.45f} + 0.45e^{-14f} + 0.074f$$

$$\angle S_{22} = 35e^{-25f} - 12.5e^{-7.5f} - 22.5 - 7.5f - 3.8f^2$$

No significance is to be assigned to the various functions. Outside the frequency range of 1 MHz to 6 GHz the functions are not valid. They are not based on physical concepts but seem to match measured data for the small signal SOT23 part reasonably well.

15.3 MATCHED AMPLIFIER DESIGN

15.3.1 Device Characterization and Mounting

The specification used for the amplifier is that it is required to have 10 dB of gain at 1.25 GHz and have simultaneous conjugate match. The device needs to be mounted on soldering or attachment lines on the printed circuit board. A completed amplifier is shown in Figure 15.1 to show that the device is mounted on 1.27 × 2.54-mm transmission lines as discussed in Section 15.1. The first thing then that needs to be done is to prepare a set of scattering parameters for the device plus the transmission lines. This is done by adding a phase shift versus frequency to the scattering parameters' angles that tracks this change. In this chapter, it is assumed that the microstrip line is dispersionless. At 6 GHz, the 2.54-mm transmission lines have a phase shift of −29.54 degrees. The transmission lines have an impedance of 61.2 ohms. The scattering parameters are first converted from 50-ohm scattering parameters to 61.2-ohm scattering parameters using the equations given in Section 2.3.3 to convert to the normalized

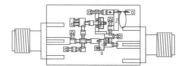

Figure 15.1 Completed 1.25-GHz amplifier.

immittance parameters. The immittance parameters are renormalized to the new impedance level and the inverse transformation is used to convert to the scattering parameters for the new impedance level. The equations in Section 2.6.3 are used to give the new scattering parameters shifted to the new reference planes. The scattering parameters are then changed back to 50-ohm scattering parameters using the reverse of the procedure in going from 50 ohms to 61.2 ohms. This is best done with a computer program and was done to form the data set used for these analyses. At the design frequency of 1.25 GHz, the input scattering parameters were:

Original S Parameters—1.25 GHz	Shifted S Parameters—1.25 GHz
$S_{11} = 0.25050$ at 174.29430 deg	$S_{11} = 0.26223$ at 158.31056 deg
$S_{21} = 2.81072$ at 63.04329 deg	$S_{21} = 2.82456$ at 50.54997 deg
$S_{12} = 0.16256$ at 72.22701 deg	$S_{12} = 0.16336$ at 59.73369 deg
$S_{22} = 0.42287$ at -37.81356 deg	$S_{22} = 0.40807$ at -49.15499 deg

If the transmission lines were 50 ohms on the device, then the shift expected for each of the parameters would be $2(-29.54)1.25/6$ degrees or -12.3 degrees. The shift is close to that except for S_{11}. This is a result of the larger mismatch between 50 ohms and S_{11} than for S_{22}. Why are the parameters shown to five or six digits? One cannot be expected to measure or synthesize circuits to that precision. The reason is that when one does self-consistent calculations between frequencies close to each other (e.g., in an oscillator or narrowband filter), then the variation of the scattering parameters is needed. Either the scattering parameters are used to this precision or one has to calculate slopes over a larger frequency range. In addition, stability equations become sensitive to the actual values when K approaches 1. Even though the parameters will vary, the relationship between the four scattering parameters varies in a self-consistent manner. This author suggests that where possible, the calculations should be carried out using more precision than is deemed necessary unless it is obvious that relationships between the various four scattering parameters do not matter.

15.3.2 Device Stabilization

The quantities K, 2-L, and μ discussed for stability considerations in Section 5.3.2 are plotted as K, L, and M respectively for the device in Figure 15.2. Notice that the device is potentially unstable over the whole frequency range under consideration. Notice as well that the K, L, and M stability factors are very close in value to each other as they approach 1 in value. The device needs to be stabilized. The values from the series and parallel resistor values from equations discussed in Section 5.3.3 are plotted versus frequency in Figures 15.3 through 15.6. These were calculated using the scattering parameters with and without the reference plane shifted values. When using the reference plane shifted scattering parameters, the resistors, either parallel or series, need to be inserted at the shifted reference plane point.

The design criterion used for this amplifier is that the stabilization is to be done on the input. That criterion needs to be determined for each instance. It was the choice made here. If the amplifier were to be designed for power generation, then the choice would clearly be to stabilize at the input if possible. If the amplifier were to be designed for low-noise operation, then the choice would clearly be to stabilize at the output if possible.

What value of resistance should be chosen? The equation given at the end of Section 5.5.4 indicates that a K value of 1.153698 is required for a gain of 10 dB at 1.25 GHz. A CAD program can be used to calculate K versus a series resistance value on port one. The calculation can be done manually by changing the scattering parameters to impedance parameters and adding the series resistance to Z_{11} and converting the resultant impedance matrix back to

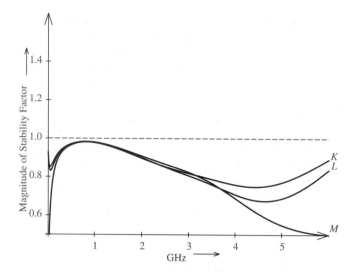

Figure 15.2 Stability factors for the device.

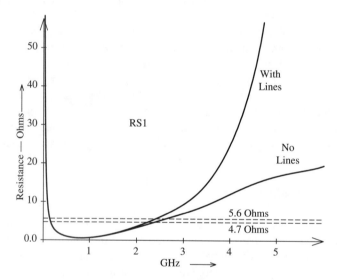

Figure 15.3 Series resistance values for stability vs. frequency on port one.

scattering parameters. *K* is then calculated from the "new" scattering parameters. Using either procedure, a series resistance of 5.6 ohms is determined. From Figure 15.3, a 5.6-ohm series resistor will stabilize the device over a wide frequency range around the design frequency but not everywhere. A value of 4.7 ohms was chosen since additional loss will come from adding additional components to stabilize the device at frequencies well removed from 1.25 GHz. Figure 15.7 shows a plot of parallel resistance values that need to be added in addition to 4.7-ohm series to give stability. Notice that nothing is needed over the region around 1.25 GHz. The device is already stable there. A minimum of 270 ohms is needed at around 4.8 GHz. At low frequencies, the value decreases rapidly with frequency decreasing to under 100 ohms at the low limit given by the scattering parameters. At frequencies below this low limit, the bias network will stabilize the circuit and bias stability will be determined when the total network is completed. One does not want to put 270 ohms in parallel with the input at 1.25 GHz. That would degrade the gain beyond what is needed. A series *R-L-C* of 180 ohms, 1.5

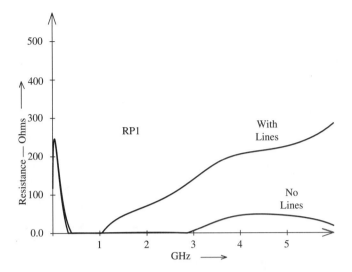

Figure 15.4 Parallel resistance values for stability vs. frequency on port one.

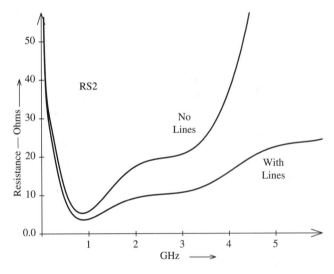

Figure 15.5 Series resistance values for stability vs. frequency on port two.

nH, and 0.25 pF was chosen to provide the stability network required for high frequencies. The 0.25-pF capacitor limits the loss effect at 1.25 GHz and the 1.5-nH inductor in series with the 0.25-pF capacitor increases the slope of the effect of the 0.25-pF capacitor so that it starts to be effective well before the 8.2-GHz series resonance frequency. The Q of the series R-L-C circuit is kept below 1 so that it is effective over a wide frequency range. A series resistance value of 180 ohms (a standard 10% resistance value) is adequate in series with the series LC to provide less than 270 ohms parallel at 5 GHz yet limit the loading effect at 1.25 GHz. In order to stabilize the low-frequency end, a resistance of 50 ohms in series with an inductance is chosen. Inductors of a 1608 physical size and in the 100-nH range have approximately 0.15 pF of parasitic capacity. Smaller values have parasitic capacitance values closer to 0.25 pF. Nominally, 0.15 pF gives a parallel inductance of 110 nH for resonance at 1.25 GHz. An inductor value of 100 nH was chosen to put in series with the 50-ohm resistor. Since the bias current required by the bipolar part is quite small, the base current will be provided through the

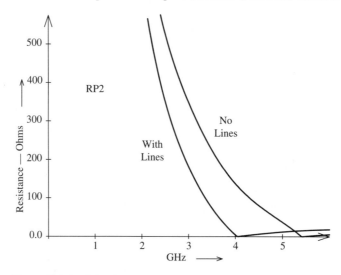

Figure 15.6 Parallel resistance values for stability vs. frequency on port two.

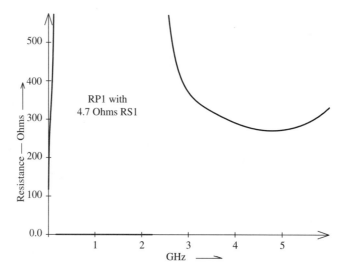

Figure 15.7 Parallel resistance in addition to 4.7-ohm series on port one.

100-nH and 50-ohm resistor. If a FET device were used, the gate voltage could be provided through such an arrangement.

Figure 15.8 shows K, 2-L, and μ values for the device with a series 4.7-ohm resistor and the two series circuits in parallel at the circuit (not the device) end of the 4.7-ohm resistance. Notice that the stability factors are all above 1 everywhere although they are close to 1 at about 3 GHz.

Figure 15.9 shows the maximum available gain calculation using the formula from Section 5.5.3. The available gain at 1.25 GHz is just above 10 dB. Notice that the gain goes through zero dB at about 5.0 GHz and continues to drop after that. The circuit would not be expected to have gain beyond 6 GHz based on this trend. In other circuits, the user needs to determine from the scattering parameters of the device or circuit whether there is the possibility of gain beyond the frequencies under consideration. For instance, when the circuit is actually fabricated, if series or parallel feedback is inadvertently added, the gain calculations may not be accurate. Often this occurs in the manner in which the common terminal of the active device

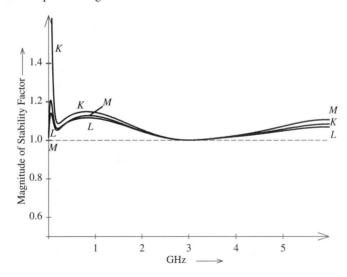

Figure 15.8 Stability factors with three stabilization networks.

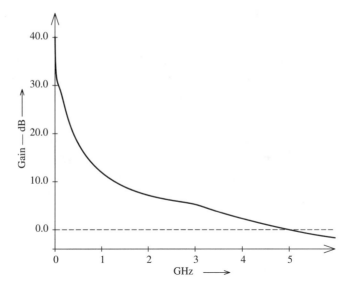

Figure 15.9 Maximum available gain from the device with stabilization networks.

is grounded. It is important that the device parameters are measured with the ground that is used or that the common mode impedance is characterized and incorporated in the analysis.

15.3.3 Matching for Simultaneous Conjugate Match

The circuit as stabilized will have a simultaneous conjugate match since the K factor is greater than 1. The scattering parameters for the stabilized amplifier at 1.25 GHz are:

$S_{11} = 0.20051$ at 157.83883 deg
$S_{21} = 2.63198$ at 50.11479 deg
$S_{12} = 0.15222$ at 59.29851 deg
$S_{22} = 0.39767$ at -47.80842 deg

Using the formulas in Section 5.4, the simultaneous conjugate match impedances and reflection coefficients are:

$$ZS1 = 15.1524 - j\,9.04772 \text{ ohms or } 20.555 \text{ ohms in parallel with } -j\,34.424 \text{ ohms}$$
$$ZS2 = 53.0627 + j\,86.6650 \text{ ohms or } 194.61 \text{ ohms in parallel with } +j\,119.15 \text{ ohms}$$

where ZS1 and ZS2 stand for the series source impedance on port one and load impedance on port two. Some CAD programs often give these as ZM1 and ZM2.

15.3.3.1 Matching for Simultaneous Conjugate Match—No Parasitic Compensation.

For the input match, the series configuration is chosen. The parallel configuration could have been chosen and the impedance level change would have been less. The matching will be done using the matching formulas in Section 6.3.1. For this particular circuit, by choosing the series configuration, a single capacitor results in the match. This capacitor can be used as the input blocking capacitor. Therefore, the input match consists of transforming a 50-ohm source to 15.1524 ohms. This match has a transformation Q of 1.5165, giving a series capacitance of 5.54 pF ($-j22.978$ ohms) and a shunt inductance of $j50$ divided by 1.5165 or $j32.97$ ohms. This is a 4.198-nH inductor at 1.25 GHz. The series capacitance of $-j22.978$ ohms must be added in series with the input resonating reactance of $-j9.0477$ ohms to give a total series reactance of $-j32.026$ ohms or a series capacitance of 3.98 pF.

For the output match, the parallel configuration was chosen. The parallel inductance required to resonate the output will be put in parallel with the inductance from the transformation. The transformation Q is 1.700. This gives a parallel inductance reactance of 194.61 divided by 1.700 or $j114.43$ ohms. This transformation reactance is in parallel with the resonating reactance of $j119.15$ ohms giving a combined inductive reactance of $j58.37$ ohms or an inductance of 7.432 nH. The series capacitive reactance is 50 ohms times 1.700 or $-j85.0$ ohms or 1.497 pF at 1.25 GHz.

Without adding parasitic components, the circuit just designed was analyzed. The responses are shown in Figure 15.10. Note that the 3-dB bandwidth extends from 1 to 1.55 GHz and that the better-than-10-dB return loss on the input and output matches extends over approximately 1.15 to 1.4 GHz. The maximum gain occurs at just below 1.25 GHz and the gain at 1.25 GHz is about 10.2 dB.

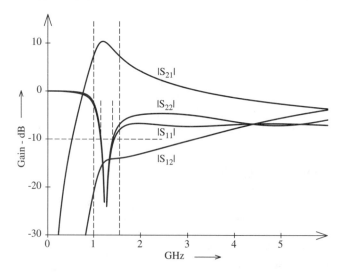

Figure 15.10 Gain and port matches for lumped-constant matched amplifier.

Figure 15.11 shows the stability factors for the lumped-constant matched amplifier. Notice again that the stability factors are too close to 1 at about 3 GHz. Some margin of stability could have been added to the initial stabilization. This exercise was given this way to show that the K stability factor does not get better with matching changes; in fact, as indicated in Section 5.3.2, K does not change when the part is matched. Compare K of Figure 15.8 with Figure 15.11. However, the μ and L factors do change as a function of matching.

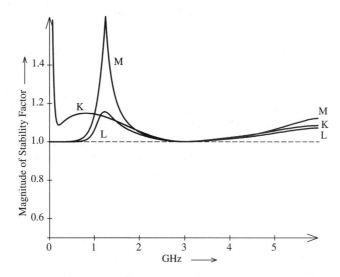

Figure 15.11 Stability factors for lumped-constant matched amplifier.

15.3.3.2 Matching for Simultaneous Conjugate Match with Parasitic Compensation.

Now that the circuit is matched with ideal capacitors and inductors, the circuit needs to be changed to accommodate real components. The gap in the transmission line for mounting the 1608 (0603) components is nominally 1 mm. As indicated in Section 15.1, in this design example, the capacitors and resistors are considered to be ideal resistors and capacitors. However, these ideal components are placed in the center of a section of 1-mm-long transmission line with a characteristic impedance of 80 ohms and an effective dielectric constant of 2.52. Using the equations in Section 2.6.2, the tee and pi models of this section of transmission line are given as shown in Figure 15.12. The component values between the two circuits are not actually in the ratio of 2 : 1 but appear to be in that circuit since the transmission line is short. Refer to Figure 15.1. The output capacitor has a parasitic inductance of about 0.424 nH. One-half of the parasitic capacitance of 0.03 pF ($-j4000$ ohms) appears across the output 50 ohms and is ignored. The other half appears across the output inductor and is incorporated in parallel with the parasitic capacitance of the output inductor. The parasitic capacitor of the output inductor is approximately 0.25 pF giving a total parasitic capacitance for the inductance of 0.28 pF. The output capacitance was originally calculated at a value of 1.497 pF. The new

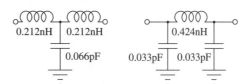

Figure 15.12 Equivalent circuit of a 1-mm length of 80-ohm TL with a 2.52 ϵr.

value compensated for by compensating for the series inductance is:

$$-jX_{c\_\text{orig}} = -j85.05 \Rightarrow 1.50 \text{ pF}$$

$$+jX_L = +j3.33$$

$$-jX_{c\_\text{new}} = -j85.05 - j3.33 = -j88.38 \Rightarrow 1.44 \text{ pF}$$

The new value for the 7.432-nH output inductance is:

$$-jB_{L\_\text{orig}} = -j0.01713 \Rightarrow 7.432 \text{ nH}$$

$$+jX_C = +j0.0022$$

$$-jX_{c\_\text{new}} = -j0.01713 - j0.0022 = -j0.01933 \Rightarrow 6.59 \text{ nH}$$

For the input series blocking capacitor, the inductance of the series 4.7-ohm resistor, the 5.54-pF inductor, and the 1.9-mm-long, 61.2-ohm TL are lumped together since each part is a small portion of a wavelength. If the total length of the line is greater than about 30 degrees, then distributed transmission line analyses should be used. In this case, the total length is about ten degrees. The inductance of the TL is about 0.63 nH and the inductance of the resistor and the capacitor are both just about 0.4 nH, giving a total series inductance of 1.48 nH. Therefore, the new value to be used for the input 3.98-pF capacitor is:

$$-jX_{c\_\text{orig}} = -j32.0 \Rightarrow 3.98 \text{ pF}$$

$$+jX_L = +j11.4$$

$$-jX_{c\_\text{new}} = -j32.0 - j11.4 = -j43.4 \Rightarrow 2.93 \text{ pF}$$

The input inductance has the parasitic parallel resonant capacity plus one-half of the parasitic TL capacity associated with the input capacitor. This is approximately 0.25 pF plus 0.03 pF or about 0.28 pF. The new value to be assigned to the 4.198-nH inductor is:

$$-jB_{L\_\text{orig}} = -j0.0303 \Rightarrow 4.198 \text{ nH}$$

$$+jX_C = +j0.002$$

$$-jX_{c\_\text{new}} = -j0.0303 - j0.002 = -j0.0323 \Rightarrow 3.94 \text{ nH}$$

The circuit shown in Figure 15.1 was analyzed with the values given here, and using transmission lines between the components, and so forth. The stability factors are shown in Figure 15.13 and the gain and match curves are shown in Figure 15.14. Notice that the

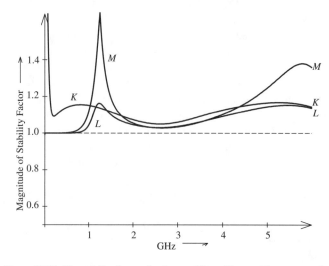

Figure 15.13 The stability factors for the amplifier with parasitic compensation.

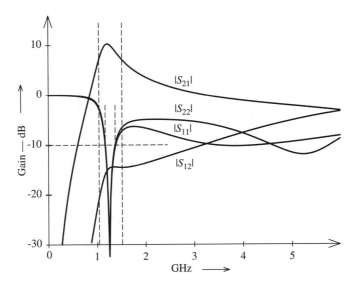

Figure 15.14 Gain and match curves for the amplifier with parasitic compensation.

stability has improved a little at 3 GHz. This is due to the movement of the shunt RL and shunt R-L-C circuits on the input due to the size of transmission line pads when those circuits were placed on the transmission lines as shown in Figure 15.1. Figure 15.15 shows a comparison between the lumped-element model and the parasitic-compensated model. The model with parasitic is just a little narrower band but the response and match are still quite good, showing that this first-order compensation is actually very good. The final circuit could be input to an optimizer CAD program for final tweaking. However, the component values used have been given to two and three decimal points for demonstration. The actual component values would be given to two decimal points and with possibly 10% component values. Therefore, at this point the reader would be encouraged to do a component variation analysis with actual component values instead of further tweaking the circuit. A broader-band response could be

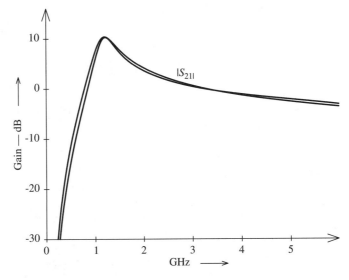

Figure 15.15 Comparison of the lumped-constant and parasitic-compensated amplifier.

procured by changing the transformation networks and using a couple of sections on the high Q transformation sections.

15.3.4 Bias Circuit Stability

After the complete circuit is designed, the bias circuit needs to be checked for oscillation. The bias circuit should be stable without the bypass capacitors and biasing inductors. Figure 15.16 shows a schematic diagram of the amplifier circuit with bias supply. The current source at the base of the PNP transistor supplies a pulse of current to give a transient change to the reference point in the regulator. This causes a transient in collector current. If the transient dies down after the pulse goes away, the circuit is transient stable. If the bias circuit would oscillate, the transient would build up rather than die down. Figure 15.17 shows the transient response of the collector current to the current transient. Notice that the transient does die out with time. The figure shows the response both with bias components and without bias components. There is a significant difference in the two responses, indicating the necessity for checking the circuit after the total rf circuit including bias components are determined. Notice that the input and output rf terminals are terminated with an open circuit. In this analysis this is appropriate since the coupling capacitors on those ports are very small and insignificant with respect to the bias supply response frequencies. However, if a designer persists in making the coupling capacitor very large (bigger is not necessarily better), then the effect of the rf port terminations would have to be investigated.

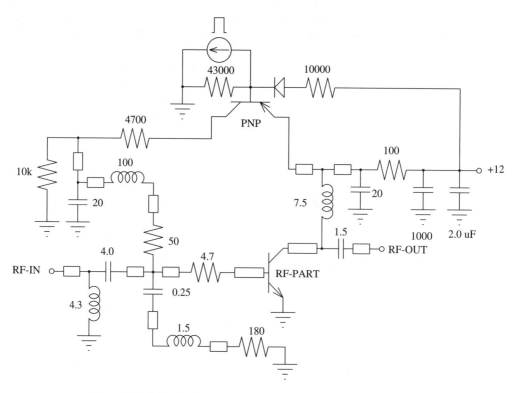

Figure 15.16 Schematic diagram of the amplifier circuit including bias components.

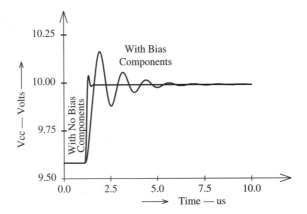

Figure 15.17 Transient response of the amplifier bias.

15.4 SHUNT OSCILLATOR DESIGN

The purpose of this section is to show the design approach for a shunt oscillator. The active device will be the same one as used in the matched amplifier except that the part will be matched for maximum power generation rather than maximum gain. Depending on the oscillator specifications, different design scenarios can be used. The part could also be matched for minimum noise figure within the oscillator bandwidth. However, in this design, maximum dc-to-rf power generation is the design criterion.

15.4.1 Determining the Active Device Load Line for the Shunt Oscillator

The oscillator part is biased at 10 volts and 20 milliamps of collector current. A collector saturation voltage of 2 volts is assumed. The minimum current is assumed to be close to zero and is ignored. This may not be accurate for some FET devices but for many bipolar devices it is not too bad an assumption. The load line analysis given in Section 7.4 shows that the parasitic free load line is:

$$R_L = \frac{V_{cc} - V_{sat}}{I_{cc} - I_{min}} = \frac{10 - 2}{0.02 - 0} = 400$$

The design considers that the load line on the part must be 400 ohms. This means that the parasitic reactances of the mounting pad and bond wire inductances must be considered. The bond wire inductance for the part would be on the order of 1 nanohenry or about 8 ohms of inductive reactance at 1.25 GHz. This design ignores that amount of inductance in relation to the 400 ohms of resistance. The designer should not forget the bond wire inductance since it is often quite important, especially in higher-power parts.

The part will not be matched for maximum gain. What is the value of resonating reactance to be put in parallel with the 400 ohm resistance? The equation given in Section 6.2 allows one to calculate this reactance. What scattering parameters are to be used? In the discussion given in Section 8.4.4 it is suggested that the magnitude of S_{21} be reduced to 80% of the small signal value. This allows for about 2 dB of loop gain saturation and also allows one to calculate the change in output resistance versus saturation. The oscillator is designed using the reduced gain value but the output resistance or conductance is checked with the small signal value to determine the sign of the rate of change of resistance or conductance for the stability calculations shown in Section 8.4.7. Three different device scattering parameter sets are used in the analysis. The first is the small signal scattering parameters. These are the parameters that are important for oscillator startup. Another set is used that has the magnitude of S_{21} gradually reduced over some range around the design frequency. This set is used to indicate the locus of points that the oscillator takes as it approaches steady-state operation. A third set of scattering parameters is used that have the magnitude of S_{21} reduced to 80% at all frequencies. This set is used to set the locus at right angles to the real axis to maximize operating point stability. This will be more evident when the oscillator examples are looked at.

The original scattering parameters and the scattering parameters with a reduced forward gain magnitude at 1.25 GHz for the part are given here.

Original S Parameters—1.25 GHz	Shifted S Parameters—1.25 GHz
$S_{11} = 0.25050$ at 174.29430 deg	$S_{11} = 0.25050$ at 174.29430 deg
$S_{21} = 2.81072$ at 63.04329 deg	$S_{21} = 2.24858$ at 63.04329 deg
$S_{12} = 0.16256$ at 72.22701 deg	$S_{12} = 0.16256$ at 72.22701 deg
$S_{22} = 0.42287$ at -37.81356 deg	$S_{22} = 0.42287$ at -37.81356 deg

These are scattering parameters right at the device. The effect of the mounting pad will be considered later. Using these parameters in the equations in Section 6.2 give the load required on the part as 400 ohms in parallel with $j151.1$ ohms and the value of the input impedance into the device with this load as $9.1925 + j3.0221$ ohms. These are the loads required right at the device plane. The input impedance will not be used directly since the oscillator input impedance will be modified when feedback is added to the part. However, the load impedance is used directly. The effect of the mounting pads will be considered only after the appropriate load on the oscillator is determined. It is important to do the shunt oscillator calculations for load line right at the active device.

15.4.2 Determining the Device Termination Impedances for the Shunt Oscillator

These values of scattering parameters and load required for the active part are used in the equations given in Section 8.4.5. However, the loss of other parts needs to be considered before those equations can be used. In this design, it is assumed that the feedback components will have 0.5 ohm of series resistance. The components are usually capacitors for this type of oscillator at this frequency and power level. If they turn out to be inductors, then the appropriate loss component will be used. The reflection coefficient circle for a 0.5-ohm series resistance is centered at 0.0099099 with a radius of .990099. These values are used in the oscillator equations given in Section 8.4.5. Substituting these values into those equations gives the following values of impedance to be presented to the device.

Z_port1: $0.5 - j9.502$

Z_port2: $43.332 + j139.2$ or 490.5 ohms in parallel with $j152.7$ ohms

Z_common: $0.5 - j36.022$

The reader needs to keep in mind that the 0.5 ohms shown on the input port and the common terminal port are not extra resistors to be added to the design. These are parasitic components expected to be in the capacitors (or inductors) added on the part. The model for these components in the CAD program used to characterize the oscillator needs to include this amount of resistance or the output impedance calculation for the oscillator will be incorrect.

15.4.3 Determining the Shunt Oscillator Circuit Configuration

The mounting pads for the oscillator in this design example are considered to be 2.54 mm long and 1.27 mm wide, just as they were in the matched amplifier design. The emitter (or common terminal) of the device is no longer at ground potential so a mounting pad needs to be provided for it as well. In this oscillator design, the emitter termination is capacitive and an inductance will need to be provided to provide for the dc bias current to return to ground. This inductance often provides an "extra" resonance in the oscillator. The output impedance of the oscillator will have to be checked for spurious resonances after the part is designed for printed circuit board or MMIC layout.

The oscillator load will be presented at the output end of the output TL. De-embedding using a negative length of transmission line as shown in Section 6.5.1 is done to transform the impedance seen at the device to the impedance presented at the end of the TL. A line with a characteristic impedance of 61.2 ohms, a line length of 2.54 mm, and an effective dielectric constant of 2.606 is used. This line transforms an impedance of $43.332 + j139.2$ at the device into a load impedance of 443.94 ohms in parallel with a load reactance of $j115.54$ at the end of the TL. The output match uses the $Q^2 + 1$ equations from Section 6.3.1. The transformation Q going from 50 ohms to 443.94 ohms is 2.807. One could use a series reactance of negative $j140.35$ ohms or a 0.907-pF capacitor and a shunt reactance of $j158.16$ ohms. This shunt reactance combined with the resonating inductance of $j115.54$ ohms gives a combined reactance of $j66.77$ ohms or a parallel inductance of 8.5 nH. However, if one tries this match, there is a parasitic output negative resistance in the 1.4 GHz region. Instead, an alternate match consisting of a series reactance of $+j140.35$ ohms and a shunt reactance of $-j158.16$ ohms is used. Combining this shunt reactance with the resonating reactance of $j115.54$ ohms gives a net shunt reactance of $j428$ ohms. Therefore, a 54.5-nH shunt inductor is used in the analysis. This inductor is also used as the collector bias inductor. This is a fairly large inductor at 1.25 GHz and the part would have to be characterized to give the correct reactance at 1.25 GHz. A typical value of 0.1 to 0.14 pF of shunt capacity for 1608 (0603) chips indicates that somewhere around a 100-nH chip would actually be used. The mounting configuration would also be critical. This inductance is a critical frequency-setting component and there will likely be some tweaking necessary to set the oscillator right on frequency. For that reason, 54.5 nH is used in the analysis. The output now needs a blocking capacitor and a 100-pF chip is used for that.

The capacitor used on the input needs to be de-embedded along the input mounting line. This changes the resultant capacity from 13.4 pF right at the device to 7.78 pF mounted on the end of the input TL. The common terminal capacitance would be 3.53 pF if mounted right at the transistor, but with de-embedding through the line on the common terminal the capacity is 2.8 pF. The common terminal also has a self-resonant 100-nH chip inductor from it to ground to provide a dc current return path.

15.4.4 Examining the Output Impedance of the Shunt Oscillator

The circuit diagram for the oscillator is given in Figure 15.18. Note that the schematic circuit shown in Figure 15.18 shows standard value components. The circuit will first be analyzed using the precise value of components derived in the analysis above.

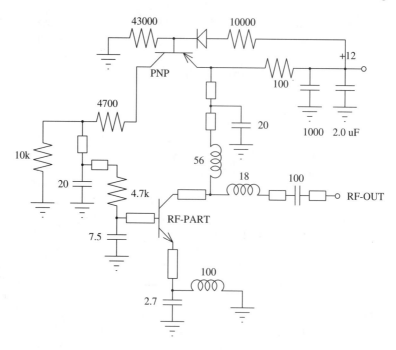

Figure 15.18 Schematic diagram of the oscillator using standard value components.

Figure 15.19 shows the output impedance of the oscillator as a function of frequency. This chart shows negative R but X in its normal arrangement (i.e., the top of the chart is inductive). The point nearest the center of the chart is at 1.2525 GHz. There are three loci on the chart. They are labeled SS, 80%, and GR. The SS locus is calculated using small signal scattering parameter values. These determine the startup conditions of the oscillator. Notice that the startup impedance is more negative than −50 ohms, ensuring startup. The GR locus uses scattering parameters that have a varying amount of reduction on the magnitude of S_{21}. These scattering parameters were graded from 1.15 GHz to 1.35 GHz. That locus helps show the locus the oscillator output impedance takes as the oscillation builds up. The third locus is the 80% locus and that locus has all the magnitudes of S_{21} reduced by 80%. That locus is used to determine whether a phase shifter needs to be put on the output of the oscillator. This particular oscillator design has the 80% locus going at right angles to the real axis, therefore eliminating the need for a phase shifter on the output.

Figure 15.20 shows the output impedance of the oscillator using standard component values. This particular oscillator works fairly well and close to the design frequency using standard component values. This is not typical. Also keep in mind that the actual component capacitance used in the oscillator would need to be compensated for the inductance in series with them to ground. That inductance is not shown in the schematic. The inductance would typically be 0.5 nH for these chip capacitors and 0.3 nH for getting through the 0.75-mm thick board. The frequency closest to the center of the chart is 1.2525 GHz.

At 1.2425 GHz, the output reactance is $-j4.03255$ ohms and at 1.2625 GHz the output reactance is $j3.86633$, giving a positive reactance slope. The output resistance varies from about −100 ohms for small signal amplitudes to −50 ohms at the operating point. Therefore, the resistance slope versus power (higher power at the operating point) is positive. Since the locus crosses the real axis at right angles, the derivative of resistance versus frequency at the operating point is zero. Therefore, the second part of the stability equation in Section 8.4.7 is

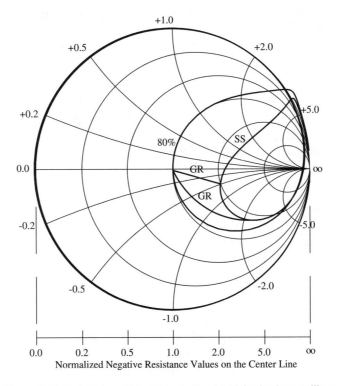

Figure 15.19 Reflection coefficient locus ($-R$ and $+X$) for the shunt oscillator.

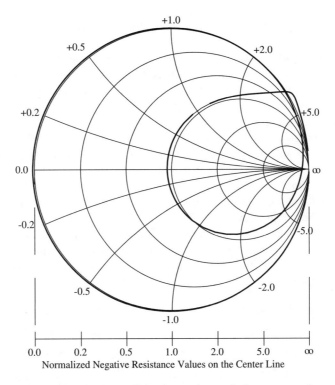

Figure 15.20 Reflection coefficient locus using standard component values.

zero. Each of the first two terms of the impedance stability equation is positive as just shown, indicating a stable operating point.

The equations in Section 8.4.8 for external Q and frequency change can be rewritten as:

$$Q_{\text{ext}} = \frac{2|\Gamma_{\text{LOAD}}|f}{\Delta f}$$

where the frequency difference in the denominator is the double-sided frequency shift and the magnitude of the load reflection coefficient is the maximum change from zero. For the data for this oscillator, the change in reflection coefficient is close to plus or minus 0.04 for the frequency shifts of 1.2425 GHz to 1.2625 GHz. Therefore, the external Q of this oscillator is approximately:

$$Q_{\text{ext}} \approx \frac{2(0.04)1.25}{(1.2625 - 1.2425)} = 5$$

In order to increase the external Q, a series resonator could be inserted in series with the output since this oscillator has a series type of resonance at the output. This would help reduce the load-pulling effects of the oscillator.

15.5 LOOP OSCILLATOR DESIGN

The loop oscillator design follows a similar approach to the shunt oscillator design. However, the device's ground is not disturbed. The load line on the part is the same as the load line of the part in the shunt oscillator. This was 400 ohms in parallel with $j151.1$ ohms, which is equivalent to $50 + j132$ ohms. It just happens that this is 50 ohms plus a reactance and is not related to the 50-ohm normalization impedance. With this value of load impedance on the device, the device has a $9.1925 + j3.0221$-ohm input impedance.

15.5.1 Loop Oscillator Design—Amplifier Design

The values of the load impedance for the amplifier need to be de-embedded to determine the values that would need to be presented at the output of a 2.54-mm long, 1.27-mm wide line. This mounting line is the same as was used in the matched amplifier design and in the shunt oscillator design. The value of the input impedance into the device will also be transformed to the value needed at the input of the input line. The TL impedance for the mounting lines is 61.2 ohms and the lines have an effective dielectric constant of 2.606.

These transformations are essentially a 12.3-degree rotation in the cw direction on a 61.2-ohm normalized reflection coefficient chart for the input and a 12.3-degree rotation in the ccw direction on a 61.2-ohm normalized reflection coefficient chart for the output impedance. Performing those rotations for the given load impedance gives $33.1 + j104$ as the impedance required at the output of the output TL.

Rather than calculating the input impedance at the input to the input transmission line, the scattering parameters for the device with input and output mounting lines will be used. These scattering parameters were discussed in Section 15.4.1 and the values at 1.25 GHz were given there and are repeated here.

Original S Parameters—1.25 GHz	Shifted S Parameters—1.25 GHz
$S_{11} = 0.25050$ at 174.29430 deg	$S_{11} = 0.25050$ at 174.29430 deg
$S_{21} = 2.81072$ at 63.04329 deg	$S_{21} = 2.24858$ at 63.04329 deg
$S_{12} = 0.16256$ at 72.22701 deg	$S_{12} = 0.16256$ at 72.22701 deg
$S_{22} = 0.42287$ at -37.81356 deg	$S_{22} = 0.42287$ at -37.81356 deg

The value of the load required at the output of the device with the output line on it has just been calculated. That load will be put on the new device. This is the device with the two mounting lines. Care needs to be taken that the correct load is put on this network in order that the device sees its correct load impedance right at the device. It is a common error to not de-embed the impedance as was just done.

The design procedures in Section 8.4.6 and Figure 8.22 are used to calculate the loop oscillator components.

Power gain for the network (device plus TLs) with a load impedance of $33.1 + j104$ ohms (362 ohms in parallel with $j115$ ohms) and the shifted scattering parameters is 9.04 dB. The loop is assumed to have a lossless phase shifter and lossless components. Loss can be included in the design by adding a loss in the feedback and reducing the loss in the branch circuit. For this design, the branch loss must be -9.04 dB to end up with unity gain around the loop. The parallel combination of the loop resistance and the load resistance needs to be 362 ohms. A loss of 9.04 dB is a power ratio of 0.125. Therefore, the loop resistance needs to be eight times as big as 362 ohms or 2896 ohms. The parallel combination of the load and loop resistances needs to be 362 ohms so the load resistance is 414 ohms.

$Q^2 + 1$ matches are used for these transformations at 1.25 GHz. The output of computer calculations for these matches is given here. Since the computation is done with more decimal points than in the previous paragraph, the numbers are slightly different.

Power Gain is 9.037075 dB

Branch Gain is -9.037075 dB

The resistance looking into the feedback is 2897.944000 ohms

The resistance looking our at the output is 413.319800 ohms

The impedance looking into the device input is $9.357 + j9.505$

For the input of the device to 50-ohm match, the following components are calculated.

Series X is $-j24.272310$ ohms toward device

Shunt B is $-j2.553354E\text{-}02$ mhos $= j39.164170$ ohms

Resonating shunt reactance on the device $= -j18.716670$ ohms

For the branch impedance transformer:

Series X is $-j377.355500$ ohms toward loop

Shunt B is $-j2.604299E\text{-}03$ mhos $= j383.980600$ ohms

For the output impedance transformer

Series X is $-j134.781300$ ohms toward load

Shunt B is $-j6.521888E\text{-}03$ mhos $= j153.329800$ ohms

Resonating shunt reactance on the device output $= j114.7471$ ohms

Calculating this gives a loop gain of 1 at an angle of 169 degrees. Therefore, a -169-degree phase shifter is used. Figure 15.21 shows a schematic of the circuit used. Note the different parts of the circuit. On the left there is a tee circuit for the -169-degree phase shifter. This circuit has a characteristic impedance of 50 ohms. Next is the input impedance $Q^2 + 1$ transformation circuit. Next are the resonating capacitor and the active device with input and output TLs. Next comes the output resonating inductor. After that the circuit branches. The next shunt L and branch series C form the branch feedback, impedance transformation circuit. Finally, the last shunt L series C form the impedance transformation circuit from the 50-ohm load to the oscillator circuit. Figure 15.22 shows the gain response of the circuit for small

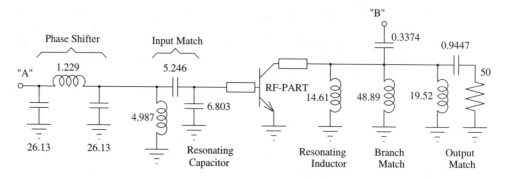

Figure 15.21 Loop oscillator amplifier circuit.

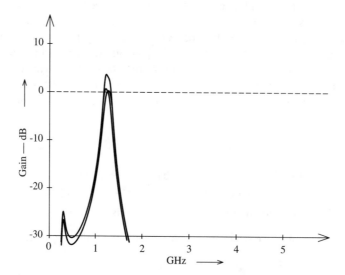

Figure 15.22 Response of the loop oscillator, amplifier circuit.

signal scattering parameters, for scattering parameters having the magnitude of S_{21} graded to 80% over a 200-MHz range, and the response with the magnitude of S_{21} set at 80% for all frequencies. Figure 15.23 shows a closeup view of the match and response of the amplifier around 1.25 GHz using the magnitude of S_{21} set at 80% for all frequencies. Notice that the part is matched as a power amplifier. The input is matched but the output is badly mismatched. This is what is expected when the part is matched for maximum dc-to-rf power conversion. The gain is 0 dB as expected. The forward gain phase of the amplifier around 1.25 GHz is shown in Figure 15.24. As expected it passes through zero degrees. The phase slope will govern the external Q of the oscillator. Notice that there is no added resonator in the oscillator. One could be added to increase this phase slope. The circuit will be reduced to fewer components before further analysis.

15.5.2 Loop Oscillator Design—Component Reduction

The circuit shown in Figure 15.21 contains more components than necessary. By combining components that are in parallel and applying tee to pi and pi to tee transformations, the circuit can be reduced to the circuits shown in Figure 15.25. The parallel inductor and capacitor in the middle of the input network are combined at 1.25 GHz to give 22.9 pF. All three inductors on the output are combined into one capacitor. The center of the input circuit is

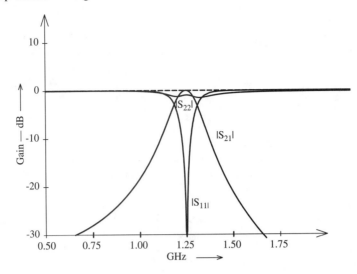

Figure 15.23 Response and match of the loop oscillator amplifier circuit.

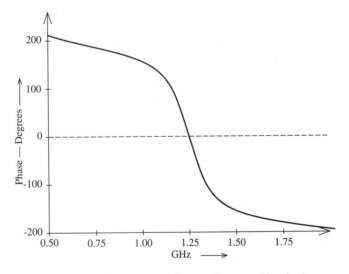

Figure 15.24 Phase response of loop oscillator amplifier circuit.

now a three-component tee circuit. That circuit is converted to a pi circuit with the result of a shunt 0.8-nH inductor, a series 3.5-nH inductor, and a shunt 8.04-pF capacitor. The first shunt inductor is combined with the leftmost shunt 26.13-pF capacitor and the two capacitors that are now on the input of the device are combined. This gives the circuit shown in Figure 15.25. The points "A" and "B" are still 50-ohm points if one wishes to insert a 50-ohm resonator at that point. The resonator would have to have zero phase shift to not disturb the oscillation conditions. Whether one inserts a series LC or a shunt LC circuit will dictate what type of phase shifter will be necessary on the output to ensure stable oscillations. In this design, another pi-to-tee conversion is made on the input side and the resulting series L on the input side at "A" is combined with the series capacitor at "B" to form a new series capacitor as shown in Figure 15.26. Another tee-to-pi transformation could be performed on the input circuit to allow the combination of another parallel component with the shunt inductor on the output, but the result necessitates the addition of a feedback blocking capacitor and the advantage is lost. The circuit shown in the bottom of Figure 15.26 without the bias supply is now analyzed.

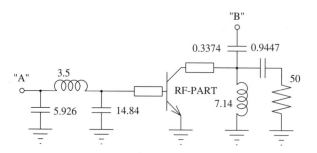

Figure 15.25 Loop oscillator circuit formed by component reduction.

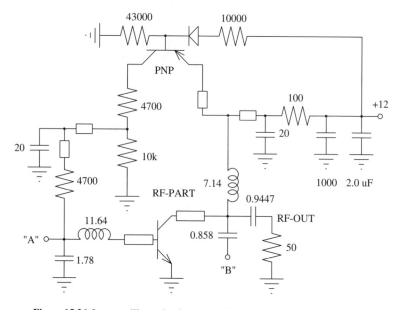

Figure 15.26 Loop oscillator circuit—a second version of component reduction.

15.5.3 Examining the Output Impedance of the Loop Oscillator

Figure 15.27 shows the output impedance locus for the circuit. Notice that the locus for this oscillator is also stable. The derivative of the output impedance versus power is positive as is the reactance slope, giving a positive first term of the stability equation in Section 8.4.7. The second part of that term is zero at the design frequency since the locus crosses the real axis at right angles. The output reactance is $-j4.05$ at 1.2475 GHz and $+j4.20$ at 1.2525 GHz. The one-sided change in reflection coefficient is approximately 0.041. This gives an approximate external Q of:

$$Q_{ext} \approx \frac{2(0.041)1.25}{(1.2525 - 1.2475)} = 21$$

This external Q is about four times higher than the shunt oscillator that was designed. The external Q of both oscillators will be improved with the use of either internal or external resonators.

There is a small possibility of an oscillation at 6 GHz. Notice that the locus comes back into the chart. There is a small negative resistance there. Likely the small losses of the capacitors and inductors that are not included in this analysis would eliminate that possibility.

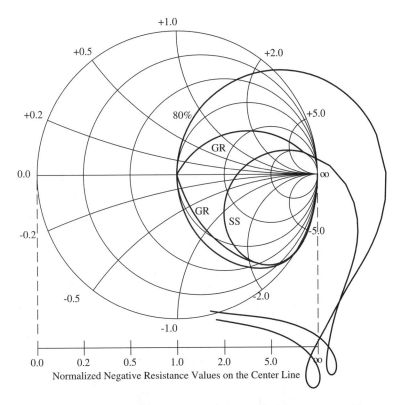

Figure 15.27 Locus of output impedance for the reduced-component loop oscillator.

The analysis was done with the 7.14-nH inductor that is on the collector grounded. The reader should attempt to do the analysis with all the parasitics included and compensated for as well as the bias supply included. This will help determine whether additional design work needs to be done to eliminate the 6-GHz possibility as well as any other resonances introduced by the bias supply connections. The oscillator should be checked to see if the oscillator has startup difficulties and whether there are any additional resonances using minimum and maximum expected losses in the reactive elements. The component values shown in Figure 15.26 are compensated for the parasitics in the same manner as was done for the amplifier circuit.

15.5.4 Shifting the Locus of an Oscillator for Maximum Stability

In order to demonstrate shifting the locus of an oscillator for maximum stability, the output impedance of the unreduced circuit shown in Figure 15.21 will be considered. Figure 15.28 shows the output impedance of the oscillator from Figure 15.21. Notice that the locus does not cross the real axis at right angles. The locus needs to be rotated about 50 degrees counterclockwise. This requires a 25-degree lead network. A tee network consisting of 11.5-pF capacitors in the series branches and a 15.1-nH inductor for the shunt component makes a 50-ohm +25-degree phase delay network. Figure 15.29 shows the output impedance of the oscillator of Figure 15.21 with that network on the output. Notice that the loci now cross the real axis at right angles. The circuit is still not acceptable since there are parasitic resonances still within the chart. However, this procedure demonstrates how to rotate a locus to get the locus to cross the real axis at right angles at the resonant frequency.

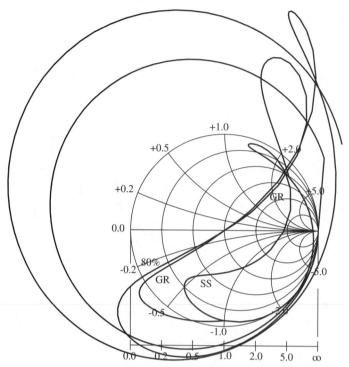

Figure 15.28 Locus of the unreduced loop oscillator circuit.

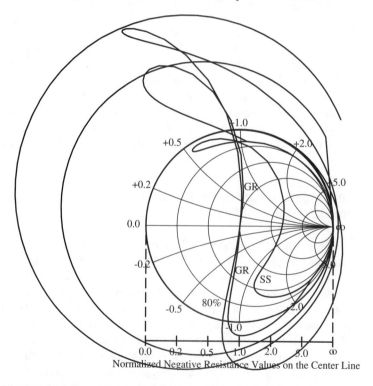

Figure 15.29 Locus of the unreduced loop oscillator circuit with a +25-degree phase shifter.

15.6 ONE-GIGAHERTZ THREE-POLE COMB LINE FILTER DESIGN

A three-pole, 3% bandwidth filter will be designed at 1.25 GHz. Quarter-wavelength resonators in a comb line filter structure will be used. A Butterworth response will be used as well. The g parameters g_0, g_1, g_2, g_3, g_4 for the three-pole Butterworth response are 1, 1, 2, 1, 1.

15.6.1 Inverter Design for a Three-Pole Comb Line Filter

In order to eliminate the input and output inverters, the J_{01} and J_{34} inverters are set equal to the source and load admittance values, 0.02, for a 50-ohm-to-50-ohm filter. Using identical design resonators and the formulas at the end of Section 9.7.2, we have:

$$J_{0,1}^2 = \frac{w\omega_o C_r}{1 \times 1 \times 50} = 0.0004$$

$$\Rightarrow w\omega_o C_r = 0.02$$

$$J_{1,2} = \frac{\omega_o w C_r}{\sqrt{1 \times 2}} = 0.014142$$

From the formula at the end of Section 9.8.1, the effective capacitor of the resonator versus tap point is:

$$w\omega C_{\text{eff}} = \frac{\pi w Y_0}{4} \csc^2(\theta_1) = 0.02$$

$$\sin(\theta_1) = \sqrt{\frac{\pi w Y_0}{0.02 \times 4}}$$

In order to find a tap point, the characteristic impedance needs to be determined. A width-to-height ratio of 5 is used for the resonator. For a dielectric constant of 3.38, the characteristic impedance is 29.3 ohms. Microstrip Q tends to go up as the resonator gets wider. However, if the resonator becomes too wide, then there is significant phase shift across it for energy going through the filter. With the given characteristic impedance, the tap point is at 11.56 degrees from the shorted end.

The inverters between the resonator need to be determined next. An equation for a transmission line inverter was given in Section 9.8.2. There are many circuits that can be used for the inverter. However, if one uses the one described in Section 9.8.2 for this filter, the characteristic impedance for the inverter is:

$$J_{1,2} = Y_{\text{inv}} \csc(11.56°) = 0.014142$$

$$\Rightarrow Z_{\text{inv}} = 353 \text{ ohms}$$

That high a level of characteristic impedance cannot be realized on a 0.76-mm-thick substrate. The line would be on the order of 30 microns wide, an impractical width for a copper printed circuit board line. The line would have a very high loss as well. Since the transmission line inverter will not work well, a lumped-constant inverter will be tried. For this filter, since the tap point is close to the ground, much less than an eighth of a wavelength, the inductance of a stub of that length will not vary very fast versus frequency. A stub that is the length of line long from the open end of the resonator to the tap point will vary significantly with frequency. For this reason, an inductive inverter will work better than a capacitive inverter for this filter.

The value of an inductor or a capacitor to use for the inverter is:

$$J_{1,2} = 0.014142 = \frac{1}{\omega L} = \omega C$$

$$\Rightarrow L = 9 \text{ nH}$$
$$\Rightarrow C = 1.8 \text{ pF}$$

If the inductor is used as the inverter, the characteristic impedance of the bottom portion of the resonator must be increased to account for the negative inductance of the inverter. The equivalent characteristic impedance of the inverter is the negative of the characteristic impedance calculated for the TL inverter. Therefore, the characteristic impedance of the lower portion of the end resonators is:

$$Z_{\text{lower}} = \left(\frac{1}{29.3} - \frac{1}{353} \right)^{-1} = 32 \text{ ohms}$$

The middle resonator has an inverter on either side of it. Therefore, the characteristic impedance of the lower portion of the middle resonator is:

$$Z_{\text{lower}} = \left(\frac{1}{29.3} - \frac{1}{353} - \frac{1}{353} \right)^{-1} = 35.1 \text{ ohms}$$

If a capacitor is used for the inverter, then the lower portion of the resonator can have their impedances reduced (using plus signs in the two equations above) or the top portion of the resonators can have the same impedances as calculated above. Since the tap point is so close to the ground, the first choice would be better, but since the first choice is really using an inductor in place of a negative capacitor, the inverter will not track well in frequency and will give a bandwidth change. The second choice uses a negative-characteristic-impedance transmission line stub to compensate for a negative capacitor. Again, since the length is long, it will not track well with frequency and will cause a change in the bandwidth of the filter.

15.6.2 Physical Design for a Three-Pole Comb Line Filter

The characteristic impedance and effective dielectric constant for the transmission line resonators for the different electrical lengths are:

	Z0	$\varepsilon_{\text{reff}}$	Length-mm
Bottom stub: 11.56 deg	32	2.81	4.59
Bottom stub: 11.56 deg	35.1	2.78	4.62
Top stub: 78.44 deg	29.3	2.83	31.06

Notice that the lower portion of the center resonator is slightly longer than the lower portion of the outer resonators. The filter is shown in Figure 15.30. Notice the difference in the width of the lower portion of the resonators. If capacitors were used for inverters, then either the lower portions of the resonators would be wider or the upper portions of the resonators would be narrower.

There are dashed lines on the tops and bottoms of the resonators. The dashed lines on the top show how much of the resonator has to be cut off to account for the fringing of the resonator as described in Section 2.7.7. It was shown there that the amount that should be cut off is 0.84 times the thickness of the dielectric for wide lines on 3.38 dielectric constant material. For

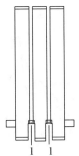

Figure 15.30 Three-pole comb line filter using inductor inverters.

0.76-mm-thick material, this is 0.64 mm. The dashed lines on the bottom show where the straps to ground would have to start. Unless a full ground plate is used on the ground end of the resonators (not a likely procedure on a printed circuit board), then the strap to ground acts like an inductance. The length of line is approximately equal to the dielectric thickness. There is hardly any capacitive energy at the shorted end of the resonator and the difference in capacity for the ground strap versus some stub length is very small. As a first-order approximation, the lower portion of the resonator would be shortened by the dielectric thickness.

15.6.3 Calculated Responses for a Three-Pole Comb Line Filter

Figure 15.31 shows the nose bandwidth response of the filter. There are three different responses on the figure. The one labeled "L" is for the filter with inductor inverters. It has a 3-dB bandwidth very close to 3%. The one labeled "C" is using capacitor inverters with the top portions of the resonators narrowed to compensate for the negative capacities. The 3-dB bandwidth at the filter is 3.35%. Not shown on the figure is the response for a filter using capacitive inverters but with the lower portion of the resonators widened to compensate for the negative capacities. That filter has a 3-dB bandwidth of 2.89%. The third curve on the figure is for a filter designed with a transmission line inverter. That filter will be discussed later.

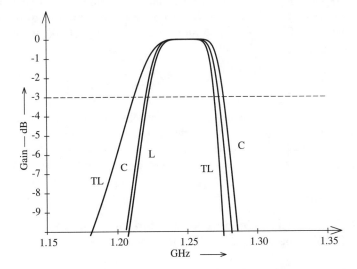

Figure 15.31 Nose response of the three-pole comb line filter.

Figure 15.32 shows the wide band response of the filters. The responses are labeled like the responses in Figure 15.31 except two different TL responses are shown. They will be

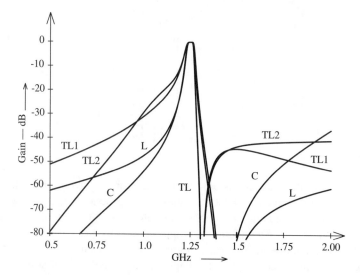

Figure 15.32 Wideband response of the three-pole comb line filter.

discussed later. Notice that the reject band responses of the "L" and "C" filters are significantly different. Notice also that there is a zero of transmission in the filters. This is a result of the distributed nature of the filter. When the length from the open end of the resonators to the inverter tap points is a quarter-wavelength, the response is shorted out and the zero appears.

When inductive inverters are used, at low frequencies the filter looks like a series L, shunt L, series L, shunt L circuit, and so on. When capacitive inverters are used, at low frequencies the filter looks like a shunt L, series C, shunt L circuit, and so on. This is a high-pass type of circuit and has a higher rejection than the inductor inverter filter at low frequencies. The opposite can be said for the high-frequency portion of the reject band. Although not shown, there is another passband when the resonators are three-quarters of a wavelength long. This is a result of the distributed nature of the transmission line resonators.

Loss due to finite resonator Q is not shown in Figure 15.31. Using the formula given in Section 9.14, the losses for a resonator can be closely approximated by placing a shunt resistor on the open end of the resonator. Assuming a resonator Q of 400, the value of the resistor is given by:

$$R_P = \frac{4 \times 400 \times 29.3}{\pi} = 14.9 \text{ k}$$

The inductor inverter filter response was calculated using that value of resistance on the ends of the resonators. Figure 15.33 shows the results of loss in the resonators. The passband loss has increased to just over 1.4 dB. Figure 15.34 shows the return loss for the filter with and without resonator loss. The return loss bandwidth is slightly wider for the filter with loss as expected.

15.6.4 Transmission Line Inverter Design for a Three-Pole Comb Line Filter

A transmission line inverter will be designed for a three-pole comb line filter. As indicated in Section 15.6.1, the characteristic impedance for a transmission line inverter for a filter without input and output inverters and a 3% bandwidth filter was too high to be realized. Assuming that a 0.2-mm line is the smallest that can be expected to be reproducible, the impedance of

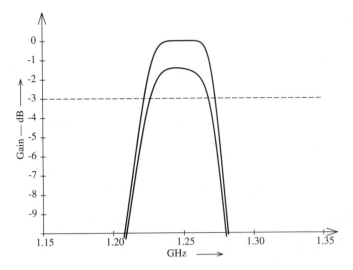

Figure 15.33 Insertion loss of the inductive inverter filter with and without resonator loss.

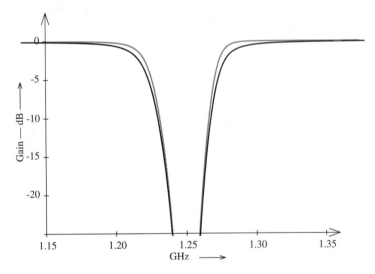

Figure 15.34 Return loss of the inductive inverter filter with and without resonator loss.

the inverter is about 133 ohms. The tap point and length of the inverter are then:

$$J_{1,2} = \frac{1}{133} \sin(\theta) = \frac{0.03\pi}{4\sqrt{2}(29.3)}$$

$$\Rightarrow \theta = 4.33°$$

This is a very short length from the tap point to the ground. In fact, it will be hard to repeatably place the inverter on the top of the board since the board is about the equivalence of 2 degrees thick. However, the design will be made and the response will be calculated. The lower end of each of the end resonators has a characteristic impedance of 37.6 ohms and the center resonator has a characteristic impedance of 52.4 ohms. The input and output inverter has a value of:

$$(J_{0,1})^2 = \frac{0.03\pi}{4(29.3)50} \frac{1}{\sin^2(4.33°)}$$

$$\Rightarrow J_{0,1} = 0.0531$$

This means that the first resonator sees:

$$R = \frac{0.02}{(J_{0,1})^2} = \frac{0.02}{0.00282} = 7.08 \text{ ohms}$$

Using a $Q^2 + 1$ match, the transformation Q is 2.46. This is a little high for using on a filter but since the bandwidth Q of the filter is just over 30, a factor of at least 10 higher, the match will be attempted. Either a shunt C series L match or a shunt L series C match will accomplish the transformation. In the passband, the responses will be similar but in the reject band they will be different. The difference is seen in Figure 15.32, where "TL1" uses the shunt C series L transformation and "TL2" uses the shunt L series C transformation. The magnitude of the reactance of the series part is 17.4 ohms and the magnitude of the reactance of the shunt part is 20.3 ohms. Notice from Figure 15.31 that the nose shape of the TL filters is shifted down and has some slope. Since the tap point is so close to ground, the zero of transmission is just above the passband.

If one desires to place the zero at some frequency just above the passband, the position of the tap point can be chosen based on that frequency. That length dictates the value of the inverters. The choice of the type of inverter will depend on whether each type is realizable.

A

An Approximate Formula for the Characteristic Impedance of a Microstrip Line

An approximate formula for the characteristic impedance as given by Wheeler [1] but written in a slightly different form follows.

$$Z_0 = \frac{42.4}{\sqrt{1.+\varepsilon_r}} \ln\left\{1.+\left(\frac{4}{\hat{W}}\right)\left[\left(\frac{14\varepsilon_r+8}{11\varepsilon_r}\right)\left(\frac{4}{\hat{W}}\right)+\sqrt{\left(\frac{14\varepsilon_r+8}{11\varepsilon_r}\right)^2\left(\frac{4}{\hat{W}}\right)^2+\frac{\varepsilon_r+1}{2\varepsilon_r}\pi^2}\right]\right\}$$

$$\hat{W} = \frac{W}{H}$$

W is the width of the line for a zero thickness line and H is the thickness of the dielectric material. W is adjusted with the use of an algorithm given in [1] when the line has finite thickness although the effect on the characteristic impedance for thin lines is not too significant if the lines are not too narrow. The formulas are stated to be accurate to under 2%.

The effective dielectric constant of the line is determined from the ratio of two characteristic impedances. First the characteristic impedance is calculated using a dielectric constant of 1. Let that characteristic impedance be designated as $Z_0(1)$. Next the characteristic impedance is calculated using the actual dielectric constant. Let that characteristic impedance be designated as $Z_0(\varepsilon_r)$. The effective dielectric constant is then given as:

$$\varepsilon_{\text{reff}} = \left(\frac{Z_0(1)}{Z_0(\varepsilon_r)}\right)^2$$

If the characteristic impedance is known and the line width is desired, then the inverse formula is used. This is written as:

$$\hat{W} = 8\frac{\sqrt{\left[\exp\left(\frac{R}{42.4}\sqrt{\varepsilon_r+1}\right)-1\right]\frac{7\varepsilon_r+4}{11\varepsilon_r}+\frac{\varepsilon_r+1}{8\varepsilon_r}\pi^2}}{\left[\exp\left(\frac{R}{42.4}\sqrt{\varepsilon_r+1}\right)-1\right]}$$

When the ratio of W to H is determined, that ratio is put back into the impedance formula using a dielectric constant of 1. Then the desired characteristic impedance and the characteristic impedance for a unity dielectric constant are used to determine the effective dielectric constant as was shown above.

One can determine the approximate electrical width used to foreshorten open stubs as discussed in Section 2.7.7 using these formulas. Two values of impedance close to the actual value of the transmission line characteristic impedance are used. These two values of characteristic impedance determine two values of the ratios of W to H. Those values are used to calculate the value Δx that the line appears to be wider on each side as shown here.

$$S = \left(\frac{\dfrac{1}{Z_0(\widehat{W}_2)} - \dfrac{1}{Z_0(\widehat{W}_1)}}{\widehat{W}_2 - \widehat{W}_1} \right)$$

$$\Delta x = \frac{1}{2}\left[\frac{1}{S}\frac{1}{Z_0(\widehat{W}_2)} - \widehat{W}_2 \right]$$

Since the slope of characteristic admittance versus width is not actually a straight line, the value of the offset varies as a function of line width. Charges across the end of the resonator are affected not only by the width of the line but by the length of the line also since the end of the line is orthogonal to the length dimension. The length of the resonator is usually much longer than its width, and one should check what the offset is for a wider (longer) line as well. For very narrowband filters and high Q resonators, some experimentation will be necessary to set the actual frequency. However, remember that the circuit layer etching tolerance is not zero and the dielectric constant of the board varies somewhat from batch to batch. All of these factors need to be considered when setting the zero point for line length.

It is interesting to note that when the width formula is expanded for very small R, by replacing the exponential term with the exponent plus 1 and ignoring the subsequent smallest term in the numerator, one gets:

$$Y_0 = \frac{1}{Z_0} = \frac{1}{\sqrt{8}}\frac{1}{42.4\pi\sqrt{\varepsilon_r}}\hat{W} = \frac{1}{376.7\sqrt{\varepsilon_r}}\hat{W}$$

This is the expected value of the characteristic impedance for a wide line. The capacity of the wide line is only the capacity directly under the line.

B

Some Complex
Variable Facts

In order to understand one of the tools of the microwave engineer, the Smith Chart™, it is helpful to consider the description of lines and circles in the complex plane. Consider the circle in the complex plane shown in Figure B.1. The equation of a circle in the complex plane is:

$$|Z - Z_c| = \rho$$

or:

$$(Z - Z_c)(Z^* - Z_c^*) = \rho^2$$
$$ZZ^* - Z_c Z^* - Z Z_c^* + Z_c Z_c^* = \rho^2$$

The equation for a straight line in the complex plane is:

$$BZ + B^* Z^* + C = 0$$

If the slope of a straight line is not infinite, the equation of a straight line expressed by $z = x + jy$ where $jy = jmx + jb$ is:

$$(m - j)z^* + (m + j)z + 2b = 0$$

For a line with infinite slope:

$$z^* + z - 2Re(z) = 0$$
$$z^* + z = 2Re(z)$$

For a line of zero slope:

$$-jz^* + jz + 2b = 0$$
$$z - z^* = 2j\,\text{Im}(z)$$

A Special Conformal Transformation

The transformation:

$$w = u + jv = \frac{Az + B}{Cz + D}$$

$$z = x + jy$$

is called a *bilinear transformation* to the w plane from the z plane. A bilinear transformation transforms a circle in one plane to a circle in another plane. A point is a circle with a zero radius and a straight line is a circle with an infinite radius. Therefore, straight lines (circles of infinite radius) in the impedance plane would transform to circles in the reflection coefficient plane.

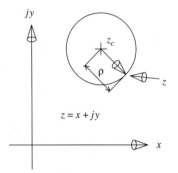

Figure B.1 Circle in the complex plane.

A bilinear transformation is one of many conformal transformations. A conformal transformation maps two curves that intersect at right angles in one plane onto another plane where the mapped curves also intersect at right angles. The reflection coefficient:

$$\Gamma = \frac{Z - Z_0}{Z + Z_0}$$

is a bilinear transformation. Therefore, straight lines in the impedance plane will map into circles in the reflection coefficient plane. Reactance lines are orthogonal to resistance lines in the impedance plane. Therefore, these lines mapped into the reflection coefficient plane will also cross at right angles. Since Y is the inverse of Z, straight lines and circles in the admittance plane also map into circles in the reflection coefficient plane.

C

Matrix Multiplication

In order to use $ABCD$ and T matrices, one needs to be able to multiply two matrices together. A review of matrix multiplication is given here. The chain matrices are 2×2 matrices. In general, in a matrix product, the number of columns in the first matrix must be the same as the number of rows in the second matrix forming the product. The resultant matrix will have the number of rows of the first matrix and the number of columns of the second matrix. For a first matrix that has r rows and n columns and a second matrix that has n rows and c columns, the multiplication of the two matrices can be expressed by the following summation. The subscript i ranges from 1 to r the number of rows in the first matrix and the subscript k ranges from 1 to c the number of columns in the second matrix.

$$(c) = (a)(b)$$

$$c_{ik} = \sum_{j=1}^{n} a_{ij} b_{jk}$$

The first subscript designates which row and the second subscript designates which column the term in the product matrix is found at. This is demonstrated by the following sequence of operations for the multiplication of two 2×2 matrices.

$$\begin{pmatrix} c_{11} & c_{12} \\ c_{21} & c_{22} \end{pmatrix} = \begin{pmatrix} a_{11} & a_{12} \\ a_{21} & a_{22} \end{pmatrix} \begin{pmatrix} b_{11} & b_{12} \\ b_{21} & b_{22} \end{pmatrix}$$

For c_{11}, take the first row of (a) times the first column of (b) as shown.

$$\begin{pmatrix} c_{11} & (\ ) \\ (\ ) & (\ ) \end{pmatrix} = \begin{pmatrix} a_{11} & a_{12} \\ (\ ) & (\ ) \end{pmatrix} \begin{pmatrix} b_{11} & (\ ) \\ b_{12} & (\ ) \end{pmatrix} \Rightarrow \frac{\begin{matrix} (a_{11} & a_{12}) \\ (b_{11} & b_{21}) \end{matrix}}{c_{11} = a_{11}b_{11} + a_{12}b_{21}}$$

In order to form a term, c_{ik}, the first vector is from the ith row, in this case the first row, and the second vector is from the kth column, in this case the first column. Notice that only the same component positions of the vectors are multiplied by each other and then added. There are no cross-product terms. This is an inner product for the two vectors. The other three terms are found in a similar manner.

$$\begin{array}{ccc} (a_{11} \quad a_{12}) & (a_{21} \quad a_{22}) & (a_{21} \quad a_{22}) \\ \underline{(b_{12} \quad b_{22})} & \underline{(b_{11} \quad b_{21})} & \underline{(b_{12} \quad b_{22})} \\ c_{12} = a_{11}b_{12} + a_{12}b_{22} & c_{21} = a_{21}b_{11} + a_{22}b_{21} & c_{22} = a_{21}b_{12} + a_{22}b_{22} \end{array}$$

If there were four columns in the first matrix and four rows in the second matrix, then the ikth term would be:

$$\begin{pmatrix} c_{ik} \end{pmatrix} = \begin{pmatrix} a_{i1} & a_{i2} & a_{i3} & a_{i4} \end{pmatrix} \begin{pmatrix} b_{1k} \\ b_{2k} \\ b_{3k} \\ b_{4k} \end{pmatrix} \Rightarrow \begin{array}{c} (a_{i1} \quad a_{i2} \quad a_{i3} \quad a_{i4}) \\ \underline{(b_{1k} \quad b_{2k} \quad b_{3k} \quad b_{4k})} \\ c_{ik} = a_{i1}b_{1k} + a_{i2}b_{2k} + a_{i3}b_{3k} + a_{i4}b_{4k} \end{array}$$

Resistor, Capacitor, and Inductor Component Modeling

Laboratory guides similar to these were used at Iowa State University during the 1990s. They were originally compiled in cooperation with the author by Dr. Paul Stucky and are included with his permission.

The structure of these laboratories is basically unchanged when going from a leaded component to a chip component. The effect of the contact on a chip is not as large as the effect of a lead on a leaded component. The reader will appreciate that there are still instances where leaded components are used. The reader may want to measure leaded versus chip components to ascertain these differences.

One key to good characterization is to have a good fixture. The discussion in Section 2.8 should be reviewed, especially that part dealing with the short. A strap to ground is not a short. A wire to ground is definitely not a short. An open circuit does not exist at the end of the conductor on a transmission line. There are good commercial fixtures available for characterization of chips. Keep in mind that these are very good for reproducible measurements from chip to chip and from batch to batch. However, the real parasitics that one desires to measure are the parasitics that are the same as those actually in use on a PCB or MMIC.

These laboratory guides can be used with MMIC components with some minor modification. The MMIC component test fixture will often use matched microprobes on a process control monitor (PCM) portion of a chip. The components to be measured will have a set of pads associated with them. These pads are included on the PCM and the calibration patterns so that they can be effectively de-embedded from the measurements. The magnitude of the parasitics for MMIC components will be different from those for PCB components but the same parasitics exist.

RESISTOR MODELING

Purpose: An accurate lumped-element model is needed for a resistor of the type that will be used in the circuits that are going to be constructed. For instance, one may wish to use one-eighth-watt carbon composition, carbon film, or metal film resistors. Resistors with different compositions and different form factors behave differently at rf and microwave frequencies. A model is required for each type of resistor to be used in the circuit designs for realistic computer simulation of the overall microwave circuit. Careful attention needs to be given to lead position, often called lead dress, when using leaded components. Many manufacturing processes require that a strain relief loop be put in the lead of leaded components. The user will want to measure the component with all of the required mechanical conditions in mind. The model will incorporate parasitic capacitance and inductance not considered at lower frequencies. The type of resistor used in any laboratory exercise is based on the type that is to be used in later exercises and designs. Leaded types that might have been used in prior laboratories in electrical engineering courses should be characterized. Chip resistors are preferable for microstrip circuit design and therefore should be characterized.

Prelab

1. Figure D.1 shows a leaded resistor in place on a microstrip circuit. Think in terms of the electric and magnetic field and answer the following questions. Consider what would be different if the resistor were a chip resistor instead of a resistor with leads. Sketches should be given where appropriate.
   (a) What is the approximate electric and magnetic field configuration in the undisturbed portion of the microstrip line?
   (b) What is the probable electric and magnetic field configuration in the portion of the T-line containing the resistor? Think about the differences that are expected depending on the value of the resistor.

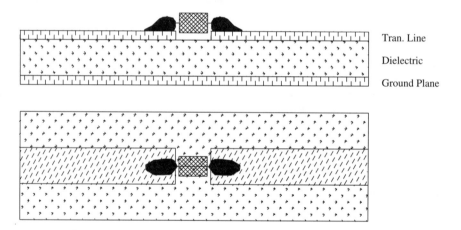

Figure D.1 Cross section of a chip resistor on a PCB.

2. One possible lumped-constant equivalent circuit for the resistor mounted on a substrate is given in Figure D.2. Some equivalent circuits with more components can also be used to model the resistor. The shunt capacitors could be positioned at different nodes but at those frequencies for which that placement is more accurate, the model should really be replaced with a distributed model. This equivalent circuit is chosen because each component is readily identifiable with a physical portion of the resistor. Each component in the model approximates (or represents) some physical aspect of the actual resistor in the microstrip circuit of Figure D.1. Be clear and complete in the answer given for the following and remember that this is only a model of the resistor as it resides on top of the microstrip line and substrate with the ground plane beneath. (*Ls* is for inductor, *R* for resistor, *Cp* for parallel capacitance, and *Cs* for shunt capacitance.)

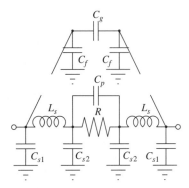

Figure D.2 Equivalent circuit of a resistor on a PCB.

(a) Why is the circuit model symmetrical?

(b) What physical parameter of the resistor, as it resides on the microstrip line, does each circuit component model and why?

3. Figure D.1 is representative of the test fixture that is used to measure the parasitic component values. It has its own equivalent circuit and must be taken into account. Consider Figure D.1 without the resistor.

(a) What would be an equivalent circuit of the gap in the microstrip?

(b) Give a sketch of the gap showing where the parasitic components of the model belong.

(c) Discuss briefly why each particular parasitic has been included.

4. In the actual lab experiment the S parameters of the resistor are measured in its test fixture. These measurements are used to estimate the values of the parasitics in Figure D.2. In order to calculate values of inductance and/or capacitance, the S parameters must be derived. It can be shown that S_{21} for the circuit in Figure D.3 is as given below. Assume Z_L is zero and Z_s is infinite ($\alpha = l$ and $\beta = 1$), then clearly show a complete derivation of S_{21}. Assume the characteristic resistance, Z_0, is the same for each port. Once S_{21} is obtained, then S_{12} is also found since the circuit is reciprocal. The circuit is also symmetrical since Z_{01} is equal to Z_{02}. Derive S_{11} by using the same assumptions, that is, assume Z_L is zero and Z_S is infinite ($\alpha = l$ and $\beta = 1$).

$$S_{21} = \frac{2Z_0}{\beta^2 Z + \alpha\beta Z_0}$$

where:

$$\alpha = 1 + \frac{Z_L}{Z_0}$$

$$\beta = 1 + \frac{Z_0 + Z_L}{Z_S}$$

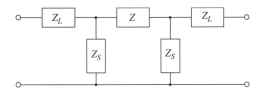

Figure D.3 S-parameter impedance model for a resistor.

Resistor Measurement

Purpose: An accurate lumped-element model is needed for a resistor of the type that will be used in the circuits that are going to be constructed. For instance, one might wish to use one-eighth-watt carbon composition, carbon film, or metal film resistors. Resistors with different compositions and different form factors behave differently at rf and microwave frequencies. A model is required for each type of resistor that is to be used in the circuit designs for realistic computer simulation of the overall microwave circuit. The model will incorporate parasitic capacitance and inductance not considered at lower frequencies. The type of resistor used in any laboratory exercise is based on the type that is to be used in later exercises and designs. Leaded types that have been used in prior laboratories in other electrical engineering courses or laboratories should be characterized. Chip resistors are preferable for microstrip circuit design and therefore should be characterized.

Measurement of Shunt Capacitance, C_s

1. Preset the Vector Network Analyzer and set the frequency range from f_L to f_H. The low-frequency limit f_L of the range often is set equal to the low-frequency limit of the Vector Network Analyzer. The high-frequency limit f_H is set to the highest frequency needed in the simulation. This is determined by the upper limit of validity for the test fixture, the highest frequency of the Vector Network Analyzer, or the highest frequency needed in the circuit simulation. Typical numbers for 1-GHz design using devices with maximum frequency of oscillations of 4 GHz might be from 0.5 MHz to 5 GHz assuming the Vector Network Analyzer and test fixture operate over that range.

2. Perform a full two-port calibration on the resistor test fixture (see Figure D.4). Refer to the user's or reference guide for the *Vector Network Analyzer* or consult with the lab instructor to answer any questions on how to do the calibration. The easiest way to learn how the instrument works is to experiment with it with the help of the *User's Guide*.

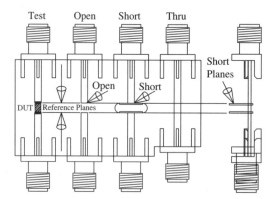

Figure D.4 Resistor test fixture.

3. The measurement of C_s is done using the open-circuit lines on the test fixture shown in Figure D.4. The test board material should be the same as the material used for the circuit that will be designed. In order to limit the interaction between one side of the fixture and the other side of the fixture, terminate the opposite side's transmission line in a load. Whether C_{s1} or C_{s2} is the best model depends on the physical configuration of the resistor.

4. The fringing capacitance at the end of the open-circuit line needs to be taken into account since a microwave open circuit is never perfect. Determine a value for the fringing capacitance from the S_{11} measurement. This value of capacitance will be small (likely less than 1 pF).

5. If the calibration and choice of the reference plane is satisfactory, then solder a 47-ohm resistor to the end of the open-circuit line of the test fixture and record the observations. Do not connect the other end to the other side of the line. Pick one frequency in the frequency range being used. (1 GHz is a good choice for common chip sizes.) Determine the capacitance from the impedance measurement. Is this capacitance relatively constant over the frequency range, and why or why not (give an opinion)? (Consider the effect of an incorrect reference plane calibration.)

6. Sketch an equivalent circuit of the fixture and resistor including the parasitics and note the position of the reference plane. Determine an approximate value for the shunt capacitance and discuss it with the lab instructor. Do not be surprised if the value of shunt capacitance is less than a picofarad. The shunt capacity of a leaded resistor will be small if it is mounted some distance from the ground plane.

Measurement of Series Inductance, L_s

1. Measure the inductance of the resistor. Connect the short-circuit line to the network analyzer and determine the position of the reference plane.

2. Solder the 47-ohm resistor across the gap and place a short circuit on the far end of the T line. The display on the network analyzer should give a locus of points that pass through the open-circuit point on the Smith chart. This is the result of the effect of standing waves on the T-line. At the load a short exists and at the reference plane at a certain frequency a high impedance (approximately an open circuit) exists.

3. Record the frequency at which there is an open circuit at the reference plane (use the marker function). This frequency should be the frequency at which the distance from the short to the reference plane is a quarter wavelength.

4. Replace the short with an open circuit. Now the short and open will switch places and a short is placed at the reference plane. Notice the impedance in question lies along the 50-ohm constant resistance circle and is in the inductive portion of the Smith chart. Use the marker to find the impedance at the quarter-wavelength frequency. The inductance of this point is approximately the inductance of the resistor. Record the inductance. Now replace the open with a 50-ohm termination. The 47-ohm resistor in series with the 50-ohms from the transmission line gives 97 ohms of resistance. The network analyzer will display approximately 100 ohms plus an inductance. This inductance is the inductance of the resistor. Compare this with the value from the short, open measurement. *Note:* If the resistor is physically small, a smaller resistance value than 47 ohms may be needed in order to accurately determine the inductance and capacitance of the resistor. Why?

5. Sketch an equivalent circuit of the resistor in the test fixture. What are the equivalent circuits when the short and open are on the line, respectively? What parasitics have been neglected in this measurement and why?

6. Estimate the inductance by measuring the length of the resistor as a whole and using the rule of thumb: 4nH per cm or 1 nH per 0.1 inches. What is the estimate of inductance and how does it compare to the measured value? Remember that the inductance depends on the total length of line from one side of the gap to the other side. This includes the lead and the resistor body!

7. Different results are often obtained if the resistor is physically shorted to ground at its far end rather than by the method described above. If there is time and an appropriate test fixture try this and compare and explain the results.

Measurement of Parallel Capacitance, C_p

1. This measurement is often difficult since the value of the capacitance is relatively small. Essentially this parasitic is due to the two leads that enter the ends of the resistor. For this measurement measure S_{21}. Also characterize the test fixture since it has a gap capacitance that is on the order of the parallel capacitance.

2. Preset the Vector Network Analyzer and set the frequency range from f_L to f_H. The low-frequency limit f_L of the range often is set equal to the low-frequency limit of the Vector Network Analyzer. The high-frequency limit f_H is set to the highest frequency needed in the simulation. This is determined by the upper limit of validity for the test fixture, the highest frequency of the Vector Network Analyzer, or the highest frequency needed in the circuit simulation. Typical numbers for 1-GHz design using devices with maximum frequency of oscillations of 4 GHz might be from 0.5 MHz to 5 GHz assuming the Vector Network Analyzer and test fixture operate over that range.

3. With no resistor in the gap on the test fixture, make a transmission measurement (S_{21}) of the gap itself. Assume the gap coupling capacitance C_g is the only parasitic and then determine the coupling capacitance from the measurement. Use the formula for S_{21} derived in the prelab with Z_L equal to zero and let Z_S approach infinity. If S_{21} was derived correctly, it should be equal to the following:

$$S_{21} = \frac{2Z_0}{2Z_0 + Z}$$

Solve this for Z and obtain an expression for the gap coupling capacitance based on the measurement of S_{21}. Discuss the results with the lab instructor.

4. Now solder a 10- to 15-kilo-ohm resistor across the gap. The reason for using such a relatively large resistor rather than using the 47-ohm resistor, is that the impedance of the capacitance to be measured is large. If a 47-ohm resistor were used, the transmission measurement would be dominated by the 47-ohm resistor rather than by the parallel capacitance. Remember that for two impedances in parallel—one small, the other large—the impedance is essentially that of the smaller impedance; therefore, the larger impedance is difficult to extract from the measurement.

5. Make the S_{21} measurement and record the observations. Now using the model of the resistor derived in the prelab and the model of the gap from the prelab, derive the expression for the unknown Z. Neglect the inductance of the leads (the impedance of the leads is small and neglecting them makes the ensuing calculation easier). The gap coupling capacitance, C_g, is then in parallel with the parallel capacitance of the resistor, C_p. For chip resistors, the parallel capacitance of the resistor essentially increases the gap coupling capacitance.

6. Determine the value of the parallel capacitance and discuss the results with the instructor. In a project at 1 GHz, this capacitance may not have any effect at all, but the point is that it exists. As the frequency increases past 3 GHz its effect will become significant and it might be important in the stability analysis of an amplifier or provide a possible spurious oscillation path in an oscillator. These effects build an intuitive sense about the factors contributing to the performance of devices and components and for that reason this measurement is important.

7. At this point the measurement portion of the resistor modeling lab is completed. Now perform the computer modeling for the resistor laboratory and then write up the lab.

Computer Modeling

1. Verify the model that was measured by characterizing it using the computer. Hopefully, all the important parameters of the test fixture and resistors have been measured. If they have not, the measurements must be done.

2. Model the gap capacitance. Alter the value of the model until it approximates the measured data. How do the model and measured value of C_g compare? Give a percent difference and plot the modeled response. Overlay the measured data on the graph of the computed data.

$$\Delta\% = \left(\frac{\text{measured } - \text{ theoretical}}{\text{theoretical}} \right) \times 100\%$$

NOTE: If an open-circuit transmission line is used in a computer-aided design program, the computer program needs to be checked to determine whether the program includes compensation for the fringing capacity of the open-circuit line. (See the manual!) If a simple transmission line is used, then the program will not compensate for the open-circuit fringing capacity. The computer model will be affected by the use of an open circuit. The model does not want to include fringing capacity twice!

3. Add the model of the resistor. Note that the values of shunt capacitance and series inductance measured are total values and should be halved when entered in the symmetrical model. Also the gap capacitance spans, or is in parallel with, the whole resistor model.

4. Compare the model to the measured data by plotting the model and the measured data together. Now try altering (either by optimization or tuning using the computer-aided-design program) the model until a better approximation is obtained.

5. Compare the data against a model constructed from a lossy transmission line. The lossy line model will likely be best for chip resistors. Leaded resistors will likely fit the lumped-constant model better. The length of the resistor is the length between the contacts on the resistor. The resistance is usually input in terms of ohms/length in a transmission line model. The inductance and capacitance of the resistor are calculated from transmission line parameters for a transmission line that is the width of the resistance element.

CHIP CAPACITOR MODELING

Purpose: In order to design a printed circuit board using chip capacitors, it is necessary to develop an accurate lumped-element model of a monolithic ceramic chip capacitor over the frequency range of interest. The model is required for realistic computer simulation of the overall microwave circuit. The model will incorporate parasitic resistance and inductance not considered at lower frequencies. The model will be specific to a capacitor of a specific physical size and construction. Each different physical model needs to be characterized.

Prelab

1. Figure D.5 shows a chip capacitor placed in a microstrip circuit. Think in terms of the electric and magnetic field and answer the following question:

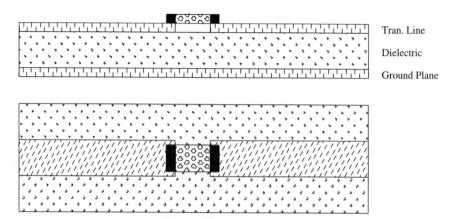

Figure D.5 Capacitor on a microstrip T-line.

(a) What is the possible electric and magnetic field configuration in the portion of the T-line containing the capacitor? Give a sketch illustrating the answer. Show a side view as well as a cross-section view.

2. Figure E.1 shows the top view of some chip capacitor sizes. These chips vary in thickness but the thickness is on the order of the width. Some vendors supply capacitors that are almost cubic in size. One possible equivalent circuit model for the chip capacitor mounted on a printed circuit board is given in Figure D.6. Think of each section of the chip capacitor individually. There are two metallic end sections and one center dielectric section, each of which has an equivalent circuit. As in the resistor modeling lab, each component in the model approximates (or represents) some physical aspect of the actual capacitor shown in Figure D.5. Be clear and complete in answering the following. (L_s and R_s are for series inductance and resistance, respectively, R_p for parallel resistance, C_p for component capacitance.) Shunt capacitance due to mounting the capacitor on a substrate is indicated by C_s. The fringing capacity C_f and gap capacity C_g of the printed circuit gap interact with these parasitics of the capacitor. The dashed lines connect with what the gap would look like without the capacitor. The capacitor as just indicated interacts with some of these elements. As a first-order approximation, if the parasitic inductance and series resistance are small, the gap capacitances are essentially just in parallel with the capacitor capacitances.

(a) Which components of the equivalent circuit coincide with the different sections of the capacitor (this is too easy!)?

(b) Explain how each circuit component models a physical parameter of the capacitor.

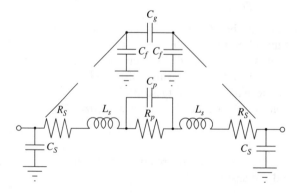

Figure D.6 Equivalent circuit of a capacitor on a PCB.

It is not shown in Figure E.1 but capacitors are often multilayered with electrodes (metal) and dielectric as alternating layers throughout the dielectric material.

Other possible equivalent circuit models that might be more valid in specific applications are discussed in Section 2.7.2. The primary difference is in where the shunt capacitors are placed.

3. Now it is necessary to perform some basic circuit analysis on the equivalent circuit. In order to save some time, refer to the *series and parallel resonance* in a circuits text that explains resonance, for example, Chapter 15 of [1].

   (a) Show that the equivalent series resistance and reactance of the model in Figure D.6 are given by the following:

   $$R_{Seq} = 2R_S + \frac{R_P}{1 + (\omega R_P C_P)^2}$$

   $$X_{Seq} = 2\omega L_S - \omega R_P C_P \left(\frac{R_P}{1 + (\omega R_P C_P)^2} \right)$$

   For this derivation, ignore the capacitance from the component to ground.

   (b) Assuming $R_p \gg R_s$, which resistance will dominate at low frequencies? at high frequencies? Which reactance, inductance or capacitance, will dominate at low frequencies? at high frequencies?

   (c) Show that the series resonant frequency is given by the following:

   $$\omega_o = \sqrt{\frac{1}{2L_S C_P} - \frac{1}{(R_P C_P)^2}}$$

   (d) Show that bandwidth, β, is given by

   $$\beta = \frac{R_{Seq}}{2L_S}$$

   and Q at resonance is given by

   $$Q = \frac{\omega_o}{\beta}$$

4. A usual assumption is that R_p is very large and therefore can be neglected at very high frequencies, but before assuming this some insight can be gained by considering its presence. Assume that there is a dial that can be turned to arbitrarily change the value of R_p from zero to infinity while everything else remains constant. Answer the following questions.

   (a) Turn the dial until R_p is equal to infinity—what does this imply about the capacitor dielectric?

   (b) Set R_p such that the resonant frequency is zero. A further decrease in R_p makes ω_o imaginary! Explain what it implies about the circuit as a whole if ω_o is zero or imaginary? (*Hint:* Suppose L_s and C_p are 1 nH and 5 pF, respectively; find R_p.)

5. Figure D.7 shows the placement of the chip capacitor in a test fixture, although not to scale. A small hole is drilled and the capacitor is soldered between the microstrip line and the ground plane. Consider the equivalent circuit as replacing the capacitor (assume R_p goes to infinity). Give a qualitative sketch showing transmission coefficient, S_{21}—log magnitude (insertion loss) and angle, as a function of frequency. Label the resonant frequency and the -3-dB frequencies.

6. Assume the capacitor is situated in the test fixture as shown in Figure D.7 and that R_p is infinite. The insertion loss in dB, abbreviated IL, is given by the equation:

$$IL_{dB} = -20\log_{10}(S_{21}) \quad (IL > 0)$$

where S_{21} is the transmission coefficient. At the resonant frequency the series resistance, R_s, can be determined directly from the insertion loss. Show that the series resistance is given by the following equation:

$$R_S = \frac{Z_0}{2(10^{\frac{IL}{20}} - 1)}$$

Draw the equivalent circuit and derive S_{21}. This derivation should be the same as the one in the inductor lab. It is helpful to sketch the test fixture circuit including the source, test fixture with capacitor, and load.

7. Typically an electrical engineering student has a reasonably good feeling for the concept of bandwidth (β, -3-dB bandwidth, half-power points, voltage or current magnitude decreased by the square root of two, etc.). On the other hand, the student likely needs to consider a little more thoroughly the definition and concept of Q or "quality factor." For these last questions, write a short paragraph or two on the subject of Q.

   (a) Find definitions of Q, be they electrical, mechanical, acoustical, etc. Do not make an exhaustive search but just see what definitions exist and write them down (equations or words). Remember that in general, Q refers to a single resonant structure.

   (b) Give an interpretation and ideas of Q in general. Refer to a paper[*] for a discussion of Q. Why is Q a useful figure of merit in electrical situations? (It might be an example; e.g., compare a high- and a low-Q circuit.)

Chip Capacitor Measurement

Purpose: In order to design a printed circuit board using chip capacitors, it is necessary to develop an accurate lumped-element model of a monolithic ceramic chip capacitor over the frequency range of interest. The model is required for realistic computer simulation of the overall microwave circuit. The model will incorporate parasitic resistance and inductance not considered at lower frequencies. The model will be specific to a capacitor of a specific physical size and construction. Each different physical version of a capacitor needs to be characterized to form a separate model for that version.

Measurement of ω_o, β, and Q

1. Select a capacitor for measurement. This capacitor can be mounted in a capacitor test fixture. Figure D.7 shows a test fixture for measuring a chip capacitor inductance.

2. Determine a proper calibration frequency range for the measurements. A quick way to determine the frequency range is to connect the capacitor test fixture to the Vector Network Analyzer and make an uncalibrated S_{21} log magnitude measurement over the analyzer's complete frequency range. An approximate value of the resonance frequency of the capacitor with its inductance should be apparent. Pick a suitable range to set the frequency range for calibration.

---

[*]Estill I. Green, "The Story of Q," *American Scientist*, vol. 43, #4, pp. 584–594, 1955.

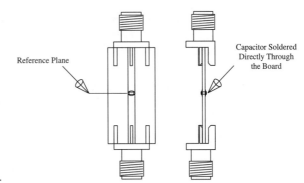

Reference Plane

Capacitor Soldered
Directly Through
the Board

Figure D.7 Capacitor inductance test fixture.

3. Preset the Vector Network Analyzer and set the frequency range as determined in part 1. Additional data can be obtained by setting the low-frequency end to a very low frequency.

4. Perform a full two-port calibration with the necessary connectors and cables attached to the Vector Network Analyzer. If there are any questions refer to the Vector Network Analyzer *User's Guide* or *Quick Reference*, or ask the lab instructor. The easiest way to learn how the network analyzer works and which connectors are needed is to experiment and refer to the *User's Guide*.

5. Record the analyzer settings for future reference. Be sure to record the position of the measurement reference plane(s). If the frequency range needs to be narrowed up in order to zoom in on the resonance, some Vector Network Analyzers have an interpolate function that can be turned on. This function is sometimes found in some analyzers by pressing the CAL key.

     The correct procedure for setting interpolation on a many Vector Network Analyzers is, (1) turn the calibration off, (2) set the low and high values of the desired frequencies of interest (or Δf, and fo), (3) turn interpolation on, (4) then turn calibration on. To reset the original calibrated frequency setting, (1) turn calibration off, (2) turn interpolation off, and (3) turn calibration on. This order allows the network analyzer to set the frequency ranges before it attempts to do a calibration interpolation.

6. Now it is time to actually measure something. If all has gone well, the transmission measurement will have a definite "suck-out." The capacitor mounted directly across from the transmission line to ground functions as a notch filter. Use the marker(s) to determine the resonant frequency, the upper and lower 3-dB bandwidth frequencies, and the insertion loss at the resonant frequency. Q can then be computed knowing ω_o and β. All this can be done manually or with some analyzers the analyzer can perform the measurement. Check the *User's Guide* to see how to set the analyzer to find a 3-dB down point from some specified value. The network analyzer often is better at finding 3 dB down than the user is.

Determining R_s and L_s

1. Assume R_p is infinite, determine L_s from ω_o and the nominal value of C_p. How does this value compare with the estimate of 4 nH per cm (1 nH per 0.1 in.)?

2. Knowing L_s and β, determine R_s. Assuming R_p is infinite, what is R_s?

3. Determine R_s from the value of insertion loss at resonance and the characteristic impedance of the transmission line. (Refer to the prelab.) Are the values obtained for

part 2 and part 3 the same? Give the percent difference. If the values are not equal then give reason(s) for the difference. Specifically, where are other sources of loss or error in the test setup? What other measurement(s), transmission or reflection, should be made to determine sources of the loss? Perform those measurements if possible.

4. Draw and label an equivalent capacitor circuit showing only the total series resistance, inductance, and capacitance. For future design purposes the only parameter that should change in the model of a given chip type is the series capacitance.

5. At this point the measurement portion of the lab is complete. Discuss the results and overall observations with the lab instructor before leaving the lab. Was the test fixture correctly calibrated? Record the test setup, etc.

Computer Modeling

1. Now perform the computer modeling of the capacitor equivalent circuit. Use the measured values of the parasitics in the CAD model and compare the S_{21} given by computer-aided-design program and the Vector Network Analyzer. Quantitatively compare the Q, bandwidth and resonant frequency of the Vector Network Analyzer model to the measured values. Give the percent difference between measured parameters and computed parameters. Try tuning the parameters of the circuit model to get a closer fit and observe the behavior of the circuit as the parameters change. Specifically, what happens when R_s, L_s, and C_p change respectively and the other two components remain constant? Give appropriate plots and circuit files (with comments) documenting these modeling efforts.

2. Discuss any "abnormalities" that may have arisen in the course of the measurements. If there were no problems, state that. The comments on measurement abnormalities are extremely helpful in identifying variations in components in different batches or different chip types.

INDUCTOR MODELING

Air-Core Inductor Modeling

Purpose: In order to design a printed circuit board using inductors, it is necessary to develop an accurate lumped-element model of a multiturn air-core inductor over the frequency range of interest. The model is required for realistic computer simulation of the overall microwave circuit. The model will incorporate parasitic capacitance and resistance not considered at lower frequencies. The model will be specific to an inductor of a specific physical size and construction. Each inductor using a different physical construction needs to be characterized. This laboratory exercise is applicable to an air-wound inductor, a chip inductor, or an inductor fabricated on magnetic core material.

Prelab

1. Figure D.8 shows an inductor mounted on a microstrip circuit. Think in terms of the electric and magnetic field and answer the following question:
   (a) What are the possible electric and magnetic field configurations in the portion of the T-line containing the inductor? Give a sketch illustrating these configurations.

2. One possible equivalent circuit model for the inductor mounted on a printed circuit board is given in Figure D.9. Think of the inductor mounted on the microstrip as shown in Figure D.8. As in the capacitor modeling lab, each component in the model

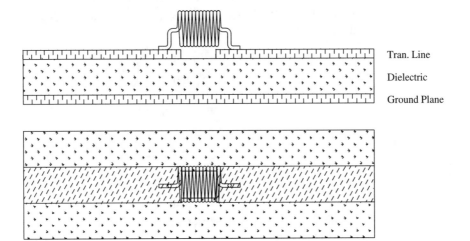

Figure D.8 Inductor on a microstrip T-line.

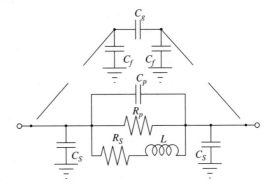

Figure D.9 Air-core inductor equivalent circuit.

approximates (or represents) some physical aspect of the actual inductor. Be clear and complete in answering the following questions and remember this is a model of the inductor as mounted on the microstrip line and substrate.

(a) What physical parameter of the inductor does each circuit component model and why?

3. Consider the four components in the center of the model, R_s, R_p, C_p, and L_s in the following analysis.

(a) Show that the equivalent parallel conductance and susceptance of the model are given by the following:

$$G = \frac{R_P + R_S(1 + \tau^2)}{R_P R_S(1 + \tau^2)}$$

$$B = \frac{\omega L_S \left(1 + \dfrac{1}{\tau^2}\right) - \dfrac{1}{\omega C_P}}{\omega L_S \left(1 + \dfrac{1}{\tau^2}\right) \left(\dfrac{1}{\omega C_P}\right)}$$

where:

$$\tau^2 = \frac{\omega^2 L_S^2}{R_S^2}$$

(b) Show that the parallel resonant frequency, bandwidth, and quality factor are given by the following:

$$\omega_o = \sqrt{\frac{1}{L_S C_P} - \left(\frac{R_S}{L_S}\right)^2}$$

$$\beta = \frac{R_S}{L_S} + \frac{1}{R_P C_P}$$

$$Q = \frac{\sqrt{\dfrac{1}{L_S C_P} - \left(\dfrac{R_S}{L_S}\right)^2}}{\dfrac{R_S}{L_S} + \dfrac{1}{R_P C_P}}$$

(c) Qualitatively, will the circuit appear inductive or capacitive at:
  (i) low frequencies?
  (ii) high frequencies?

4. Figure D.8 is representative of the inductor placed on the test fixture, although not to scale. Vary the components in the equivalent circuit. Assume that R_s goes to zero and R_p is relatively large but not infinite—say 10 to 20 kΩ). Give a qualitative sketch showing transmission coefficient, S_{21}, as a function of frequency. Label the resonant frequency and the 3-dB frequencies.

5. Assume the inductor is situated in the test fixture as shown in Figure D.8. The insertion loss in dB, abbreviated IL, is given by the equation

$$IL = -20 \log_{10} |S_{21}|$$

where S_{21} is the transmission coefficient and IL is defined to be positive. At the resonant frequency, the equivalent series resistance, R_{seq}, can be determined directly from the insertion loss. Show that the R_{seq} is given by the following:

$$R_{\text{Seq}} = \frac{\dfrac{R_P L_S}{R_S C_P}}{R_P + \dfrac{L_S}{R_S C_P}} = 2Z_0(10^{+\frac{IL}{20}} - 1)$$

In order to derive this expression, assume that the capacitors to ground in Figure D.9 are eliminated. There has to be a plus sign in the exponent of 10; why? What would R_{seq} be if there were not one?

6. One very important capability in microwave engineering (and all of engineering for that matter) is the ability to approximate. Engineering is often referred to as the "art of approximation." For this lab there are at least three parameters for which an *order of magnitude* approximation can be made to check the reasonableness of the measurements. The following three questions require some approximations and assumptions to be made. The assumptions that can be made (not necessarily all of them) and the material parameters are given. The derivation and assumptions should be very clear. Use illustrative sketches and note where assumptions are made—including the ones given.

Material Parameters:

AWG #28 copper wire: 12.6 mils Cu diameter, approx. 15 mils overall diameter including insulation.

Length: approx. 1000 mils of conductor for a two-turn air-core inductor with solder leads.

Conductivity σ of Cu: 58 MS/m.

(a) Estimate the series resistance, R_s, of the inductor wire at a frequency of 3 GHz. The resistance will increase with frequency due to the "skin effect." Assume the skin depth, δ_s (in meters), of the round wire is the same as that for an infinite half-space filled with copper:

$$\delta_S = \frac{1}{\sqrt{\pi f \mu \sigma}}$$

This is a good approximation when the skin depth is less than about one-fifth the radius of the wire. Assume the permeability of copper is that of free space. Assume the current density magnitude J is solely in the longitudinal direction of the wire, is constant, and is all concentrated within one skin depth of the surface of the conductor. An important constitutive relation that will facilitate deriving the expression for series resistance is:

$$\overline{J} = \sigma \overline{E}$$

where E is the electric field within the wire.

(b) Estimate the interwinding capacitance of a two-turn inductor with the turns as close together as possible (insulation of the wires is touching). The thickness of the insulation needs to be determined from the wire specification or measured using a micrometer. Use the parallel plate capacitance approximation:

$$C = \frac{\varepsilon A}{d}$$

This assumes no field fringing at the edges. Assume the dielectric constant is air. The dielectric constant of the insulation is greater than 1 but the distance between the turns is not constant either. This is a way of approximating the capacity only. This approximation also assumes that capacity is only between the nearest-neighbor windings.

(c) Find an approximate upper bound for the inductance of a finite turn air-core inductor by deriving an expression for the inductance of a closely wound infinite-length inductor (a solenoid). Use the dimensions of the #28 Cu wire given under the material parameters and refer to Figure D.10 for the coil dimensions. If there is no #28 wire available, use a wire size close to #28 and adjust the calculations accordingly.

The inductance L of an air-filled long solenoid of length len, cross sectional area A, and number of turns N is:

$$L = \frac{\mu_o N^2 A}{\text{len}}$$

A good place to find help with deriving this formula is in an introductory physics or fields book. Be sure to clearly and briefly explain each step of the derivation including the assumptions made.

Compare this value with the value from the formula given below. Assume N is a large number and calculate the inductance per unit length.

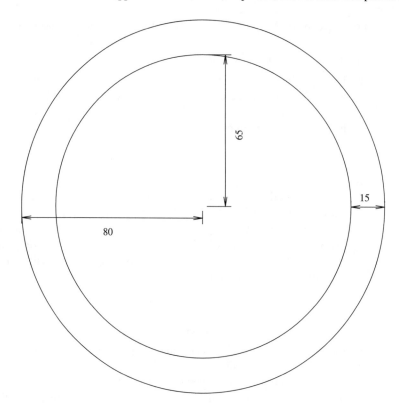

Figure D.10 Coil cross-section dimensions, #28 Cu wire with insulation.

Air Wound Inductors

The approximate inductance of an air-wound inductor is [2]:

$$L = N^2 \frac{R^2}{9R + 10D}$$

where L is in microhenries, R is in inches, and D is in inches. N is the number of turns and R and D are as shown in Figure D.11. The value of inductance is said to be accurate to about 1%. As is well known, the total length of the wire must be shorter than a small portion of a wavelength for this formula to hold and the parasitic capacity of the windings will dictate to how high a frequency the coil acts solely like an inductor.

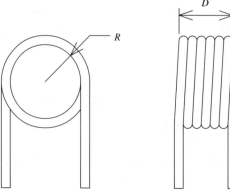

Figure D.11 Drawing of a single-layer wound inductor.

The inductance will be at its maximum when the turns are closely packed (the enamel coating keeps the windings from shorting) and will decrease as the inductor is stretched apart. This procedure is often used to tune an inductor. In addition, the inductance of the inductor will decrease if the space inside the winding (the core) is partially filled with a highly conductive metal and will increase if the space is filled with a high-permeability material. Tunable inductors are often tuned by using a screw made of a magnetic material such as ferrite. The screw is used to fill a variable portion of the core. This in effect puts two inductors in series, one primarily air filled and one primarily ferrite filled.

Data have been measured on several different-sized inductors. These are given for guidance. These should be used to check the measurement that are made on the inductors that are fabricated for measurement.

The inductors were made using #26 AWG(0.0159 inch) copper wire single belsol coated wire, close wound on a 0.125-inch mandrel. The following data were measured.

Values Measured on Fabricated Inductors

Number of Turns	Inductance (nH)	Parallel Resonant Freq. (GHz)	Equivalent C_p (pF)	Measured Length (mils)	Misc. Notes
0 (thru)	0.5				
1	5.6	5.75	0.137	12.5	
2	16.8	2.66	0.213	34	
3	31	1.79	0.255	55	fs = 5.46 GHz fp = 5.475 GHz
4	49	1.30	0.306	70	fs = 5.05 GHz fp = 5.93 GHz
5	66	1.18	0.227	90	fs, fp, fs 3.0, 3.15, 5.13 GHz
6	85	1.11	0.242	106	
7	103	1.02	0.236	120	
8	125	1.02	0.195	138	

Several of the inductors had other resonances at higher frequencies. Therefore, a simple parallel L-C circuit is inadequate to model the circuit above the first parallel resonant frequency. This is due to the distributed nature of the winding. When the winding approaches a multiple of a quarter wavelength long, the inductor will come close to another resonance.

Each turn was about 0.017 inch long (wire plus insulation) and the inner radius of each turn was 0.0625 inches. The inductance was calculated using:

$$L_{\mu H} = \frac{N^2 R^2}{9R + 10D}$$

Measured versus calculated values are shown in Figure D.12. The x points are measured values and the $+$ points are calculated values. The ordinate is inductance in nH and the abscissa is number of turns.

Air-Core Inductor Measurement

Purpose: In order to design a printed circuit board using inductors, it is necessary to develop an accurate lumped-element model of a multiturn air-core inductor over the frequency range of interest. The model is required for realistic computer simulation of the overall microwave circuit. The model will incorporate parasitic capacitance and resistance not considered at lower frequencies. The model will be specific to an inductor of a specific physical size and construction. Each inductor using a different physical construction needs to be characterized.

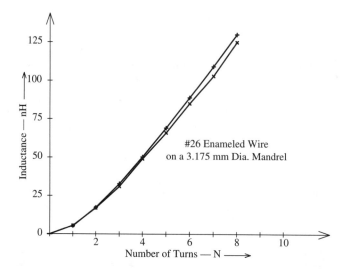

Figure D.12 Number of turns vs. measured ($\times$) and calculated ($+$) inductance.

This laboratory exercise is applicable to an air-wound inductor, a chip inductor, or an inductor fabricated on magnetic core material.

Measurement of ω_o, β, and Q

1. Make an inductor for measurement if one is not already available. Solder the inductor across the gap of the resistor test fixture in order to make a preliminary measurement.

2. Determine a proper calibration frequency range for the measurements. A quick way to determine this is to connect the test fixture with the inductor soldered on it to the network analyzer and make an uncalibrated S_{21} magnitude measurement over the analyzer's complete frequency range. The approximate frequency where the resonance occurs should be apparent. Pick a suitable frequency range. Include sufficiently low frequencies so that a "good" inductance measurement can be made.

3. Preset the Vector Network Analyzer and set the frequency range according to part 2. Additional data can be obtained by using frequencies down to very low frequencies.

4. Perform a full two-port calibration with the necessary connectors and cables attached to the Vector Network Analyzer. If there are any questions, refer to the *Vector Network Analyzer User's Guide*, *Vector Network Analyzer Quick Reference*, or the lab instructor. The easiest way to learn how the machine works and which connectors are needed is to experiment and refer to the *User's Guide*.

5. Record the analyzer settings for future reference. Be sure to record the position of the measurement reference plane(s). If it is necessary to zoom in on a small frequency range, use the interpolation function if it is available.

   The correct procedure for setting interpolation on a many Vector Network Analyzers is: (1) turn the calibration off, (2) set the low and high values of the desired frequencies of interest (or Δf, and f_o), (3) turn interpolation on, and (4) then turn calibration on. To reset the original calibrated frequency setting: (1) turn calibration off, (2) turn interpolation off, and (3) turn calibration on. This order allows the network analyzer to set the frequency ranges before it attempts to do a calibration interpolation.

6. Now it is time to actually make measurements. If all has gone well the transmission measurement has a definite "suck-out." The inductor functions as a poor notch filter

in the given arrangement. Use the marker(s) to determine the resonant frequency, the upper and lower 3-dB bandwidth frequencies, and the insertion loss at the resonant frequency. Q can then be computed knowing ω_o and β. All this can be done manually or the analyzer can be set to do it. It is suggested that one let the Vector Network Analyzer do it!

7. At the resonant frequency "something" ought to approach zero. Determine if that "something" does indeed approach zero. What measurement verified this claim?

8. Remember to get all the plots and data needed from the resonance measurements before moving on to the rest of the lab.

Determining L_s, R_s, C_p, and Q

1. Measure the series resistance, R_s, at d.c. with the digital voltmeter. The resistance of the voltmeter (ohmmeter) leads needs to be compensated for since it will be significantly higher than the resistance of the inductor itself will. In addition, the contact resistance from the ohmmeter leads to the wire is likely higher than the wire resistance. It is better to do a four-point measurement. Run about one amp through the wire and measure the millivolt voltage drop across the wire. Place the voltmeter leads on the wire, not on the clip leads that come from the power supply to the wire. Why? Also measure R_s using a low-frequency range on the network analyzer. This can be done by assuming that the terminating impedance of the network analyzer is a constant 50Ω. How do the measured values compare to the predicted value?

2. The inductance will be least affected by the parasitics at low frequencies; therefore, measure L_s in the low-frequency range. Does the inductance appear fairly constant? Is it less than the infinite solenoid prediction? Knowing the total length of the conductor, how does the inductance compare with an estimate of 1 nH per 0.1 in.? The estimate can be used to determine the coupling between turns of the inductor.

3. From the knowledge of the resonant frequency, inductance, and series resistance, determine the shunt capacitance, C_p. Do this now and discuss the result with the lab instructor. How does this value compare with the approximation calculated (percent difference)? Is the approximation at least within one order of magnitude of the calculated value at present?

4. At the parallel resonance, the approximate value of R_p can be measured. Remember that loci near the open circuit (infinite) position on the Smith Chart are subject to inaccuracy. Small movement in the gamma plane at the right-hand side of the chart represents a large relative change in impedance. Use the averaging function if it is available to reduce the error and increase the precision. Note that the magnitude of gamma is largest at resonance. If the network analyzer phase calibration is slightly off, the largest magnitude of gamma might not be at zero degrees. Determine R_p from the value of gamma at its largest magnitude. The network analyzer needs to have a good calibration or the open-circuit calibration will be off and then the parallel resistance value will be off. The value of R_p should be very high for an air-wound inductor. It is largely determined by the amount of radiation from the coil. For a coil wound on a core, R_p is largely determined by loss in the core material.

5. Using the measured values for bandwidth, series resistance, and the calculated value of shunt capacitance the corresponding value for the shunt resistance, R_p, can be determined.

6. From the data for insertion loss at resonance, the characteristic impedance of the transmission line(s), and the values of R_s, L_s, and C_p determine R_p (refer to the

prelab). Compare the values obtained for R_p in part 4, part 5 and the present value. Give the percent difference between the measured value and each of the two calculated values. If the values are not equal then give a reason(s) (opinions) for the difference. Specifically, where are at least two other possible sources of loss in the circuit arrangement? Do not be exhaustive here, but think carefully about the inductor as it resides on the substrate.

7. Draw and label the equivalent inductor circuit showing all values and the value of R_p obtained for part 5 since this will make the model self-consistent.

8. At this point the measurement portion of the inductor lab is done. If a chip inductor is available, the measurements can be repeated although the calculations will not be possible unless the details of the chip inductor fabrication are known. Discuss the results and overall observations with the lab instructor before leaving the lab.

Computer Modeling

1. Perform the computer modeling of the equivalent circuit. Use the measured and calculated values of the parasitics in a computer-aided-design program model and compare the S_{21} given by the program with data from the Vector Network Analyzer. Quantitatively compare the Q, bandwidth, and resonant frequency of the computer program model, using the component values given in part 7 of the previous section, and the measured values. For the purposes of this lab, the bandwidth and resonant frequency are the most important parameters. These values should match the computer program model values. If they do not match, try tuning the parameters to get a closer fit or observe the behavior of the circuit as the parameters change. Specifically, what happens when R_s, L_s, C_p, and R_p change respectively while the other components remain constant? Give appropriate plots and circuit files (with comments) documenting this modeling effort.

2. Record any "abnormalities" that may have arisen in the course of the measurement. If there were no problems, then so state. This information is very helpful when comparing measurements from different inductor types, etc.

Chip Resistor Sizes— Nominal Sizes Only

For actual sizes, consult the governing specifications.

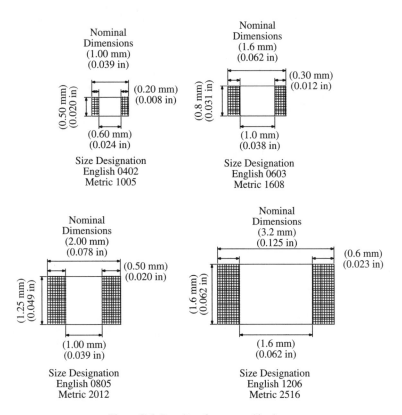

Figure E.1 Drawing of common chip sizes.

S Parameters
(Scattering Parameters—
Current Referenced)

The scattering parameter derivation can be made using a current source with the current source phase as the phase reference. An equally valid set of scattering parameters will result. The transformations to the immittance parameters could also be made and some of the equations would show more symmetry than the equations that are related to voltage-phase-referenced scattering parameters.

Scattering parameters are derived from a superposition of voltage and current. The parameters used in the superposition can vary. A dual circuit to that shown in the circuit in Figure 2.3 can be used to derive an alternative definition of scattering parameters as shown in Figure F.1. The phase reference will be the phase of the current source rather than the phase of the voltage source. In a manner similar to that used with the series voltage source, the maximum power is:

$$P = \frac{I_S I_S^*}{4G_S} = \frac{I_S I_S^*}{4(\sqrt{G_S})^2} = \left(\frac{I + VY_S}{2\sqrt{G_S}}\right)\left(\frac{I + VY_S}{2\sqrt{G_S}}\right)^* = aa^*$$

One can identify the square root of power with the left term on the right-hand side of the equation. Then:

$$a = \left(\frac{I + VY_S}{2\sqrt{G_S}}\right)$$

It can be shown that the reflected power is:

$$P_{\text{refl}} = \left(\frac{I - VY_S^*}{2\sqrt{G_S}}\right)\left(\frac{I - VY_S^*}{2\sqrt{G_S}}\right)^* = bb^*$$

A legitimate conclusion might be to associate the b term with the left side of the middle terms. Then:

$$b = \left(\frac{I - VY_S^*}{2\sqrt{G_S}}\right)$$

A correct set of scattering parameters will result. These scattering parameters, however, will not be the same as one gets by using the voltage definition. These would be scattering parameters referenced to the phase of the current source. Using this set of choices, the reflection

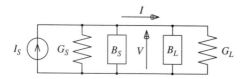

Figure F.1 Current source used to derive current-based scattering parameters.

coefficient is then:

$$\Gamma = \frac{b}{a} = \frac{I - VY_S^*}{I + VY_S} = \frac{Y_L - Y_S^*}{Y_L + Y_S} = -\frac{Z_L - Z_S^*}{Z_S + Z_L}\frac{Z_S}{Z_S^*}$$

Notice that when the source is purely real, the reflection coefficient described using a current source is the negative of the reflection coefficient using a voltage source. This is to be expected when one considers that the current reflection coefficient on a transmission line is the negative of the voltage reflection coefficient on a transmission line.

Convention seems to be to reference the scattering parameters to the phase of the voltage source. Notice that the power terms aa^* and bb^* are the same in both definitions. It can be shown that the value of the Norton equivalent G_s is related to the Thevenin equivalent R_s by:

$$G_S = \mathrm{Re}\left[\frac{1}{Z_S}\right] = \mathrm{Re}\left[\frac{Z_S^*}{Z_S Z_S^*}\right] = \frac{R_S}{Z_S Z_S^*} = R_S Y_S Y_S^*$$

Substituting this in the aa^* equation using the current definition one gets:

$$P = \left(\frac{I + VY_S}{2\sqrt{G_S}}\right)\left(\frac{I + VY_S}{2\sqrt{G_S}}\right)^* = \frac{Y_S(V + IZ_S)(V^* + I^*Z_S^*)Y_S^*}{4R_S}Z_S Z_S^*$$

$$= \frac{(V + IZ_S)(V^* + I^*Z_S^*)}{4R_S} = aa^*$$

and similarly for the reflected quantity bb^*.

$$P_{\mathrm{refl}} = \left(\frac{I - VY_S^*}{2\sqrt{G_S}}\right)\left(\frac{I - VY_S^*}{2\sqrt{G_S}}\right)^* = \frac{-Y_S(V - IZ_S)(-1)(V^* - I^*Z_S^*)Y_S^*}{4R_S}Z_S Z_S^*$$

$$= \frac{(V - IZ_S)(V^* - I^*Z_S^*)}{4R_S} = bb^*$$

The right side of each of the equations is the same as when using the voltage definitions. A modification of the choices for a and b derived when using a current source will result in the scattering parameters being the same between the two choices. It should be emphasized that there is nothing wrong with the first choices shown above. They just are not the convention used in the industry. If the following choices are made for a and b, then the current and voltage scattering parameters are the same. Notice that the magnitude of the ratio of Y and Y conjugate is 1 and therefore power computations are not affected. When the source is purely real, the only difference is in the negative sign for b. Only the phases of the current scattering parameters are affected. A negative sign and the square root of the appropriate admittance ratios correct the phases. The a and b quantities are then:

$$a = \left(\frac{I + VY_S}{2\sqrt{G_S}}\right)\sqrt{\frac{Y_S^*}{Y_S}}$$

$$b = -\left(\frac{I - VY_S^*}{2\sqrt{G_S}}\right)\sqrt{\frac{Y_S}{Y_S^*}}$$

G

Modeling Using an Equivalent Mechanical Model

This section is provided for those interested in determining capacitance and inductance by relating these parameters to equivalent mechanical parameters.

It is sometimes helpful to use the analogy between force and voltage and current and velocity to investigate parasitic elements and thus develop an equivalent circuit for circuit components. Consider an inertial particle of mass m moving with acceleration a and velocity v. The force and velocity are related as:

$$F = ma = m\frac{dv}{dt}$$

The kinetic energy E_k of this particle is:

$$E_k = \frac{1}{2}mv^2$$

Now consider the force F on a spring with spring constant K having a displacement x. The force is:

$$F = Kx = K\int v dt$$

The spring is compressed or expanded a distance x from its no-force position. The potential energy stored in the spring is:

$$E_p = \frac{1}{2}Kx^2 = \frac{1}{2}\frac{F^2}{K}$$

Now consider the voltage V induced across an inductor of magnitude L by a current I changing through it:

$$V = L\frac{dI}{dt}$$

The energy E_L stored in the inductor is:

$$E_L = \frac{1}{2}LI^2$$

Also consider the current I into a capacitor C caused by a changing voltage V across it:

$$V = \frac{1}{C} \int I \, dt$$

The energy E_C stored in the capacitor is:

$$E_C = \frac{1}{2} C V^2$$

Notice that by using the force-voltage and current-velocity analogy, the kinetic energy of a mass has the same mathematical form as the magnetic energy of an inductor circuit and that the potential energy of a spring has the same mathematical form as the electric energy of a capacitance circuit. The inductance L corresponds to the mass m. The capacitance C corresponds to the inverse of the spring constant $1/K$. There are some physical limitations to this analogy. However, if one considers that moving a charge from one point to another is similar to moving a mass and that separating two charges is similar to compressing a spring, one is able to identify where one needs to incorporate inductance or capacitance in an equivalent circuit.

Bibliography

Chapter 1

[1] Robert A. Chipman, *Theory and Problems of Transmission Lines*, New York, McGraw-Hill, 1968.

[2] K. C. Gupta, Ramesh Garg, Inder Bahl, Prakash Bhartia, *Microstrip Lines and Slotlines*, Second Edition, Dedham, MA, Artech House, 1996.

Chapter 2

[1] *Reference Data for Radio Engineers*, Fourth Edition, Copyright ©1956, New York, International Telephone and Telegraph Corporation, p. 508.

[2] Bernard C. DeLoach, "A New Microwave Measurement Technique to Characterize Diodes and an 800-Gc Cutoff Frequency Varactor at Zero Volts Bias," *IEEE Transactions on Microwave Theory and Techniques*, vol. MTT-12, no. 1, 1964, pp. 15–20.

[3] Guillermo Gonzalez, *Microwave Transistor Amplifiers*, Second Edition, Upper Saddle River, Prentice-Hall, 1997.

[4] Hewlett Application Note 154, *S-Parameter Design*, April 1972 (May 1973 revision), p. 9.

[5] *Reference Data for Radio Engineers*, Fourth Edition, Copyright ©1956, New York, International Telephone and Telegraph Corporation, p. 508.

[6] P. H. Smith, "Transmission Line Calculator," *Electronics*, vol. 12, January 1939, pp. 29–31.

[7] P. S. Carter, "Charts for Transmission-Line Measurements and Computations," *Radio Corporation of America Review*, vol. III, no. 3, January 1939, pp. 355–368.

[8] B. R. Hallford, "A 90-dB Microstrip Switch on a Plastic Substrate (Correspondence)," *IEEE Transactions on Microwave Theory and Techniques*, vol. MTT-19, no. 7, 1971, pp. 654–657.

[9] *Reference Data for Radio Engineers*, Fourth Edition, Copyright ©1956, New York, International Telephone and Telegraph Corporation, p. 112.

[10] Z. M. Hejazi, P. S. Excell, and Z. Jiang, "Accurate Distributed Inductance of Spiral Resonator," *IEEE Microwave and Guided Wave Letters*, vol. 8, No. 4, April 1998, p. 164–166.

[11] J. M. Lopez-Villegas, J. Samitier, C. Cane, and P. Losantos, "Improvement of the Quality Factor of RF Integrated Inductors by Layout Optimization," *1998 Radio Frequency Integrated Circuits (RFIC) Symposium 98* (1998 [RFIC]), pp. 169–172.

Chapter 3

[1] *Hewlett Packard Journal*, vol. 17, No. 9, May 1966, p. 10.

[2] Robert Weber, "Modified Pulsed Two-Port Load Pull *S*-Parameter Derivation," *1994 High-Power Microwave Workshop*, Dorsey Center, Baltimore, John Hopkins University, Oct. 7, 1994.

[3] Robert J. Weber, "Oscillator Design Techniques Using Calculated and Measured *S* Parameters," Copyright ©1991, Short Course, *45th Annual Symposium on Frequency Control*, Los Angeles, May 1991.

[4] Jerome L. Altman, *Microwave Circuits*, Princeton, Van Nostrand, 1964, p. 117.

Chapter 4

[1] E. M. T. Jones and J. T. Bolljahn, "Coupled-Strip-Transmission-Line Filters and Directional Couplers," *IRE Transactions on Microwave Theory and Techniques*, 1956, pp. 75–81.

[2] Gerorge I. Zysman and Kent A. Johnson, "Coupled Transmission Line Networks in an Inhomogeneous Dielectric Medium," *IEEE Transactions on Microwave Theory and Techniques*, vol. MTT-17, No. 10, 1969, pp. 753–759.

[3] D. E. Bockelman and W. R. Eisenstadt, "Combined Differential and Common-mode Scattering Parameters: Theory and Simulation," *IEEE Transactions on Microwave Theory and Techniques*, vol. 43, No. 7, 1995, pp. 1530–1539.

Chapter 5

[1] John G. Linville and James F. Gibbons, *Transistors and Active Circuits*, New York, McGraw-Hill, 1961.

[2] J. M. Rollett, "Stability and Power Gain Invariants of Linear Two-Ports," *IRE Transactions on Circuit Theory*, vol. CT-9, 1962, pp. 29–32.

[3] Robert J. Weber, "Some Design Considerations for L-Band Power MMICs," *Proceedings RF Expo EAST*, November 1990, pp. 187–199.

[4] R. J. Weber, RF/Microwave Class, University of Iowa, 1987.

[5] Marion Lee Edwards and Jeffrey H. Sinsky, "A New Criterion for Linear 2-Port Stability Using a Single Geometrically Derived Parameter," *IEEE Transactions on Microwave Theory and Techniques*, vol. 40, no. 12, 1992, pp. 2303–2311.

[6] Robert J. Weber, "Even Mode versus Odd Mode Stability," *40th Midwest Symposium on Circuits and Systems*, August 1997, pp. 607–610.

[7] K. Kurokawa, "Power Waves and the Scattering Matrix," *IEEE Transactions on Microwave Theory and Techniques*, vol. 13, no. 2, 1965, pp. 194–202.

[8] George E. Bodway, "Two Port Power Flow Analysis Using Generalized Scattering Parameters," *Microwave Journal*, vol. 10, no. 6, May 1967.

Chapter 6

[1] Wai-Kai Chen, *Theory and Design of Broadband Matching Networks*, Oxford, Pergamon Press, 1976.

[2] R. M. Fano, "Theoretical Limitations on the Broadband Matching of Arbitrary Impedances," *Journal of the Franklin Institute*, January 1950, pp. 57–83, February 1950, pp. 139–154.

[3] R. Weber, *Microwave Engineering Course Notes*, ©1987–1997, August 1997 Edition.

[4] *Reference Data for Radio Engineers*, Fourth Edition, Copyright ©1956, International Telephone and Telegraph Corporation, p. 122.

[5] G. L. Matthaei, "Short-step Chebyshev Impedance Transformers," *IEEE Transactions of Microwave Theory and Techniques*, vol. MTT-14, no. 8, August 1966, pp. 372–383.

[6] K. Kuroda, "Derivation Methods of Distributed Constant Filters from Lumped Constant Filters," test for lectures at *Joint Meeting of Konsoi Branch of Institute of Elec. Commun., of Elec., and of Illumin. Engrs. of Japan*, October 1952, p. 32. (In Japanese.) (Ref. taken from [7].)

[7] R. J. Wenzel, "Exact Design of TEM Microwave Networks Using Quarter-Wave Lines," *IEEE Transactions on Microwave Theory and Techniques*, vol. MTT-12, no. 1, 1964, pp. 94–111.

Chapter 7

[1] Herbert L. Krauss, Charles W. Bostian, and Frederick H. Raab, *Solid State Radio Engineering*, New York, John Wiley & Sons, 1980.

[2] Steve C. Cripps, *RF Power Amplifiers for Wireless Communications*, Boston, MA, Artech House, 1999.

[3] U.S. Patent 3,919,656, "High-Efficiency Tuned Switched Power Amplifier," 1975.

[4] N. O. Sokal and A. D. Sokal, "Class E, a New Class of High-Efficiency Tuned Single-Ended Power Amplifiers," *IEEE Journal of Solid State Circuits*, SC-10, June 1975, pp. 168–176.

[5] J. M. Early, "Structure-Determined Gain-Band Product of Junction Triode Transistors," *Proceedings of the IEEE*, Dec. 1958, pp. 1924–1927.

Chapter 8

[1] "Dictionary of Electronic Terms," Published by Allied Radio, Chicago, edited by Robert E. Beam, Ph.D., Seventh Edition, Fourth Printing, March 1965.

[2] G. L. Matthaei, Leo Young, and E. M. T. Jones, *Microwave Filters, Impedance Matching Networks, and Coupling Structures*, New York, New York, McGraw-Hill, 1964.

[3] Edward L. Ginzton, *Microwave Measurements*, New York, McGraw-Hill, 1957.

[4] Hewlett Packard Application Note 339-1, "Impedance Characterization of Resonators Using the HP 4194A Impedance/Gain-Phase Analyzer," March 1988.

[5] G. R. Basawaptna and R. B. Stancliff, "A Unified Approach to the Design of Wide-band Microwave Solid State Oscillators," *IEEE Transactions on Microwave Theory and Techniques*, vol. MTT-27, no. 5, May 1979, pp. 379–385.

[6] R. J. Weber, "Oscillator Design Using S-Parameters and a Predetermined Source or Load," *Proceedings 45th Annual Symposium on Frequency Control*, May 1991, pp. 364–367.

[7] R. J. Weber, R. Huisinga, and D. Ripley, "A Microwave Oscillator S-Parameter Design Procedure for Loop Oscillators," *Proceedings 35th Midwest Symposium on Circuits and Systems*, August 1992, pp. 1341–1344.

[8] Robert J. Weber, "Oscillator Design Using S Parameters," Copyright ©1993, Short Course, *47th Annual International Symposium on Frequency Control*, Salt Lake City, June 1993.

Chapter 9

[1] Herman J. Blinchikoff and Anatol Zverev, *Filtering in the Time and Frequency Domains*, New York, Wiley, 1976.

[2] T. W. Parks and C. S. Burns, *Digital Filter Design*, New York, Wiley, 1987.

[3] S. B. Cohn, "Dissipation Loss in Coupled Resonator Filters," *Proceedings of the IRE*, August 1959, pp. 1342–1348.

[4] S. B. Cohn, "Direct-Coupled-Resonator Filters," *Proceedings of the IRE*, vol. 45, February 1957, pp. 187–196.

[5] G. L. Matthaei, "Design of Wide-Band (and Narrow-Band) Band-Pass Microwave Filters on the Insertion Loss Basis," *IRE Transactions on Microwave Theory and Techniques*, vol. MTT-8, November 1960, pp. 580–593.

[6] R. M. Fano, "Theoretical Limitations on the Broadband Matching of Arbitrary Impedances," *Journal of the Franklin Institute*, January 1950, pp. 57–83, February 1950, pp. 139–154.

[7] P. I. Richards, "Resistor Transmission-Line Circuits," *Proceedings of the IRE*, vol. 36, February 1948, pp. 217–220.

[8] S. Butterworth, "On the Theory of Filter Amplifiers," *Wireless Engineer* (London), vol. 7, October 1930, pp. 536–541.

[9] Yale Jay Lubkin, *Filter Systems and Design, Electrical, Microwave, and Digital*, Reading, MA, Addison-Wesley, 1970.

[10] M. Dishal, "Alignment and Adjustment of Synchronously Tuned Multiple-Resonant-Circuit Filters," *Proceedings of the IRE*, vol. 39, November 1951, pp. 1448–1455.

Chapter 10

[1] Robert E. Collin, *Foundations for Microwave Engineering*, New York, McGraw-Hill, 1966.

[2] H. Rothe and W. Dahlke, "Theory of Noisy Fourpoles," *Proceedings of the IRE*, vol. 44, June 1956, pp. 811–818.

[3] Robert V. Pound, *Microwave Mixers*, New York, McGraw-Hill, 1948.

[4] "IRE Standards on Methods of Measuring Noise in Linear Twoports, 1959," *Proceedings of the IRE*, January 1960, pp. 60–68.

[5] "Representation of Noise in Linear Twoports," *Proceedings of the IRE*, January 1960, pp. 69–74.

Chapter 11

[1] Robert V. Pound, *Microwave Mixers*, New York, McGraw-Hill, 1948.

[2] Henry C. Torrey and Charles A. Whitmer, *Crystal Rectifiers*, New York, McGraw-Hill, 1948.

[3] Stephen A. Maas, *Microwave Mixers*, Boston, MA, Artech House, 1986, Second Edition, 1993.

Chapter 12

[1] Jerome L. Altman, *Microwave Circuits*, Princeton, Van Nostrand, 1964.

[2] S. M. Sze, *High Speed Semiconductor Devices*, New York, Wiley, 1990.

[3] S. M. Sze, *Semiconductor Devices*, New York, Wiley, 1985.

[4] Kai Chang, *Handbook of Microwave and Optical Components, Vol. 1, Microwave Passive & Antenna Components*, New York, Wiley, 1989.

[5] Kai Chang, *Handbook of Microwave and Optical Components, Vol. 2, Microwave Solid-State Components*, New York, Wiley, 1990.

[6] James D. Brobst and Robert J. Weber, "L-X Band 200us Coherent Power Sources," *IEEE International Solid-State Circuits Conference Digest*, 1981, pp. 134.

[7] J. B. Gunn, "Microwave Oscillations of Current in III-V Semiconductors," *Solid State Commun.*, vol. 1, September 1963, p. 88.

[8] E. M. T Jones and J. T. Bolljahn, "Coupled-Strip-Transmission-Line Filters and Directional Couplers," *IRE Transactions on Microwave Theory and Techniques*, 1956, pp. 75–81.

[9] Gerorge I. Zysman and Kent A. Johnson, "Coupled Transmission Line Networks in an Inhomogeneous Dielectric Medium," *IEEE Transactions on Microwave Theory and Techniques*, vol. MTT-17, no. 10, 1969, pp. 753–759.

[10] S. March, "A Wideband Stripline Hybrid Ring," *IEEE Transactions on Microwave Theory and Techniques*, vol. MTT-16, no. 6, June 1968, p. 361.

[11] J. Smit and H. P. S. Wijn, *Ferrites*, N. V. Philips' Gloeilampenfabrieken, Eindhoven, 1959, p. 289.

[12] J. Smit and H. P. J. Wijn, *Ferrites*, N. V. Philips' Gloeilampenfabrieken, Eindhoven, 1959, p. 275.

[13] William R. Smythe, *Static and Dynamic Electricity*, Second Edition, New York, McGraw Hill, 1950, p. 234.

Chapter 13

[1] J. C. Rodrigues Paulo, *Computer-Aided Analysis of Nonlinear Microwave Circuits*, Boston, MA, Artech House, 1998.

[2] Stephen A. Maas, *Nonlinear Microwave Circuits*, Norwood, MA, Artech House, 1988.

Chapter 14

[1] Paul Penfield and Robert P. Rafuse, *Varactor Applications*, Cambridge, M.I.T. Press, 1962.

[2] J. C. Rodrigues Paulo, *Computer-Aided Analysis of Nonlinear Microwave Circuits*, Boston, MA, Artech House, 1998.

[3] Stephen A. Maas, *Nonlinear Microwave Circuits*, Norwood, MA, Artech House, 1988.

Appendix A

[1] Harold A. Wheeler, "Transmission-Line Properties of a Strip on a Dielectric Sheet on a Plane," *IEEE Transactions on Microwave Theory and Techniques*, vol. MTT-25, no. 8, August 1977, pp. 631–647.

Appendix D

[1] James W. Nilsson, *Electric Circuits*, Reading MA, Addison-Wesley, 1990.

[2] *Reference Data for Radio Engineers*, Fourth Edition, Copyright ©1956, International Telephone and Telegraph Corporation, p. 112.

Index